Silicon in organic synthesis

Butterworths Monographs in Chemistry and Chemical Engineering

Butterworths Monographs in Chemistry and Chemical engineering is a series of occasional texts by internationally acknowledged specialists, providing authoritative treatment of topics of current significance in chemistry and chemical engineering

Forthcoming titles:

Alkaloid Biosynthesis
Particulate Systems
Fluidized Bed Reactors
Crystallisation Process Design

Butterworths Monographs in Chemistry and Chemical Engineering

Silicon in Organic Synthesis

Ernest W. Colvin, BSc, PhD
Lecturer in Chemistry,
University of Glasgow

Butterworths
London Boston Sydney Wellington Durban Toronto

First published 1981

British Library Cataloguing in Publication Data

Colvin, Ernest
Silicon in organic synthesis. — (Butterworths monographs in chemistry and chemical engineering)
1. Organosilicon
I. Title
547′.08 QD412.S6

ISBN 0-408-10619-0
ISBN 0-408-10831-2 Pbk.

Typeset by Tunbridge Wells Typesetting Services
Printed and bound by Mackays of Chatham

Preface

The past fifteen years have witnessed a truly explosive growth in the organic chemistry of silicon. It is my hope that this volume will serve as a timely introduction to the subject for students and practitioners of synthetic organic chemistry, as well as providing a source of useful information and possibly of new ideas to those already experienced in the area.

Particular emphasis is placed throughout the book on the concept of silicon as a 'ferryman', mediating the transformation of one wholly organic molecule into another. For this reason, most of the more silicon-orientated areas of organosilicon chemistry, such as those of low-valent silicon species and sila-heterocycles, are not discussed.

Reviews and leading references have normally been given precedence over earlier reports. Consequently, the bibliography is heavily biased in favour of more recent publications, an effort having been made to provide thorough and current (through June, 1980) references on all topics covered.

Work was originally planned to commence during nine months of sabbatical leave from the University of Glasgow in 1978, as a Senior CIBA-GEIGY Fellow at the ETH, Zürich. Bench chemistry, however, overtook this intention, and I thank Professor Dr Dieter Seebach for his hospitality throughout a most enjoyable stay. Writing began seriously in Spring, 1979, and reached completion in October, 1980.

I acknowledge with pleasure the helpful comments and constructive criticisms of Dr Ian Fleming and Professor Gordon Kirby. Their advice has had a material effect on the final form of the book. Any errors or omissions are, of course, entirely my responsibility, and for these I apologize in advance.

Ernest W. Colvin
University of Glasgow
November, 1980

To My Family

Abbreviations

The following abbreviations have been used throughout:

Ac	Acetyl
acac	Acetylacetonyl
Cp	Cyclopentadienyl
DBN	1,5-Diazabicyclo[4.3.0]non-5-ene
DBU	1,5-Diazabicyclo[5.4.0]undec-5-ene
DCC	Dicyclohexylcarbodiimide
DDQ	2,3-Dichloro-5,6-dicyano-1,4-benzoquinone
DME	1,2-Dimethoxyethane
DMF	*NN*-Dimethylformamide
DMSO	Dimethyl sulphoxide
$E(E^+)$	Electrophile
HMPA	Hexamethylphosphoramide (Hexamethylphosphoric triamide)
Im	1-Imidazolyl
LDA	Lithium di-isopropylamide
MCPBA	*m*-Chloroperbenzoic acid
NBS	*N*-Bromosuccinimide
$Nu(Nu:^-)$	Nucleophile
Ph	Phenyl
py	Pyridine
TFA	Trifluoroacetic acid
THF	Tetrahydrofuran
THP	Tetrahydropyranyl
TMEDA	*NNN'N'*-Tetramethylethylenediamine
Ts	Toluene-*p*-sulphonyl (tosyl)

Contents

1 Introduction 1

2 Physical properties of organosilicon compounds 4
2.1 Relative bond strengths 4
2.2 Electronegativity 5
2.3 *p*-Orbitals 7
2.4 *d*-Orbitals 10

3 The β-effect 15

4 α-Metallated organosilanes 21

4.1 Organometallic addition to vinylsilanes 21
4.2 Proton abstraction 22
4.3 Metal–halogen exchange 25
4.4 Transmetallation 25

5 Rearrangement reactions with migration of silicon 30
5.1 1,2-Rearrangements 30
5.2 1,3-Rearrangements 33
5.3 1,4-Rearrangements 37
5.4 1,5-Rearrangements 37

6 Organohalogenosilanes and substitution at silicon 40

7 Vinylsilanes 44
7.1 Preparation 45
7.2 Geometric differentiation 60
7.3 Reactivity 62
7.4 Some other reactions of vinylsilanes 76
7.5 Addendum 79

8 $\alpha\beta$-Epoxysilanes as precursors of carbonyl compounds and heteroatom-substituted alkenes 83
8.1 Preparation 83
8.2 Isomerization 84

9 Allylsilanes 97
9.1 Preparation 98
9.2 Electrophilic substitution 104
9.3 Other selected examples of silyl control of carbonium ion formation and collapse 117
9.4 Some reactions not involving carbonium ions 118
9.5 Addendum 119

10 Arylsilanes 125
10.1 Preparation 125
10.2 Electrophile-induced desilylation 126

11 Organosilyl anions 134
11.1 Preparation 134
11.2 Reactions 136

12 Alkene synthesis by 1,2-elimination reactions of β-functional organosilanes 141
12.1 β-Hydroxyalkylsilanes 142
12.2 β-Halogenoalkylsilanes and related species 152
12.3 Addendum 160

13 Alkynylsilanes and Allenylsilanes 165
13.1 Alkynylsilanes 165
13.2 Allenylsilanes 170

14 Silylketenes 174

15 Alkyl silyl ethers 178
15.1 Solvolysis 178
15.2 Trimethylsilyl ethers 179
15.3 t-Butyldimethylsilyl ethers 184
15.4 Addendum 189

16 Acyloxysilanes (silyl carboxylates) 193

17 Silyl enol ethers and silyl ketene acetals 198
17.1 Preparation of silyl enol ethers 198
17.2 Preparation of silyl ketene acetals 213
17.3 Reactions 215
17.3.1 Generation of specific enolate anions 217
17.3.2 Lewis acid-catalysed alkylation 221
17.3.3 Alkylation and α-methylenation 223
17.3.4 Hydroxyalkylation and related reactions 226
17.3.5 Silyl dienol ethers and bis(silylenol) ethers 229
17.3.6 Acylation 232

17.3.7 Hydroboration–oxidation 235
17.3.8 Oxidation 236
17.3.9 Cycloaddition 243
17.3.10 Sigmatropic rearrangements and related processes 260
17.3.11 Modified acyloin condensations and related reactions 267
17.3.12 1-Trimethylsilyl trimethylsilyl enol ethers 273
17.3.13 Silyl nitronates 275
17.4 Addendum 276

18 Trimethylsilyl-based reagents 288
18.1 Reagent preparation 289
18.2 Ester and ether cleavage 291
18.3 Carbonyl addition processes 295
18.4 Other addition processes 304
18.5 Other functionalized silane reagents 304
18.6 Addendum 307

19 Nitrogen-substituted silanes 314
19.1 Amino-silanes 314
19.2 Oxidative decyanation 316
19.3 Amides and amide bond formation 316

20 Silicon-substituted bases and ligands 321

21 Silanes as reducing agents 325
21.1 Hydrosilylation 325
21.2 Ionic hydrogenation 329
21.3 Reductive silylation 331
21.4 Deoxygenation processes 333

Index 337

Chapter 1

Introduction

Organosilicon compounds are defined as those species which possess carbon–silicon bonds; the nomenclature used in this book is very simple, such compounds being described, with few exceptions, as derivatives of silane, SiH_4.

Organosilanes have a long pedigree, although early investigations gave little clue to the potential now being realized. Silicon is the second most abundant element in the Earth's crust. It does not occur in the free state, being found always in combination with the most abundant element, oxygen, as silica or the metal silicates. Since no organosilicon compound is known to occur naturally, these intractable sources of silicon presented early investigators with a great obstacle. Berzelius overcame this hurdle in 1823, when he successfully prepared tetrachlorosilane, $SiCl_4$. In 1857, Buff and Wöhler synthesized trichlorosilane, Cl_3SiH. These two compounds, $SiCl_4$ and Cl_3SiH, were the prototypes of the halogeno- and the hydro-silanes, the two classes of silane which were to become intimately involved in the major methods for carbon–silicon bond formation.

The first organosilane, tetraethylsilane, was prepared by Friedel and Crafts in 1863. Coupled with the newly discovered Grignard reaction, the availability of tetrachlorosilane allowed Kipping, one of the major early investigators, to prepare and to investigate the properties of a range of organosilanes. He concluded[1] rather depressingly that 'as . . . the few (organic derivatives of silicon) which are known are very limited in their reactions, the prospects of any immediate and important advance in this section of organic chemistry does not seem to be very hopeful'. Fortunately, he was wrong!

The discovery and development of the silicone polymers led to the now giant silicone industry, and also ensured a continuing availability of organosilicon monomers for research. The explosive growth of organosilicon chemistry over the past fifteen years has created a growing awareness of its considerable utility to the organic chemist. By formally substituting a silicon moiety in place of a hydrogen atom of an organic substrate or of a reagent, one can activate the substrate to reaction, one can direct the course of reaction, or one can afford temporary protection to the substrate from unwanted reaction. These facets will all be revealed in subsequent chapters, which are organized mainly in terms of the carbon functionality. Throughout, emphasis is placed upon those sequences in which silicon is absent from the final products, it having acted as 'ferryman' in mediating the course of the particular transformation. No specific chapter deals with formation or cleavage reactions of carbon–silicon bonds; instead, such processes are treated individually in appropriate chapters.

In addition to the primary literature, many books and review articles concentrating on the organic chemistry of organosilanes have been of immense help in the preparation of this treatise. These indispensable sources include general accounts of organosilicon chemistry[2–17] and the more recent compilations[18,19] of an on-going survey of its synthetic applications. Other more specific sources are referred to in the relevant subject chapters.

The spectral characteristics of organosilicon compounds are not discussed in any detail, since they are normally similar to those of the corresponding all-carbon compounds. The NMR spectral characteristics of silicon itself have been tabulated[20,21]. Of more use to the organic chemist is the fact that the system H–C–Si (with C bearing two further bonds) is readily detected by NMR spectroscopy, normally appearing in the 'clean' region between 0 and 1 ppm. For this reason, it is advisable to determine the NMR spectrum of the substrate prior to the addition of tetramethylsilane as internal standard.

Many simple organosilicon compounds are now commercially available. Major 'fine chemical' companies which can supply from an extensive range include: Fluka AG, CH-9470 Buchs, Switzerland; Aldrich Chemical Co., Milwaukee, Wisconsin, U.S.A., and Gillingham, Dorset SP8 4JL, U.K., and subsidiary companies; Cambrian Chemicals, Croydon CRO 4XB, U.K.; Petrarch Systems, Inc., Levittown, Pennsylvania, U.S.A., whose catalogue for 1979 includes a good précis of organosilicon chemistry; P.C.R. Research Chemicals, Inc., Gainesville, Florida, U.S.A. and Ventron GmbH, Karlsruhe, West Germany; Pierce Chemical Co., Rockford, Illinois, U.S.A. and Pierce and Warriner (UK) Ltd., Chester, Cheshire, U.K.

References

1 KIPPING, F. S., *Proc. R. Soc. A,* **159,** 139 (1937)
2 EABORN, C., *'Organosilicon Compounds',* Butterworths, London (1960)
3 PETROV, A. D., MIRONOV, B. F., PONOMARENKO, V. A. and CHERNYSHEV, E. A., *'Synthesis of Organosilicon Monomers',* Heywood, London (1964)
4 SOMMER, L. H., *'Stereochemistry, Mechanism, and Silicon',* McGraw-Hill, New York (1965)
5 BAZANT, V., CHVALOVSKY, V. and RATHOUSKY, J., *'Organosilicon Compounds',* vols. 1 and 2, Academic Press, New York (1965)
6 VOORHOEVE, R. J. H., *'Organohalosilanes',* Elsevier, Amsterdam (1967)
7 EABORN, C. and BOTT, R. W., 'Synthesis and Reactions of the Silicon–Carbon Bond', in *'Organometallic Compounds of the Group IV Elements',* Ed. MacDiarmid, A. G., vol. 1, part 1, p. 105, Marcel Dekker, New York (1968)
8 PIERCE, A. E., *'Silylation of Organic Compounds',* Pierce Chemical Co., Rockford, Illinois (1968)

9 BIRKOFER, L. and RITTER, A., 'The Use of Silylation in Organic Synthesis', in *'Newer Methods in Preparative Organic Chemistry'*, Ed. Foerst, W., vol. 5, p. 211, Academic Press, New York (1968)
10 KLEBE, J., 'Silyl–Proton Exchange Reactions', *Accts chem. Res.* **3**, 299 (1970)
11 CHVALOVSKY, V., 'Cleavage Reactions of the Carbon–Silicon Bond', *Organometallic Reactions* **3**, 191 (1972)
12 KLEBE, J., 'Silylation in Organic Synthesis', *Adv. org. Chem.* **8**, 97 (1972)
13 FLEMING, I., 'Bond Formation Controlled by Silicon', *Chemy Ind.* 449 (1975)
14 HUDRLIK, P. F., 'Organosilicon Compounds in Organic Synthesis', *J. Organometallic Chem. Libr.* Ed. Seyferth, D., vol. 1, p. 127, Elsevier, Amsterdam (1976)
15 COLVIN, E. W., 'Silicon in Organic Synthesis', *Chem. Soc. Rev.* **7**, 15 (1978)
16 FLEMING, I., 'Organic Silicon Chemistry', in *'Comprehensive Organic Chemistry'*, Eds. Barton, D. H. R. and Ollis, W. D., vol. 3, p. 539, Pergamon, Oxford (1979)
17 BIRKOFER, L. and STUHL, O., 'Silylated Synthons', *Topics curr. Chem.* **88**, 33 (1980)
18 WASHBURNE, S. S., *J. organometal. Chem.* **83**, 155 (1974); **123**, 1 (1976); *Organometal. Chem. Rev.* **4**, 263 (1977)
19 RUBOTTOM, G. M., *Organometal. Chem. Rev.* **8**, 263 (1979); **10**, 277 (1980)
20 EBSWORTH, E. A. V., 'Physical Basis of the Chemistry of the Group IV Elements', in *'Organometallic Compounds of the Group IV Elements'*, Ed. MacDiarmid, A. G., vol. 1, part 1, p. 1, Marcel Dekker, New York (1968)
21 HARRIS, R. K., KENNEDY, J. D. and McFARLANE, W., in *'NMR and the Periodic Table'*, Eds. Harris, R. K. and Mann, B. E., p. 310, Academic Press, New York (1978)

Chapter 2

Physical properties of organosilicon compounds

Silicon's utility in organic synthesis depends upon three main factors: its relative bond strengths to other elements, its relative electronegativity, and the involvement or lack of involvement of its valence *p*- and empty *d*-orbitals.

2.1 Relative bond strengths

Table 1[1-3] indicates the approximate bond dissociation energies for Si–X and C–X containing compounds. The energies are those required to bring about

Table 2.1 Approximate bond dissociation energies (*D*) and bond lengths (*r*) for Si–X and C–X

Bond	*Compound*	*D*/kJ mol^{-1}	*r*/nm	*Bond*	*D*/kJ mol^{-1}	*r*/nm
Si–C	Me_4Si	318	0.189	C–C	334	0.153
Si–H	Me_3SiH	339		C–H	420	0.109
	Cl_3SiH	378, 382				
	D_3SiH		0.148			
Si–O	Me_3SiOMe	531		C–O	340	0.141
	$(Me_3Si)_2O$	812				
	$(H_3Si)_2O$		0.163			
Si–S		*ca.* 293		C–S	313	0.180
	$(H_3Si)_2S$		0.214			
Si–N	$(Me_3Si)_2NH$	320(*E*)		C–N	335	0.147
	$(H_3Si)_3N$		0.174			
Si–F	Me_3SiF	807		C–F	452	0.139
	H_3SiF		0.16			
Si–Cl	Me_3SiCl	471				
	H_3SiCl		0.205	C–Cl	335	0.178
Si–Br	Me_3SiBr	403				
	H_3SiBr		0.221	C–Br	268	0.194
Si–I	Me_3SiI	322		C–I	213	0.214
	H_3SiI		0.244			

homolytic fission, and as such do not give an accurate picture of the ease of heterolytic fission, which is the normal mode of behaviour. Further, the value for the silicon–nitrogen bond is a thermochemical bond energy term, i.e., is derived from the sum of the formation of all the bonds from gaseous atoms.

In spite of these limitations, certain generalities of profound significance can be made. In particular, it can be seen that whereas silicon's bonds to oxygen and fluorine are stronger than the corresponding bonds between carbon and these elements, its bonds to carbon and hydrogen are weaker.

2.2 Electronegativity

Relative electronegativity can be established on several scales, all of which are empirical to some extent. The values shown in *Table 2.2* are non-empirical, having been derived[4] from *ab initio* FSGO wave functions.

Table 2.2 Relative electronegativity

H	B	C	N	O	F
2.79	1.84	2.35	3.16	3.52	4.0
	Al	Si	P	S	Cl
	1.40	1.64	2.11	2.52	2.84
		Ge	As	Se	Br
		1.69	1.99	2.4	2.52

Regardless of the scale used, silicon always appears as being markedly more electropositive than carbon, resulting in strong polarization (1) of C–Si bonds and in a tendency for nucleophilic attack to occur at silicon. Such attack leads to bond heterolysis[5], especially when the carbon-containing moiety is a good leaving group and the attacking nucleophile is oxygen or halogen (*Scheme 2.1*); similar generalizations hold for bonds from silicon to oxygen, nitrogen and halogen.

To concentrate for example on the carbon–silicon bond, this bond is relatively stable towards homolytic fission, but is more readily cleaved by ionic reagents, either by nucleophilic attack at silicon or electrophilic attack at carbon. Since carbon–hydrogen bonds break in the same direction, C^-H^+, as do carbon–silicon bonds, C^-Si^+, a good indication of the likely behaviour of a C–Si bond can be predicted by consideration of an analogous C–H bond. Just as Ar–H bonds are broken by treatment with electrophiles such as bromine, so also are Ar–Si bonds. Similarly, the β-elimination reactions displaced by H–C–C–X systems can occur even more readily in the fragmentation reactions of Si–C–C–X systems.

Indeed, as a broad generality, it is usually true that when a C–H bond is cleaved by a particular ionic reagent, the corresponding $C–SiMe_3$ bond will be broken by the same reagent even more readily. This statement has to be

$\overset{\delta+}{Si}$—$\overset{\delta-}{C}$

(1)

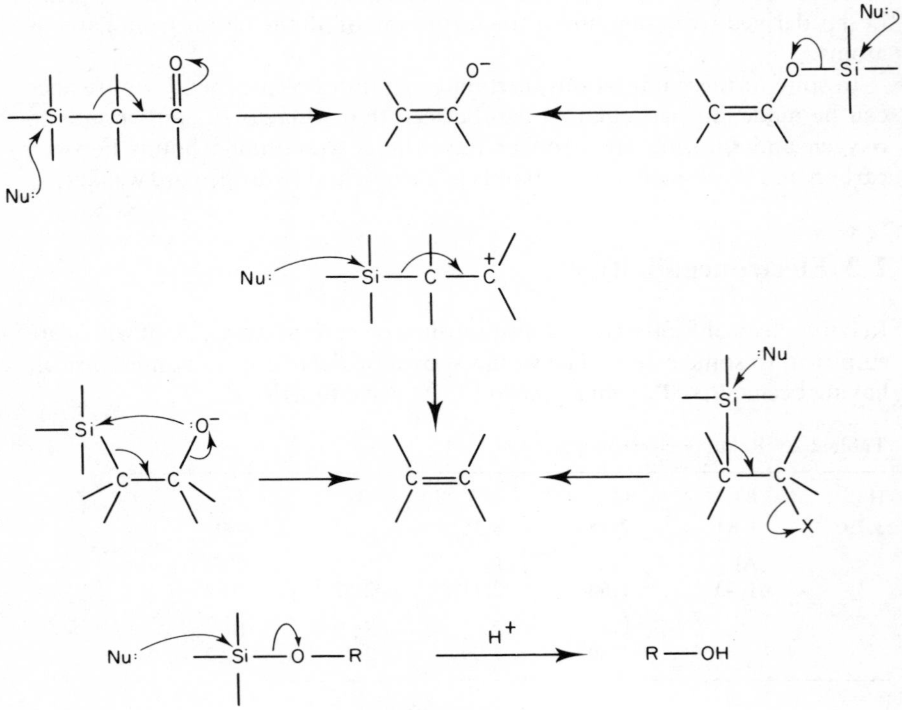

Scheme 2.1

qualified by comparison of the reactivities of C–Si and C–H bonds, in similar chemical environments, towards nucleophiles/bases. In such situations, the C–Si bond is the more reactive towards oxygen and halogen nucleophiles/bases, whereas the C–H bond is the more reactive towards carbon and nitrogen nucleophiles/bases.

On the other hand, a polarity reversal is observed when C–H and Si–H bonds are compared, with a resulting contrast in some aspects of chemical reactivity. On treatment with organometallic, i.e., carbon, bases, the C–H bond breaks as $C^{-}H^{+}$, whereas the Si–H bond breaks as $Si^{+}H^{-}$ (*Scheme 2.2*). Hydride bases are surprisingly non-selective, cleaving Si–H bonds as Si H^{+} as discussed in Chapter 18.

Such cleavages must be seen in perspective. Silicon's bond to carbon, while certainly polarized, is only weakly so when compared with those of other organometallic compounds. In general, organosilicon compounds can be handled readily, without the necessity for inert atmospheres or moisture exclusion. C–Si bonds can withstand a wide variety of reaction conditions and reagents, yet they have a latent lability which can be revealed at the appropriate moment. Reactions such as catalytic hydrogenation, hydroboration, hydroalumination, alcohol halogenation and free radical

$$Ph_3SiH + MeLi \longrightarrow Ph_3SiMe + LiH$$

$$Ph_3CH + MeLi \longrightarrow Ph_3CLi + MeH$$

$$R_3SiH + KH \longrightarrow R_3SiK + H_2$$

Scheme 2.2

halogenation, alcohol oxidation and reduction, epoxidation, base-induced nucleophilic addition and organometallic nucleophilic addition reactions have all been performed on C–Si compounds without C–Si bond cleavage.

2.3 *p*-Orbitals

Despite the great abundance of stable $(2p\text{–}2p)_\pi$ multiply bonded compounds which carbon provides, analogous compounds with Si–C, Si–N, or Si–O $(3p\text{–}2p)_\pi$ multiple bonds have proven to be elusive[6–8].

At first sight, this is rather surprising, since *ab initio* MO calculations have indicated a surprisingly strong Si–C $(p\text{–}p)_\pi$ bond in silaethene (2) with a π-bond strength of *ca.* 185 kJ mol^{-1}; this value is *ca.* two thirds of the π-bond energy of ethene itself, viz., 230–270 kJ mol^{-1}. The alternative formulation as a 1,2-diradical has been computed[9] to be *ca.* 120 kJ mol^{-1} higher in energy than the π-bonded ground state. On the other hand, the π-bond energy of 1,1-dimethylsilaethene (3), as determined[10] by ion cyclotron resonance spectroscopy, is significantly less than this, being *ca.* 142 kJ mol^{-1}. The latter value is in somewhat better accord with other thermochemically derived data[6–8] which suggest a π-bond energy of between 125 and 170 kJ mol^{-1}. Although relatively strong, the bond is highly reactive, owing to strong polarization and to the presence of a rather low-lying π^* antibonding MO. The best representation is probably as a π-bonded closed-shell (singlet) with a polarized bond, as depicted in (4).

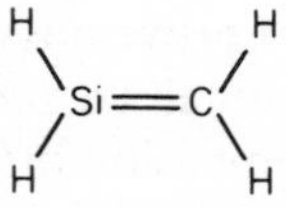

$$H_2Si{=}CH_2 \quad (2) \qquad Me_2Si{=}CH_2 \quad (3) \qquad \overset{\delta+}{Si}{=}\overset{\delta-}{C} \quad (4)$$

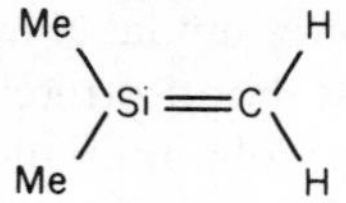

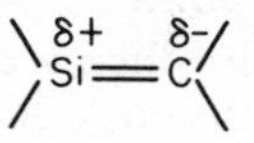

Silaethenes have been invoked many times[6,11] as reaction intermediates, and indeed iron complexes of η^3-1-silapropenyl have been claimed[12,13] to have been isolated (*Scheme 2.3*), but simple silaethenes have been isolated and characterized only recently.

Trimethylsilyldiazomethane (5) has been observed[14] to give both thermolysis and ambient temperature photolysis products suggestive of the intermediacy of 1,1,2-trimethylsilaethene (6). More recently, irradiation[15,16] of the diazoalkane (5), matrix isolated in argon at 8 K, resulted in a

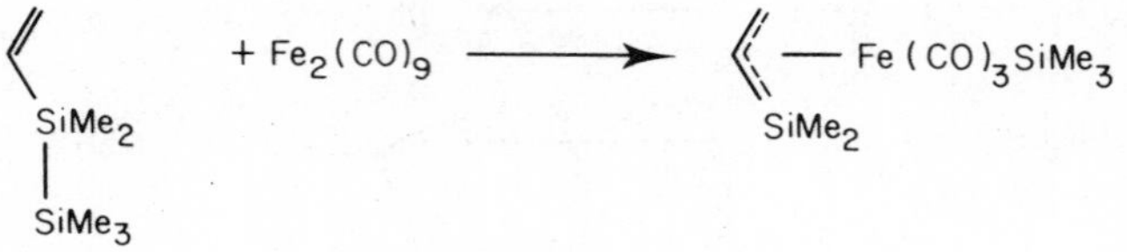

Scheme 2.3

photostationary state involving the diazirine (7). Continued irradiation gave the silaethene (6) via the carbene (8), the ESR spectrum of which was determined (*Scheme 2.4*).

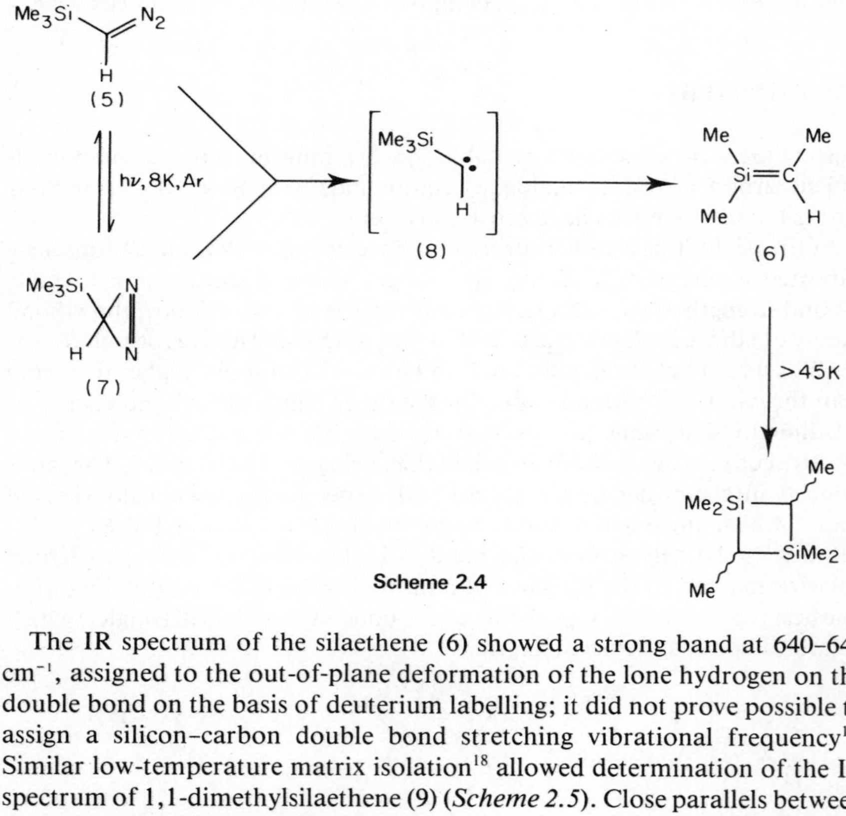

Scheme 2.4

The IR spectrum of the silaethene (6) showed a strong band at 640–645 cm^{-1}, assigned to the out-of-plane deformation of the lone hydrogen on the double bond on the basis of deuterium labelling; it did not prove possible to assign a silicon–carbon double bond stretching vibrational frequency[17]. Similar low-temperature matrix isolation[18] allowed determination of the IR spectrum of 1,1-dimethylsilaethene (9) (*Scheme 2.5*). Close parallels between these spectra and that of 2-methylbut-2-ene suggest an overall similarity in geometry of these molecules.

Silaethene analogues of silyl enol ethers (Chapter 17) are produced by the photolysis of acyl polysilanes, as evidenced by trapping experiments[19]; in

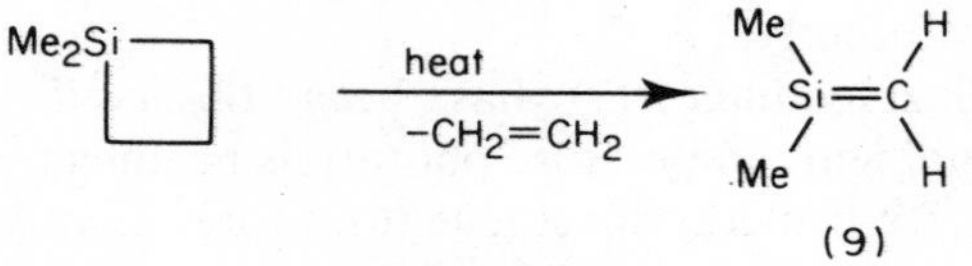

Scheme 2.5

the absence of trapping agents, these silaethenes dimerize to 1,2-disilacyclobutanes (*Scheme 2.6*). The highly hindered silaethene derived from pivaloyltris(trimethylsilyl)silane has a moderate life-time at room temperature, and is in mobile equilibrium with its dimer (11).

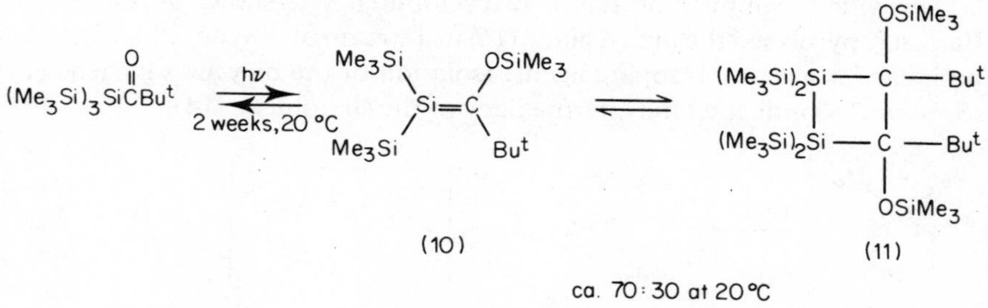
(Me3Si)3SiCBut
hν
2 weeks, 20 °C
Me3Si
OSiMe3
Si=C
Me3Si
But
(10)
(Me3Si)2Si — C — But
(Me3Si)2Si — C — But
OSiMe3
OSiMe3
(11)
ca. 70:30 at 20 °C

Scheme 2.6

A variety of highly reactive silaethenes can be readily if transiently obtained under mild conditions in organic solvents by 1,2-elimination reactions of α-lithiosilyl halides[20] and silanol[21] derivatives (*Scheme 2.7*).

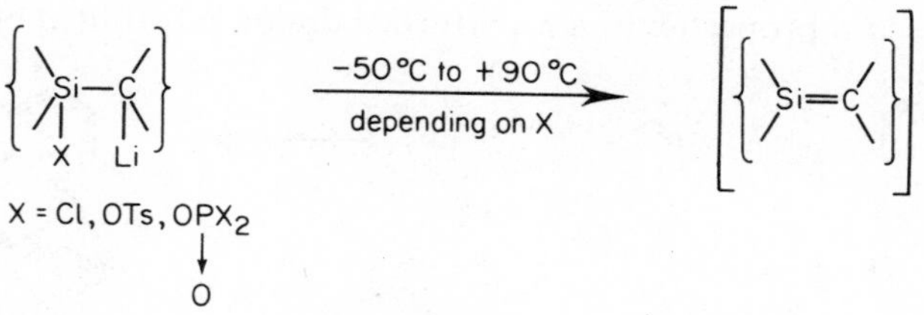
Si—C
X
Li
X = Cl, OTs, OPX2
O
−50 °C to +90 °C
depending on X
Si=C

Scheme 2.7

Disilenes (12) have been generated as transient species, and their chemical behaviour has been studied[22].

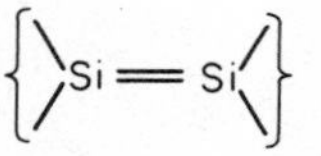

(12)

Much effort has also been devoted to the generation of Si–C aromatic π-bonds. *ab initio* MO Calculations[23] on silabenzene (13) have indicated that its resonance energy should be two thirds that of benzene itself; these calculations also showed that, although other isomers of C_5SiH_6 may be similar in energy, silabenzene itself should have all the attributes expected of an analogue of benzene. In spite of such predictions, silabenzene has proven

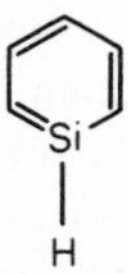

(13)

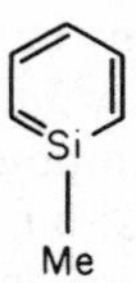

(14)

to be most elusive, many unsuccessful attempts[24] having been made to prepare it. Success was finally achieved by Barton[25a] and his group, who generated and trapped the silatoluene (14).

Following a report by Block[26] that flash vacuum pyrolysis of diallylsilanes (15) provided a simple synthesis of silacyclobutenes, as shown in *Scheme 2.8*, Barton[25a] pyrolysed the diallylsilane (16) in a stream of ethyne, which acted as both carrier gas and trapping agent. Isolation of the bicyclic silatriene (17) (*Scheme 2.9*) indicated the intermediacy of the silatoluene (14).

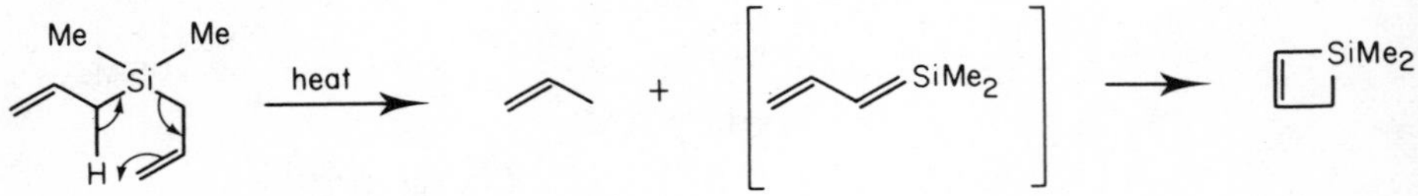

Me Me Si H heat + SiMe$_2$ SiMe$_2$

(15)

Scheme 2.8

The same chemical methodology, coupled to a special short-path pyrolysis apparatus, has allowed Maier and Reisenauer and their groups to prepare *and* characterize silabenzene[25b] itself. Although very labile, the molecule is stable in an argon matrix at 10 K, when it shows the IR, UV, and photoelectron spectral properties of a π-perturbed donor-substituted benzene.

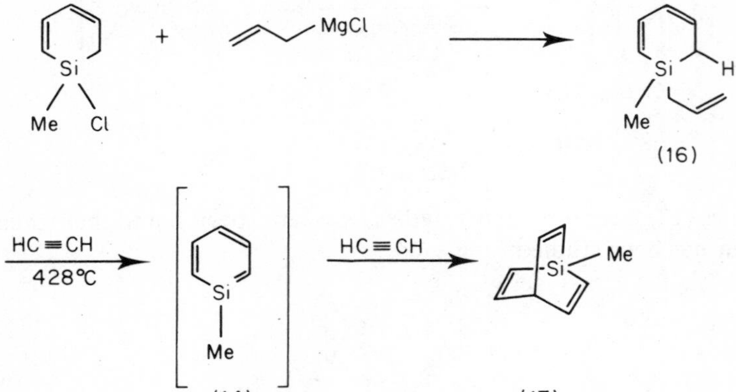

Si Me Cl + MgCl Si H Me (16)

HC≡CH 428°C Si Me HC≡CH Si Me

(14) (17)

Scheme 2.9

2.4 *d*-Orbitals

Organic compounds of silicon are normally quadricovalent, with silicon making tetrahedral sp^3-hybridized bonds. This is in spite of silicon differing from carbon in having vacant *d*-orbitals, with the outer electronic configuration, $3s^2 3p^2 3d^0$. Certain physical and chemical properties of organosilicon compounds raise the question of possible involvement of these vacant *d*-orbitals, but a detailed discussion of this topic is quite beyond the scope or intention of this book. Several excellent expositions[1,27–30] are recommended for such detail. Discussion will be limited here to those areas and examples which may have relevance to the synthetic utility of organosilicon compounds.

The three main areas of interest which involve possible *d*-orbital participation are the formation of additional σ-bonds, the stabilization of reaction intermediates and transition states, and the formation of internal π-bonds.

One of the best studied examples of octet expansion is the hexafluorosilicate ion, SiF_6^{2-}; an *X*-ray crystallographic analysis[31] revealed that the six fluorine atoms are arranged octahedrally about silicon, implying $3sp^3d^2$ hybrid bonds. Species such as $Na_2Si(CH_3)F_5$ probably contain related anions[32]. There is a wealth of spectroscopic and other evidence for a large degree of transannular interaction between nitrogen and silicon in compounds such as the biologically potent silatranes[33,34] (18), which have N–Si separations of only 0.219–0.234 nm. The quinquecovalent silicon species (19) appears[35] to have an only slightly elongated Si–O bond length of 0.1918 nm, and syntheses of the quinquecovalent compounds (20) have been reported[36] recently.

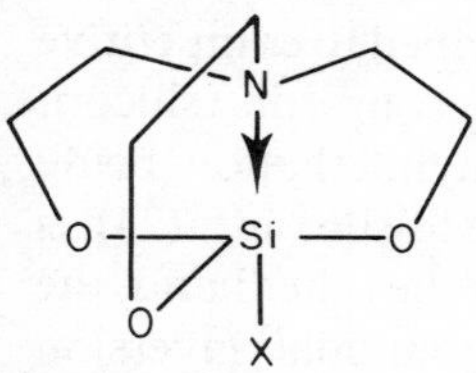

(18) 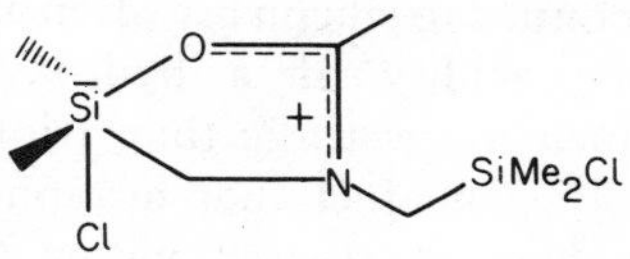 (19) 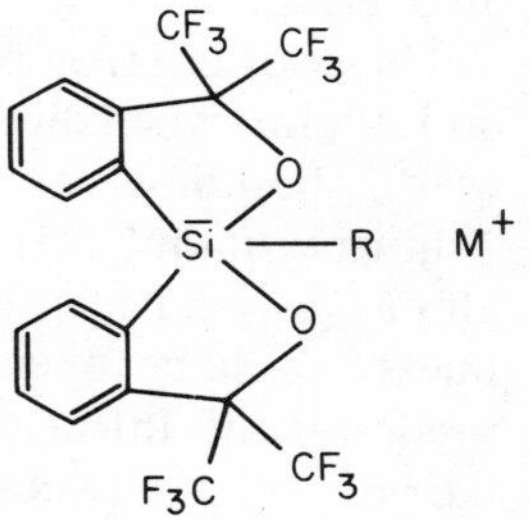 (20) R = Me, Ph
M = Li, Me_4N

Sommer's elegant and detailed studies[28] on chiral silicon compounds have shown that S_N2 reactions at silicon are often extremely fast, and often proceed with retention of stereochemistry. It is reasonable to propose that the involvement of *d*-orbitals will lower both those transition state energies involved in frontal attack (resulting in retention), and those involved in backside attack (resulting in inversion); these observations are at least consistent with *d*-orbitals taking part in transition state development.

Although some *d*-orbital involvement in coordination expansion by σ-bond formation is generally accepted, controversy becomes more heated over proposals of $(p\text{–}d)_\pi$ bonding[1,27-30]. UV spectroscopic studies of alkyl derivatives of both the six-membered cyclic siloxane (21), a 6π-system with a nodal plane (a Möbius anti-aromatic) and the seven-membered cyclic siloxane (22), an 8π-system with a nodal plane (a Möbius aromatic), have indicated[37] the normal properties associated with the linear analogues. Any degree of overlap of the empty $3d$ orbital on silicon with an adjacent filled $2p$ orbital on oxygen is apparently unimportant, even in cyclic systems where

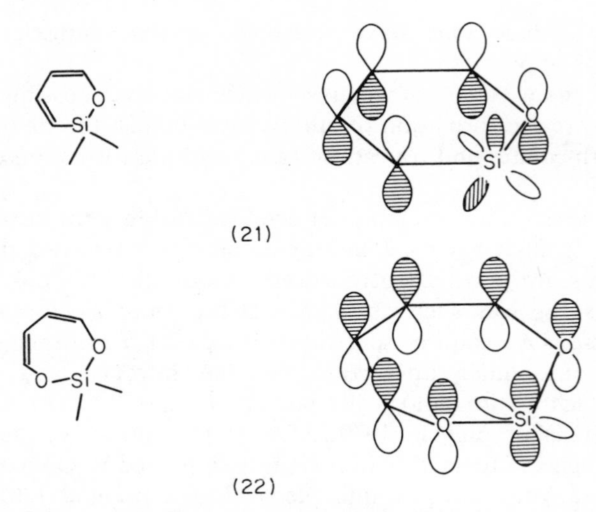

(21)

(22)

symmetry and theory both predict that stabilization would result from such an overlap.

The poor electron pair donor properties of nitrogen, phosphorus, oxygen and sulphur when directly bonded to silicon are often ascribed to competitive $(p\text{–}d)_\pi$ bonding, as is the acidity of a hydroxyl group on silicon. Triphenylsilanol (23) is much more acidic than triphenylmethanol, being almost as acidic as phenol[38]. The fact that hexaphenyldisiloxane (24) is linear[39] could be taken as evidence for $(p\text{–}d)_\pi$ bonding. On the other hand, the predominant influence on pyramidal stability and pyramidal inversion barriers of silyl-substituted amines, phosphines and arsines has been ascribed[40] entirely to substituent electronegativity, without the need of involving $(p\text{–}d)_\pi$ overlap.

Ph_3SiOH (23) $\qquad$ $Ph_3Si{-}O{-}SiPh_3$ (24)

It is well established, and it will shortly be exemplified, that a simple silyl substituent strongly favours carbonium ion development at the β-carbon atom (the β-effect; Chapter 4) and that it exerts a weak electron-attracting effect at the α-carbon atom. To explain the transmission of such electronic effects, a $(p\text{–}d)_\pi$ model was initially favoured over a $(p\text{–}\sigma^*)_\pi$ hyperconjugative one, but support[41] for the latter has increased. CNDO/2 calculations[42] have reproduced this effect of silyl substituents *without* inclusion of *d*-orbitals in the basis set. More recent calculations[43], and ^{19}F and ^{13}C NMR studies[44], have been interpreted in favour of the $(p\text{–}d)_\pi$ model, $SiH_n(CH_3)_{3-n}$ substituent groups being described as possessing $+F-I+M$ effects[45], with opposing field and inductive properties, when attached to a π-system.

Similar arguments can arise over α-anion or α-metalloid stabilization by silicon (Chapter 4). Such stabilization (*Scheme 2.10*) can be accounted for without having to involve $(p\text{–}d)_\pi$ overlap if analogy is drawn with the corresponding α-thio species. An important mechanism of stabilization of carbanions or polarized metalloids by adjacent sulphur is polarization of the electron distribution, dispersing the charge over the whole molecule. Additionally, perturbational MO calculations[46] have indicated that (n_C–σ^*_{SR}) charge transfer interactions of the carbon lone pair with the antibonding σ^* orbital of the adjacent antiperiplanar SR bond can contribute strongly to carbanion stabilization. The polarization of the C–Si bond in the direction $C^{\delta-}$–$Si^{\delta+}$ will ensure a relatively high coefficient on silicon in the σ^* level, further enhancing the stabilizing effect of such overlap. This suggests the stabilization (25) shown in *Scheme 2.10* for α-silyl carbanions and the related metalloids.

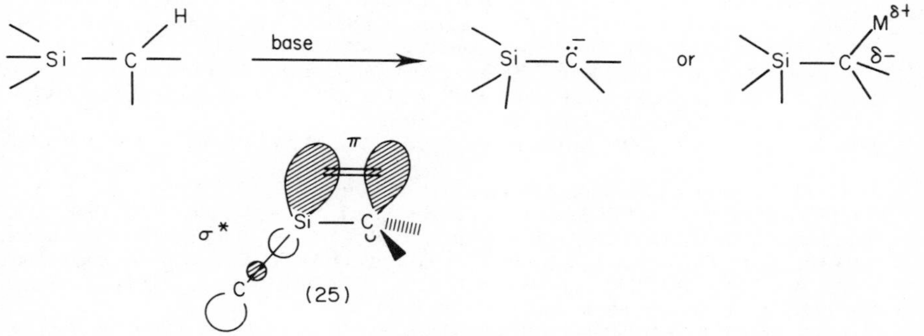

Scheme 2.10

References

1 EBSWORTH, E. A. V., in *'Organometallic Compounds of the Group IV Elements'*, Ed. MacDiarmid, A. G., vol. 1, part 1, Marcel Dekker, New York (1968)
2 COTTRELL, T. L., *'The Strengths of Chemical Bonds'*, Butterworths, London (1958)
3 WALSH, R. and WELLS, J. M., *J. chem. Soc. Faraday I* **72,** 100, 1212 (1976); DONCASTER, A. M. and WALSH, R., *J. chem. Soc. chem. Communs.* 904 (1979); *J. phys. Chem.* **83,** 3037 (1979)
4 SIMONS, G., ZANDLER, M. E. and TALATY, E. R., *J. Am. chem. Soc.* **98,** 7869 (1976)
5 EABORN, C. and BOTT, R. W., *'Organometallic Compounds of the Group IV Elements'*, Ed. MacDiarmid, A. G., vol. 1, part 1, pp. 105–536, Marcel Dekker, New York (1968)
6 GUSEL'NIKOV, L. E., NAMETKIN, N. S. and VDOVIN, V. M., *Accts chem. Res.* **8,** 18 (1975); GUSEL'NIKOV, L. E. and NAMETKIN, N. S., *Chem. Rev.* **79,** 529 (1979)
7 BALLARD, R. W. and WHEATLEY, P. J., *Organometal. Chem. Rev.* **2,** 1 (1976)
8 ISHIKAWA, M., *Pure appl. Chem.* **50,** 11 (1978)
9 AHLRICHS, R. and HEINZMANN, R., *J. Am. chem. Soc.* **99,** 7452 (1977)
10 PIETRO, W. J., POLLACK, S. K. and HEHRE, W. J., *J. Am. chem. Soc.* **101,** 7126 (1979)
11 *See,* for example, BERTRAND, G., MANUEL, G. and MAZEROLLES, P., *Tetrahedron Lett.* 2149 (1978); ISHIKAWA, M., FUCHIKAMI, T. and KUMADA, M., *J. Am. chem. Soc.* **99,** 245 (1977); BROOK, A. G. and HARRIS, J. W., *J. Am. chem. Soc.* **98,** 3381 (1976); BARTON, T. J. and KILGOUR, J. A., *J. Am. chem. Soc.* **98,** 7746 (1976); SAKURAI, H., KAMIYAMA, Y. and NAKADAIRA, Y., *J. Am. chem. Soc.* **98,** 7424 (1976); ANCELLE, J., BERTRAND, G., JOANNY, M. and MAZEROLLES, P., *Tetrahedron Lett.* 3153 (1979)
12 SAKURAI, H., KAMIYAMA, Y. and NAKADAIRA, Y., *J. Am. chem. Soc.* **98,** 7453 (1976); for refutation, *see* RADNIA, P. and McKENNIS, J. S., *J. Am. chem. Soc.* **102,** 6349 (1980)
13 *See also* BULKOWSKI, J. E., MIRO, N. D., SEPELAK, D. and Van DYKE, C. H., *J. organometal. Chem.* **101,** 267 (1974); MALISCH, W. and PANSTER, P., *J. organometal. Chem.* **64,** C5 (1974)

14 KREEGER, R. L. and SCHECHTER, H., *Tetrahedron Lett.* 2061 (1975)
15 CHAPMAN, O. L., CHANG, C.-C., KOLC, J., JUNG, M. E., LOWE, J. A., BARTON, T. J. and TUMEY, M. L., *J. Am. chem. Soc.* **98,** 7844 (1976)
16 CHEDEKEL, M. R., SKOGLUND, M., KREEGER, R. L. and SCHECHTER, H., *J. Am. chem. Soc.* **98,** 7846 (1976)
17 For a theoretical prediction of the IR spectrum of silaethene, *see* SCHLEGEL, H. B., WOLFE, S. and MISLOW, K., *J. chem. Soc. chem. Communs* 246 (1975)
18 MAL'TSEV, A. K., KHABASHESKY, V. N. and NEFODOV, O. M., *Izv. Akad. Nauk SSSR, Ser.khim.* 113 (1976); *Dokl. Akad. Nauk SSSR* **223,** 421 (1977); for an electron diffraction study, *see* MAHAFFY, P. G., GUTOWSKY, R. and MONTGOMERY, L. K., *J. Am. chem. Soc.* **102,** 2854 (1980)
19 BROOK, A. G., HARRIS, J. W., LENNON, J. and SHEIKH, M. El., *J. Amer. chem. Soc.* **101,** 83 (1979); BROOK, A. G., NYBURG, S. C., REYNOLDS, W. F., POON, Y. C., CHANG, Y.-M., LEE, J.-S. and PICARD, J.-P., *J. Am. chem. Soc.* **101,** 6750 (1979)
20 JONES, P. R. and LIM, T. F. O., *J. Am. chem. Soc.* **99,** 8447 (1977); JONES, P. R., LIM, T. F. O. and PIERCE, R. A., *J. Am. chem. Soc.* **102,** 4970 (1980)
21 WIBERG, N. and PREINER, G., *Angew. Chem. int. Edn* **16,** 328 (1977)
22 WULFF, W. D., GOURE, W. F. and BARTON, T. J., *J. Am. chem. Soc.* **100,** 6236 (1978); HENGGE, E., *Topics curr. Chem.* **51,** 1 (1974)
23 SCHLEGEL, H. B., COLEMAN, B. and JONES, Jr., M., *J. Am. chem. Soc.* **100,** 6499 (1978)
24 JUTZI, P., *Angew. Chem. int. Edn* **14,** 232 (1975)
25 (*a*) BARTON, T. J. and BURNS, G. T., *J. Am. chem. Soc.* **100,** 5246 (1978); BARTON, T. J. and BANASIAK, D. S., *J. Am. chem. Soc.* **99,** 5199 (1977); for matrix isolation, *see* KREIL, C. L., CHAPMAN, O. L., BURNS, G. T. and BARTON, T. J., *J. Am. chem. Soc.* **102,** 841 (1980); BOCK, H., BOWLING, R. A., SOLOUKI, B., BARTON, T. J. and BURNS, G. T., *J. Am. chem. Soc.* **102,** 429 (1980)
(*b*) SOLOUKI, B., ROSMUS, P., BOCK, H. and MAIER, G., *Angew. Chem. int. Edn* **19,** 51 (1980); MAIER, G., MIHM, G. and REISENAUER, H. P., *Angew. Chem. int. Edn* **19,** 52 (1980)
26 BLOCK, E. and REVELLE, L. K., *J. Am. chem. Soc.* **100,** 1630 (1978)
27 EABORN, C., *'Organosilicon Compounds',* pp. 91–113, Butterworths, London (1960)
28 SOMMER, L. H., *'Stereochemistry, Mechanism, and Silicon',* McGraw-Hill, New York (1965)
29 BÜRGER, H., *Angew. Chem. int. Edn* **12,** 474 (1973)
30 KWART, H. and KING, K., 'd-*Orbital Involvement in the Organo-Chemistry of Silicon, Phosphorus, and Sulphur',* Springer Verlag, Berlin (1977)
31 KETEELAR, J. A. A., *Z. Kristallogr.* **92,** 155 (1935)
32 MÜLLER, R. and DATHE, C., *Chem. Ber.* **98,** 235 (1965)
33 VORONKOV, M., *Pure appl. Chem.* **13,** 35 (1966); *Chemy Brit.* **9,** 411 (1973); *Topics curr. Chem.* **84,** 77 (1979); TACKE, R. and WANNAGAT, U., *Topics curr. Chem.* **84,** 1 (1979)
34 TURLEY, J. W. and BOER, F. P., *J. Am. chem. Soc.* **90,** 4026 (1968); BOER, F. P., TURLEY, J. W. and FLYNN, J. J., *J. Am. chem. Soc.* **90,** 5102 (1968)
35 ONAN, K. D., McPHAIL, A. T., YODER, C. H. and HILLARD, R. W., *J. chem. Soc. chem. Communs* 209 (1978)
36 PEROZZI, E. F. and MARTIN, J. C., *J. Am. chem. Soc.* **101,** 1591 (1979)
37 CHILDS, M. E. and WEBER, W. P., *J. org. Chem.* **41,** 1799 (1976)
38 WEST, R. and BANEY, R. H., *J. inorg. nucl. Chem.* **7,** 297 (1958); ALLRED, L., ROCHOW, E. G. and STONE, F. G. A., *J. inorg. nucl. Chem.* **2,** 416 (1956)
39 GLIDEWELL, C. and LILES, D. C., *J. chem. Soc. chem. Communs* 632 (1977)
40 BAECHLER, R. D., ANDOSE, J. D., STACKHOUSE, J. and MISLOW, K., *J. Am. chem. Soc.* **94,** 8060 (1972); BAECHLER, R. D., CASEY, J. P., COOK, R. J., SENKLER, G. H. and MISLOW, K., *J. Am. chem. Soc.* **94,** 2859 (1972)
41 EABORN, C., *J. chem. Soc. chem. Communs* 1255 (1972)
42 PITT, C. G., *J. organometal. Chem.* **61,** 49 (1973)
43 RAMSEY, B. G., *J. organometal. Chem.* **135,** 307 (1977)
44 ADCOCK, W., ALDOUS, G. L. and KITCHING, W., *Tetrahedron Lett.* 3387 (1978)
45 DEWAR, M. J. S., *'The Molecular Orbital Theory of Organic Chemistry',* p. 410, McGraw-Hill, New York (1969)
46 EPIOTIS, N. D., YATES, R. L., BERNARDI, F. and WOLFE, S., *J. Am. chem. Soc.* **98,** 5435 (1976); LEHN, J. M. and WIPFF, G., *J. Am. chem. Soc.* **98,** 7498 (1976)

Chapter 3

The β-effect

Trialkylsilyl substituents behave in a dichotomous manner, showing the properties of both electron donor and acceptor groups. Reactions which involve carbonium ion formation or development α to silicon are disfavoured, although not unknown[1], whereas carbonium ion formation or development β to silicon is positively encouraged.

Halogenoalkyltrialkylsilanes[2a] are very unreactive under S_N1 conditions. This can be illustrated by the following observations: neither chloromethyl- nor iodomethyl-trimethylsilane[2b] react with silver(I) ions; solvolysis[3] of 2-bromo-2-trimethylsilylpropane gives *only* the vinylsilane (1) by what is believed to be an *E*2 process; the bromide (2) has been reported[4] to be 110 times less reactive towards aqueous acetone at 50 °C than is the bromide (3) at 0 °C (*Scheme 3.1*). *ab initio* Molecular orbital calculations[5] have indicated that, when compared with CH_3, H_3Si has a stronger inductive (σ) but weaker hyperconjugative (π) stabilizing donor interaction with an adjacent carbonium ion.

Indeed, when directly linked to a π-electron system, the trialkylsilyl group can often behave as a weak electron acceptor[6]. Although only a slight effect is

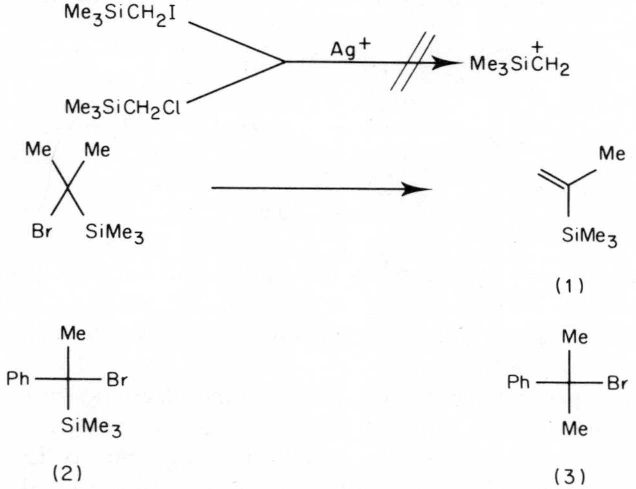

Scheme 3.1

Table 3.1 Electron-attracting effect of the Me_3Si group

Arylsilane		pK_a at 25 °C
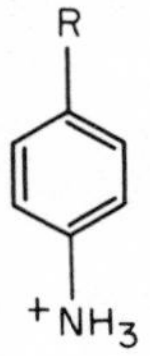 p-R-C$_6$H$_4$-NH_3^+	R = H	4.62
	= $SiMe_3$	4.36
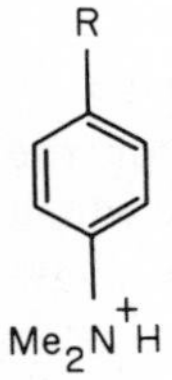 p-R-C$_6$H$_4$-N^+HMe_2	R = H	4.35
	= $SiMe_3$	3.98
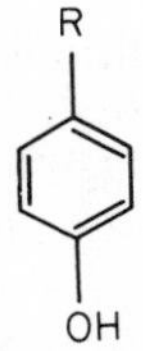 p-R-C$_6$H$_4$-OH	R = H	10.85
	= $SiMe_3$	10.64

detectable with aromatic carboxylic acids, *p*-trimethylsilyl-substituted anilines are noticeably *less* basic than their unsubstituted analogues, as are *p*-trimethylsilyl-substituted phenols (*Table 3.1*) *more* acidic[7].

Another example of this electron-attracting effect can be seen in the low temperature nitration of phenyltrimethylsilane, when a slight statistical excess of the *meta*-substituted product was obtained[8] (*Scheme 3.2*).

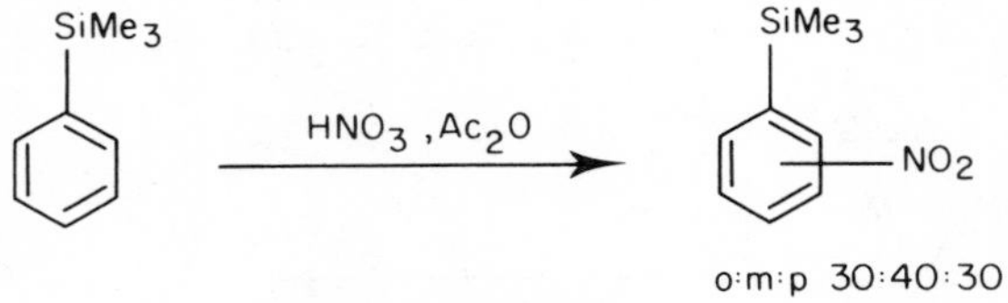

$$PhSiMe_3 \xrightarrow{HNO_3,\ Ac_2O} O_2N\text{-}C_6H_4\text{-}SiMe_3$$

o:m:p 30:40:30

Scheme 3.2

In direct contrast to this, the Me_3Si–C moiety exerts quite a strong electron-releasing effect. This results, for example, in the relative strengthening of aliphatic amine bases[9] and the weakening of carboxylic acids[10], as shown in *Table 3.2*.

In a more general sense, the Me_3Si–C unit offers a strong stabilizing, electron-donating effect to an adjacent electron-deficient centre. The mechanism by which this electron donation occurs is now believed[11] to be by hyperconjugation, that is, by $(p\text{–}\sigma)_\pi$ conjugation. The proposed involvement

Table 3.2 Electron-releasing effect of the Me_3Si–C moiety

	pK_a, 25 °C		pK_a, 25 °C
Me_3Si–CH_2NH_2	10.96	Me_3Si–CH_2CO_2H	5.22
Me_3C–CH_2NH_2	10.21	Me_3C–CH_2CO_2H	5.0

of the Si–C linkage can be understood by consideration of the high degree of polarization of the Si–C σ-bond; this ensures a high coefficient on carbon, and results in an enhanced ability to stabilize an adjacent electron-poor centre by orbital overlap. In addition, and perhaps of more importance, the Si–C bonding orbital is higher in energy than an H–C or C–C bonding orbital, and the energy match with the empty p (or unfilled π) orbital is therefore better.

Such vertical $(p\text{–}\sigma)_\pi$ hyperconjugation (4) is preferred over the alternative of neighbouring nucleophilic participation (5) to account for features in the charge-transfer spectra[12] of a range of benzylic organometallics with tetracyanoethene. Hyperconjugation, like all forms of π-conjugation, requires coplanarity, in this case of the interacting σ-bond and the axis of the electron-deficient π- or p-orbital, as illustrated in structure (6). Further evidence of such a coplanar requirement is provided by the observation that the organostannane (7) and its unsubstituted analogue (8) have identical charge-transfer maxima; the stannane (7) cannot achieve coplanarity of the tin–carbon bond and the induced electron-deficient orbital.

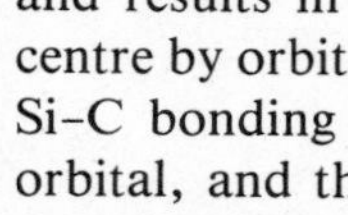

(4), M = Si, Sn, Pb

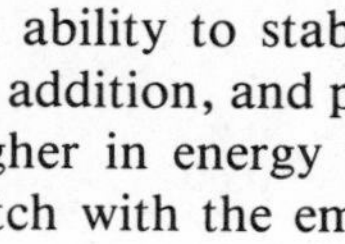

(5)

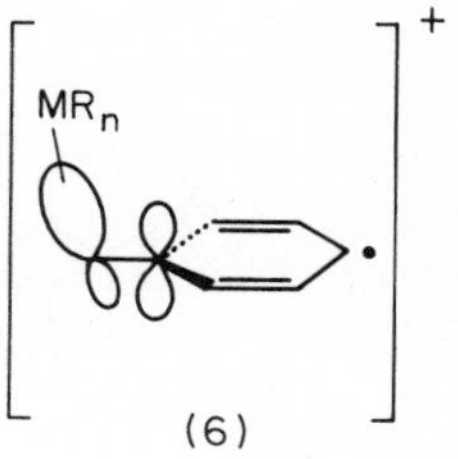

(6)

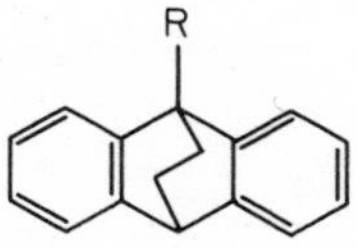

(7), R = $SnMe_3$

(8), R = H

From reaction rate[13] and charge-transfer spectral data[12], values for the Hammett *para*-substitution constants have been derived for the range of substituents shown in *Table 3.3. Table 3.3* shows that the electron-releasing effect of a Me_3Si–CH_2 group lies somewhere between that of a methyl and of a methoxy group, being roughly comparable with that of an acetamido group.

Table 3.3 Some Hammett *para*-substitution constants, σ_p^* for (4-methylphenyl)–R

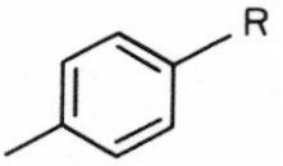

Substituent R	σ_p^* Reaction rates	σ_p^* Charge-transfer spectra
Me	−0.31	
Et	−0.58	−0.25
$(Me_3Si)_3C$	−0.52	
Me_3Si–CH_2	−0.54	−0.62
CH_3CONH	−0.58	−0.60
$(Me_3Si)_2CH$	−0.62	
MeO	−0.78	−0.74

Many aliphatic β-functionalized organosilanes exhibit abnormally high reactivity[14], which is often accompanied by anomalous electronic spectra. As discussed in Chapter 12, β-halogenoalkylsilanes are much more reactive under a wide variety of conditions than are the corresponding α- or γ-substituted analogues, with β-fragmentation processes leading to the production of alkenes (*Scheme 3.3*); similarly, β-hydroxyalkylsilanes are converted into alkenes by treatment with either acid or base. β-Ketosilanes and the related esters are readily cleaved by nucleophiles and electrophiles[15].

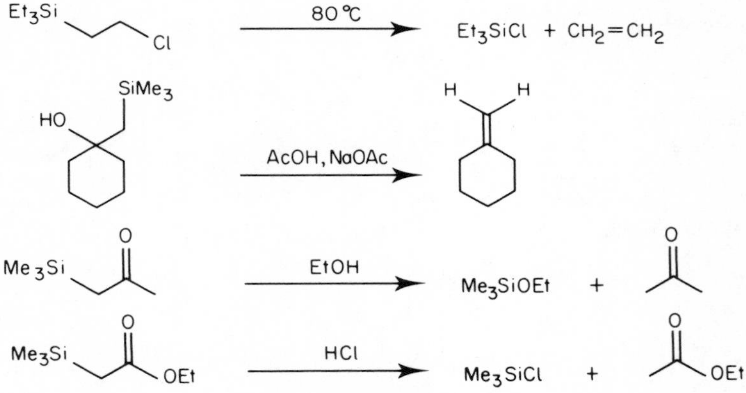

$Et_3SiCH_2CH_2Cl \xrightarrow{80\,^\circ C} Et_3SiCl + CH_2{=}CH_2$

1-(trimethylsilylmethyl)cyclohexanol (HO, CH_2SiMe_3) $\xrightarrow{AcOH,\ NaOAc}$ methylenecyclohexane ($H_2C{=}C_6H_{10}$)

$Me_3SiCH_2COCH_3 \xrightarrow{EtOH} Me_3SiOEt + CH_3COCH_3$

$Me_3SiCH_2CO_2Et \xrightarrow{HCl} Me_3SiCl + CH_3CO_2Et$

Scheme 3.3

Allyl-, vinyl-, and benzyl-silanes possess a useful degree or reactivity towards electrophilic reagents. Indeed, electrophilic attack on aryl, vinyl, allyl, alkynyl and cyclopropyl groups can often be controlled by the presence of a silyl moiety in the starting material[16,17]. The silyl group usually encourages attack at that site which will generate a carbonium ion, or at least an electron deficiency, β to silicon, and it is usually the silyl group which is lost in the second step of the reaction. These processes are treated in more detail in their appropriate functionality Chapters.

To return to the mechanism involved in creating the β-effect, while several factors may be involved, the most important of these appears to be $(p-\sigma)_\pi$ conjugation between the silicon–carbon bond and the developing positive charge in the transition state for reaction (9). It is additionally possible that this $(p-\sigma)_\pi$ conjugation may play a significant part in the weakening of the silicon–carbon bond and thus *promote* the cleavage processes which are so frequently encountered in the reactions of β-functionalized organosilanes.

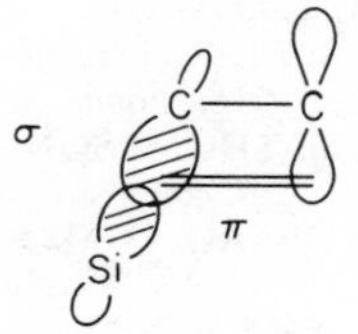

σ C—C π Si

(9)

An interesting case in which silicon is retained in the product was reported independently by Jarvie and Eaborn and their respective co-workers. Treatment of the specifically deuterium-labelled β-hydroxyethylsilane (10) with phosphorus(III) bromide produced[18] a mixture of the two β-bromoethylsilanes shown (*Scheme 3.4*). Analogously, re-isolation of starting material from the partial solvolysis[19] of the specifically deuterium-labelled β-bromoethylsilane (11) yielded material in which the deuterium had been extensively scrambled between the two carbon atoms of the ethyl group. Detailed study of these results led to the proposal of a mechanism which involves anchimerically-assisted ionization of the β C–Br or C–O bond, with a pseudosymmetrical transition state (12); the 1,2-silyl shifts linking the two isomeric types must be fast relative to loss of the silyl moiety.

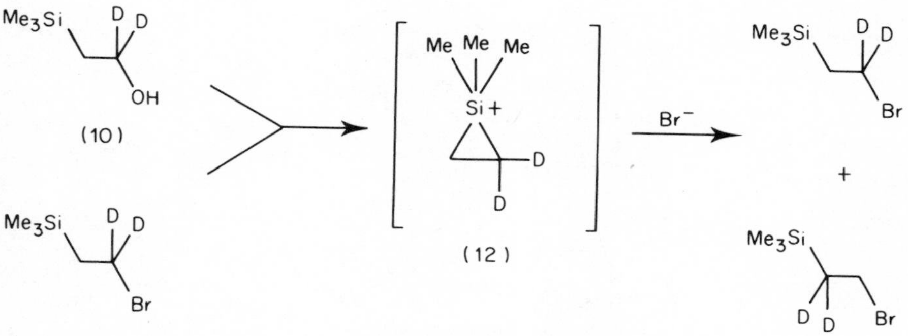

Me_3Si D D OH (10)

Me_3Si D D Br (11)

Me Me Me Si+ D D (12)

Br^-

Me_3Si D D Br + Me_3Si D D Br

Scheme 3.4

References

1 WHITMORE, F. C., SOMMER, L. H. and GOLD, J., *J. Am. chem. Soc.* **69,** 1976 (1947)
2 (*a*) For some interconversions, *see* AMBASHT, S., CHIU, S. K., PETERSON, P. E. and QUEEN, J., *Synthesis* 318 (1980)
(*b*) EABORN, C., *'Organosilicon Compounds',* Butterworths, London (1960)

3 CARTLEDGE, F. K. and JONES, J. P., *Tetrahedron Lett.* 2193 (1971)
4 COOK, M. A., EABORN, C. and WALTON, D. R. M., *J. organometal. Chem.* **29,** 389 (1971)
5 CLARK, T. and SCHLEYER, P. v. R., *Tetrahedron Lett.* 4641 (1979); APELOIG, Y., SCHLEYER, P. v. R. and POPLE, J. A., *J. Am. chem. Soc.* **99,** 1291 (1977); EABORN, C., FEICHTMAYR, F., HORN, M. and MURRELL, J. N., *J. organometal. Chem.* **77,** 39 (1974)
6 BOCK, H., BRÄHLER, G., FRITZ, G. and MATERN, E., *Angew. Chem. int. Edn* **15,** 699 (1976)
7 BENKESER, R. A. and KRYSIAK, H. R., *J. Am. chem. Soc.* **75,** 2421 (1953)
8 SPEIER, J. L., *J. Am. chem. Soc.* **75,** 2930 (1953)
9 NOLL, J. E., DAUBERT, B. F. and SPEIER, J. L., *J. Am. chem. Soc.* **73,** 3871 (1951); SOMMER, L. H. and ROCKETT, J., *J. Am. chem. Soc.* **73,** 5130 (1951); SHEA, K. J., GOBEILLE, R., BRAMBLETT, J. and THOMPSON, E., *J. Am. chem. Soc.* **100,** 1611 (1978)
10 SOMMER, L. H., GOLD, J. R., GOLDBERG, G. M. and MARANS, N. S., *J. Am. chem. Soc.* **71,** 1509 (1949)
11 BOTT, R. W., EABORN, C. and GREASLEY, P. M., *J. chem. Soc.* 4804 (1964); BERWIN, H. J., *J. chem. Soc. chem. Communs* 237 (1972); EABORN, C., *J. chem. Soc. chem. Communs* 1255 (1972); EABORN, C., *J. organometal. Chem.* **100,** 43 (1975); FLEMING, I., *'Frontier Orbitals and Organic Chemical Reactions',* pp. 81–83, Wiley, London (1977)
12 HANSTEIN, W., BERWIN, H. J. and TRAYLOR, T. G., *J. Am. chem. Soc.* **92,** 829, 7476 (1970); HARTMAN, G. D. and TRAYLOR, T. G., *J. Am. chem. Soc.* **97,** 6147 (1975)
13 COOKE, M. A., EABORN, C. and WALTON, D. R. M., *J. organometal. Chem.* **24,** 293 (1970)
14 JARVIE, A. W. P., *Organometal. Chem. Rev. A* **6,** 153 (1970)
15 BAUKOV, Yu. I. and LUTSENKO, I. F., *Organometal. Chem. Rev. A* **6,** 355 (1970)
16 EABORN, C. and BOTT, R. W., *'Organometallic Compounds of the Group IV Elements',* Ed. MacDiarmid, A. G., vol. 1, part 1, Marcel Dekker, New York (1972)
17 CHAN, T. H. and FLEMING, I., *Synthesis* 761 (1979)
18 JARVIE, A. W. P., HOLT, A. and THOMPSON, J., *J. chem. Soc. B* 852 (1969); BOURNE, A. J. and JARVIE, A. W. P., *J. organometal. Chem.* **24,** 335 (1970)
19 COOKE, M. A., EABORN, C. and WALTON, D. R. M., *J. organometal. Chem.* **24,** 301 (1970)

Chapter 4

α-Metallated organosilanes

α-Metallated organosilanes play a fundamental role in preparative organosilicon chemistry. This Chapter is devoted entirely to methods of preparation of these species; their stability is discussed in Chapter 2, and their utility is displayed in many chapters, in particular in Chapter 12.

There are basically four methods[1] of achieving α-metallation of an organosilane. These comprise addition of an organometallic species to a vinylsilane, proton abstraction, metal–halogen exchange, and transmetallation of an already α-metallated organosilane.

4.1 Organometallic addition to vinylsilanes

This process of nucleophilic addition to a vinyl silane was first observed by Cason and Brooks[2], and can be depicted as shown in *Scheme 4.1*. Some representative examples of this addition are given in *Table 4.1*.

Table 4.1 Organometallic addition to vinylsilanes

R^1_3Si–CH=CH$_2$ + R^2M ⟶ R^1_3Si–CH(M)–CH$_2$–R^2

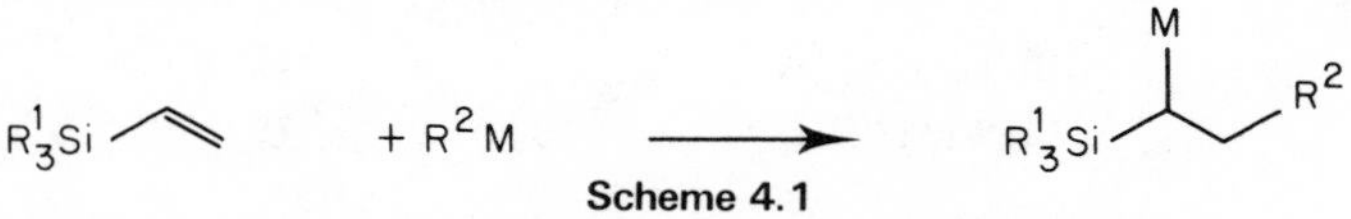

Scheme 4.1

Vinylsilane	*Organometallic reagent*	*Notes*	*References*
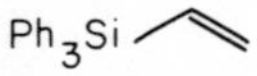 Ph_3Si–CH=CH$_2$	R^2Li	1	2, 3
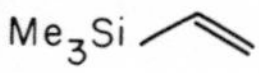 Me_3Si–CH=CH$_2$	EtLi		4
	Bu^tLi		5
$(Me_3Si)_2C{=}CH_2$	R^2Li	2	6
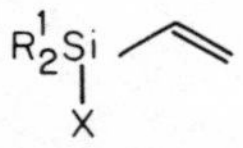 $R^1_2Si(X)$–CH=CH$_2$	R^2MgX	3	7

[1] R^2 = Me, 0%; Pr^n, 0%; Bu^n, 67%, Ph, 84%
[2] R^2 = Bu^n, Bu^s, Bu^t.
[3] R^2 = Me, Ph, CH_2Ph; X = Cl, OEt, which exert an activating effect, with often quantitative yields of adducts; R^1 = range of alkyl groups, with reactivity order of tertiary > secondary > primary.

4.2 Proton abstraction

A wide range of variously functionalized organosilanes has been deprotonated α to silicon (*Scheme 4.2*). The bases most commonly used are either alkyl-lithiums, often in the activating presence of TMEDA, or LDA; this illustrates the propensity of carbon and nitrogen bases/nucleophiles to attack at hydrogen rather than at silicon. Yields are highest when additional carbanion or carbanionoid stabilizing groups are present, and a selection of substrates which have been deprotonated successfully is shown in *Table 4.2*; the substrates have been drawn to indicate the proton being removed.

Table 4.2 α-Proton abstraction

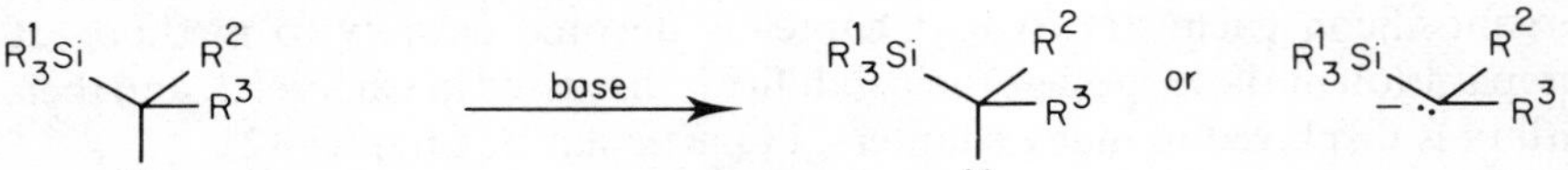

R^1_3Si–CR^2R^3–H $\xrightarrow{\text{base}}$ R^1_3Si–CR^2R^3–M or R^1_3Si–$\bar{C}$:R^2R^3

Scheme 4.2

Organosilane	*Base*	*Notes*	*References*
Me_3SiCH_2–H	Bu^nLi/TMEDA	1	8
$Bu^nMe_2SiCH_2$–H	Bu^nLi/TMEDA		8
$Me_2Si(X)CH_2$–H	Bu^nLi/TMEDA	2	9
$(Me_3Si)_3C$–H	MeLi		10
	Bu^tLi/HMPA		11
Me_3SiCHR–H	Bu^nLi/TMEDA	3	12
	Bu^nLi	4	13
$Ph_3SiCHPh$–H	Bu^nLi		14
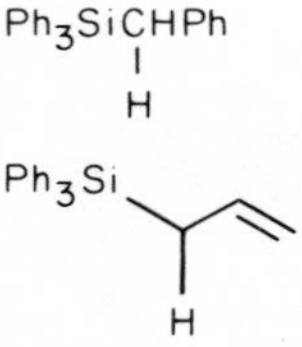 $Ph_3SiCH(H)CH{=}CH_2$	Ph_3SiLi		15
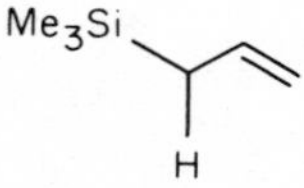	Bu^nLi/TMEDA		16
$Me_3SiCH(H)CH{=}CH_2$	Bu^tLi/HMPA		17
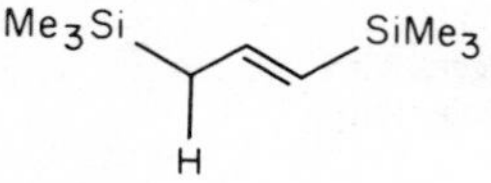	Bu^nLi		18
$Me_3SiCH(H)CH{=}CHSiMe_3$	Bu^nLi/TMEDA		19

Table 4.2—*cont.*

Organosilane	*Base*	*Notes*	*References*
$Me_3SiCH_2CH{=}NR$	LDA		22
$Ph_3SiCH_2N{=}CPh_2$	LDA or Bu^nLi		23
2-Me_3Si-1,3-dithiane (H at C-2)	Bu^nLi	5	24–27
$Me_3SiCH_2SiMe_3$	Bu^tLi/HMPA		11
$Me_3SiCH{=}N_2$	$KOBu^t$ or Bu^tLi		20, 21
$Me_3SiCH_2\overset{+}{S}Me_2$	$KOBu^t$	6	28, 29
$Me_3SiCH_2\overset{+}{P}Ph_3$	Bu^sLi		30
	PhLi		31, 32
Me_3SiCH_2SePh	Bu^sLi/TMEDA		33
$Me_3SiCH(Ph)SeAr$	LDA	7	34
	$LiNEt_2$		33
Me_3SiCH_2CN	LDA		35
$Me_2PhSiCH(Me)CN$	LDA		35
$Me_3SiCH_2C(O)OH$	LDA		36

Table 4.2—*cont.*

Organosilane	*Base*	*Notes*	*References*
Me_3Si–C(H)–CO_2Et	LDA		37
Me_3Si–C(H)–CO_2Bu^t	LDA		38
$(Me_3Si)_2C(H)CO_2Bu^t$	LDA		38
Me_3Si–C(H)(Cl)–CO_2Bu^t	LDA		39
Me_3Si–C(H)–$CONR_2$	LDA		40, 41
Me_3Si–C(H)–COSR	LDA		42
3-H-3-Me_3Si-5-R-lactone (O, O, R)	$LiCPh_3$	8	36, 43
dihydro-oxazine (N, O)–C(H)(X)$SiMe_3$	Bu^nLi	9	44
benzothiazole (N, S)–$CH(H)SiMe_3$	Bu^nLi		45
3-H-3-Me_3Si-β-lactam (O, NPh)	LDA		46
Me_3Si–C(H)–Cl	Bu^sLi		47

Table 4.2—*cont.*

Organosilane	*Base*	*Notes*	*References*
Me_3Si, Me, Cl, H	Bu^sLi		47
Ph_3Si, O, H	Bu^nLi		48
Ph, O, $SiMe_3$, H	Bu^tLi		48
Me_3Si, OMe, H	Bu^sLi		49

[1] 36% yield
[2] X = Cl, F, OMe, OEt
[3] R = Ph, PPh_2, $P(S)Ph_2$, SMe
[4] R = SPh, S(O)Ph, $P(O)(OEt)_2$
[5] In the presence of 1–2 equivalents HMPA, this anion adds almost exclusively (1,4) to cyclohexenone; in the absence of HMPA, 1,2-addition takes place (ref. 26).
[6] For the preparation of the corresponding sulphoxonium ylid, *see* ref. 28.
[7] According to ref. 33, LDA is completely ineffective at removing this proton.
[8] For the preparation of such species, *see* ref. 36.
[9] X = H, $SiMe_3$.

4.3 Metal–halogen exchange

Many α-halogenoalkyl- and α-halogenoalkenylorganosilanes undergo metal–halogen exchange on reaction with either a metal or an organometallic reagent (*Scheme 4.3*). Yields are normally excellent, and some examples are given in *Table 4.3*.

4.4 Transmetallation

This process can be represented as shown in *Scheme 4.4*. As yet, there are comparatively few examples of this process. A particularly interesting application is shown in *Table 4.4*; the cleavage of silicon–carbon bonds in polysilylated methanes can be readily achieved by use of a good nucleophile for silicon, in this case methoxide ion.

Table 4.3 Metal–halogen exchange

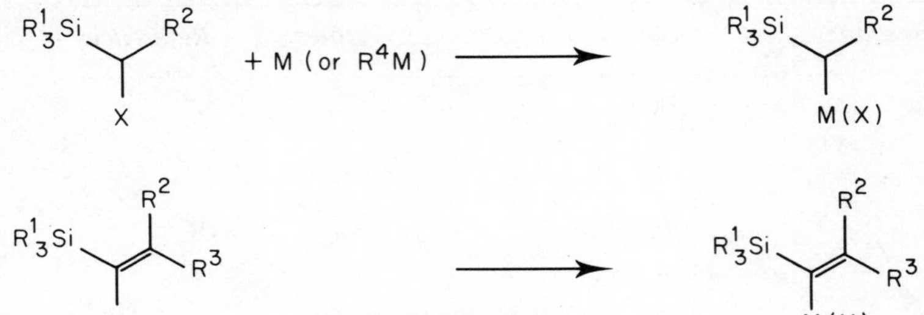

$R^1_3SiCH(R^2)X$ + M (or R^4M) ⟶ $R^1_3SiCH(R^2)M(X)$

$R^1_3SiC(X)=C(R^2)R^3$ ⟶ $R^1_3SiC(M(X))=C(R^2)R^3$

Scheme 4.3

α-Halogeno-organosilane	*Metal/Organometallic*	*Notes*	*References*
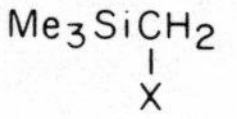 Me_3SiCH_2X	Mg	1	50, 51
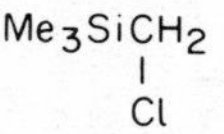 Me_3SiCH_2Cl	Li	2	52
 Me_3SiCH_2Br	Bu^nLi		53
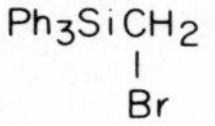 Ph_3SiCH_2Br	Bu^nLi		53
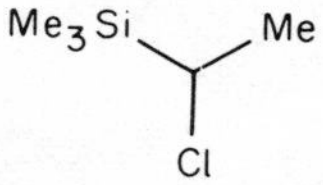 $Me_3SiCH(Me)Cl$	Mg		50
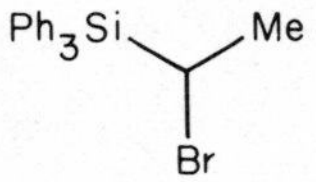 $Ph_3SiCH(Me)Br$	Bu^nLi		53
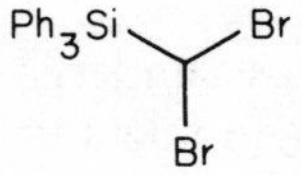 $Ph_3SiCHBr_2$	Bu^nLi		53
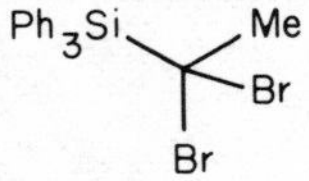 $Ph_3SiC(Me)Br_2$	Bu^nLi		53
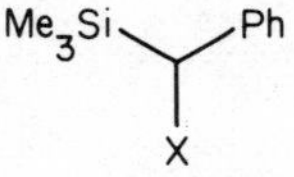 $Me_3SiCH(Ph)X$	Mg	1	50, 51
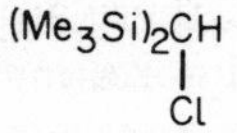 $(Me_3Si)_2CHCl$	Li		54
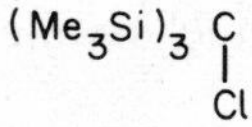 $(Me_3Si)_3CCl$	MeLi		10

Table 4.3—*cont.*

α-Halogeno-organosilane	*Metal/Organometallic*	*Notes*	*References*
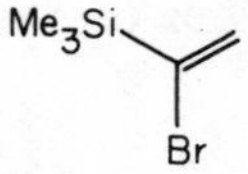 Me_3Si–C(Br)=CH_2	Bu^tLi		55
Et_3Si–C(Br)=CH_2	Bu^nLi		56
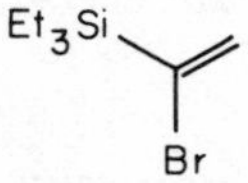	Mg		57
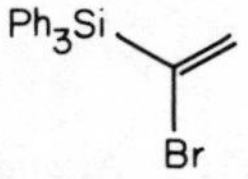 Ph_3Si–C(Br)=CH_2	Bu^nLi		53
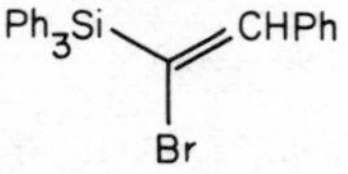 Ph_3Si–C(Br)=CHPh	Bu^nLi		53

[1] X = Cl, Br
[2] Me_3SiCH_2Li is reported to be a white crystalline solid, m.p. 112 °C

Table 4.4 Transmetallation

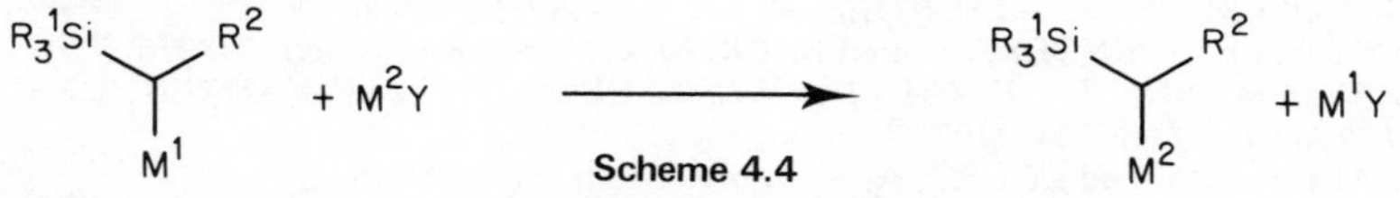

$$R^1_3Si\text{–}CH(R^2)\text{–}M^1 + M^2Y \longrightarrow R^1_3Si\text{–}CH(R^2)\text{–}M^2 + M^1Y$$

Scheme 4.4

α-Metallated organosilane	*Transmetallating reagent*	*References*
Me_3SiCH_2–$SiMe_3$	NaOMe/HMPA	58
$Me_3SiCHPh$–$SiMe_3$	NaOMe/HMPA	58
$(Me_3Si)_2CH$–$SiMe_3$	LiOMe/HMPA	58
$(Me_3Si)_3C$–$SiMe_3$	NaOMe/HMPA	58
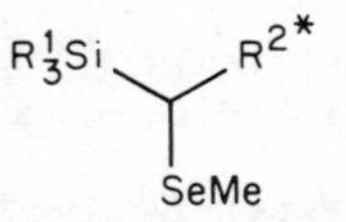 R^1_3Si–CH(R^2)–SeMe *	Bu^nLi	59
Me_3SiCH_2–$SnBu^n_3$	Bu^nLi	60

*R^1 = Me, Et; R^2 = Me, C_6H_{13}, SeMe

References

1 PETERSON, D. J., *Organometal. Chem. Rev. A* **7**, 295 (1972)
2 CASON, L. F. and BROOKS, H. G., *J. Am. chem. Soc.* **74**, 4582 (1952); *J. org. Chem.* **19**, 1278 (1954)
3 CHAN, T. H., CHANG, E. and VINOKUR, E., *Tetrahedron Lett.* 1137 (1970); CHAN, T. H. and CHANG, E., *J. org. Chem.* **39**, 3264 (1974)
4 HUDRLIK, P. F. and PETERSON, D., *Tetrahedron Lett.* 1133 (1974)
5 MULVANEY, J. E. and GARDLUND, Z. G., *J. org. Chem.* **30**, 917 (1965)
6 GRÖBEL, B.-Th. and SEEBACH, D., *Angew. Chem. int. Edn* **13**, 83 (1974)
7 BUELL, G. R., CORRIU, R., GUERIN, C. and SPIALTER, L., *J. Am. chem. Soc.* **92**, 7424 (1970)
8 PETERSON, D. J., *J. organometal. Chem.* **9**, 373 (1967)
9 GORNOWICZ, G. A. and WEST, R., *J. Am. chem. Soc.* **90**, 4478 (1968)
10 COOK, M. A., EABORN, C., JUKES, A. E. and WALTON, D. R. M., *J. organometal. Chem.* **24**, 529 (1970)
11 GRÖBEL, B.-Th. and SEEBACH, D., *Chem. Ber.* **110**, 852 (1977)
12 PETERSON, D. J., *J. org. Chem.* **33**, 780 (1968)
13 CAREY, F. A. and COURT, A. S., *J. org. Chem.* **37**, 939 (1972); CAREY, F. A. and HERNANDEZ, O., *J. org. Chem.* **38**, 2670 (1973)
14 WU, T. C., WITTENBERG, D. and GILMAN, H., *J. org. Chem.* **25**, 596 (1960)
15 GILMAN, H. and AOKI, D., *J. organometal. Chem.* **2**, 44 (1964)
16 CORRIU, R. J. P., MASSE, J. and SAMATE, D., *J. organometal. Chem.* **71**, 93 (1975)
17 LAU, P. W. K. and CHAN, T. H., *Tetrahedron Lett.* 2383 (1978)
18 AYALON-CHASS, D., EHLINGER, E. and MAGNUS, P., *J. chem. Soc. chem. Communs* 772 (1977)
19 CARTER, M. J. and FLEMING, I., *J. chem. Soc. chem. Communs* 679 (1976)
20 COLVIN, E. W. and HAMILL, B. J., *J. chem. Soc. chem. Communs* 151 (1973); *J. chem. Soc. Perkin I* 869 (1977)
21 SCHÖLLKOPF, U. and SCHOLZ, H.-U., *Synthesis* 271 (1976)
22 COREY, E. J., ENDERS, D. and BOCK, M. G., *Tetrahedron Lett.* 7 (1976)
23 KAUFFMANN, T., KOCH, U., STEINSEIFER, F. and VAHRENHORST, A., *Tetrahedron Lett.* 3341 (1977)
24 CAREY, F. A. and COURT, A. S., *J. org. Chem.* **37**, 1926 (1972)
25 JONES, P. F. and LAPPERT, M. F., *J. chem. Soc. chem. Communs* 526 (1972)
26 SEEBACH, D., GRÖBEL, B.-Th., BECK, A. K., BRAUN, M. and GEISS, K.-H., *Angew. Chem. int. Edn* **11**, 443 (1972); SEEBACH, D., KOLB, M. and GRÖBEL, B.-Th., *Tetrahedron Lett.* 3171 (1974); GRÖBEL, B.-Th., BURSTINGHAUS, R. and SEEBACH, D., *Synthesis* 121 (1976)
27 BROWN, C. A. and YAMAICHI, A., *J. chem. Soc. chem. Communs* 100 (1979)
28 FLEISCHMANN, C. and ZBIRAL, E., *Tetrahedron* **34**, 317 (1978); *see also* COOPER, G. D., *J. Am. chem. Soc.* **76**, 3713 (1954)
29 SCHMIDBAUR, H. and KAPP, W., *Chem. Ber.* **105**, 1203 (1972)
30 COOKE, F., MAGNUS, P. and BUNDY, G. L., *J. chem. Soc. chem. Communs* 714 (1978)
31 GILMAN, H. and TOMASI, R. A., *J. org. Chem.* **27**, 3647 (1962)
32 SCHMIDBAUR, H. and STÜHLER, H., *Angew. Chem. int. Edn* **12**, 321 (1973)
33 REICH, H. J. and SHAH, S. K., *J. Am. chem. Soc.* **97**, 3250 (1975); *J. org. Chem.* **42**, 1773 (1977)
34 SACHDEV, K. and SACHDEV, H. S., *Tetrahedron Lett.* 4223 (1976)
35 OJIMA, I., KUMAGAI, M. and NAGAI, Y., *Tetrahedron Lett.* 4005 (1974)
36 GRIECO, P. A., WANG, C-L. J. and BURKE, S., *J. chem. Soc. chem. Communs* 537 (1975)
37 SHIMOJI, K., TAGUCHI, H., OSHIMA, K., YAMAMOTO, H. and NOZAKI, H., *J. Am. chem. Soc.* **96**, 1620 (1974); TAGUCHI, H., SHIMOJI, K., YAMAMOTO, H. and NOZAKI, H., *Bull. chem. Soc. Japan* **47**, 2529 (1974)
38 HARTZELL, S. L., SULLIVAN, D. F. and RATHKE, M. W., *Tetrahedron Lett.* 1403 (1974); HARTZELL, S. L. and RATHKE, M. W., *Tetrahedron Lett.* 2737, 2757 (1976)
39 CHAN, T. H. and MORELAND, M., *Tetrahedron Lett.* 515 (1978)
40 WOODBURY, R. P. and RATHKE, M. W., *Tetrahedron Lett.* 709 (1978); *J. org. Chem.* **43**, 1947 (1978)
41 HART, D. J., CAIN, P. A. and EVANS, D. A., *J. Am. chem. Soc.* **100**, 1548 (1978)

42 LUCAST, D. H. and WEMPLE, J., *Tetrahedron Lett.* 1103 (1977)
43 HUDRLIK, P. F., PETERSON, D. and CHOU, D., *Synth. Communs* **5,** 359 (1975)
44 SACHDEV, K., *Tetrahedron Lett.* 4041 (1976)
45 COREY, E. J. and BOGER, D. L., *Tetrahedron Lett.* 5, 9, 13 (1978)
46 KANO, S., EBATA, T., FUNAKI, K. and SHIBUYA, S., *Synthesis* 746 (1978)
47 COOKE, F. and MAGNUS, P., *J. chem. Soc. chem. Communs* 513 (1977); BURFORD, C., COOKE, F., EHLINGER, E. and MAGNUS, P., *J. Am. chem. Soc.* **99,** 4536 (1977); MAGNUS, P. and ROY, G., *J. chem. Soc. chem. Communs* 297 (1978)
48 EISCH, J. J. and GALLE, J. E., *J. Am. chem. Soc.* **98,** 4646 (1976)
49 MAGNUS, P. and ROY, G., *J. chem. Soc. chem. Communs* 822 (1979)
50 WHITMORE, F. C. and SOMMER, L. H., *J. Am. chem. Soc.* **68,** 481, 485 (1946)
51 HAUSER, C. R. and HANCE, C. R., *J. Am. chem. Soc.* **74,** 5091 (1952)
52 CONNOLLY, J. W. and URRY, G., *Inorg. Chem.* **2,** 645 (1963)
53 BROOK, A. G., DUFF, J. M. and ANDERSON, D. G., *Can. J. Chem.* **48,** 561 (1970)
54 DAVIDSON, P. J., HARRIS, D. H. and LAPPERT, M. F., *J. chem. Soc. Dalton* 2268 (1976)
55 GRÖBEL, B.-Th. and SEEBACH, D., *Chem. Ber.* **110,** 867 (1977)
56 BOECKMAN, R. K. and BRUZA, K. J., *Tetrahedron Lett.* 3365 (1974)
57 STORK, G. and GANEM, B., *J. Am. chem. Soc.* **95,** 6152 (1973)
58 SAKURAI, H., NISHIWAKI, K. and KIRA, M., *Tetrahedron Lett.* 4193 (1973)
59 DUMONT, W. and KRIEF, A., *Angew Chem. int. Edn* **15,** 161 (1976); Van ENDE, D., DUMONT, W. and KRIEF, A., *J. organometal. Chem.* **149,** C10 (1978); DUMONT, W., Van ENDE, D. and KRIEF, A., *Tetrahedron Lett.* 485 (1979)
60 Ref. 27 cited in ref. 1; for a quantitative method, *see* SEITZ, D. E. and ZAPATA, A., *Tetrahedron Lett.* 3481 (1980)

Chapter 5

Rearrangement reactions with migration of silicon

Discussion in this chapter is restricted to those migration/rearrangement reactions[1,2] which have demonstrable or obvious potential synthetic utility.

5.1 1,2-Rearrangements

The best-known example of this rearrangement is the migration of silicon from carbon to oxygen which can occur[3] when silylmethanols (1) are treated with catalytic amounts of base (*Scheme 5.1*). This general process has been

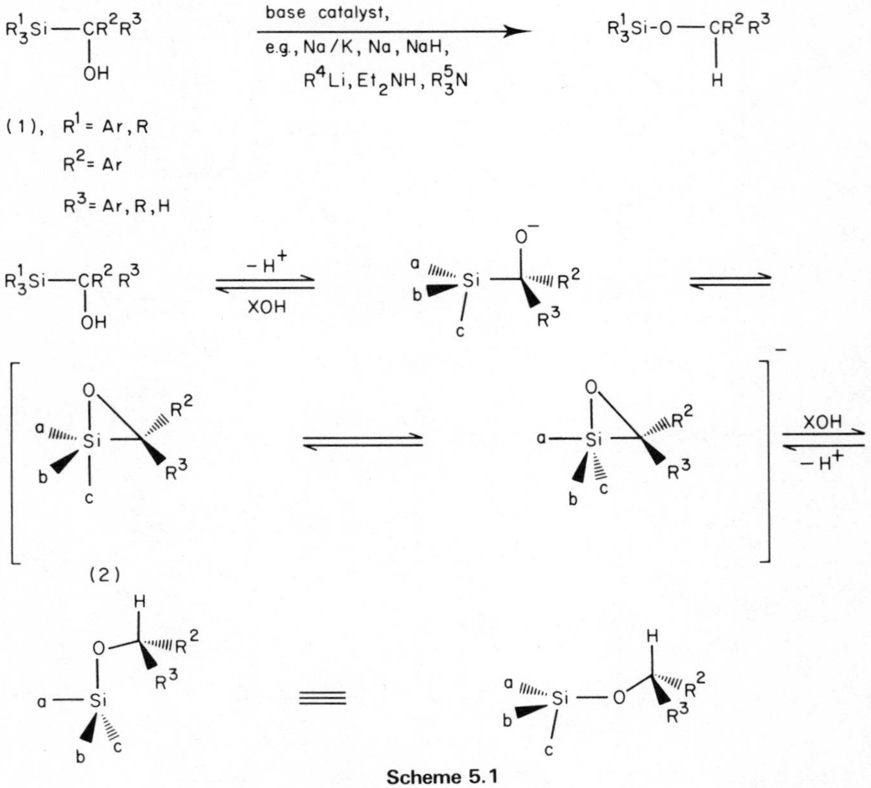

$$R^1_3Si{-}CR^2R^3(OH) \xrightarrow[\text{e.g., Na/K, Na, NaH, } R^4Li,\ Et_2NH,\ R^5_3N]{\text{base catalyst,}} R^1_3Si{-}O{-}CR^2R^3(H)$$

(1), R^1 = Ar, R
R^2 = Ar
R^3 = Ar, R, H

$$R^1_3Si{-}CR^2R^3(OH) \underset{XOH}{\overset{-H^+}{\rightleftharpoons}} (a)(b)(c)Si{-}CR^2R^3(O^-) \rightleftharpoons [(2)] \underset{-H^+}{\overset{XOH}{\rightleftharpoons}} (a)(b)(c)Si{-}O{-}CHR^2R^3$$

Scheme 5.1

studied extensively by Brook[1], with whose name the rearrangement is now associated. Extensive kinetic studies have indicated that the rearrangement proceeds through the series of equilibria shown. Intramolecular attack by oxygen produces the hypervalent pentacoordinate species (2), which by pseudorotation and expulsion of the then apical carbon leaving group leads to the observed result. As expected from such a mechanism, this rearrangement has a large negative entropy of activation. When chiral substrates are employed, *retention* at silicon and *inversion* at carbon are observed to take place.

Further evidence in favour of such a sequence includes the observation that substituents R^2 and R^3 which can delocalize negative charge, such as phenyl or allyl, accelerate the rate of rearrangement: simple alkyl substituted methanols rearrange at an extremely slow rate, too slow to be of any utility. The reversibility of these processes, while not normally observable (but see *Scheme 11.11*, p. 137), can be deduced from the fact that when *stoichiometric* amounts of base are used, equilibrium ratios are determined by the relative stabilities of the anions. This can be illustrated by the contrasting cases of benzyloxysilanes and diphenylmethoxysilanes shown in *Scheme 5.2*. Under catalytic conditions, on the other hand, equilibrium ratios are determined by the relative stabilities of the neutral compounds, where formation of the much stronger Si–O bond in the rearranged product gives the observed outcome.

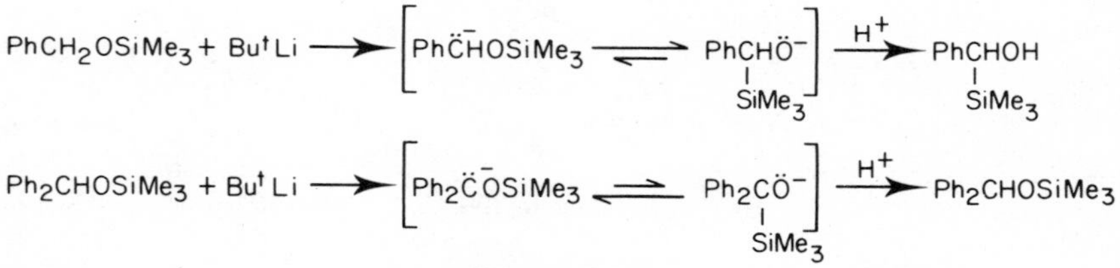

$$PhCH_2OSiMe_3 + Bu^tLi \longrightarrow \left[Ph\ddot{C}^{-}HOSiMe_3 \rightleftharpoons PhCH(SiMe_3)\ddot{O}^{-} \right] \xrightarrow{H^+} PhCH(SiMe_3)OH$$

$$Ph_2CHOSiMe_3 + Bu^tLi \longrightarrow \left[Ph_2\ddot{C}^{-}OSiMe_3 \rightleftharpoons Ph_2C(SiMe_3)\ddot{O}^{-} \right] \xrightarrow{H^+} Ph_2CHOSiMe_3$$

Scheme 5.2

It is interesting to note at this point that benzylthiosilanes rearrange on treatment with excess of base to give α-silylbenzylthiols (*Scheme 5.3*) by migration from sulphur to carbon. The reverse rearrangement can be induced by heat or free radical catalysis, but *not* by treatment with catalytic quantities of base[4].

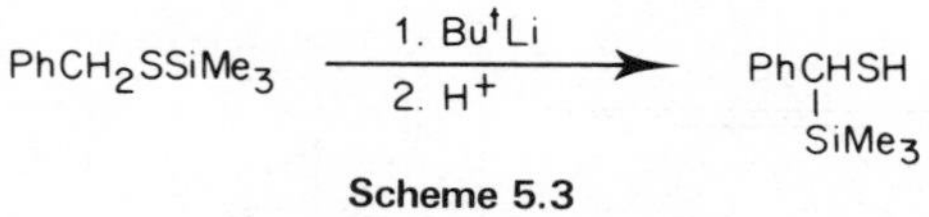

$$PhCH_2SSiMe_3 \xrightarrow[2.\ H^+]{1.\ Bu^tLi} PhCH(SiMe_3)SH$$

Scheme 5.3

Allyl silylmethyl ethers (3) undergo a related rearrangement[5] on being heated (*Scheme 5.4*); with fluoride-ion catalysis[6] this rearrangement and a similar rearrangement of benzyl silylmethyl ethers occurs at room temperature.

α-Ketosilanes react with Wittig ylides by two distinct pathways. With aliphatic carbonyl substrates, normal Wittig olefination takes place to give

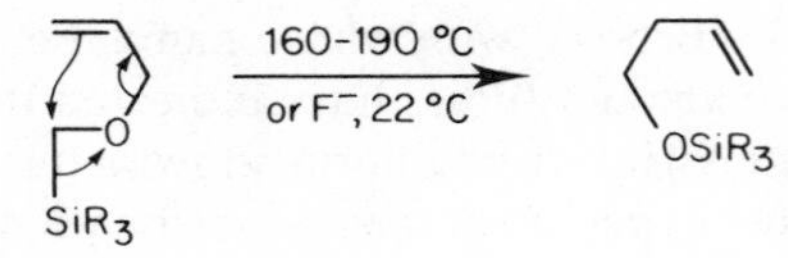

160–190 °C or F^-, 22 °C

(3) SiR_3 → $OSiR_3$

Scheme 5.4

vinylsilanes (*Scheme 5.5*); aromatic carbonyl substrates, on the other hand, produce silyl enol ethers, the intermediate (4) having undergone Brook rearrangement[7].

α-Ketosilanes react with diazoalkanes and derived species to give products arising from similar rearrangements occurring on the initially produced adducts[8].

A variety of other 1,2-rearrangements have been described[2] involving the migration of silicon between nitrogen and other elements.

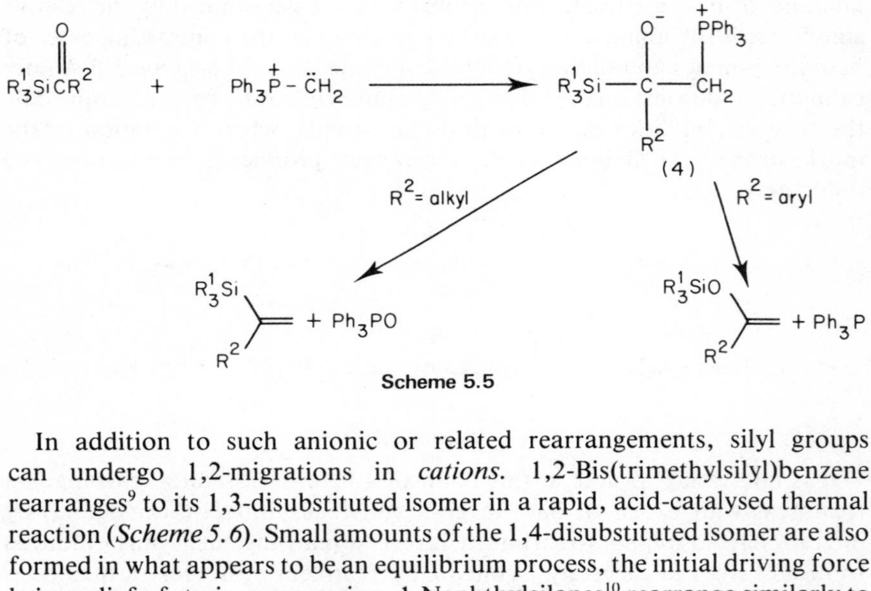

$R^1_3SiCOR^2$ + $Ph_3\overset{+}{P}-\ddot{C}H_2^-$ → $R^1_3Si-C(O^-)(R^2)-CH_2-\overset{+}{P}Ph_3$ (4)

R^2 = alkyl → $R^1_3Si(R^2)C{=}CH_2$ + Ph_3PO

R^2 = aryl → $R^1_3SiO(R^2)C{=}CH_2$ + Ph_3P

Scheme 5.5

In addition to such anionic or related rearrangements, silyl groups can undergo 1,2-migrations in *cations.* 1,2-Bis(trimethylsilyl)benzene rearranges[9] to its 1,3-disubstituted isomer in a rapid, acid-catalysed thermal reaction (*Scheme 5.6*). Small amounts of the 1,4-disubstituted isomer are also formed in what appears to be an equilibrium process, the initial driving force being relief of steric compression. 1-Naphthylsilanes[10] rearrange similarly to

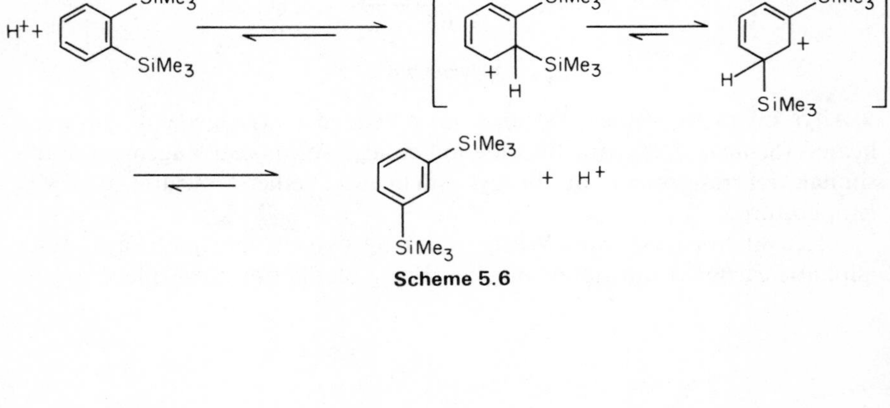

H^+ + 1,2-$C_6H_4(SiMe_3)_2$ ⇌ [cation (H, $SiMe_3$, $SiMe_3$) ⇌ cation (H, $SiMe_3$, $SiMe_3$)] ⇌ 1,3-$C_6H_4(SiMe_3)_2$ + H^+

Scheme 5.6

their 2-substituted isomers; competitive protiodesilylation occurs in the acidic conditions employed. A synthetic application of this rearrangement is described in Chapter 10.

5.2 1,3-Rearrangements

In 1968, Brook and Anderson described[11] a novel variation of the Pummerer rearrangement[12] of sulphoxides. Trimethylsilylmethyl phenyl sulphoxide (5) was observed to undergo a remarkably smooth, low-temperature rearrangement to give a Pummerer-like product (*Scheme 5.7*). The homologue (6) was found[13] to rearrange similarly.

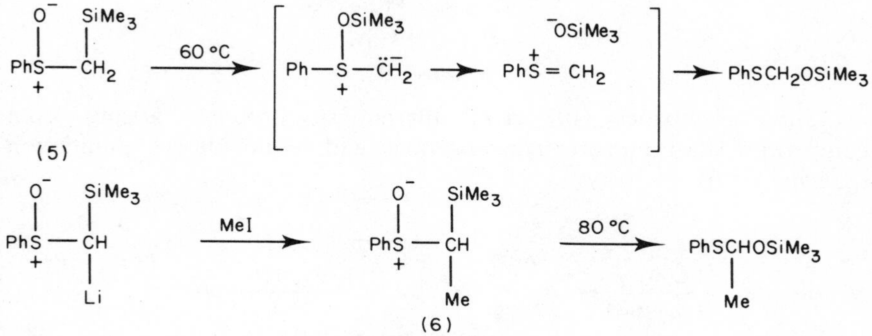

$PhS^{+}(O^{-})—CH_2SiMe_3$ (5) —60 °C→ [$Ph—S^{+}(OSiMe_3)—\ddot{C}H_2^{-}$ → $^{-}OSiMe_3$ $Ph\overset{+}{S}=CH_2$] → $PhSCH_2OSiMe_3$

$PhS^{+}(O^{-})—CH(SiMe_3)Li$ —MeI→ $PhS^{+}(O^{-})—CH(SiMe_3)Me$ (6) —80 °C→ $PhSCH(Me)OSiMe_3$

Scheme 5.7

The full potential of this very mild version of the Pummerer rearrangement has still to be realized. A recent improvement[14] in the preparation of the requisite α-silyl sulphoxides involves the inverse addition of the sulphoxide anion to trimethylsilyl chloride (*Scheme 5.8*).

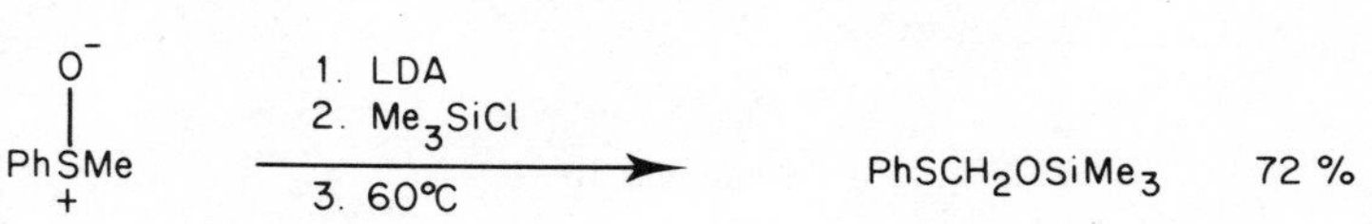

$PhS^{+}(O^{-})Me$ —1. LDA; 2. Me_3SiCl; 3. 60°C→ $PhSCH_2OSiMe_3$ 72 %

Scheme 5.8

The same authors[14] provide evidence substantiating Brook's mechanistic proposals. Silylation of the sulphoxide (7) gave a mixture of diastereoisomers, one of which underwent rearrangement below room temperature, whereas the other had to be heated to 70 °C before rearrangement took place. This can be rationalized by proposing that the diastereoisomer which rearranged at low temperature is (8), with a relatively unhindered conformation for intramolecular silyl transfer; the other diastereoisomer (9) will suffer adverse interaction between the phenyl and t-butyl substituents when it takes up the required conformation for silyl transfer (*Scheme 5.9*).

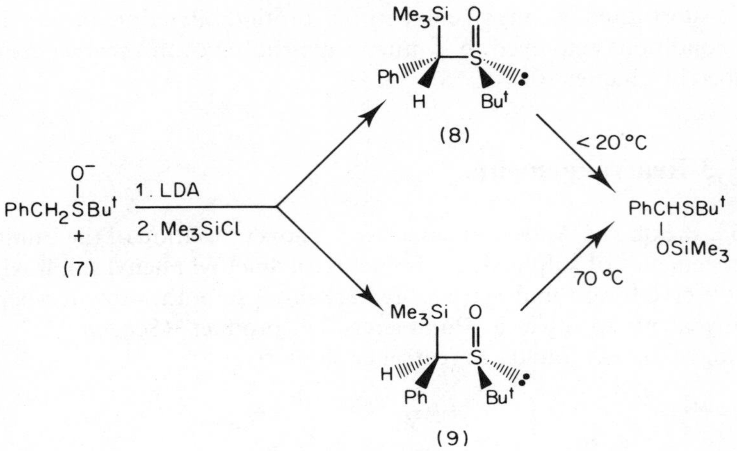

Scheme 5.9

α-Silyl selenoxides (10) give[15] thermolysis products arising from competitive sila-Pummerer rearrangement and selenoxide *syn* elimination (*Scheme 5.10*).

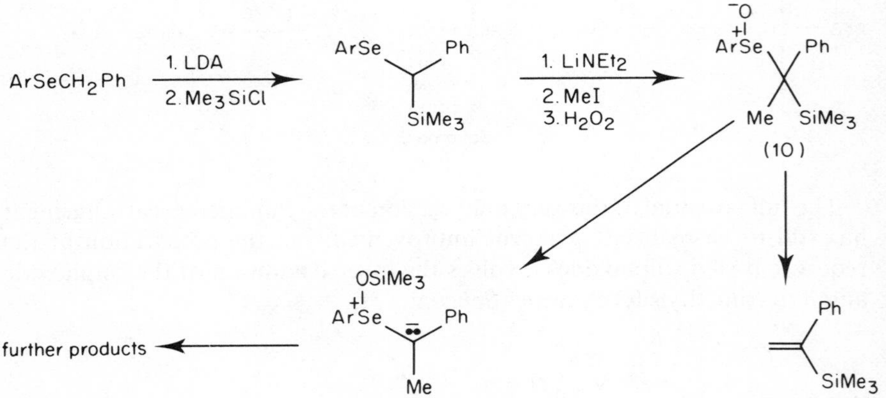

Scheme 5.10

The occurrence of 1,2-elimination in this instance is undoubtedly due to selenoxides having a greater tendency to eliminate than do the corresponding sulphoxides. In suitable cases, elimination can be suppressed and this silicon/selenium version of the Pummerer rearrangement can then be used[16] to convert primary alkyl halides into the homologous aldehydes (*Scheme 5.11*).

β-Ketosilanes rearrange[1,17] thermally to silyl enol ethers (*Scheme 5.12*). Although yields are good, and only moderate temperatures are required, it is unlikely that this route to silyl enol ethers would ever be competitive with the more standard methods detailed in Chapter 17. Kinetic and thermodynamic data are in full accord with an intramolecular four-centred concerted mechanism, as shown. Silicon-chiral silanes rearrange with *retention* of

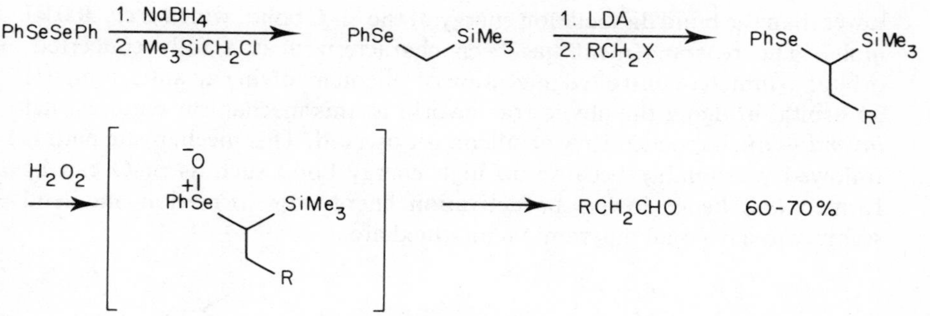

$$PhSeSePh \xrightarrow[\text{2. } Me_3SiCH_2Cl]{\text{1. } NaBH_4} PhSe\text{—}CH_2\text{—}SiMe_3 \xrightarrow[\text{2. } RCH_2X]{\text{1. LDA}} PhSe\text{—}CH(CH_2R)\text{—}SiMe_3$$

$$\xrightarrow{H_2O_2} \left[PhSe^{+}(O^{-})\text{—}CH(CH_2R)\text{—}SiMe_3 \right] \longrightarrow RCH_2CHO \quad 60\text{-}70\%$$

Scheme 5.11

configuration at silicon, in agreement with the postulated intermediacy of the hypervalent structure (11), which by pseudorotation and collapse will give the observed stereochemical result.

This mechanism has been challenged on the basis of substituent Hammett ρ values; these can be interpreted[18] as indicating that little or no charge development occurs on either the silicon or the carbonyl carbon in the rate-determining step. A more recent investigation[19] utilizing $^{29}Si/^{28}Si$ isotope effects vindicates the original mechanism, and establishes that an intermediate trigonal bipyramid or tetragonal pyramid is formed in the rate-determining step. In other words, the central silicon atom is capable of undergoing octet expansion because of the highly favourable creation of a strong Si–O bond.

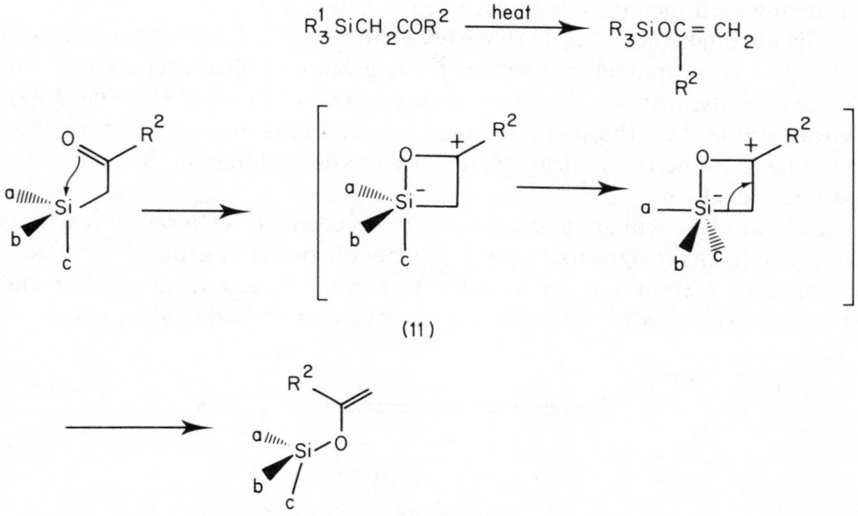

$$R^1_3SiCH_2COR^2 \xrightarrow{\text{heat}} R_3SiOC(R^2){=}CH_2$$

(11)

Scheme 5.12

Simple allylsilanes are normally regiostable at room temperature, but can be induced[20] to undergo thermal 1,3-sigmatropic rearrangement (*Scheme 5.13*); relatively high temperatures are required. The activation energy for this process is in the order of 210 kJ mol^{-1}, an energy considerably lower than the bond dissociation energy of the Si–C bond, which is *ca.* 300 kJ mol^{-1}. The rearrangement has been characterized as a fully concerted, orbital-symmetry controlled migration of silicon involving an antisymmetric 3*p* orbital bridging the allylic framework; as this mechanism requires, net *inversion* of stereochemistry at silicon is observed. This mechanistic path is followed presumably because no high energy bond such as Si–O can be formed and hence lower the activation energy for formation and lend stability to a trigonal bipyramidal intermediate.

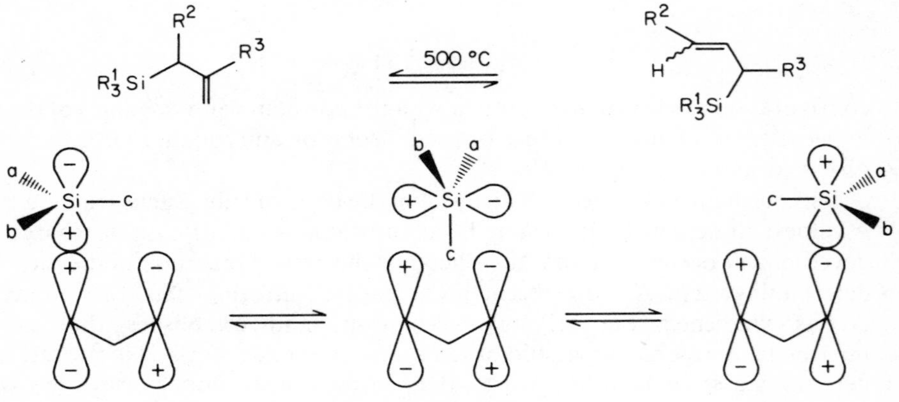

Scheme 5.13

Trimethylsilylcyclopentadiene is much more labile to heat, 1,5-silyl shifts occurring with facility; this is discussed in Chapter 9.

Silyl nitronates (*see* 17.3.13) have been shown[21,22] by NMR spectroscopy to undergo a very rapid intramolecular 1,3-migration of silicon between the two oxygen atoms, with an activation energy of *ca.* 40 kJ mol^{-1} (*Scheme 5.14*). *X*-Ray studies[22] on the silyl nitronates (12) and (13) show structures whose bonding parameters indicate some progression along an S_N2 *retention* pathway at silicon.

Such bond-switching has considerable precedent. 1,3-Migration reactions of silicon from nitrogen to oxygen[23-26], from nitrogen to nitrogen[24,27,28], from nitrogen to carbon[29], from nitrogen to sulphur[24], and from sulphur and selenium to oxygen[30,31] have all been reported and studied in some detail.

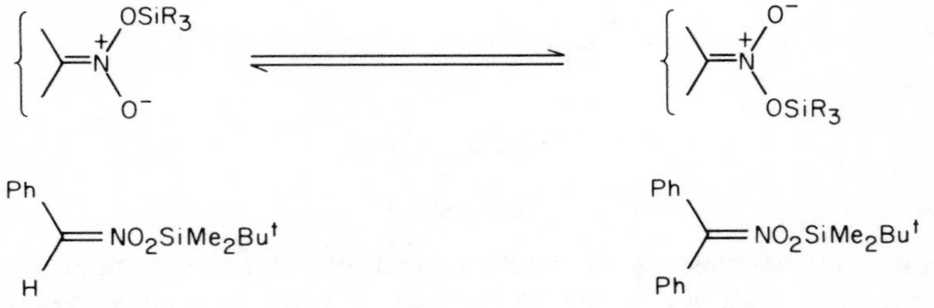

(12) (13)

Scheme 5.14

5.3 1,4-Rearrangements

Anionic 1,4-silyl group shifts are quite common. For example, 1,1-bis(trimethylsilyl)-1,2-diphenylethane undergoes[32] consecutive 1,2- and 1,4-shifts[33] from carbon to carbon when treated with excess of strong base (*Scheme 5.15*). Several cases[34-36] of 1,4-silyl group shifts from carbon to oxygen have been reported; remarkably, the reverse migration (*Scheme 5.16*) has also been described[37].

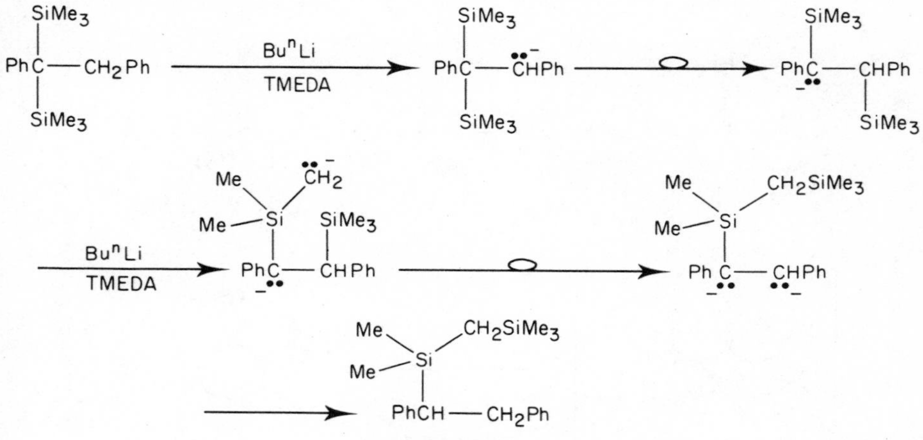

Scheme 5.15

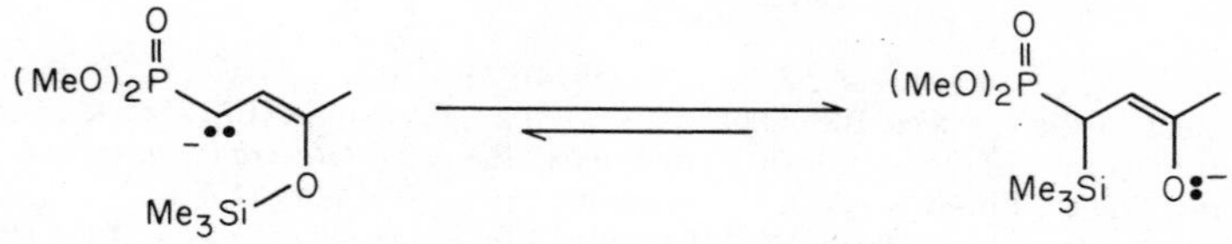

Scheme 5.16

5.4 1,5-Rearrangements

β-Diketone silyl enol ethers can exhibit bond-switching. For the case shown in *Scheme 5.17,* an activation energy of *ca.* 55 kJ mol^{-1} was calculated[38] from ^{1}H-NMR coalescence temperatures. This particular intramolecular reaction has been shown[39] to proceed with *retention* of stereochemistry at silicon.

Trimethylsilyl *β*-keto esters undergo thermal silatropic rearrangement[40] to produce silyl enol ethers (Chapter 17) regiospecifically (*Scheme 5.18*). This general process is the silicon analogue of the ready prototropic decarboxylation of *β*-keto acids to enols. Such analogy can be extended further to provide a silatropic version of the Carroll reaction, whereby trimethylsilyl enol ethers of allyl *β*-keto esters, e.g. (14), are transformed into *α*-allyl trimethylsilyl enol ethers (15); this sequence is also exemplified in *Scheme 5.18.*

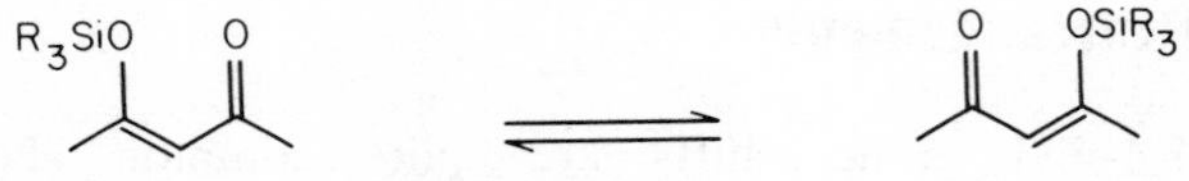

Scheme 5.17

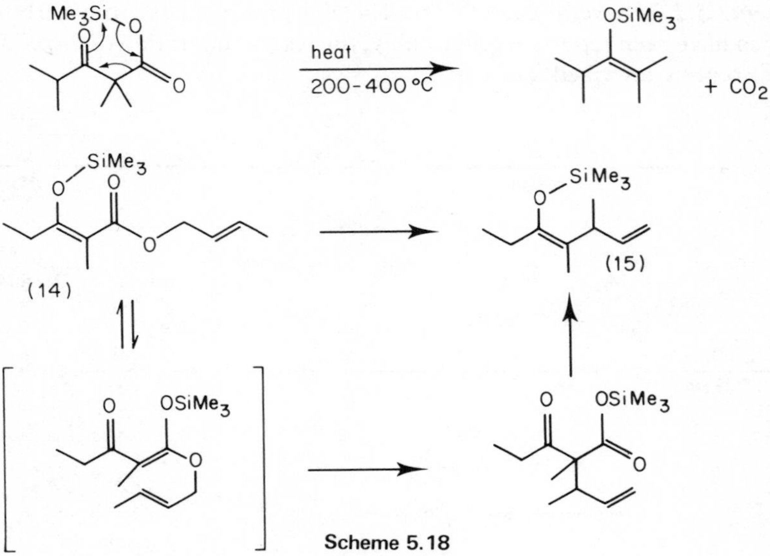

heat

200–400 °C

$+ CO_2$

(14)

(15)

Scheme 5.18

References

1 BROOK, A. G., *Accts chem. Res.* **7,** 77 (1974); *Pure appl. Chem.* **13,** 215 (1966); for an excellent review, *see* BROOK, A. G. and BASSINDALE, A. R., 'Molecular Rearrangements of Organosilicon Compounds', Essay 9 in '*Rearrangements in Ground and Excited States',* Ed. de MAYO, P., Academic Press, New York (1980)
2 WEST, R., *Pure appl. Chem.* **19,** 291 (1969); *Adv. organometal. Chem.* **16,** 1 (1977)
3 GILMAN, H. and WU, T. C., *J. Am. chem. Soc.* **75,** 2935 (1953); **76,** 2502 (1954); GILMAN, H. and LICHTENWALTER, G. D., *J. Am. chem. Soc.* **80,** 2680 (1958)
4 WRIGHT, A., LING, D., BOUDJOUK, P. and WEST, R., *J. Am. chem. Soc.* **94,** 4784 (1972)
5 REETZ, M. T., *Angew. Chem. int. Edn* **18,** 173 (1979); *Chem. Ber.* **110,** 954, 965 (1977); *Adv. organometal. Chem.* **16,** 33 (1977)
6 REETZ, M. T. and GREIF, N., *Chem. Ber.* **110,** 2958 (1977)
7 BROOK, A. G. and FIELDHOUSE, S. A., *J. organometal. Chem.* **10,** 235 (1967)
8 *See,* e.g., BROOK, A. G., LIMBURG, W. W., MacRAE, D. and FIELDHOUSE, S. A., *J. Am. chem. Soc.* **89,** 704 (1967); SEKIGUCHI, A., KABE, Y. and ANDO, W., *J. chem. Soc. chem. Communs* 233 (1979)
9 SEYFERTH, D. and WHITE, D. L., *J. Am. chem. Soc.* **94,** 3132 (1972)
10 BECKER, B., HERMAN, A. and WOJNOWSKI, W., *J. organometal. Chem.* **193,** 293 (1980); *see also* WROCZYNSKI, R. J., BAUM, M. W., KOST, D., MISLOW, K., VICK, S. C. and SEYFERTH, D., *J. organometal. Chem.* **170,** C29 (1979)
11 BROOK, A. G. and ANDERSON, D. G., *Can. J. Chem.* **46,** 2115 (1968)
12 *See,* e.g., BLOCK, E., *'Reactions of Organosulphur Compounds',* Academic Press, New York (1978)
13 CAREY, F. A. and HERNANDEZ, O., *J. org. Chem.* **38,** 2670 (1973)
14 VEDEJS, E. and MULLINS, M., *Tetrahedron Lett.* 2017 (1975)
15 REICH, H. J. and SHAH, S. K., *J. org. Chem.* **42,** 1773 (1977)

16 SACHDEV, K. and SACHDEV, H. S., *Tetrahedron Lett.* 4223 (1976); *see also* DUMONT, W., Van ENDE, D. and KRIEF, A., *Tetrahedron Lett.* 485 (1979); KOCIENSKI, P. J., *Tetrahedron Lett.* 1559 (1980); AGER, D. J. and COOKSON, R. C., *Tetrahedron Lett.* 1677 (1980)
17 BROOK, A. G., MacRAE, D. M. and LIMBURG, W. W., *J. Am. chem. Soc.* **89,** 5493 (1967)
18 BROOK, A. G., *J. organometal. Chem.* **86,** 185 (1975); LARSON, G. L. and FERNANDEZ, Y. V., *J. organometal. Chem.* **86,** 193 (1975)
19 KWART, H. and BARNETTE, W. E., *J. Am. chem. Soc.* **99,** 614 (1977)
20 KWART, H. and SLUTSKY, J., *J. Am. chem. Soc.* **94,** 2515 (1972); SLUTSKY, J. and KWART, H., *J. Am. chem. Soc.* **95,** 8678 (1973)
21 IOFFE, S. L., SHITKIN, V. M., KHASAPOV, B. N., KASHUTINA, M. V., TARTAKOVSKII, V. A., MYAGI, M. Y. and LIPPMAA, E. T., *Izv. Akad. Nauk SSSR, Ser. khim.* 2146 (1973); English translation, p. 2100
22 COLVIN, E. W., BECK, A. K., BASTANI, B., SEEBACH, D., KAI, Y. and DUNITZ, J., *Helv. chim. Acta* **63,** 697 (1980)
23 ITOH, K., KATSUDA, M. and ISHII, Y., *J. chem. Soc. (B)* 302 (1970)
24 ITOH, K., KATSUURA, T., MATSUDA, A. and ISHII, Y., *J. org. Chem.* **34,** 63 (1972)
25 FUKUI, M., ITOH, K. and ISHII, Y., *J. chem. Soc. Perkin II* 1043 (1972)
26 PUMP, J. and ROCHOW, E. G., *Chem. Ber.* **97,** 627 (1964)
27 SCHERER, O. J. and HORNIG, P., *Chem. Ber.* **101,** 2533 (1968)
28 KLEBE, J. F., *J. Am. chem. Soc.* **90,** 5246 (1968)
29 WILBURN, J. C. and NEILSON, R. H., in preparation
30 EBSWORTH, E. A. V., ROCKTÄSCHEL, G. and THOMPSON, J. C., *J. chem. Soc. (A)* 362 (1967)
31 CRADOCK, S., EBSWORTH, E. A. V. and JESSUP, H. F., *J. chem. Soc. Dalton* 359 (1972)
32 EISCH, J. J. and TSAI, M.-R., *J. Am. chem. Soc.* **95,** 4065 (1973)
33 DANEY, M., LAPOUYADE, R., LABRANDE, B. and BOUAS-LAURENT, H., *Tetrahedron Lett.* 153 (1980)
34 WOODBURY, R. P. and RATHKE, M. W., *J. org. Chem.* **43,** 1947 (1978)
35 ISOBE, M., KITAMURA, M. and GOTO, T., *Tetrahedron Lett.* 3465 (1979)
36 MATSUDA, I., MURATA, S. and IZUMI, Y., *Bull. chem. Soc. Japan* **52,** 2389 (1979); MATSUDA, I., MURATA, S. and ISHII, Y., *J. chem. Soc. Perkin I* 26 (1979)
37 EVANS, D. A., TAKACS, J. M. and HURST, K. M., *J. Am. chem. Soc.* **101,** 371 (1979)
38 PINNAVAIA, T., COLLINS, W. T. and HOWE, J. J., *J. Am. chem. Soc.* **92,** 4544 (1970); REICH, H. J. and MURCIA, D. A., *J. Am. chem. Soc.* **95,** 3418 (1973)
39 KUSNEZOWA, I. K., RÜHLMANN, K. and GRÜNDEMANN, E., *J. organometal. Chem.* **47,** 53 (1973)
40 COATES, R. M., SANDEFUR, L. O. and SMILLIE, R. D., *J. Am. chem. Soc.* **97,** 1619 (1975)

Chapter 6

Organohalogenosilanes and substitution at silicon

Organohalogenosilanes are by far the most important intermediates in organosilicon chemistry, and their involvement *via* nucleophilic displacement reactions is displayed in almost every chapter of this book. The particular synthetic utility of bromo- and iodo-trimethylsilane and of trimethylsilyl halogenoids is discussed in Chapter 18. It is beyond the scope or intention of this book to discuss the stereochemistry and mechanism[1-3] of substitution at silicon in any great detail, as emphasis here is on the 'ferryman' ability of silicon to mediate in the conversion of a silicon-free precursor into a silicon-free product. Similarly, since most of the commonly used halogenosilanes are commercially available, preparative routes will not be discussed; if desired, several excellent literature preparative surveys[4-88] are available.

However, some broad and useful generalities can be made. Organohalogenosilanes form part of an empirical conversion series[7,9], illustrated in an abbreviated manner in *Scheme 6.1.* Any compound type in the series can be converted into any other on its right-hand side by heating with an appropriate silver salt, either in a non-polar solvent or without solvent at all; attempted conversions in the other direction are normally unsuccessful. For example, silver chloride will convert iodotrimethylsilane into the corresponding chloride, but silver iodide will *not* effect the reverse transformation upon chlorotrimethylsilane; *see,* however, Chapter 18.1.

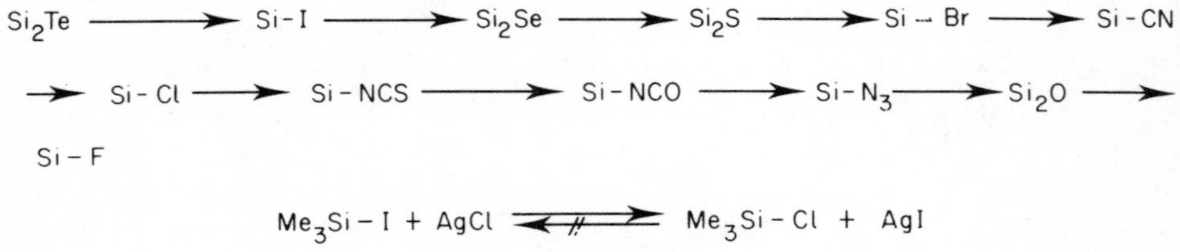

$$Si_2Te \longrightarrow Si{-}I \longrightarrow Si_2Se \longrightarrow Si_2S \longrightarrow Si{-}Br \longrightarrow Si{-}CN$$

$$\longrightarrow Si{-}Cl \longrightarrow Si{-}NCS \longrightarrow Si{-}NCO \longrightarrow Si{-}N_3 \longrightarrow Si_2O \longrightarrow$$

$$Si{-}F$$

$$Me_3Si{-}I + AgCl \rightleftharpoons Me_3Si{-}Cl + AgI$$

(the reverse arrow in this equilibrium is struck through)

Scheme 6.1

The substituents of this series are arranged in order of increasing 'hardness'[10,11]; the 'hard' silicon prefers to bond to a 'hard' nucleophile, with the displaced 'soft' nucleophile bonding well with the 'soft' silver cation. It is not always necessary to use silver salts. Fluorosilanes can be prepared from chlorosilanes by direct treatment[12,13] with hydrofluoric acid (*Scheme 6.2*).

$$R_3SiCl \xrightarrow[\text{aqueous or alcoholic}]{HF} R_3SiF$$

Scheme 6.2

Fluorosilanes can themselves be converted into the corresponding chlorides, bromides or iodides by heating with the appropriate aluminium(III) halide; in such cases the direction of reaction is governed[9,14] by the gain in lattice energy on formation of aluminium(III) fluoride.

When compared with carbon centres of organic compounds, the reaction rates of chloro-, bromo- and iodo-silanes are frequently too fast to measure by conventional techniques. Fluorides react sufficiently slowly to allow kinetic study, and considerable amounts of additional data have been obtained from silanes with other poor leaving groups such as alkoxy, aryloxy, and hydrogen. In a comparable series of halogenosilanes, the general reactivity of the silicon–halogen bond decreases in the expected order, as shown in *Scheme 6.3*.

$$Si-I > Si-Br > Si-Cl > Si-F$$

Scheme 6.3

A useful example of such high reactivity is illustrated in the preparation of chloromethyltrimethylsilane, a most useful synthon for further transformation. Chloromethyltrichlorosilane reacts[15,16] with an excess of methylmagnesium bromide with exclusive replacement of the chlorine atoms on silicon (*Scheme 6.4*).

$$Cl_3SiCH_2Cl + \text{excess } MeMgI \longrightarrow Me_3SiCH_2Cl \quad 90\%$$

Scheme 6.4

Cleavage of the silicon–halogen bond with water, alcohols, amines, and similar nucleophiles is actually reversible, with various equilibrium positions. For all apart from fluorosilanes, hydrolysis is essentially complete. For example, under appropriate conditions, hydrolysis of chlorotriethylsilane produces triethylsilanol; on the other hand, exposure of triethylsilanol to concentrated hydrochloric acid gives chlorotriethylsilane[17], the immiscibility of which in the medium results in the observed displacement of equilibrium (*Scheme 6.5*).

$$H_2O + Et_3SiCl \rightleftharpoons Et_3SiOH + HCl$$

Scheme 6.5

Equilibrium positions in the alcoholyses of organohalogenosilanes are not displaced so far towards the products as are those in the corresponding hydrolyses[18]. It is an unfavourable equilibrium, rather than a slow rate of reaction, which lies behind the apparent inertness of the silicon–fluorine bond towards solvolysis. For example, both fluorotriphenylsilane and fluorotributylsilane are stable in ethanol at 25 °C; fluorotriphenylsilane can even be crystallized from ethanol[13], because of an equilibrium which favours only 2 per cent alcoholysis.

Such equilibria can, of course, be readily displaced by conventional means such as acid neutralization, volatilization, phase separation, etc.

The main stereoelectronic features of substitution at silicon can be summarized as follows:

1. The most common mechanism for polar reactions of organohalogenosilanes is S_N2–Si; in acyclic cases, this leads to *inversion* of configuration at silicon.
2. *Retention* of configuration is favoured when relatively poor leaving groups, such as H, OR, and sometimes F, are involved;

S_N2 – Si inversion

S_N2 – Si retention

3. This duality of stereochemical outcome can be explained, at least in part, by consideration[19] of frontier orbital HOMO–LUMO interactions between the Si–X bond and the incoming nucleophile. The more electronegative X becomes, the more is the σ^* Si–X antibonding orbital concentrated between the two atoms; this situation favours 'frontside' attack, and hence *retention* of stereochemistry. With less electronegative X groups, the σ^* Si–X orbital is more concentrated on the 'backside', thus favouring substitution with *inversion*. Additionally, soft, delocalized and polarizable anions tend to attack with overall *inversion* of stereochemistry.
4. Certain, but not all, intramolecular migration reactions of silicon result in *retention* of configuration at silicon by an S_Ni-Si process:

S_Ni – Si retention

5. *d*-Orbital participation is probably small in most cases, but in some it could be making a significant contribution.
6. S_N1 reactions at silicon are extremely rare, if they occur at all, largely because the rate of the alternative S_N2-Si reaction is so high.

7. Electron-withdrawing substituents on silicon increase the rate of substitution.
8. Oxygen and halogen ion species are powerfully nucleophilic towards silicon, whereas nitrogen and organometallic carbon nucleophiles are less so.
9. Steric hindrance to substitution at silicon can be pronounced, and can be put to good use.

References

1 SOMMER, L. H., *'Stereochemistry, Mechanism, and Silicon'*, McGraw-Hill, New York (1965)
2 EABORN, C., *'Organosilicon Compounds'*, Butterworths, London (1960)
3 FLEMING, I., in *'Comprehensive Organic Chemistry'*, Eds. Barton, D. H. R. and Ollis, W. D., vol. 3, Pergamon Press, Oxford (1979)
4 VOORHOEVE, R. J. H., *'Organohalosilanes'*, Elsevier, Amsterdam (1967)
5 BAZANT, V., CHVALOVSKY, V. and RATHOUSKY, J., *'Organosilicon Compounds'*, Academic Press, New York and London (1965)
6 PETROV, A. D., MIRONOV, B. F., PONOMARENKO, V. A. and CHERNYSHEV, E. A., *'Synthesis of Organosilicon Monomers'*, Heywood, London (1964)
7 Van DYKE, C. H., in *'Organometallic Compounds of the Group IV Elements'*, Ed. MacDiarmid, A. G., vol. 2, part 1, Marcel Dekker, New York (1972)
8 GEORGE, P. D., PROBER, M. and ELLIOTT, J. R., *Chem. Rev.* **56,** 1065 (1956)
9 EABORN, C., *J. chem. Soc.* 3077 (1950)
10 PEARSON, R. G., *Chemy Brit.* **3,** 103 (1967)
11 HO, T.-L., *Chem. Rev.* **75,** 1 (1975)
12 MARANS, N. S., SOMMER, L. H. and WHITMORE, F. C., *J. Am. chem. Soc.* **73,** 5127 (1951)
13 EABORN, C., *J. chem. Soc.* 2846 (1952)
14 EABORN, C., *J. chem. Soc.* 494 (1953)
15 WHITMORE, F. C., SOMMER, L. H. and GOLD, J., *J. Am. chem. Soc.* **69,** 1976 (1947)
16 MIRONOV, V. F., *Dokl. Akad. Nauk SSSR* **108,** 266 (1956)
17 SOMMER, L. H., PIETRUSZA, E. W. and WHITMORE, F. C., *J. Am. chem. Soc.* **68,** 2282 (1946)
18 ALLEN, A. D., CHARLTON, J. C., EABORN, C. and MODENA, G., *J. chem. Soc.* 3668 (1957); ALLEN, A. D. and MODENA, G. *J. chem. Soc.* 3671 (1957)
19 CORRIU, R., *Organometal. Chem. Rev.* **9,** 357–373 (1980); *see also,* ANH, N. T., *Topics curr. Chem.* **88,** 145 (1980); ANH, N. T. and MINOT, C., *J. Am. chem. Soc.* **102,** 103 (1980)

Chapter 7

Vinylsilanes

The great synthetic utility[1] of vinylsilanes shortly to be discussed is critically dependent on their availability. Fortunately, there are now many suitable preparative routes, several being highly stereoselective. Most existing

Table 7.1 Preparation of vinylsilanes

Substitution type	*Alkynes*	*Availability* *Carbonyl compounds*	*Vinyl halides*
Monosubstitution			
(H)(H)C=C(Si–)(R)	✓	✓	✓
(H)(R)C=C(Si–)(H)	✓	✓	✓
(H)(R)C=C(H)(Si–)	✓	✓	
Disubstitution			
(R^1)(H)C=C(Si–)(R^2)	✓	✓	
(H)(R^1)C=C(Si–)(R^2)	✓	✓	✓
(R^1)(R^2)C=C(Si–)(H)	✓	✓	
Trisubstitution			
(R^1)(R^2)C=C(Si–)(R^3)	✓	✓	✓

methodologies utilize alkynes, carbonyl compounds, or vinyl halides as starting materials (*Table 7.1*), and these will be considered in turn.

7.1 Preparation

7.1.1 From alkynes

Two distinct routes are possible from terminal alkynes, differing in the timing of introduction of the silyl group (*Scheme 7.1*). The alkyne can either be converted via its anion into the alkynylsilane, and then reacted with a range of organometallic species or reduced catalytically, or it can be treated directly with a silyl hydride (Hydrosilylation, Chapter 21.1); obviously, only the second alternative is possible for internal alkynes.

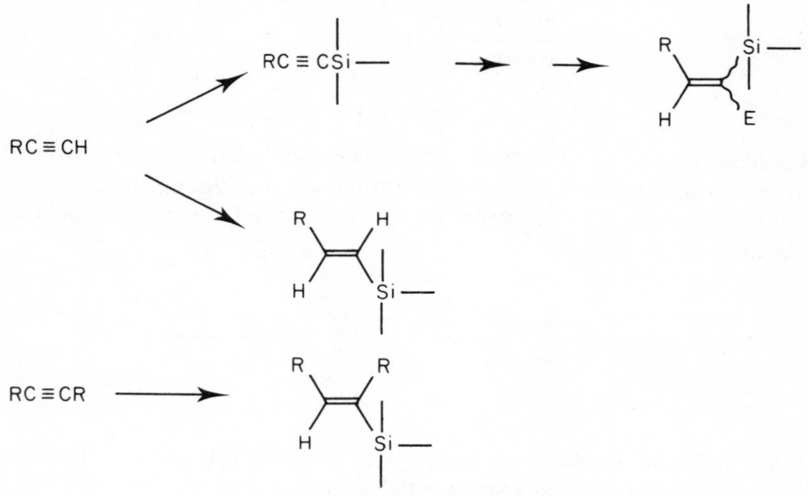

RC≡CH → RC≡CSi— → → (R)(H)C=C(Si—)(E)

RC≡CH → (R)(H)C=C(H)(Si—)

RC≡CR → (R)(H)C=C(R)(Si—)

Scheme 7.1

A simple illustration is the separate preparations[2] of (*Z*)- and (*E*)-*β*-trimethylsilylstyrene; in practice, the (*Z*)-isomer was obtained more easily by photochemical isomerization, as shown in *Scheme 7.2*.

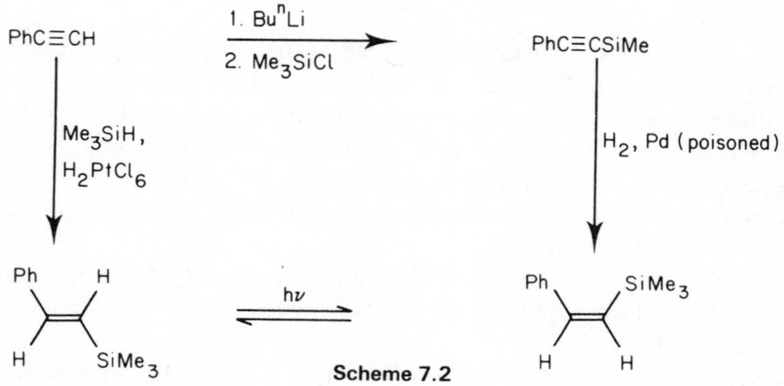

PhC≡CH → (1. Bu^nLi; 2. Me_3SiCl) → PhC≡CSiMe

PhC≡CH → (Me_3SiH, H_2PtCl_6) → (Ph)(H)C=C(H)($SiMe_3$)

PhC≡CSiMe → (H_2, Pd (poisoned)) → (Ph)(H)C=C($SiMe_3$)(H)

(Ph)(H)C=C(H)($SiMe_3$) ⇌ (hν) (Ph)(H)C=C($SiMe_3$)(H)

Scheme 7.2

Hydrosilylation of alkynes

The catalysed addition of silyl hydrides across multiple bonds will be discussed in some detail later (Chapter 21.1). Suffice it to say that such addition can be accomplished with a variety of catalysts, the best being chloroplatinic acid, when *cis*-addition with terminal regioselectivity[3] is normally observed (*Scheme 7.3*); peroxide initiated addition can result in the products of *trans*-addition.

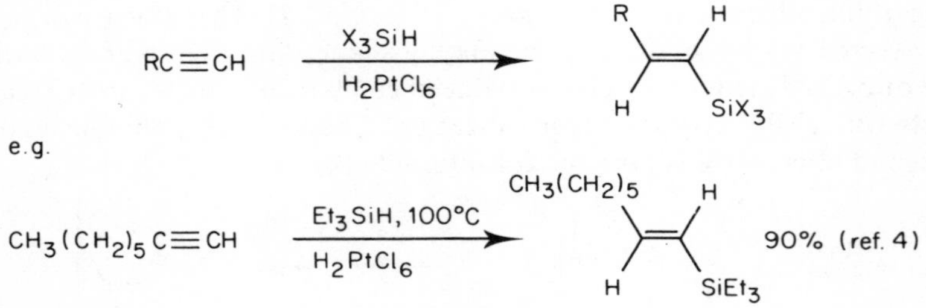

$RC{\equiv}CH \xrightarrow[H_2PtCl_6]{X_3SiH}$ (R, H / H, SiX_3 alkene)

e.g.

$CH_3(CH_2)_5C{\equiv}CH \xrightarrow[H_2PtCl_6]{Et_3SiH,\ 100°C}$ ($CH_3(CH_2)_5$, H / H, $SiEt_3$ alkene) 90% (ref. 4)

Scheme 7.3

Regiospecificity is reversed when 1-trimethylsilylalkynes are the substrates; the product 1,2-disilyl-substituted alkynes undergo regiospecific protiodesilylation[5] and so provide a good route[6] to monosubstituted vinylsilanes of the type (1).

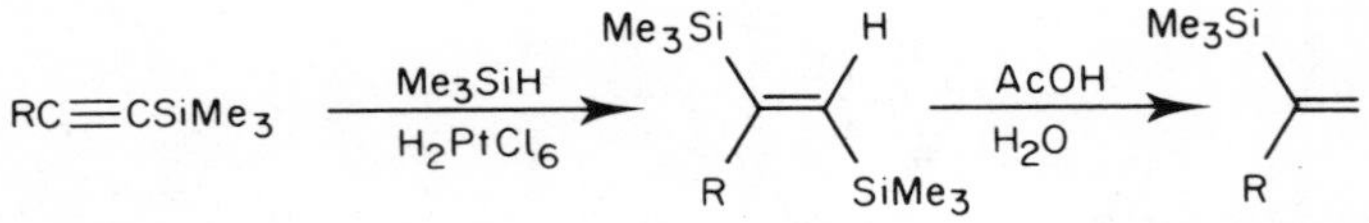

$RC{\equiv}CSiMe_3 \xrightarrow[H_2PtCl_6]{Me_3SiH}$ (Me_3Si, H / R, $SiMe_3$ alkene) $\xrightarrow[H_2O]{AcOH}$ (Me_3Si / R alkene) (1)

Hydrosilylation is rarely of utility with internal alkynes as substrates, unless they are symmetrically substituted[7] (*Scheme 7.4*)[7].

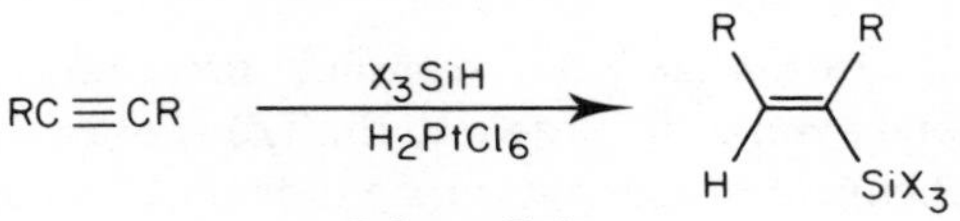

$RC{\equiv}CR \xrightarrow[H_2PtCl_6]{X_3SiH}$ (R, R / H, SiX_3 alkene)

Scheme 7.4

One such case[8] can be seen in the preparation of 2-triethylsilyl-1,4-butadiene (2); the silyl group of this diene exerted only a weak directing effect in its Diels–Alder reactions (see pp. 102), although boron trifluoride etherate catalysis proved beneficial in improving the regioselectivity of its reaction with ethyl acrylate (*Scheme 7.5*).

Use of a soluble Ni(II) catalyst leads stereoselectively[9] to the products of *cis* double silylation (*Scheme 7.6*).

An alternative route[10] to 1,2-disilylethenes is the palladium(0)-catalysed addition–elimination sequence shown in *Scheme 7.7*, the (*E*)-alkene giving the (*E*)-disilylethene (3).

Occasionally, it has proven possible hydrosilylate to regioselectively

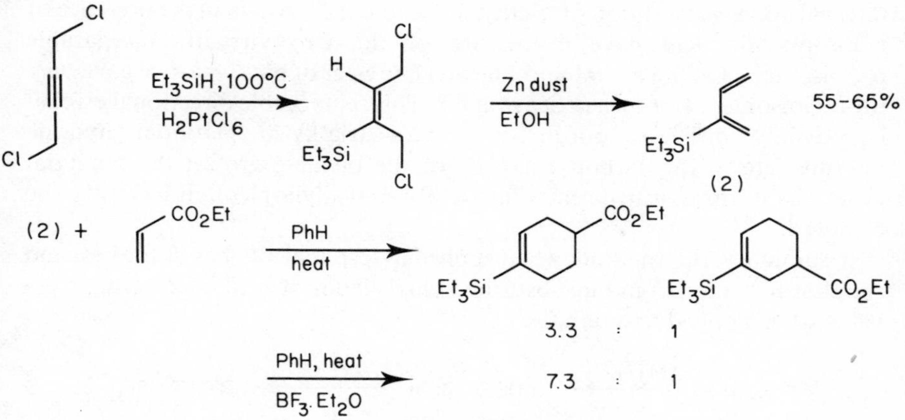

Et_3SiH, 100°C; H_2PtCl_6; Zn dust, EtOH; 55–65%; (2)

(2) + CO_2Et; PhH, heat; CO_2Et; Et_3Si; CO_2Et

3.3 : 1

PhH, heat, $BF_3.Et_2O$ 7.3 : 1

Scheme 7.5

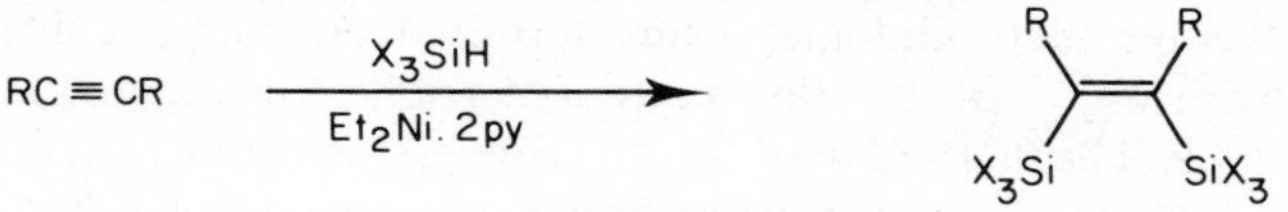

RC≡CR $\xrightarrow[Et_2Ni.\,2py]{X_3SiH}$ R, R, X_3Si, SiX_3

Scheme 7.6

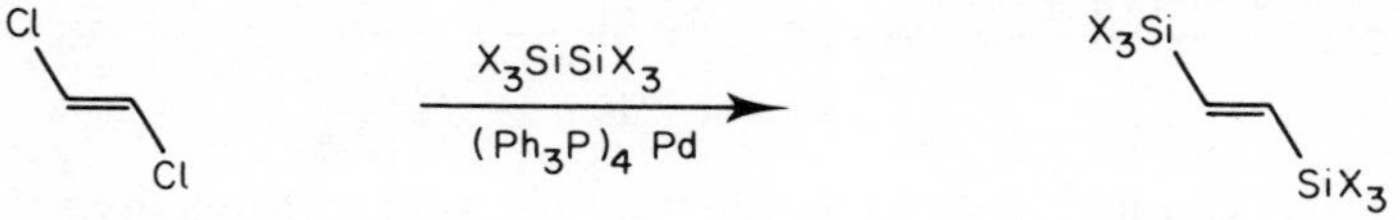

Cl, Cl $\xrightarrow[(Ph_3P)_4Pd]{X_3SiSiX_3}$ X_3Si, SiX_3 (3)

Scheme 7.7

$CH_3C{\equiv}CCH_2OCOBu^t$ $\xrightarrow[H_2PtCl_6]{Et_3SiH}$ [CH_3, H, Pt, O, Et_3Si, O, Bu^t]

→ CH_3, CH_2OCOBu^t, H, $SiEt_3$ (4) → CH_3, CH_2OH, H, $SiEt_3$

$CH_3C{\equiv}CCH_2OH$ $\xrightarrow[H_2PtCl_6]{Et_3SiH}$ CH_3, CH_2OH, H, $SiEt_3$ + CH_3, CH_2OH, Et_3Si, H

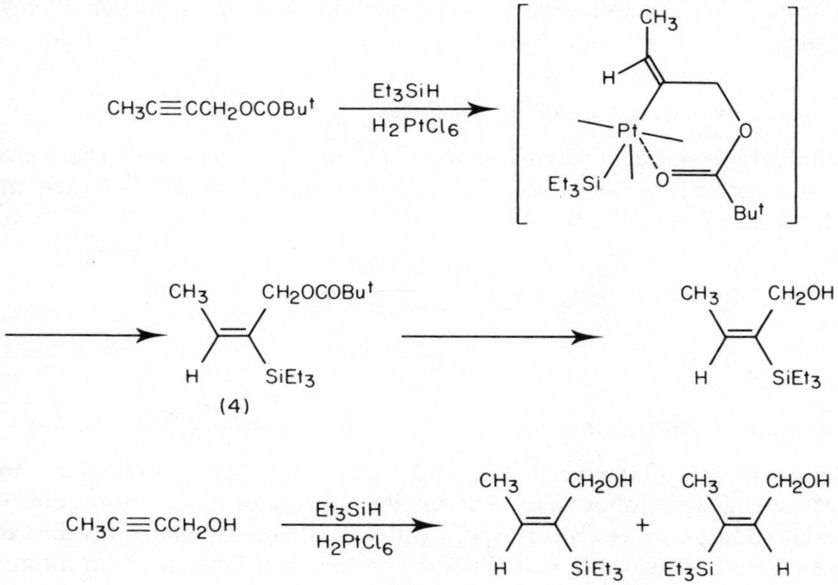

Scheme 7.8

internal alkynes. Addition of triethylsilane to but-3-yn-1-ol in the presence of chloroplatinic acid gave a mixture of the two virtually inseparable regioisomers. On the other hand, the pivalate ester of but-3-yn-1-ol gave *only* the regioisomer (4) shown in *Scheme 7.8.* This remarkable directional effect[11] is possibly due to coordination of a trialkylsilylplatinum hydride intermediate to the carbonyl oxygen of the pivalate group; the potential generality of this synthetic entry into α-silylated allylic alcohols has still to be explored.

In summary, this method, when applied to terminal alkynes, is the best and simplest route to (*E*)-monosubstituted vinylsilanes; it is of considerably less value when applied to internal alkynes.

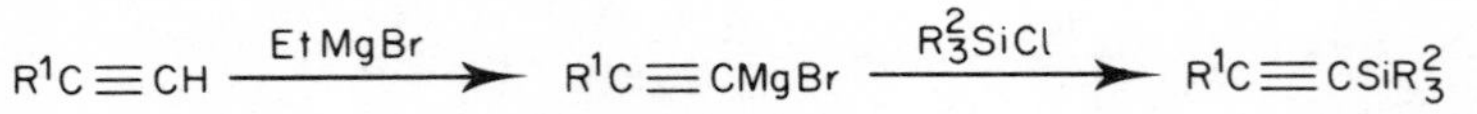

$$R^1C{\equiv}CH \xrightarrow{EtMgBr} R^1C{\equiv}CMgBr \xrightarrow{R^2_3SiCl} R^1C{\equiv}CSiR^2_3$$

Scheme 7.9

More flexible methodologies using terminal alkynes devolve from the initial conversion[12,13] of the alkyne into its 1-trialkylsilyl analogue, as shown in *Scheme 7.9.* The method used for this conversion is quite general, and has led to the preparation of a wide range of alkynylsilanes (Chapter 13),

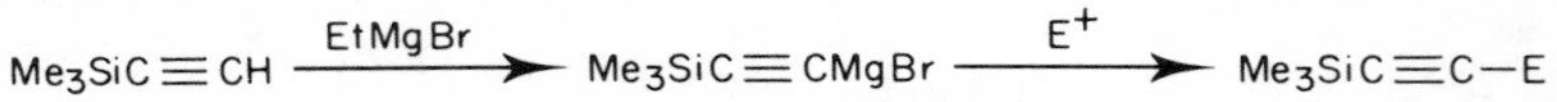

$$Me_3SiC{\equiv}CH \xrightarrow{EtMgBr} Me_3SiC{\equiv}CMgBr \xrightarrow{E^+} Me_3SiC{\equiv}C{-}E$$

Scheme 7.10

including the simplest and generally useful, trimethylsilylethyne[14]. This alkyne, on sequential treatment[4] with ethylmagnesium bromide and electrophiles, can give rise to yet another range of alkynylsilanes, as shown in *Scheme 7.10.* The following series of methods all start with such alkynyl silanes.

Catalytic hydrogenation of alkynylsilanes

Semihydrogenation of alkynylsilanes[12,13] can be readily achieved. The degree of stereoselectivity is variable, although the product is normally largely the (*Z*)-alkene (*Scheme 7.11*).

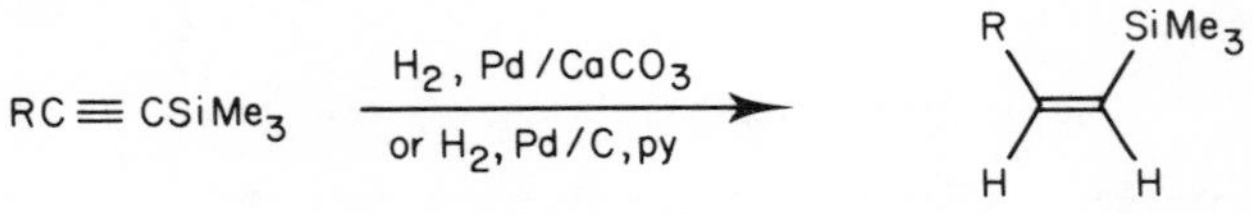

$$RC{\equiv}CSiMe_3 \xrightarrow[\text{or } H_2,\ Pd/C,\ py]{H_2,\ Pd/CaCO_3} (Z)\text{-}RCH{=}CHSiMe_3$$

Scheme 7.11

Hydrometallation/protio-, carbo-, or halo-demetallation of alkynylsilanes

Considerable stereoselectivity and flexibility are attainable by hydrometallation followed by electrophilic cleavage of the resulting adduct. Initial addition proceeds regiospecifically, the metal becoming attached to the silicon-bearing carbon. Protiodesilylation then leads to 2-substituted vinylsilanes.

Hydroboration[15] results in overall *cis*-addition, producing the (*Z*)-alkene (*Scheme 7.12*); final basic peroxide treatment, while mechanistically unnecessary, greatly simplified isolation procedures by consuming residual organoborane species.

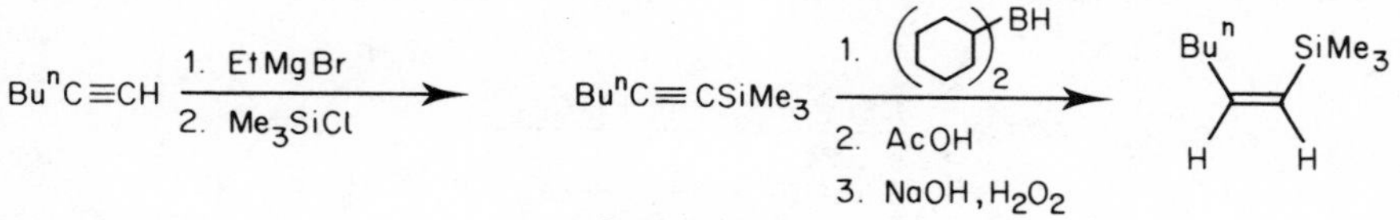

$Bu^nC{\equiv}CH \xrightarrow[2.\ Me_3SiCl]{1.\ EtMgBr} Bu^nC{\equiv}CSiMe_3 \xrightarrow[2.\ AcOH;\ 3.\ NaOH, H_2O_2]{1.\ (C_6H_{11})_2BH}$ (Bu^n, H)C=C(H, $SiMe_3$)

Scheme 7.12

In contrast, hydroalumination[16] takes place in a *trans* manner in hydrocarbon solvents (*Scheme 7.13*), whereas in donor solvent mixtures, clean *cis*-addition is observed (*Scheme 7.14*).

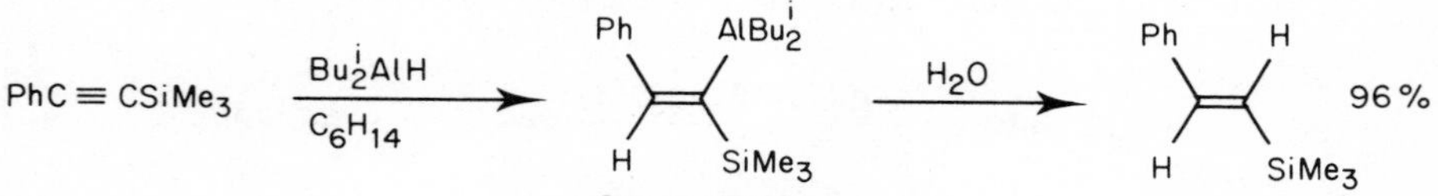

$PhC{\equiv}CSiMe_3 \xrightarrow[C_6H_{14}]{Bu^i_2AlH}$ (Ph, H)C=C($AlBu^i_2$, $SiMe_3$) $\xrightarrow{H_2O}$ (Ph, H)C=C(H, $SiMe_3$) 96%

Scheme 7.13

The resulting vinylalanes, as the corresponding alanates, can be carbodemetallated[17] with allyl bromide or iodide. Strict retention of alkene geometry was observed in the alane (5) arising from *trans*-addition, but the isomer (6) reacted sluggishly with allyl bromide, and some double bond

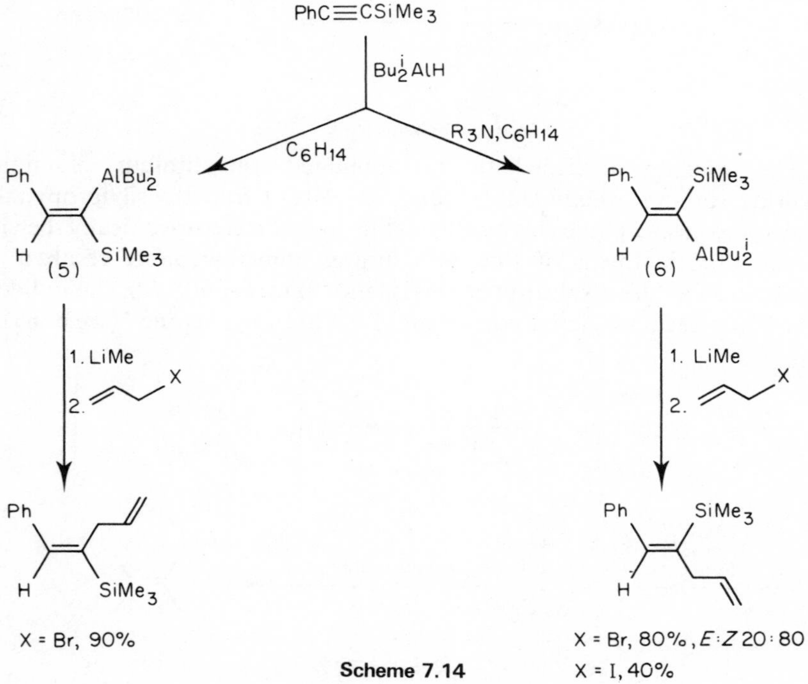

$PhC{\equiv}CSiMe_3$, Bu^i_2AlH

C_6H_{14}: (Ph, H)C=C($AlBu^i_2$, $SiMe_3$) (5) → 1. LiMe; 2. $CH_2{=}CHCH_2X$ → (Ph, H)C=C($CH_2CH{=}CH_2$, $SiMe_3$); X = Br, 90%

R_3N, C_6H_{14}: (Ph, H)C=C($SiMe_3$, $AlBu^i_2$) (6) → 1. LiMe; 2. $CH_2{=}CHCH_2X$ → (Ph, H)C=C($SiMe_3$, $CH_2CH{=}CH_2$); X = Br, 80%, *E*:*Z* 20:80; X = I, 40%

Scheme 7.14

isomerization took place; allyl iodide reacted in better stereochemical but poorer chemical yield. However, with alkylalkynylsilanes, the (*E*)-isomers formed in hexane by *trans*-addition isomerize rapidly, and the initially high stereoselectivity is lost; when a 1:1 ether–hexane solvent system was used, *cis*-addition[18] was observed (*Scheme 7.15*), producing (*Z*)-alkenes.

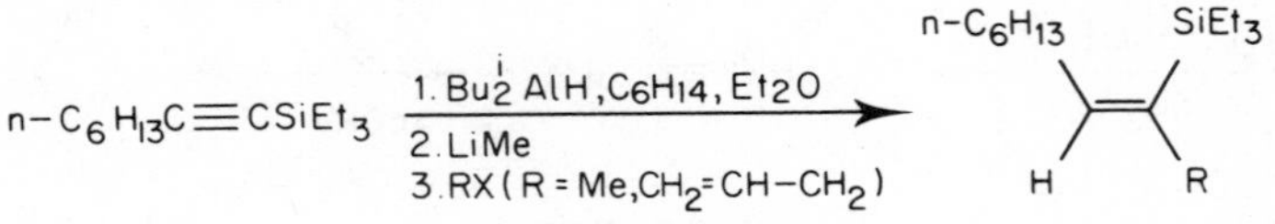

$$n\text{-}C_6H_{13}C{\equiv}CSiEt_3 \xrightarrow[\text{2. LiMe; 3. RX (R = Me, } CH_2{=}CH{-}CH_2)]{\text{1. } Bu^i_2AlH, C_6H_{14}, Et_2O} (n\text{-}C_6H_{13})(H)C{=}C(SiEt_3)(R)$$

Scheme 7.15

Regiospecific and stereoselective direct alkylative reduction can be achieved[19] using Ziegler–Natta alkylating agents. With alkylalkynylsilanes as substrates, *trans*-addition is preferred, whereas with aryl- or alkenyl-alkynylsilanes, no stereoselectivity is observed (*Scheme 7.16*).

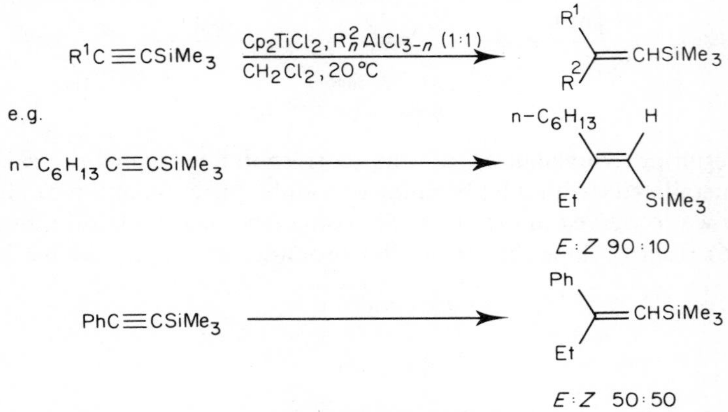

$$R^1C{\equiv}CSiMe_3 \xrightarrow[CH_2Cl_2,\ 20\,°C]{Cp_2TiCl_2,\ R^2_nAlCl_{3-n}\ (1:1)} R^1R^2C{=}CHSiMe_3$$

e.g.

$$n\text{-}C_6H_{13}C{\equiv}CSiMe_3 \longrightarrow (n\text{-}C_6H_{13})(Et)C{=}C(H)(SiMe_3)$$

E:*Z* 90:10

$$PhC{\equiv}CSiMe_3 \longrightarrow (Ph)(Et)C{=}CHSiMe_3$$

E:*Z* 50:50

Scheme 7.16

The alanate intermediate (7) obtained on lithium aluminium hydride–sodium methoxide reduction of 1-trimethylsilylpropyn-3-ol undergoes electrophilic cleavage by iodine to give stereospecifically the vinyl iodide (8); treatment of this with lithium dimethylcopper resulted[11] in production of the disubstituted vinylsilane (9) (*Scheme 7.17*), the iodide of which acts as an alkylative equivalent of methyl vinyl ketone[20] (see p. 85).

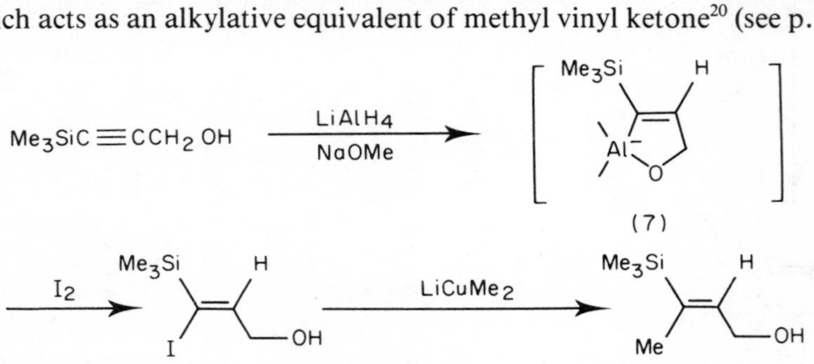

$$Me_3SiC{\equiv}CCH_2OH \xrightarrow[NaOMe]{LiAlH_4} [\text{(7)}] \xrightarrow{I_2} (Me_3Si)(I)C{=}C(H)CH_2OH\ \text{(8)} \xrightarrow{LiCuMe_2} (Me_3Si)(Me)C{=}C(H)CH_2OH\ \text{(9)}$$

(7) (8) (9)

Scheme 7.17

Indeed, one of the most generally applicable routes to the disubstituted vinylsilanes (10) and (11) is provided[21] by sequential combination of hydroalumination, halogen cleavage to the vinyl halide, then alkylation or coupling techniques. Regio- and stereo-selective monohydroalumination of alkynylsilanes in ethereal solvent, followed by cleavage of the resulting (*Z*)-1-alumino-1-alkenylsilanes with halogen, gave the corresponding (*E*)-vinyl halides in high yield and high isomeric purity. Bromine-catalysed photochemical isomerization afforded the corresponding (*Z*)-isomers. Individual treatment of these separate isomers, as illustrated in *Scheme 7.18*, constitutes one of the simplest stereoselective entries to 1,2-dialkylvinylsilanes.

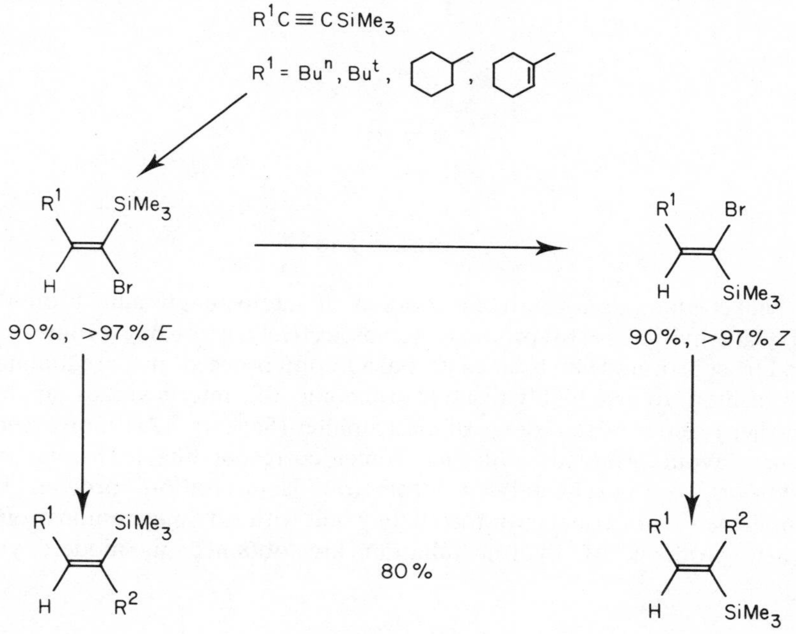

$R^1C{\equiv}CSiMe_3$

$R^1 = Bu^n, Bu^t$,

R^1 $SiMe_3$ / H Br — 90%, >97% *E*

R^1 Br / H $SiMe_3$ — 90%, >97% *Z*

R^1 $SiMe_3$ / H R^2 — (10) >99% *Z*

80%

R^1 R^2 / H $SiMe_3$ — (11) >99% *E*

Scheme 7.18

In a similar fashion, stereo- and regio-selective hydroboration[22] of alkynylsilanes gave the vinylboranes (12). Conversion *via* the corresponding borates into vinyl cuprates then allowed reaction with a wide variety of primary alkyl halides and tosylates, stereoselectively producing (*Z*)-1,2-dialkylvinylsilanes. The intermediate borates react directly with reactive halides such as methyl iodide and allyl halides (*Scheme 7.19*).

Ethynylsilanes themselves undergo direct *cis*-addition when treated with organocoppers or organocuprates; the intermediate species (13) are cleaved electrophilically by primary alkyl iodides and allyl bromide, ultimately yielding[23] (*E*)-1,2-dialkylvinylsilanes in high stereochemical purity (*Scheme 7.20*).

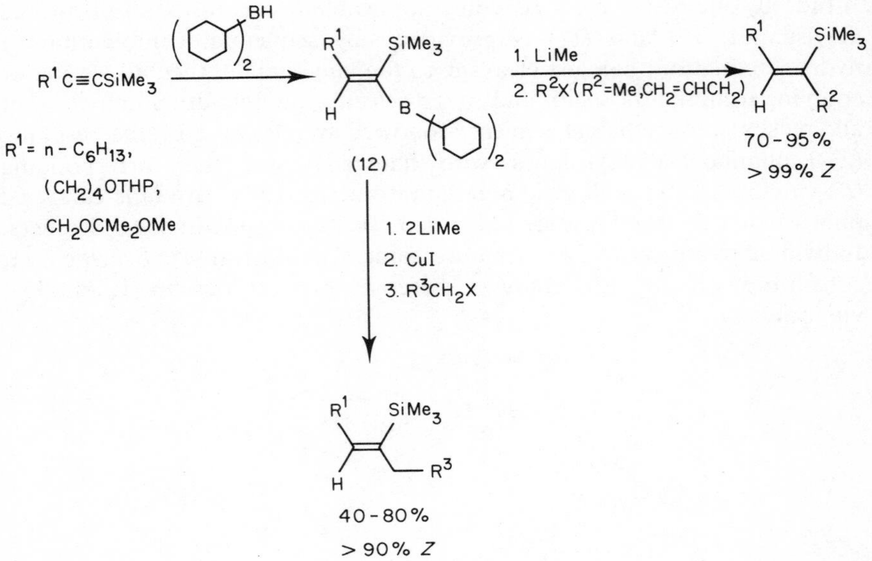

Scheme 7.19

Nickel–aluminium-catalysed addition of methylmagnesium bromide to alkylalkynylsilanes (14) proceeds stereoselectively, affording a route[24] to di- and tri-substituted vinylsilanes. Initial addition proceeds in a predominantly *cis* fashion, to give highly reactive organometallic intermediates capable of further reaction with a range of electrophiles (*Scheme 7.21*). Long reaction times favour production of that isomer corresponding to formal *trans*-addition, a nickel-catalysed geometric isomerization process being implicated. This sequence unfortunately fails with ethylmagnesium bromide, when products of hydrometallation are obtained in modest yields;

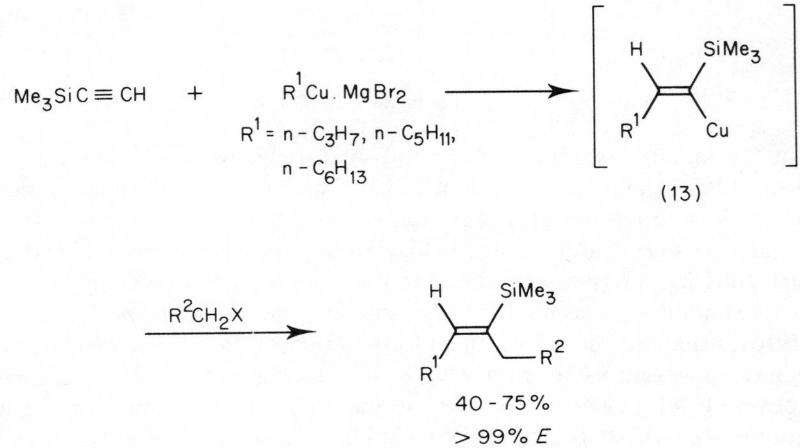

Scheme 7.20

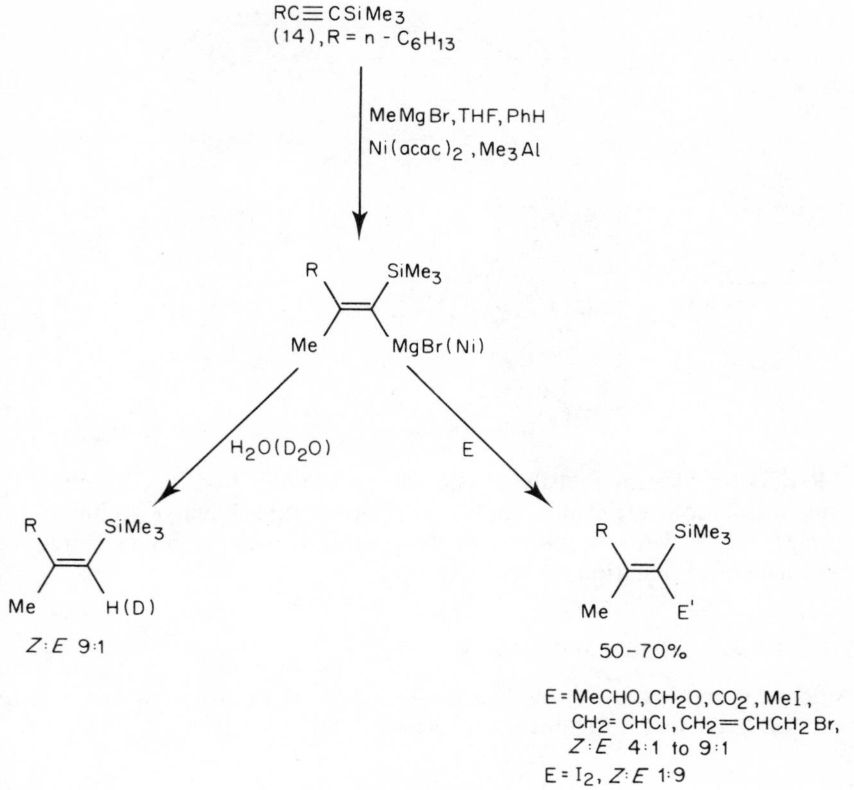

$RC{\equiv}CSiMe_3$
(14), $R = n\text{-}C_6H_{13}$

$MeMgBr$, THF, PhH
$Ni(acac)_2$, Me_3Al

R, $SiMe_3$, Me, MgBr(Ni)

$H_2O(D_2O)$ — E

R, $SiMe_3$, Me, H(D)
Z:*E* 9:1

R, $SiMe_3$, Me, E'
50–70%

E = MeCHO, CH_2O, CO_2, MeI,
$CH_2{=}CHCl$, $CH_2{=}CHCH_2Br$,
Z:*E* 4:1 to 9:1
E = I_2, *Z*:*E* 1:9

Scheme 7.21

presumably this is due to β-elimination of the intermediate ethylnickel species to give ethene and nickel hydride, which will then hydrometallate the alkyne.

Alkynyl trialkylborates undergo 1,2-alkyl migration from boron to carbon on treatment with electrophiles to produce alkenyl boranes[25], the major isomer of which corresponds to the formal *anti*-addition of boron and the electrophile; such alkenylboranes can undergo further electrophile-induced 1,2-migration. This sequence, when applied[26] to trimethylsilylethyne, is of great utility in the preparation of trisubstituted vinylsilanes symmetrically substituted at the β-carbon (*Scheme 7.22*).

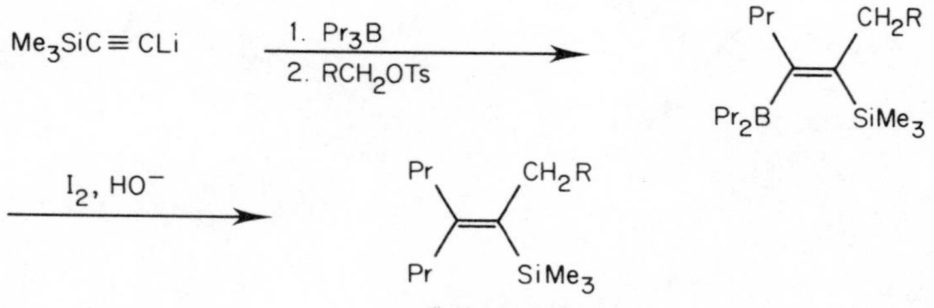

$Me_3SiC{\equiv}CLi$ —(1. Pr_3B; 2. RCH_2OTs)→ Pr, CH_2R, Pr_2B, $SiMe_3$

—(I_2, HO^-)→ Pr, CH_2R, Pr, $SiMe_3$

Scheme 7.22

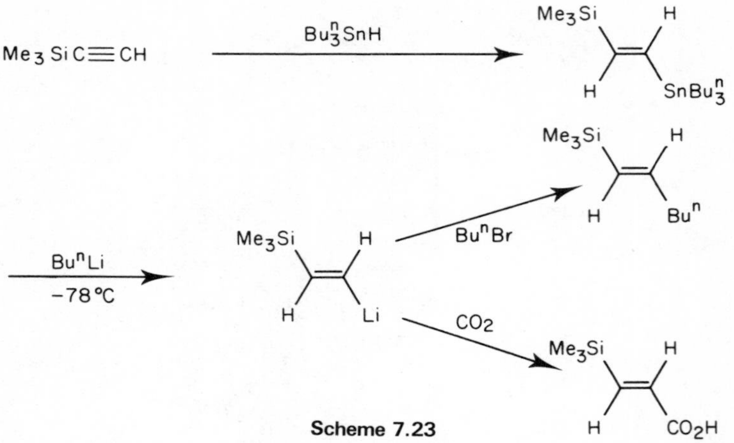

Scheme 7.23

Reduction of trimethylsilylethyne with tributyltin hydride, followed by transmetallation, efficiently leads[27] to (*E*)-trimethylsilylvinyl-lithium. This species reacts with a range of electrophiles with a high degree of stereochemical retention (*Scheme 7.23*).

7.1.2 From aldehydes and ketones

Aldehydes and ketones provide a fertile source of vinylsilanes, a variety of reaction sequences being possible (*Scheme 7.24*).

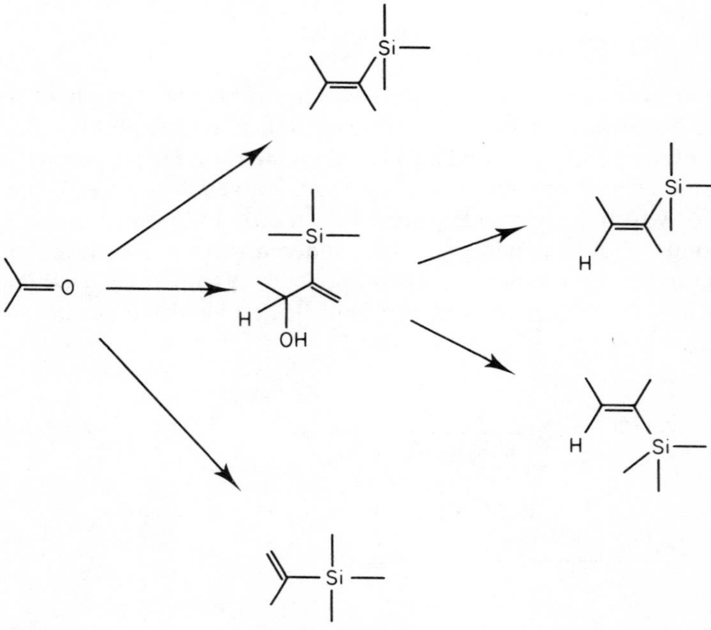

Scheme 7.24

By 1,2-elimination

Here, the elimination reaction most frequently employed is the silyl-Wittig–Peterson Reaction (Chapter 12). Metal salts of bis(trimethylsilyl)methane react with aldehydes and ketones to produce, after elimination of trimethylsilanol or its equivalent, vinylsilanes. The requisite metal salts can be prepared in two separate ways, utilizing either the propensity for oxygen bases/nucleophiles to attack at silicon[28], or the preferential removal of a proton by hindered carbon bases[29] (*Scheme 7.25*).

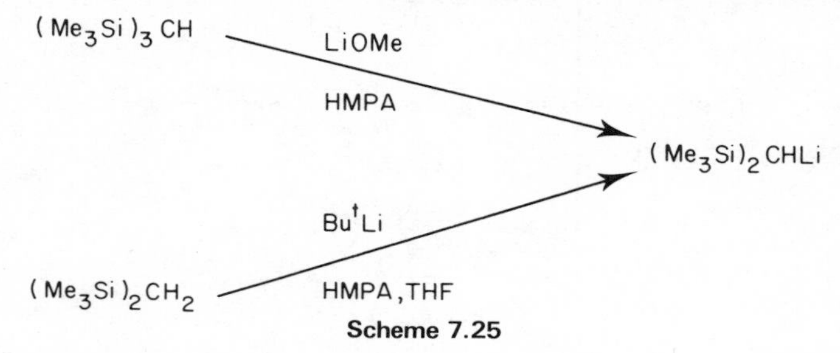

Scheme 7.25

Aldehyde substrates react smoothly to produce vinylsilanes as (*Z,E*)-mixtures (*Scheme 7.26*); yields are, however, rather poor if the aldehyde is enolizable.

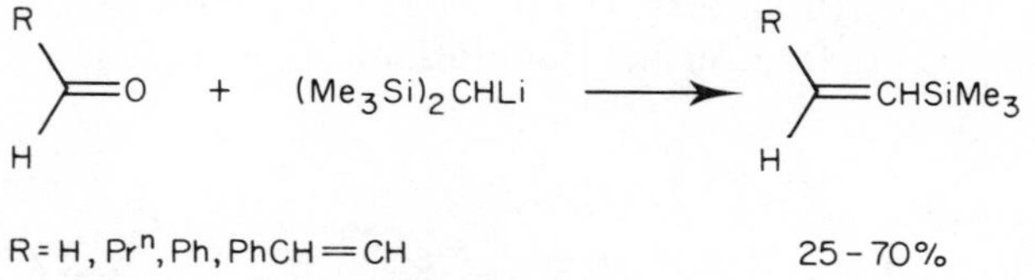

$RCH{=}O + (Me_3Si)_2CHLi \longrightarrow RCH{=}CHSiMe_3$

$R = H, Pr^n, Ph, PhCH{=}CH$ 25–70%

Scheme 7.26

Ketones, on the other hand, must be non-enolizable to be of any utility; here, stereoselectivity has been determined in only one case, that of t-butyl phenyl ketone, when the isomer (15) was sole product, presumably because of the greater eclipsing interactions in the alternative rotamer (16) (*Scheme 7.27*).

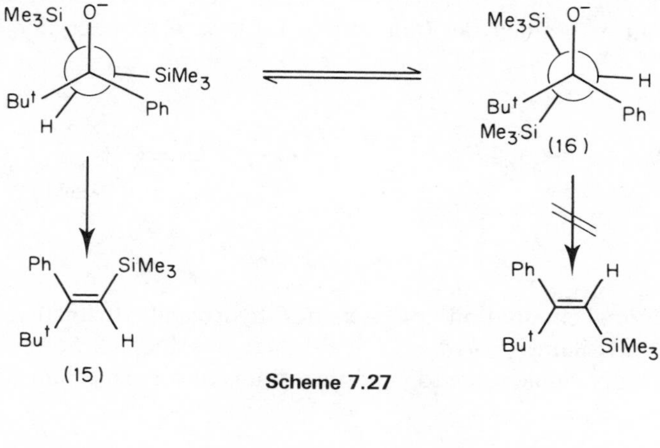

(15) (16)

Scheme 7.27

Interestingly, 1-hydroxy-2-silyl-2-selenyl species can be induced[30] to undergo a stereoselective *anti*-elimination of the hydroxyl and selenyl moieties. Thus, α-lithio-α-silylselenides react with aldehydes to give readily separable mixtures of diastereoisomers, which, in turn, lead stereoselectively to mono- or di-substituted vinylsilanes (*Scheme 7.28*).

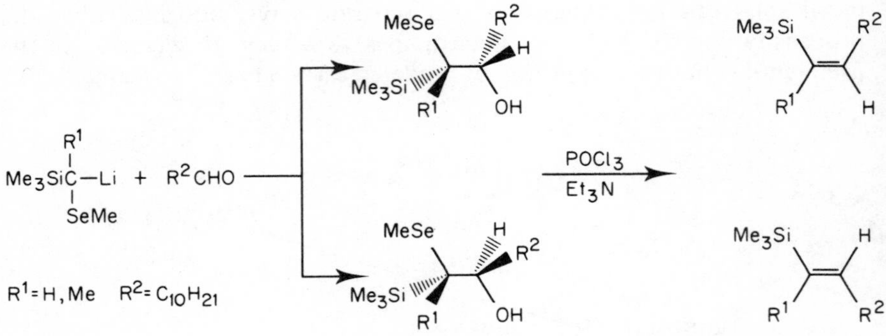

Scheme 7.28

The silyl-Wittig reaction can also be used in the production of highly functionalized vinylsilanes. The lithium salt of the dihydro-1,3-oxazine (17) reacts[31] smoothly with aldehydes (though not with ketones), stereoselectively producing the vinylsilanes (18) (*Scheme 7.29*); such species afford yet unexplored potential for further elaboration.

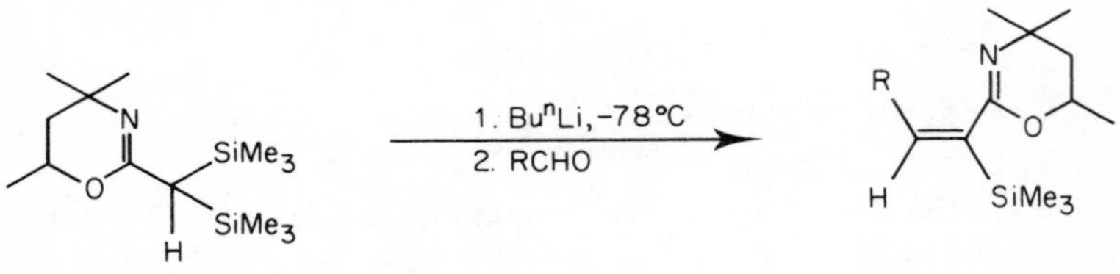

Scheme 7.29

Analogously, the bis(trimethylsilyl)acetic acid ester (19) reacts[32] as its lithium salt with aldehydes (not ketones) to give α-silyl acrylate esters (20) (*Scheme 7.30*).

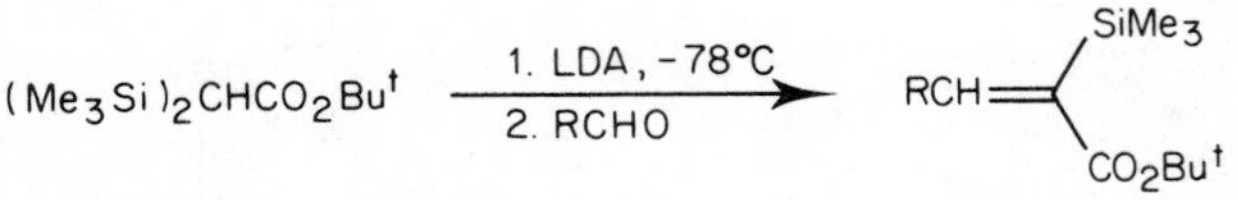

Scheme 7.30

The thermal elimination[33] of esters of β-hydroxyalkyltrimethylsilanes can produce vinylsilanes (*Scheme 7.31*); although this does not normally provide a synthetically viable route to vinylsilanes, it is of some mechanistic interest.

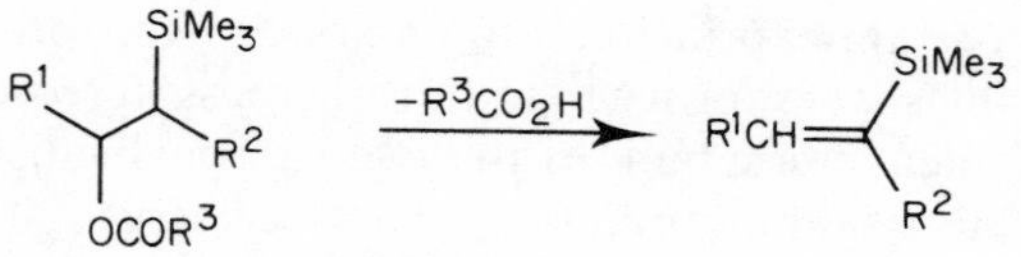

R^1, $SiMe_3$, R^2, $OCOR^3$ — $-R^3CO_2H$ → $R^1CH{=}C(SiMe_3)R^2$

Scheme 7.31

Using metallated vinylsilanes

α-Bromotrimethylsilylethene (21) is readily available on a large scale, the best route[34] starting from vinyltrimethylsilane itself (*Scheme 7.32*).

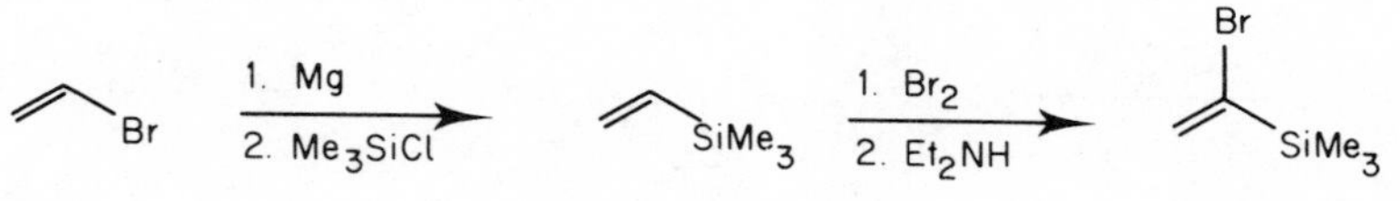

$CH_2{=}CHBr$ — 1. Mg, 2. Me_3SiCl → $CH_2{=}CHSiMe_3$ — 1. Br_2, 2. Et_2NH → $CH_2{=}C(Br)SiMe_3$ (21)

Scheme 7.32

It is readily metallated to the corresponding lithium or magnesium vinyl, the reactions of which with electrophiles will now be discussed.

In a sequence which illustrates an alternative route to α-bromotrimethylsilylethene, Seebach and Gröbel[35] showed that the derived vinyl-lithium (22) reacts readily with carbonyl compounds and pentyl iodide, as shown in *Scheme 7.33*. Reaction of the vinyl-lithium (22) with aldehydes results

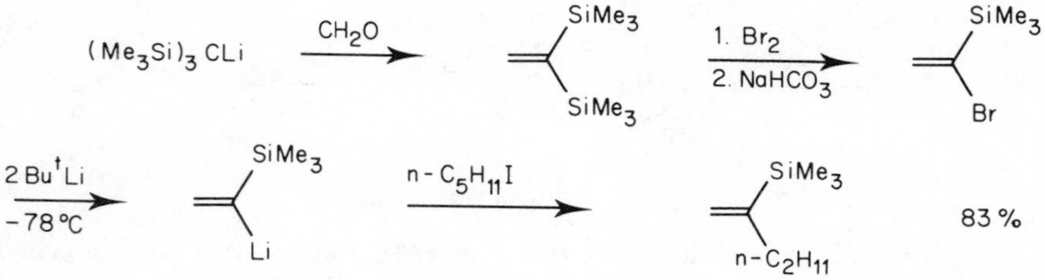

$(Me_3Si)_3CLi$ — CH_2O → $CH_2{=}C(SiMe_3)_2$ — 1. Br_2, 2. $NaHCO_3$ → $CH_2{=}C(SiMe_3)Br$

— $2\,Bu^tLi$, $-78\,°C$ → $CH_2{=}C(SiMe_3)Li$ (22) — $n\text{-}C_5H_{11}I$ → $CH_2{=}C(SiMe_3)(n\text{-}C_2H_{11})$ 83 %

Scheme 7.33

in a stereoselective route[36] to either (*Z*)- or (*E*)-1,2-dialkylvinylsilanes (*Scheme 7.34*).

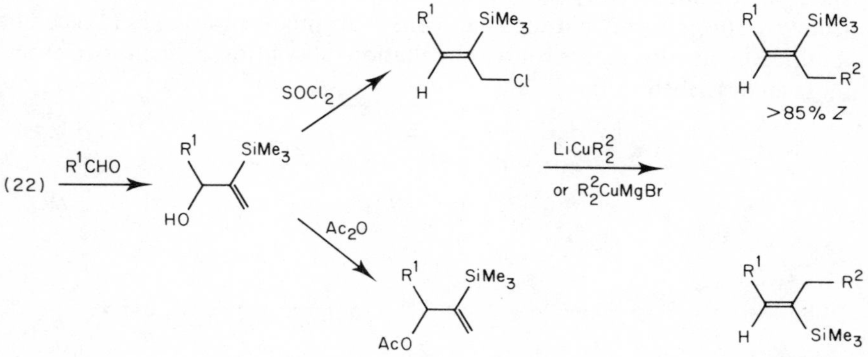

(22) — R^1CHO → $R^1CH(HO)C(SiMe_3){=}CH_2$

— $SOCl_2$ → R^1, H, $SiMe_3$, Cl

— Ac_2O → $R^1CH(AcO)C(SiMe_3){=}CH_2$

— $LiCuR^2_2$ or $R^2_2CuMgBr$ → R^1, H, $SiMe_3$, R^2 >85% *Z* ; R^1, H, R^2, $SiMe_3$ >85% *E*

Scheme 7.34

Extension of this principle to ketones, to provide routes to 1,2,2-trialkylvinylsilanes, proved[37] to be rather less stereoselective, owing to the more equivalent steric bulk of the two carbonyl substituents; the only method[38] of acetylation found to be successful is that shown in *Scheme 7.35*.

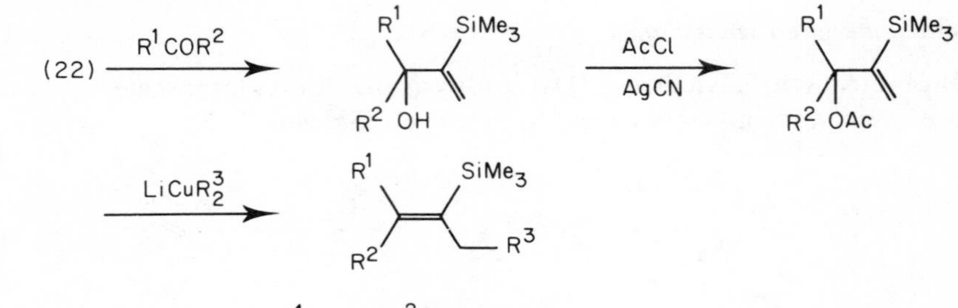

$R^1 = Me$, $R^2 = Et$, $E:Z$ 2:1
$= Bu^i$, $E:Z$ 4:1
$= Pr^i$, $E:Z$ 11:1

Scheme 7.35

1-Hydroxyalkylvinylsilanes (23) can be oxidized[39] to α-trialkylsilylvinyl ketones (24) (*Scheme 7.36*), the use of which in annelation reactions is discussed shortly (p. 76).

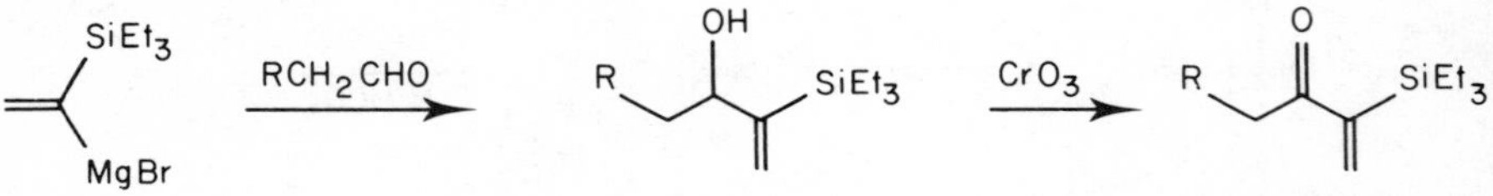

Scheme 7.36

A potentially more direct approach to (24), that of reacting the vinyl organometallic with an acid anhydride[40], is not feasible for anhydrides other than acetic anhydride.

The lithio species (25) transforms aldehydes into 1-bromoalkenylsilanes (26), no stereoselectivity being observed; such vinyl bromides provide a route[41] to 1,2-dialkylvinylsilanes, as shown in *Scheme 7.37*. Ketones do not behave in this manner, but the analogous 1-bromoalkenylsilanes (27) can be obtained by an alternative route; metallation–alkylation of such species has not been described so far.

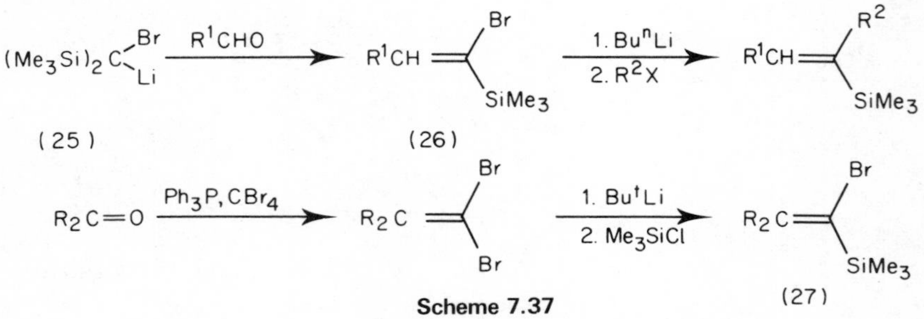

Scheme 7.37

Without change in carbon skeleton

When treated with four equivalents of n-butyl-lithium, aryl sulphonylhydrazones are converted[42] into vinyl carbanions/carbanionoids; these can be trapped with good electrophiles including trimethylsilyl chloride, whereupon the silyl group becomes bonded[43] to the original carbonyl carbon atom (*Scheme 7.38*). This reaction shows a considerable degree of regio- and stereo-selectivity, as illustrated, and allows access to a range of vinylsilanes otherwise obtainable only with difficulty.

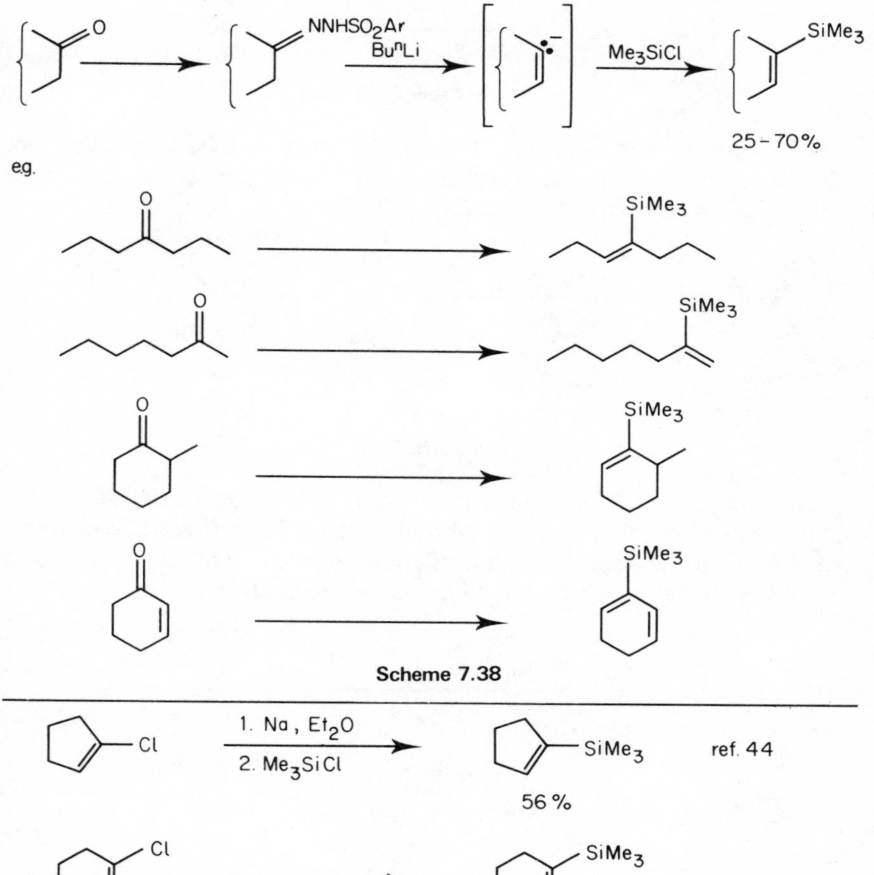

NNHSO$_2$Ar

Bu^{n}Li

Me$_3$SiCl

SiMe$_3$

25–70%

e.g.

O

SiMe$_3$

O

SiMe$_3$

O

SiMe$_3$

O

SiMe$_3$

Scheme 7.38

Cl

1. Na, Et$_2$O

2. Me$_3$SiCl

SiMe$_3$

ref. 44

56%

Cl

SiMe$_3$

ref. 45

65%

Ph

Cl

Ph

Me$_3$Si

ref. 45

40%

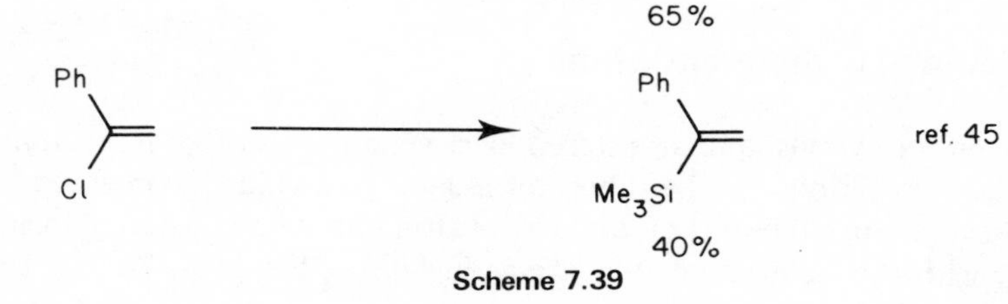

Scheme 7.39

7.1.3 From vinyl halides

One of the simplest routes to vinylsilanes starts from vinyl halides. Metal–halogen exchange, followed by electrophilic attack by trialkylsilyl chloride, can provide the vinylsilane quickly and in good yield[12,13], as utilized earlier (p. 57) in the preparation of 1-bromovinylsilanes. Some other examples are shown in *Scheme 7.39.*

Terminal vinyl bromides have been shown[46] to proceed through this sequence with geometric retention (*Scheme 7.40*).

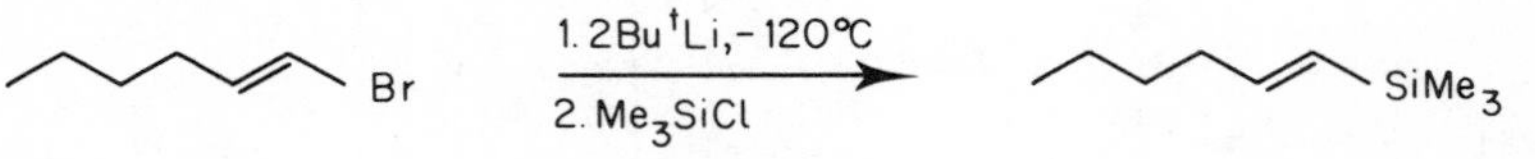

$$\text{C}_5\text{H}_{11}\text{CH=CHBr} \xrightarrow[\text{2. Me}_3\text{SiCl}]{\text{1. 2Bu}^t\text{Li, -120°C}} \text{C}_5\text{H}_{11}\text{CH=CHSiMe}_3$$

Scheme 7.40

As discussed earlier (p. 46), 1,2-dihaloethenes can be converted into 1,2-disilylethenes. A similar transformation[10] can be performed on simple vinyl halides, as shown in *Scheme 7.41.*

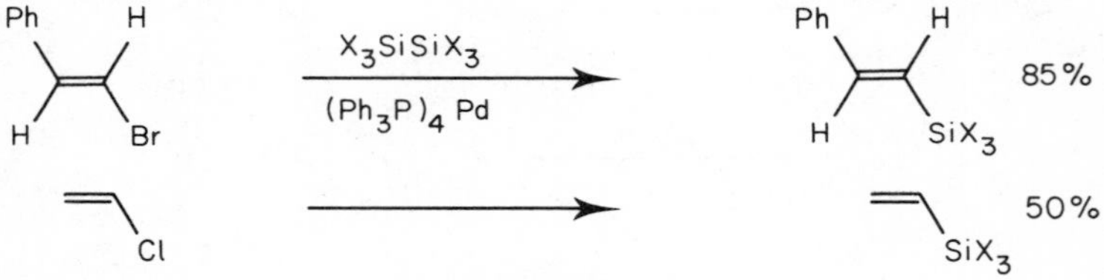

$$\text{PhCH=CHBr} \xrightarrow[(\text{Ph}_3\text{P})_4\text{Pd}]{\text{X}_3\text{SiSiX}_3} \text{PhCH=CHSiX}_3 \quad 85\%$$

$$\text{CH}_2\text{=CHCl} \longrightarrow \text{CH}_2\text{=CHSiX}_3 \quad 50\%$$

Scheme 7.41

Reversing the sequence of this methodology, 1-bromotrimethylsilylethene and its readily available[47] isomer, 2-bromotrimethylsilylethene (28) (*Scheme 7.42*), undergo a process of metallation–carbodemetallation[48] to yield 1-alkylvinylsilanes and (*E*)-2-alkylvinylsilanes, respectively.

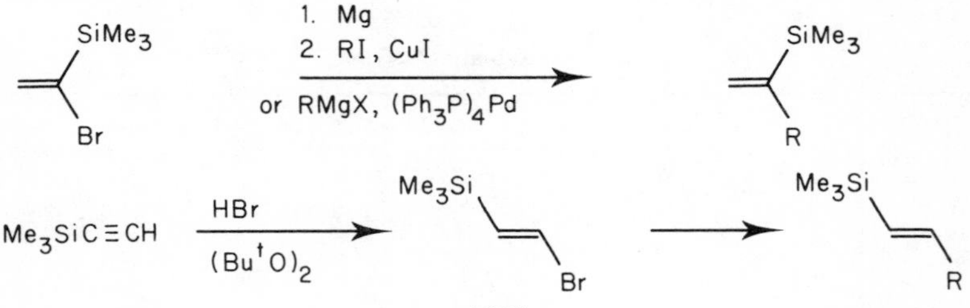

$$\text{CH}_2\text{=C(SiMe}_3\text{)Br} \xrightarrow[\text{or RMgX, (Ph}_3\text{P)}_4\text{Pd}]{\text{1. Mg; 2. RI, CuI}} \text{CH}_2\text{=C(SiMe}_3\text{)R}$$

$$\text{Me}_3\text{SiC}\equiv\text{CH} \xrightarrow[(\text{Bu}^t\text{O})_2]{\text{HBr}} \text{Me}_3\text{SiCH=CHBr}\ (28) \longrightarrow \text{Me}_3\text{SiCH=CHR}$$

Scheme 7.42

7.2 Geometric differentiation

Many routes to vinylsilanes are stereoselective, and it is often important for subsequent reactions to be able to assign particular geometries. For 2-substituted vinylsilanes, (*Z*)- and (*E*)-proton coupling constants normally suffice, with the caveat that they are rather large, being *ca.* 14 and 19 Hz respectively. The geometries of more highly substituted vinylsilanes are not so easy to assign. One frequently used method is to convert the isomeric vinylsilanes into the corresponding alkenes by protiodesilylation; in suitable cases the alkene geometry can then be determined spectroscopically. Since protiodesilylation proceeds with strict retention of configuration (p. 63), the geometry of the starting vinylsilane can then be deduced with confidence.

In a study of 1,2-dialkylvinylsilanes, Chan[49] and co-workers reported that a combination of NMR spectroscopy and GLC retention time ratios could be applied profitably to the qualitative and quantitative estimation of geometric isomers of such vinylsilanes.

In the ^{1}H NMR spectra, the chemical shifts of the vinyl protons of the (*E*)-isomers are consistently at *ca.* 0.3 ppm higher field than the corresponding (*Z*)-isomers. The chemical shifts of the methyl protons of the trimethylsilyl groups of the (*E*)-isomers are also at higher field than those of the (*Z*)-isomers, though here the difference is much smaller, *ca.* 0.08 ppm, and cannot be used with confidence if only one isomer is present. However, since they resonate in a 'clean' part of the spectrum, they can be useful for the quantitative determination of the relative proportions of the isomers in a given mixture.

In the ^{13}C NMR spectra, the methyl carbons of the trimethylsilyl groups of the (*E*)-isomers resonate at higher field than do those of the (*Z*)-isomers; the difference is large, *ca.* 1.5 ppm, and consistent enough to be useful as a method for structural differentiation (*Table 7.2*).

Table 7.2 NMR data for 1,2-dialkylvinylsilanes; ^{1}H NMR spectra were taken in CCl_4 and ^{13}C NMR spectra were taken in $CDCl_3$

$Me_3Si(R^1)C{=}CHR^2$

Vinylsilane	*Stereo-chemistry*	^{1}H, δ/ppm C=CH	^{1}H, δ/ppm CH_3Si	^{13}C, δ/ppm CH_3Si	^{13}C, δ/ppm =CSi	^{13}C, δ/ppm =CH
$R^1 = C_2H_5$	*E*	5.52	0.05	−1.08	140.05	145.87
R^2 = cyclohexyl	*Z*	5.71	0.13	0.56	138.79	147.81
$R^1 = n\text{-}C_5H_{11}$	*E*	5.51	0.05	−1.035	137.89	147.68
$R^2 = i\text{-}C_3H_7$	*Z*	5.80	0.13	0.604	136.29	150.40
$R^1 = n\text{-}C_5H_{11}$	*E*	5.53	0.05	−1.035	140.90	140.47
$R^2 = n\text{-}C_{10}H_{21}$	*Z*	5.83	0.13	0.431	139.18	143.11
$R^1 = (CH_2)_4CHMe_2$	*E*	5.53	0.05	−1.035	140.86	140.43
$R^2 = n\text{-}C_{10}H_{21}$	*Z*	5.83	0.13	0.431	139.25	143.15

In the four (*Z,E*)-geometric pairs studied by GLC, it was observed that the ratio of the retention time of the (*Z*)-isomer relative to that of the (*E*)-isomer was greater than unity. This is not sufficiently rigorous to be diagnostic on its own, but a combination of ^{1}H NMR, ^{13}C NMR, and GLC should allow unambiguous structural assignment, especially since in most cases stereoselectivity rather than absolute stereospecificity is observed.

7.3 Reactivity

Vinylsilanes react readily with a wide range of electrophiles to give products of substitution or addition; in the latter case, such products can be induced to undergo subsequent elimination, either thermally, when *syn*-elimination is observed, or by treatment with a good nucleophile for silicon, when *anti*-elimination is preferred, both processes resulting again in overall substitution (*Scheme 7.43*).

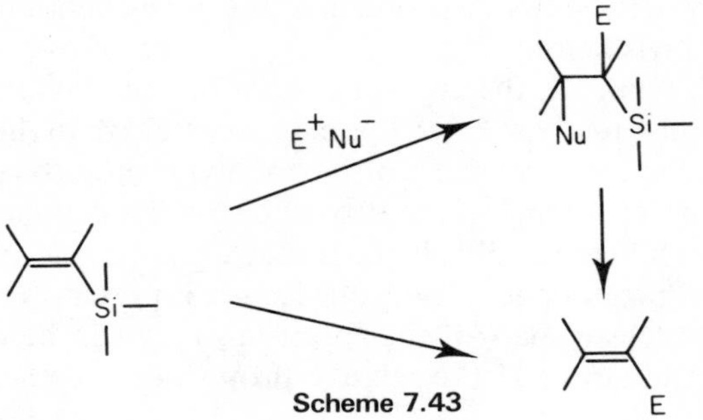

E^+Nu^- Nu Si E Si E

Scheme 7.43

The overall stereochemical result of such substitution will depend on a number of factors, including the stereochemistry of addition and that of subsequent elimination; the stereoselectivity of such processes will be discussed shortly. A most important point to bear in mind is that the *regiochemistry* of substitution/addition is normally unambiguous, the β-effect (p. 15) ensuring that carbonium ion development will occur at the carbon terminus β to silicon, with overall replacement of the C–Si bond with a new C–electrophile bond (*Scheme 7.44*); this obviously confers great synthetic utility on such processes, the starting vinylsilanes being readily available by the variety of routes just discussed.

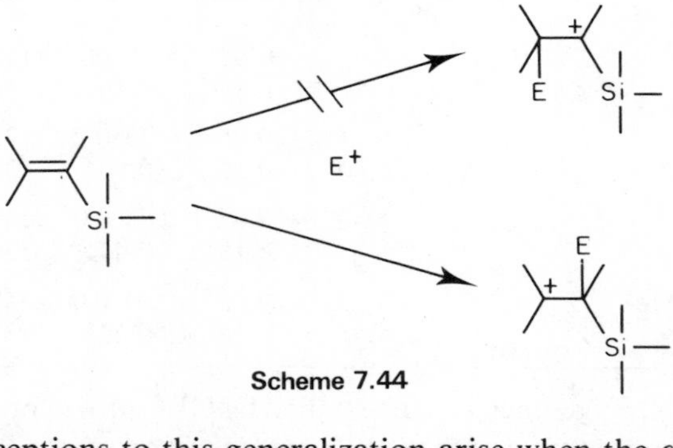

Si E^+ E Si E Si

Scheme 7.44

The few exceptions to this generalization arise when the α-carbon atom carries a substituent which can stabilize α-carbonium ion development. One substituent in such a category is oxygen; 1-trimethylsilyl trimethylsilyl enol ethers (29) (p. 273) give[50] products derived from attack of the electrophile at the β-carbon atom, as exemplified in *Scheme 7.45*.

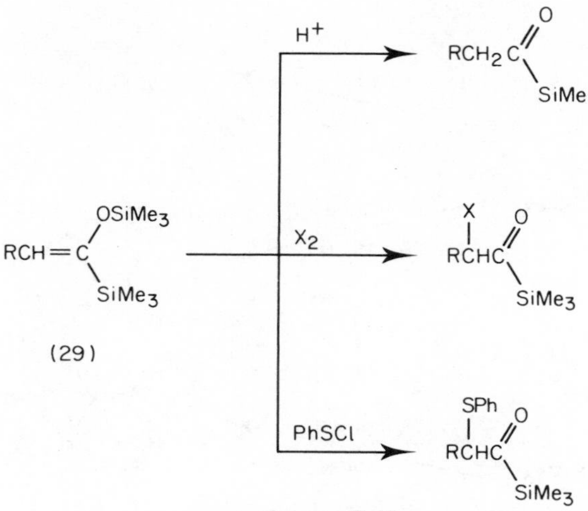

$$RCH{=}C(OSiMe_3)(SiMe_3) \;(29) \xrightarrow{H^+} RCH_2C(=O)SiMe_3$$

$$\xrightarrow{X_2} RCH(X)C(=O)SiMe_3$$

$$\xrightarrow{PhSCl} RCH(SPh)C(=O)SiMe_3$$

Scheme 7.45

In 1954, Sommer[51] and his co-workers reported that β-trimethylsilylstyrene was converted into styrene when treated with acid (*Scheme 7.46*). Little further work was described until 1971, when Eisch[16a] commented that β-trimethylstyrene reacted with bromine to give β-bromostyrene with an extremely high degree of retention of double bond geometry (*Scheme 7.47*).

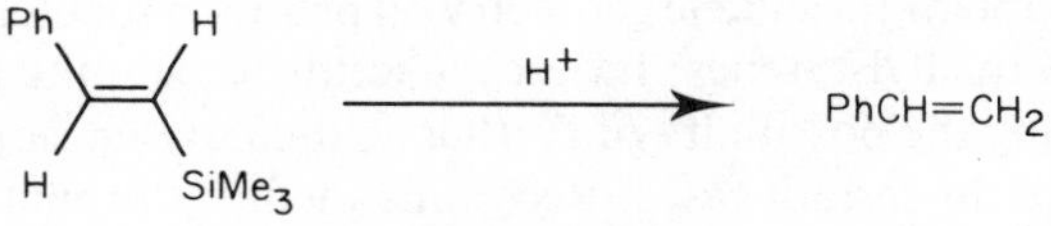

$$PhCH{=}CHSiMe_3 \xrightarrow{H^+} PhCH{=}CH_2$$

Scheme 7.46

In two definitive papers published in 1973, Koenig and Weber[52,53] reported the stereospecific substitutive deuteriation and bromination of (*Z*)- and (*E*)-β-trimethylsilylstyrene (*Scheme 7.48*).

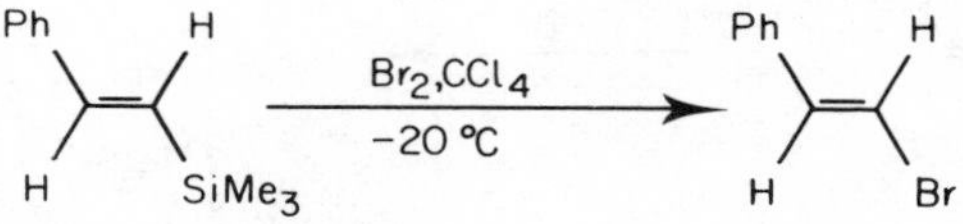

$$PhCH{=}CHSiMe_3 \xrightarrow[-20\,^{\circ}C]{Br_2, CCl_4} PhCH{=}CHBr$$

E:*Z* 95:5 → *E*:*Z* 88:12

Scheme 7.47

To rationalize the deuteriation results (factors controlling the stereoselectivity of bromination will be discussed separately) it is proposed that, simultaneously with deuteron addition to the double bond, rotation occurs about the developing carbon–carbon single bond in such a direction as to permit the C–Si bond to stabilize continuously the incipient benzylic carbonium ion; rotation in the opposite direction would bring the C–Si bond into the nodal plane of the developing ion, and preclude such continuity of overlap. This is illustrated in *Scheme 7.49* for the case of the (*Z*)-isomer, the argument being equally applicable to the (*E*)-isomer.

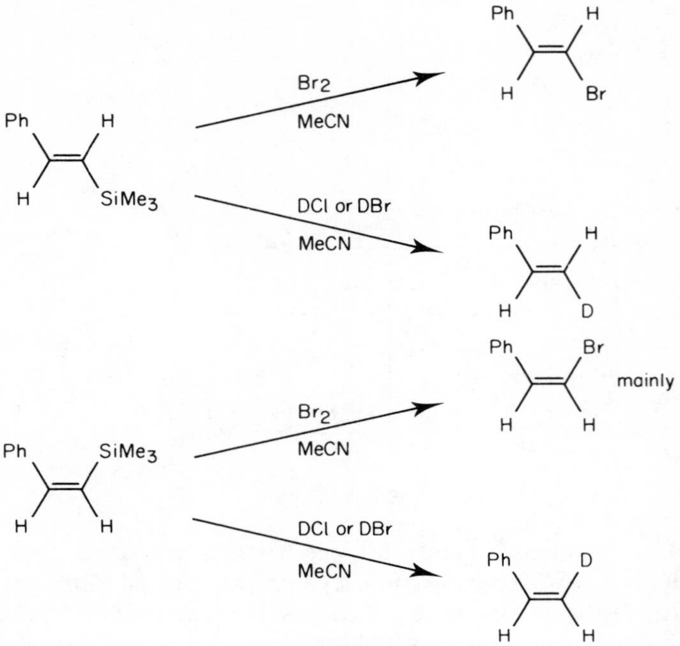

Scheme 7.48

Such retention of the alkene geometry on protio- or deuterio-desilylation is not restricted to silylstyrenes. Indeed, it seems to be quite general, the only limitation being the possibility of further acid-catalysed isomerization of the alkene product in certain cases. Recommended[26a] protonation/deuteriation systems include HI, or water/D_2O containing a catalytic amount of iodine. Some simple examples[26a] are shown in *Scheme 7.50.*

p-Toluenesulphinic acid has advantages[54] when acid-sensitive groups such

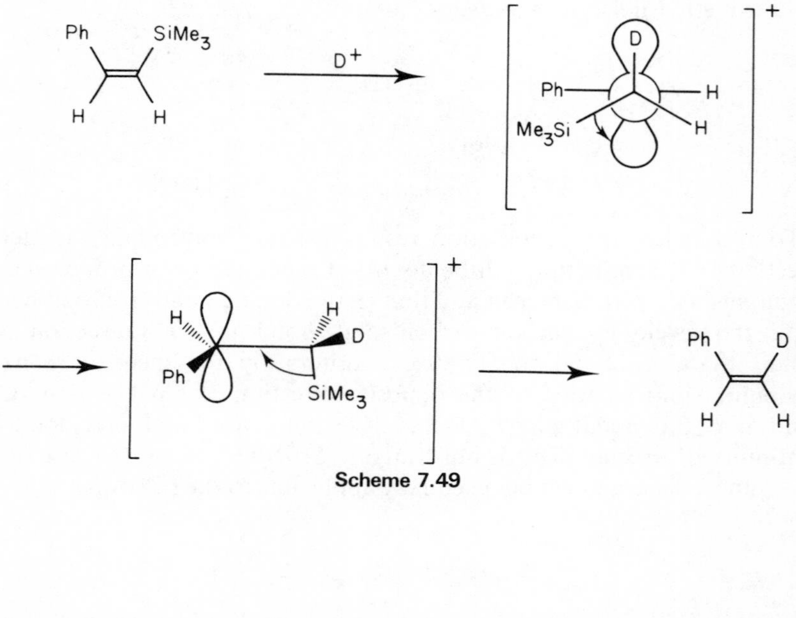

Scheme 7.49

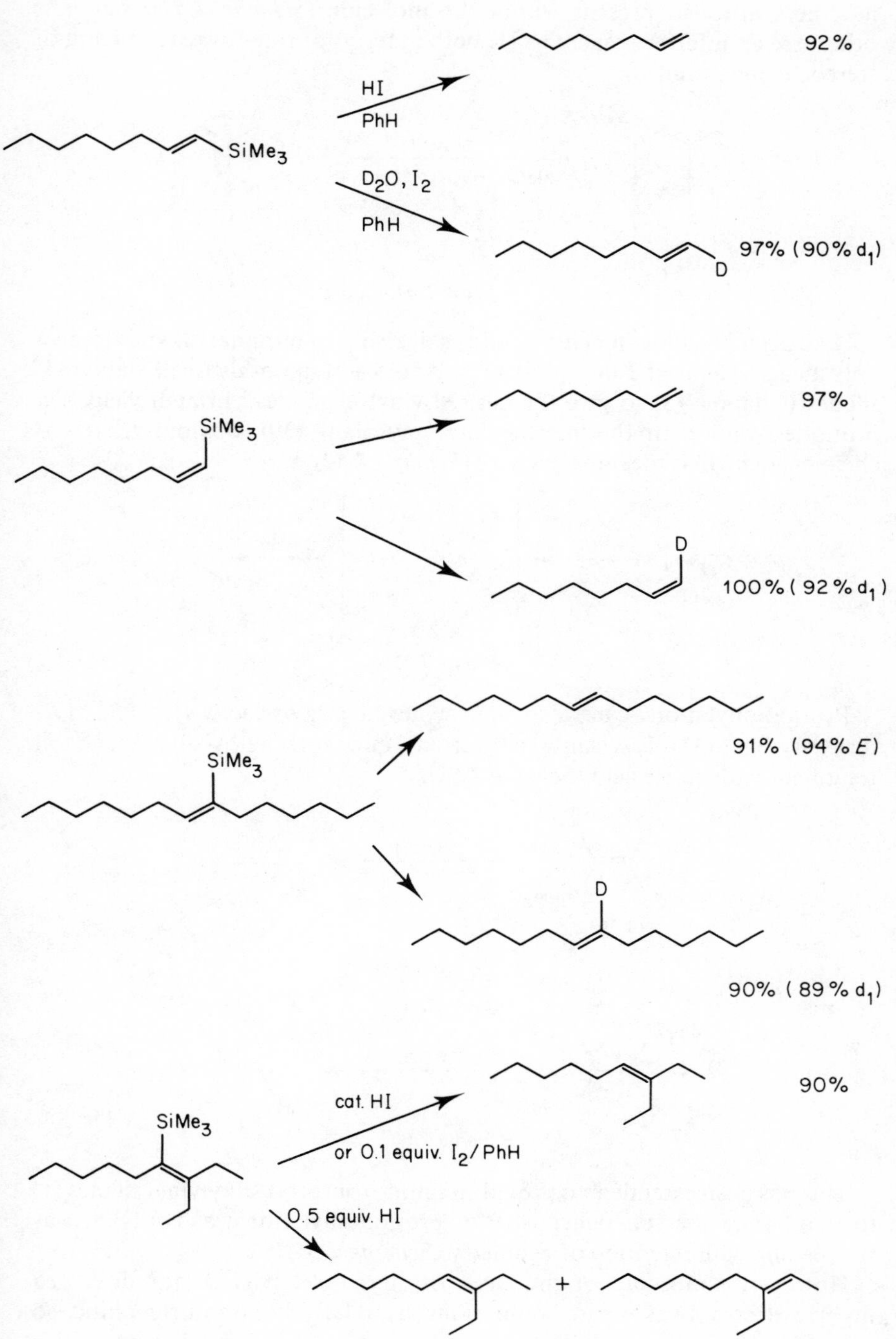

HI
PhH

92%

SiMe3

D_2O, I_2
PhH

D
97% (90% d_1)

SiMe3

97%

D
100% (92% d_1)

SiMe3

91% (94% *E*)

D
90% (89% d_1)

SiMe3

cat. HI
or 0.1 equiv. I_2/PhH

90%

0.5 equiv. HI

+

Scheme 7.50

as ethers are also present within the molecule (*Scheme 7.51*), but it is otherwise an inferior reagent to HI both in terms of rates of reaction and of stereochemical control.

p-$CH_3C_6H_4SO_2H$

MeCN, H_2O

$SiMe_3$

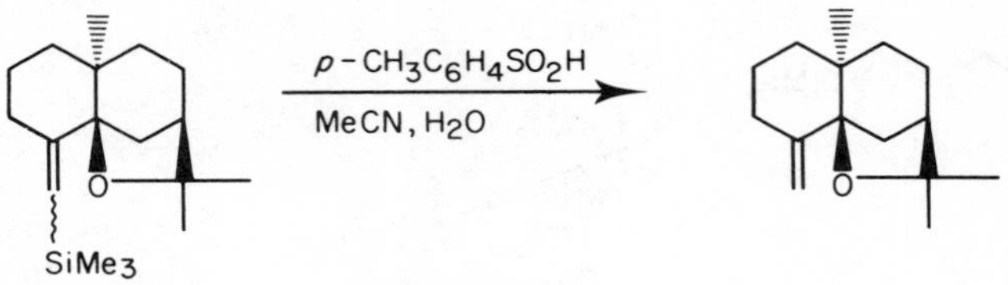

Scheme 7.51

Hydrogen bromide in pentane adds sluggishly to terminal alkynes to give only meagre yields of 2-bromoalkenes. It does add smoothly to alkylalkynylsilanes (Chapter 13) to give the desired vinyl bromides[55] in high yield, *via* protiodesilylation of the intermediate vinylsilane (30); obviously, stereochemical control is meaningless here (*Scheme 7.52*).

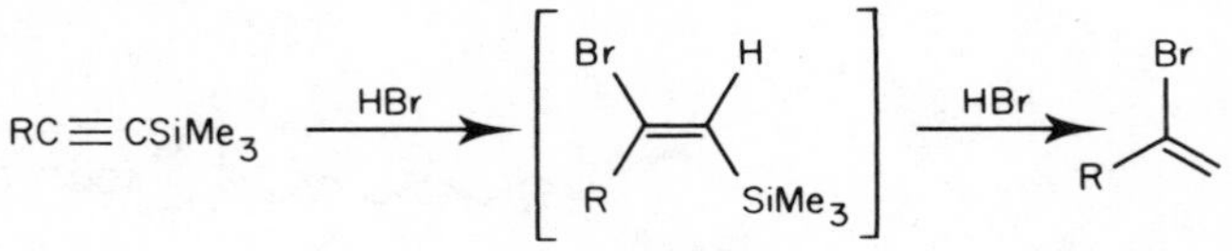

$RC{\equiv}CSiMe_3$ —HBr→ [Br, H, R, $SiMe_3$ (30)] —HBr→ Br, R

Scheme 7.52

Protiodesilylation can also be achieved *regioselectively*. The 1,2-disilylstyrene (31) is cleanly converted[5] into α-trimethylsilylstyrene on treatment with acetic acid (*Scheme 7.53*).

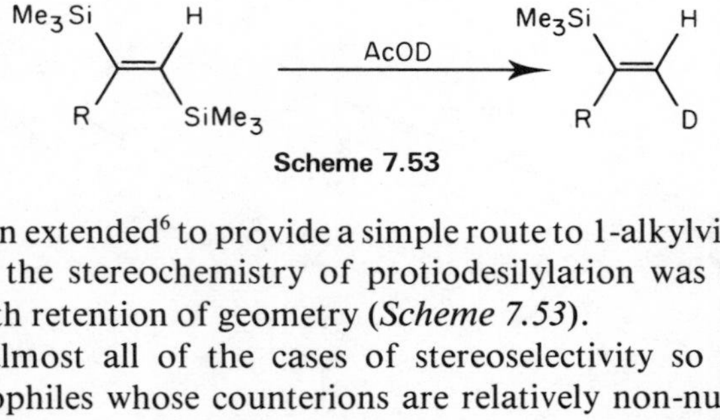

Me_3Si, H, Ph, $SiMe_3$ (31) —AcOH→ Me_3Si, Ph

Me_3Si, H, R, $SiMe_3$ —AcOD→ Me_3Si, H, R, D

Scheme 7.53

This has been extended[6] to provide a simple route to 1-alkylvinyl silanes (1) (p. 46), when the stereochemistry of protiodesilylation was confirmed as proceeding with retention of geometry (*Scheme 7.53*).

However, almost all of the cases of stereoselectivity so far discussed involve electrophiles whose counterions are relatively non-nucleophilic, so that addition to the developing carbonium ion is not kinetically competitive with collapse of the C–Si bond, thus ensuring retention of alkene geometry.

This is generally true; the nucleophilicity of the counterion can be reduced if necessary by complexation and/or by increasing its steric bulk. If, however, an adduct is formed, then the overall stereochemical outcome will depend on both the mode of addition *and* the mode of elimination. In 1969, Jarvie[56] and her co-workers reported that solvolysis of the dibromide obtained by addition of bromine to (*E*)-1-trimethylsilylpropene gave (*Z*)-1-bromopropene almost exclusively; the result is consistent with initial *anti*-addition[53] to produce the *erythro*-dibromide (32), followed by stereospecific *anti*-elimination of trimethylsilyl bromide, initial cleavage of the C–Br bond being assisted by antiperiplanar participation by the C–Si bond (*Scheme 7.54*).

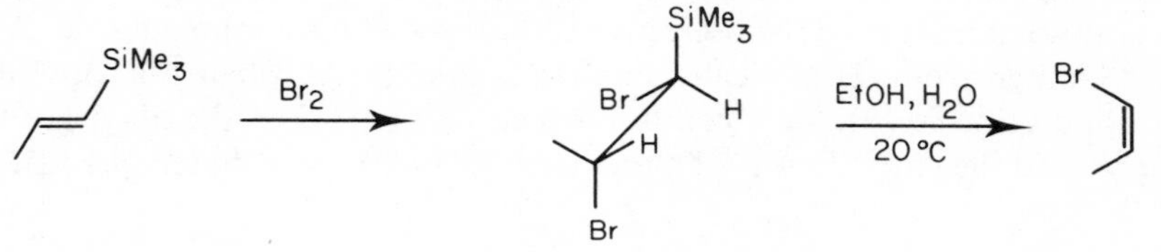

$SiMe_3$ Br_2 Br H H Br $SiMe_3$ EtOH, H_2O 20 °C Br

(32)

Scheme 7.54

More recently, many further examples of such behaviour by 2-alkylvinylsilanes have been reported; di- and tri-substituted vinylsilanes will be discussed separately.

In general, 2-alkylvinylsilanes of defined stereochemistry undergo *anti*-addition of bromine or chlorine to give intermediate 1,2-dihalogeno-compounds, which can be induced[15,21a,57] to undergo *anti*-elimination of trimethylsilyl halide, to give the requisite vinyl halide, with overall *inversion* of configuration of alkene geometry (*Scheme 7.55*).

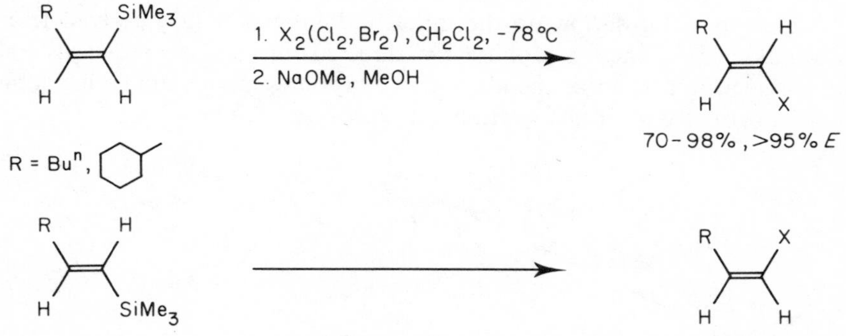

R $SiMe_3$ H H — 1. $X_2(Cl_2, Br_2)$, CH_2Cl_2, −78 °C 2. NaOMe, MeOH → R H H X

70–98%, >95% *E*

R = Bu^n, cyclohexyl

R H H $SiMe_3$ → R X H H

Scheme 7.55

66–93%, >95% *Z*

Suitable conditions for silicodehalogenation include sodium methoxide in methanol, $KF \cdot 2H_2O$ in dimethyl sulphoxide (DMSO), and alumina/pentane, of which the first proved to be the most reliable in terms of yield and stereoselectivity.

Iodination studies were initially much less successful. Direct reaction with iodine gave vinyl iodides with *retention* of alkene geometry (by collapse of the intermediate iodonium ion), but in poor yield. A more successful procedure is shown in *Scheme 7.56*; serious drawbacks to this addition–elimination sequence include low yields of adduct formation and the use of an expensive silver reagent.

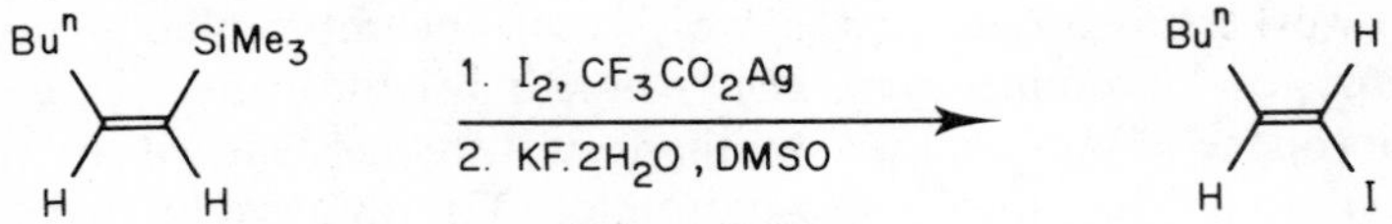
Bu^n
SiMe$_3$
H
H
1. I$_2$, CF$_3$CO$_2$Ag
2. KF.2H$_2$O , DMSO
Bu^n
H
H
I

Scheme 7.56

Halogenation with iodine monochloride overcame[58] these problems, but raised new ones of its own. Successful production of vinyl iodides using this reagent requires initial regioselective addition of iodine monochloride. When the 2-alkyl group is sterically non-demanding, and when the vinylsilane configuration is (*E*), such conditions can be met; when the alkyl group is large, and the vinylsilane is (*Z*), regioselectivity is reversed (*Scheme 7.57*).

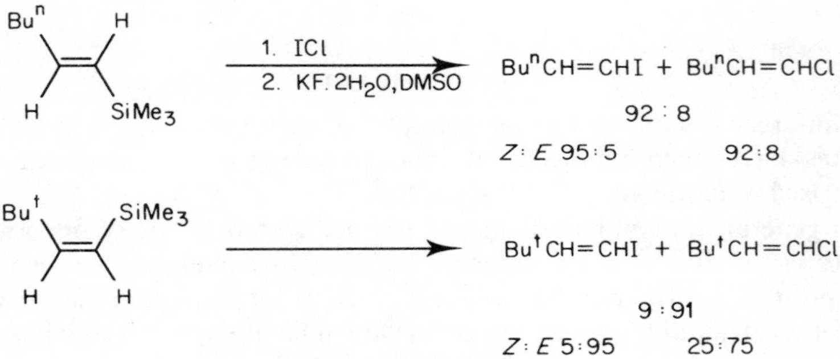
Bu^n
H
H
SiMe$_3$
1. ICl
2. KF.2H$_2$O,DMSO
Bu^{n}CH=CHI + Bu^{n}CH=CHCl
92 : 8
Z : E 95 : 5
92 : 8
Bu^t
SiMe$_3$
H
H
Bu^{t}CH=CHI + Bu^{t}CH=CHCl
9 : 91
Z : E 5 : 95
25 : 75

Scheme 7.57

Attack by chloride ion on the initial iodonium ion (33) derived from (*Z*)-2-alkylvinylsilanes is apparently directed increasingly towards the silicon-bearing carbon as the alkyl group becomes larger, indicating some balancing of electronic and steric effects (*Scheme 7.58*).

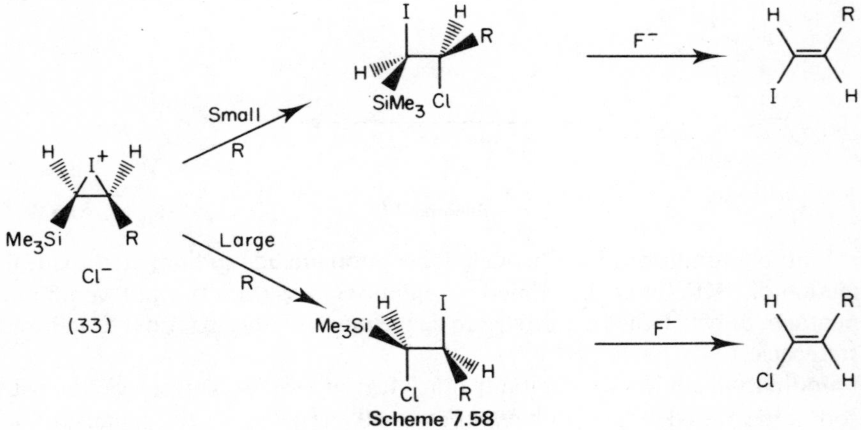
I
H
R
H
SiMe$_3$
Cl
F$^-$
H
R
I
H
Small
R
H
I$^+$
H
Me$_3$Si
R
Cl$^-$
(33)
Large
R
H
I
Me$_3$Si
H
Cl
R
F$^-$
H
R
Cl
H

Scheme 7.58

In agreement with this generalization, it has been found[59] that addition of a variety of iodine-based electrophilic reagents to 1-trimethylsilylcyclohexene and to its 4-t-butyl derivative occurs with a high degree of regio- and stereo-selectivity to give adducts in which the iodine atom is attached to C-1. Such regioselectivity is most conveniently interpreted in terms of an intermediate iodonium ion which undergoes nucleophilic diaxial opening at the carbon atom not bonded to silicon (*Scheme 7.59*).

Scheme 7.59

This orientation contrasts with the mode of opening postulated for protonated epoxysilanes (p. 87), where nucleophilic attack occurs at the carbon atom bonded to silicon. However, the transition state for iodonium ion opening is possibly less symmetric and more product-like, and can benefit from the β-effect's lending some stability to charge development at the β-carbon atom; the transition state for epoxide opening may be more reactant-like, and hence more symmetrical.

To summarize, sterically unhindered 2-alkylvinylsilanes can be converted regiospecifically and stereoselectively into the corresponding vinyl chlorides and bromides, a mechanism involving *anti*-addition and *anti*-elimination ensuring net *inversion* of alkene geometry; the addition of iodine mono-chloride to (*E*)-vinylsilanes followed by fluoride ion induced elimination can provide a viable route to (*Z*)-vinyl iodides.

As was stated earlier, when silylstyrenes are carried through such reaction sequences, products of *retention* of alkene geometry are obtained. For example, (*E*)-β-trimethylsilylstyrene reacts with bromine in carbon disulphide at -100 °C to give a dibromo adduct, which, although relatively stable, undergoes a rapid elimination reaction when dissolved in acetonitrile and produces trimethylsilyl bromide and (*E*)-β-bromostyrene; the isomeric (*Z*)-β-trimethylsilylstyrene reacts in a complementary fashion. In all cases discussed up until now, the addition of halogen has taken place with *anti*-stereochemistry, as expected *via* a bridging halonium ion. In the present case, the intermediate ion is probably much less symmetric owing to the phenyl group also stabilizing the developing carbonium ion. Attack by the nucleophilic bromide ion can now take place from the topside of the molecule, since the stabilizing C–Si bond and its terminal substituents are blocking the bottom face. This will result in overall *syn*-addition[53] of bromine; subsequent *anti*-elimination results in the observed stereochemistry (*Scheme 7.60*).

Some other electrophile-induced desilylations of (mainly) 2-alkyl- and 2-phenyl-vinylsilanes are given in *Scheme 7.61*.

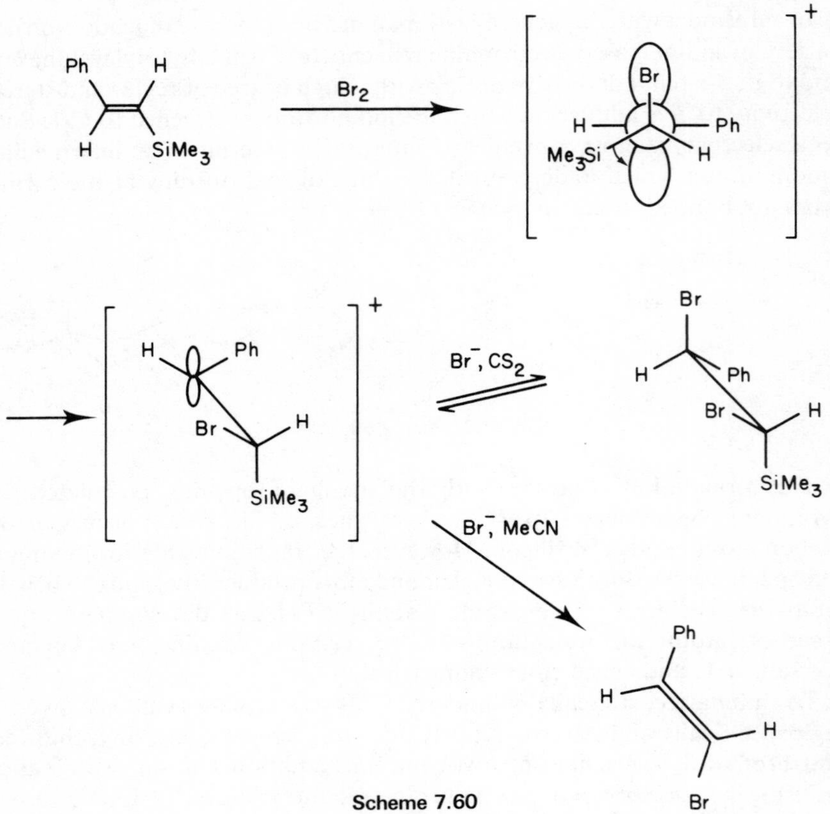

Scheme 7.60

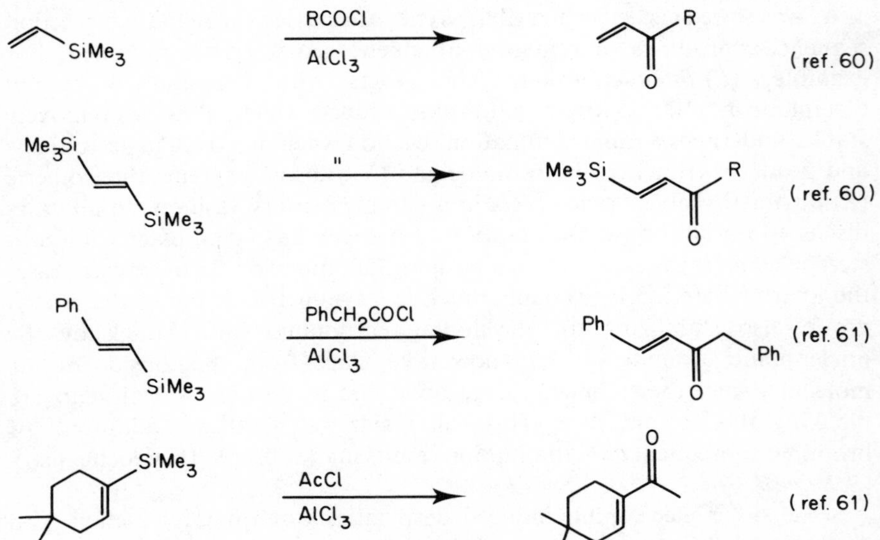

(ref. 60)

(ref. 60)

(ref. 61)

(ref. 61)

Scheme 7.61

continues

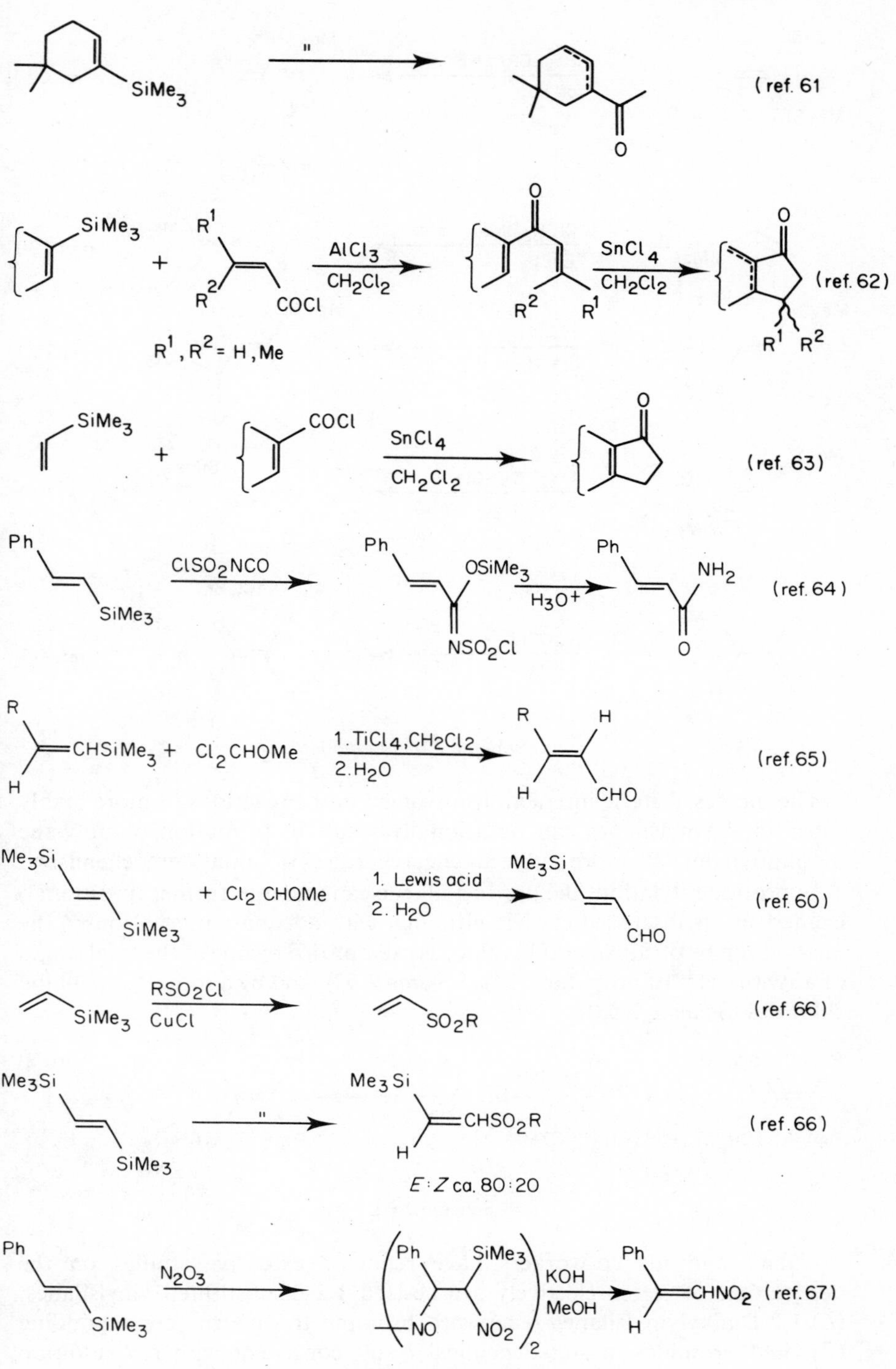

$SiMe_3$ → " (ref. 61)

$SiMe_3$ + R^1, R^2, COCl → $AlCl_3$, CH_2Cl_2 → O, R^2, R^1 → $SnCl_4$, CH_2Cl_2 → O, R^1 R^2 (ref. 62)

R^1, R^2 = H, Me

$SiMe_3$ + COCl → $SnCl_4$, CH_2Cl_2 → O (ref. 63)

Ph, $SiMe_3$ → $ClSO_2NCO$ → Ph, $OSiMe_3$, NSO_2Cl → H_3O^+ → Ph, NH_2, O (ref. 64)

R, H, $=CHSiMe_3$ + Cl_2CHOMe → 1. $TiCl_4$, CH_2Cl_2 2. H_2O → R, H, H, CHO (ref. 65)

Me_3Si, $SiMe_3$ + Cl_2 CHOMe → 1. Lewis acid 2. H_2O → Me_3Si, CHO (ref. 60)

$SiMe_3$ → RSO_2Cl, CuCl → SO_2R (ref. 66)

Me_3Si, $SiMe_3$ → " → Me_3Si, H, $=CHSO_2R$ (ref. 66)

E:*Z* ca. 80:20

Ph, $SiMe_3$ → N_2O_3 → (Ph, $SiMe_3$, NO, NO_2)$_2$ → KOH, MeOH → Ph, H, $=CHNO_2$ (ref. 67)

Scheme 7.61 *(continued)*

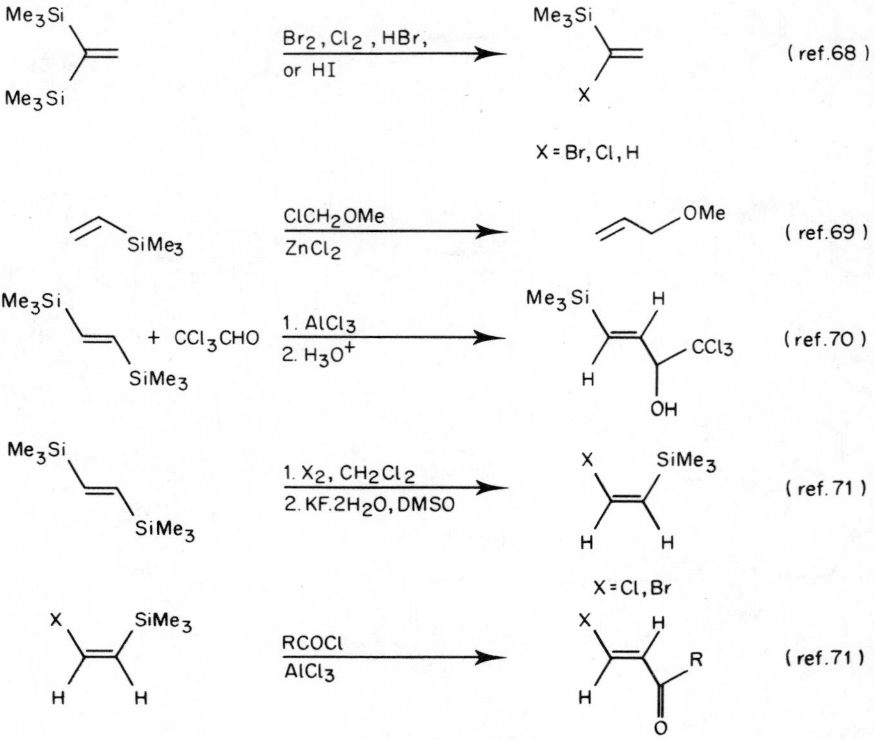

Scheme 7.61 *(continued)*

The increased steric interactions involved in the reactions of more highly substituted vinylsilanes can occasionally result in formation of only the thermodynamically more stable alkene, regardless of initial stereochemistry.

In protiodesilylation, the product alkene can undergo further (positional) isomerism, as discussed earlier, although with adequate precautions[26a] the method can be of considerable value, as exemplified by one of the final stages of a synthesis[26b] of propylure (34) (*Scheme 7.62*), and by a route[36] to (*E*)-allyl chlorides (*Scheme 7.63*).

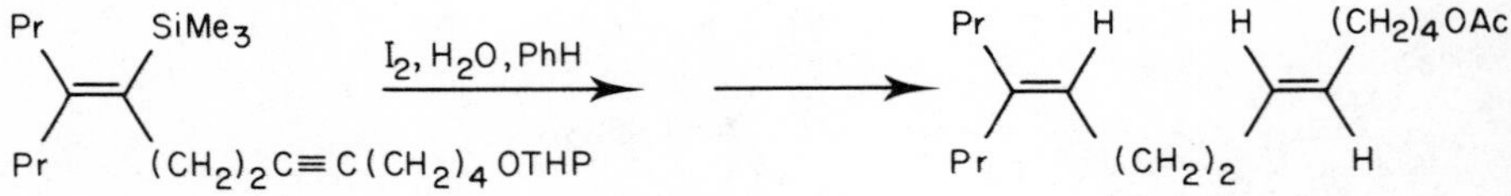

Scheme 7.62

Chan[72] and his co-workers have reported extensive studies on the desilylation of stereoselectively synthesized 1,2-disubstituted vinylsilanes. (*E*)-1,2-Dialkylvinylsilanes react with bromine to give the corresponding (*Z*)-vinyl bromides, a stereochemical result consistent with *anti*-addition followed by *anti*-elimination. (*Z*)-1,2-Dialkylvinylsilanes do not react cleanly

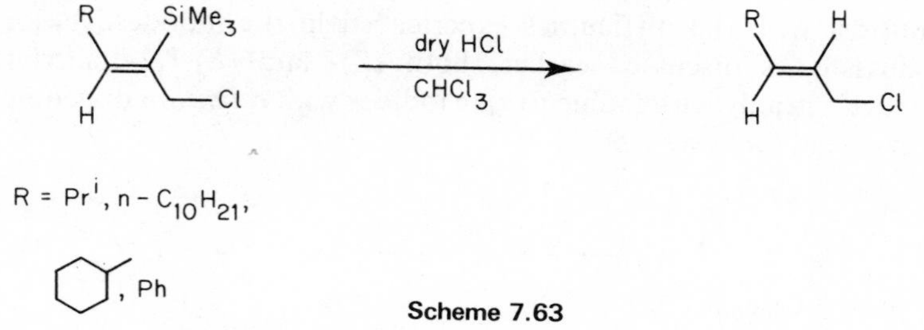
R
SiMe3
Cl
H
dry HCl
CHCl3
R
H
H
Cl
R = Pr^i, n-C10H21,
, Ph

Scheme 7.63

with bromine under such conditions. However, the (*E*)-isomers do react smoothly with cyanogen bromide–aluminium trichloride to give the (*E*)-vinyl bromides; such stereochemical retention must reflect the poor nucleophilicity of the aluminium-complexed counterion, elimination of the trimethylsilyl moiety from the intermediate carbonium ion being more rapid than its capture by nucleophile (*Scheme 7.64*).

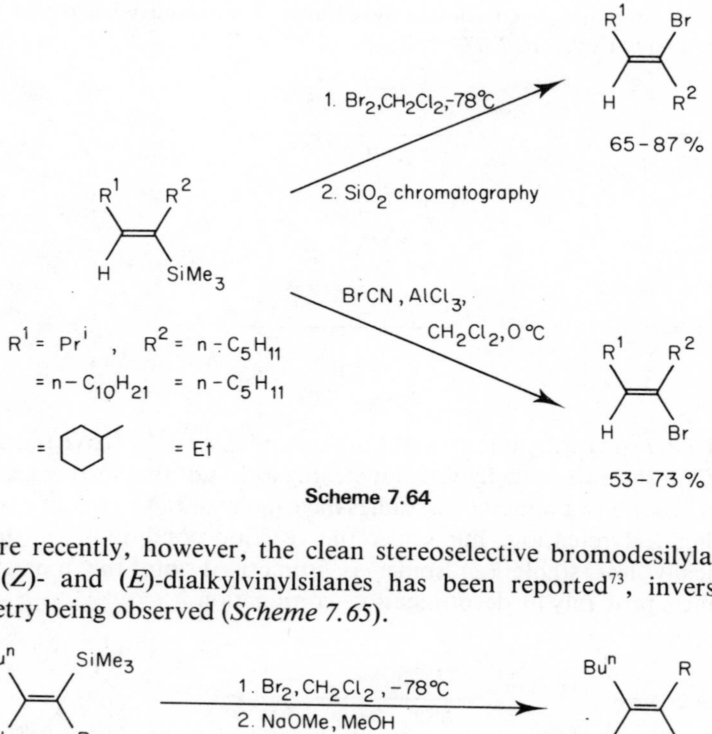
R1
Br
H
R2
65-87 %
1. Br2,CH2Cl2,-78°C
2. SiO2 chromatography
R1
R2
H
SiMe3
BrCN, AlCl3,
CH2Cl2, 0 °C
R1 = Pr^i , R2 = n-C5H11
= n-C10H21 = n-C5H11
= = Et
R1
R2
H
Br
53-73 %

Scheme 7.64

More recently, however, the clean stereoselective bromodesilylation of both (*Z*)- and (*E*)-dialkylvinylsilanes has been reported[73], inversion of geometry being observed (*Scheme 7.65*).

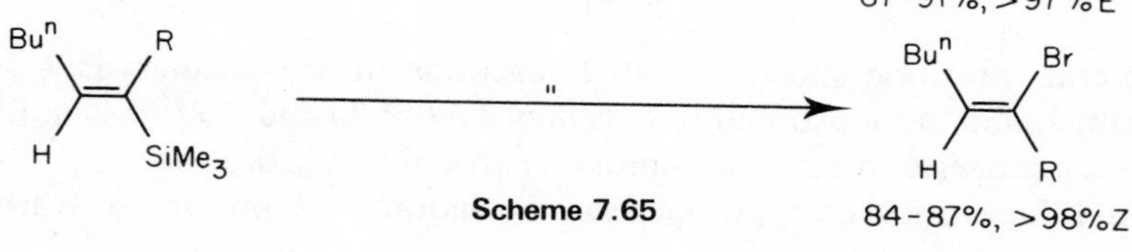
Bu^n
SiMe3
H
R
R = Me, Et
1. Br2, CH2Cl2, -78°C
2. NaOMe, MeOH
Bu^n
R
H
Br
87-91%, >97% E
Bu^n
R
H
SiMe3
"
Bu^n
Br
H
R
84-87%, >98% Z

Scheme 7.65

In contrast with the difficulties experienced in the iododesilylation of 2-alkylvinylsilanes discussed earlier, both (*Z*)- and (*E*)-1,2-dialkylvinylsilanes react[72] cleanly with iodine to give iodides with *retention* of geometry, as now expected (*Scheme 7.66*).

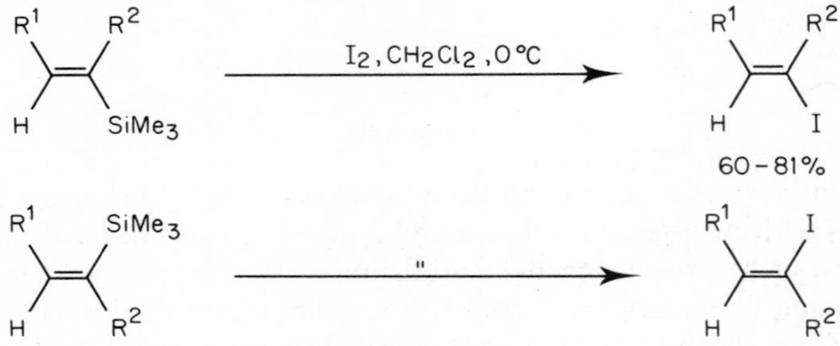

I_2, CH_2Cl_2, 0 °C

60–81%

Scheme 7.66

Friedel–Crafts acylation proceeds well with 1-alkyl-2-cyclohexylvinylsilanes, and has been shown by Chan[72] to proceed with *retention* of stereochemistry (*Scheme 7.67*).

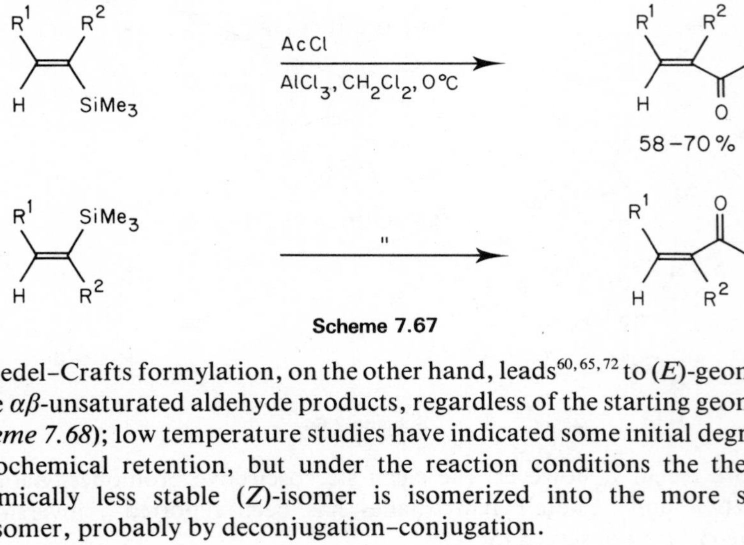

AcCl

$AlCl_3$, CH_2Cl_2, 0 °C

58–70%

Scheme 7.67

Friedel–Crafts formylation, on the other hand, leads[60,65,72] to (*E*)-geometry in the *αβ*-unsaturated aldehyde products, regardless of the starting geometry (*Scheme 7.68*); low temperature studies have indicated some initial degree of stereochemical retention, but under the reaction conditions the thermodynamically less stable (*Z*)-isomer is isomerized into the more stable (*E*)-isomer, probably by deconjugation–conjugation.

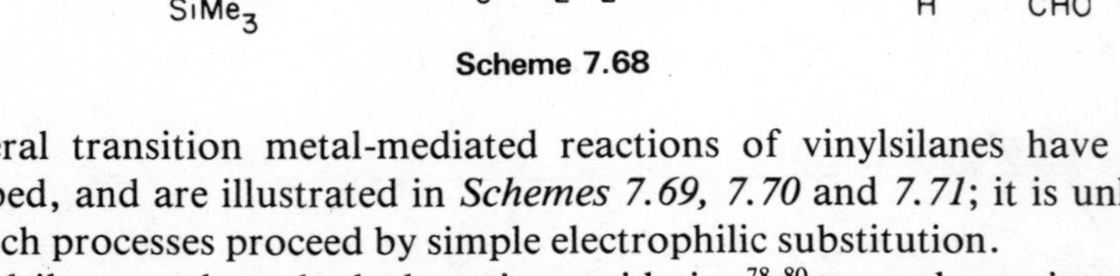

Cl_2CHOMe

$AlCl_3$, CH_2Cl_2, 0 °C

Scheme 7.68

Several transition metal-mediated reactions of vinylsilanes have been described, and are illustrated in *Schemes 7.69, 7.70* and *7.71*; it is unlikely that such processes proceed by simple electrophilic substitution.

Vinylsilanes undergo hydroboration–oxidation[78–80] to produce mixtures of

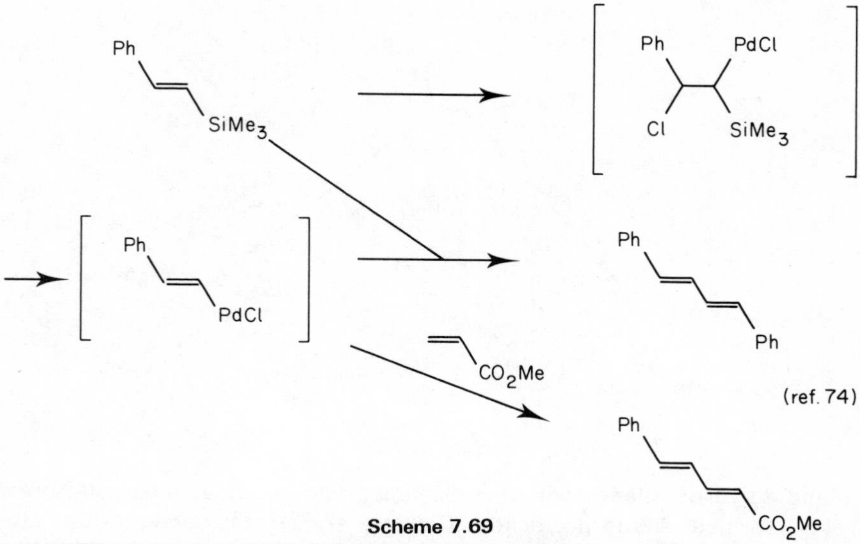

Ph SiMe$_3$ → [Ph, PdCl, Cl, SiMe$_3$] → [Ph, PdCl] → Ph, Ph ; CO$_2$Me → Ph, CO$_2$Me

(ref. 74)

Scheme 7.69

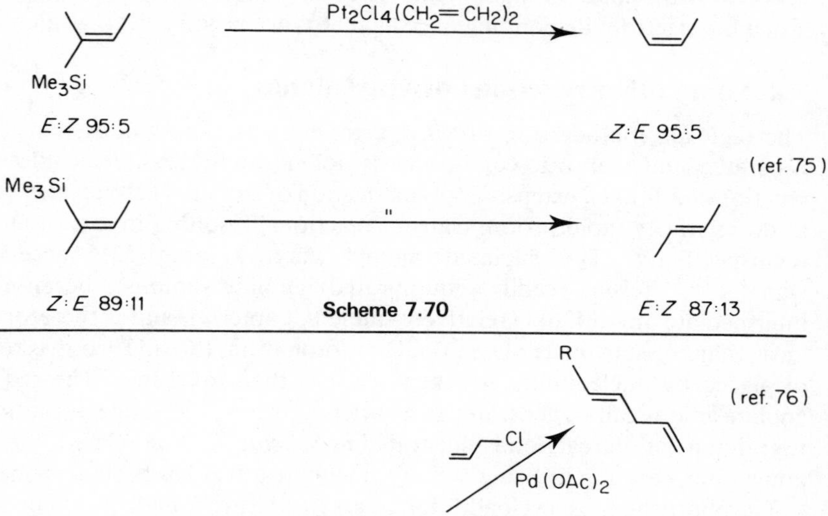

$Pt_2Cl_4(CH_2{=}CH_2)_2$

Me$_3$Si

E:*Z* 95:5

Z:*E* 95:5

(ref. 75)

Me$_3$Si

"

Z:*E* 89:11

Scheme 7.70

E:*Z* 87:13

R

(ref. 76)

Cl

Pd(OAc)$_2$

R H H SiCl$_3$

KF, H$_2$O

K$_2$ [R H H SiF$_5$]

Ag(I)

R

(ref. 77)

R

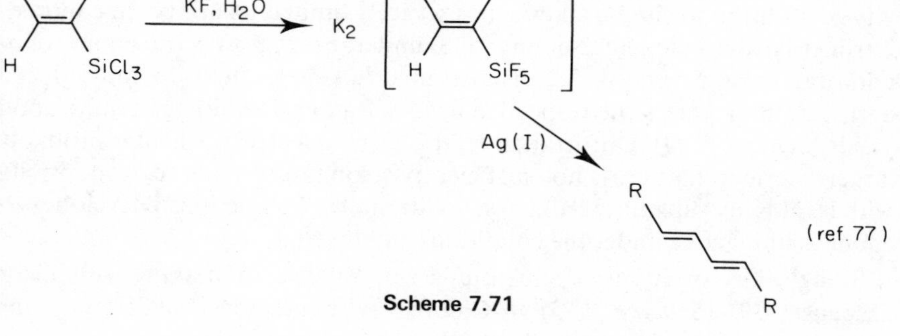

Scheme 7.71

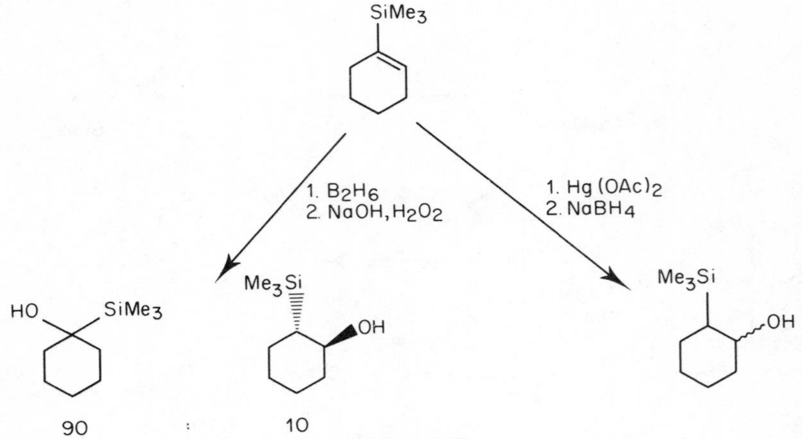

Scheme 7.72

α- and β-hydroxysilanes, with the α-isomer predominating, and the β-isomer being formed stereospecifically (*Scheme 7.72*). Oxymercuration also proceeds with similar or higher regioselectivity[81], but no stereoselectivity, the attacking electrophile bonding once again to the carbon carrying silicon.

7.4 Some other reactions of vinylsilanes

The regiospecific addition of vinyl ketones to enolate anions (kinetically generated under aprotic conditions) is not normally feasible. Under such reaction conditions, extensive polymerization of the vinyl ketone takes place, as do relatively rapid proton transfer reactions, resulting in loss of enolate regiospecificity. The silylated methyl vinyl ketone (35) successfully traps[39,82,83,84] even readily equilibrated enolate anions, because the intermediate anion (36) is relatively stable (Chapter 2), and is therefore less basic than the starting enolate (37). On protonation, the silyl group is readily displaced by nucleophilic attack, since it is then α-ketonic. The requisite enolate anions can be generated in a variety of ways. The scope of this highly useful annelation reaction is illustrated in *Scheme 7.73*. A related, but non-annelating, reaction utilizing α-silylpropenoate esters has been described[85].

The photochemical cycloaddition of alkenes to cyclic enones is in principle a most valuable reaction, but in practice the low or nonexistent regioselectivity of addition to simple alkenes has severely limited its utility. In contrast, 2-trimethylsilylcyclopent-2-enone[86] (38) undergoes regiospecific photocycloaddition to a range of 2,2-disubstituted alkenes; fluoride ion-induced desilylation produces the respective head-to-tail cycloadducts in quite good yields (*Scheme 7.74*). Unfortunately, this directing effect, which is probably largely steric in nature, is not sufficiently strong to provide regiospecificity with less highly substituted alkenes; additionally, 2-trimethylsilylcyclohex-2-enone is unreactive under the conditions employed.

Stang[87] has often used substituted vinylsilanes in making vinylidene carbenes (39) (*Scheme 7.75*). α-Chlorovinylsilanes react similarly giving vinylidene carbene derivatives in good yield (*Scheme 7.76*).

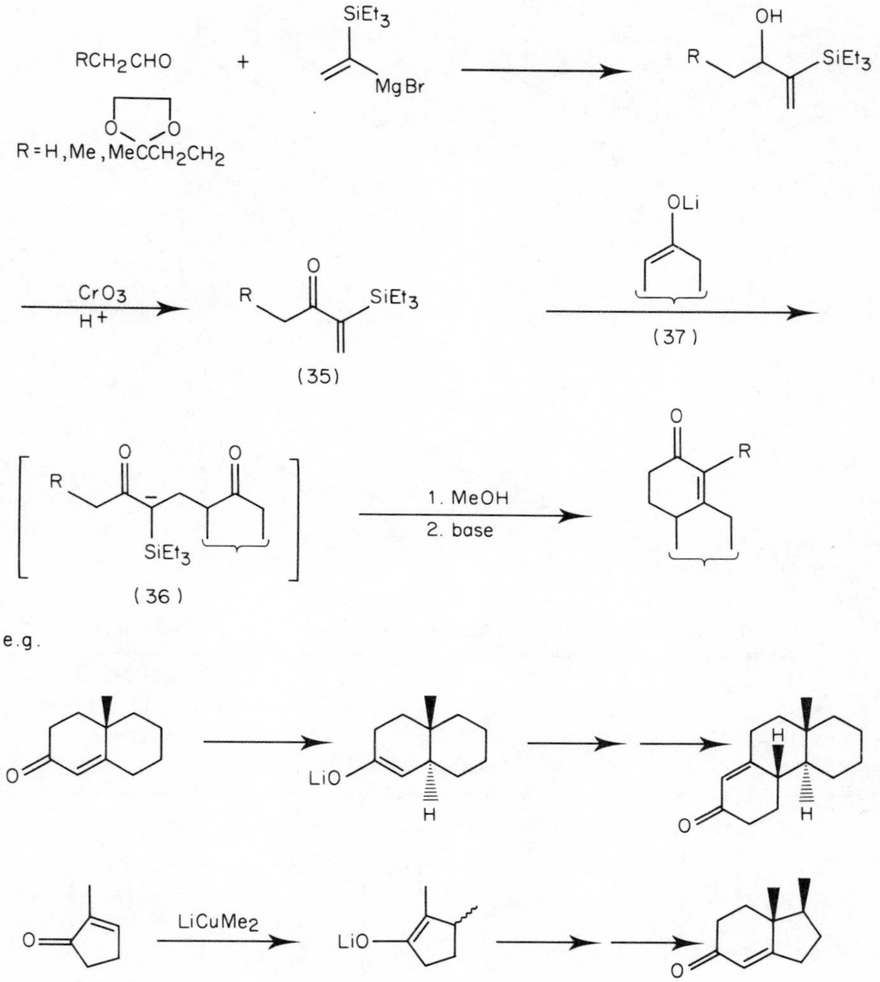

RCH_2CHO + [$CH_2=C(SiEt_3)MgBr$] → $RCH_2CH(OH)C(=CH_2)SiEt_3$

R = H, Me, $MeC(OCH_2CH_2O)CH_2CH_2$

$\xrightarrow[H^+]{CrO_3}$ (35) $\xrightarrow{(37)}$

OLi

(36)

$\xrightarrow[\text{2. base}]{\text{1. MeOH}}$

e.g.

$\xrightarrow{LiCuMe_2}$ LiO

LiO

O

Scheme 7.73

Br

$\xrightarrow[\text{2. } Me_3SiCl \quad \text{3. } H_3O^+]{\text{1. } Bu^nLi}$

(38) SiMe$_3$

$\xrightarrow{h\nu}$ R^1, R^2

SiMe$_3$, R^2, R^1

$\xrightarrow[DMSO]{KF.\ 2H_2O}$

R^2, R^1

50–55%

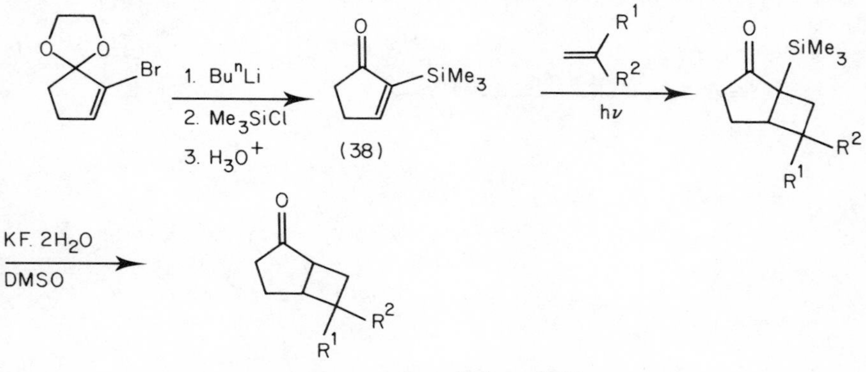

Scheme 7.74

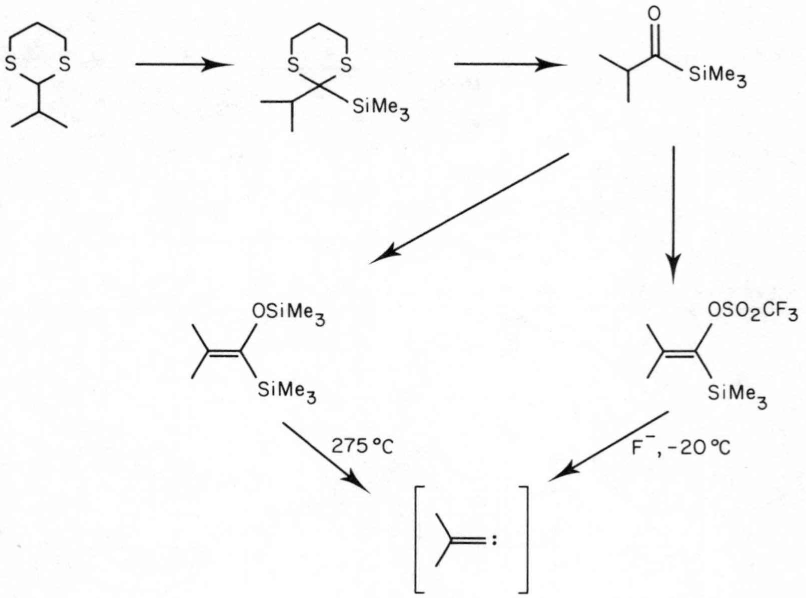

S S

S S

$SiMe_3$

O

$SiMe_3$

$OSiMe_3$

$SiMe_3$

OSO_2CF_3

$SiMe_3$

275 °C

F^-, −20 °C

(39)

Scheme 7.75

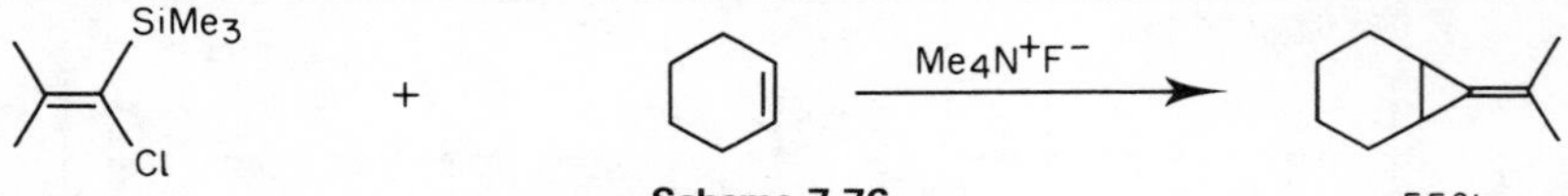

$SiMe_3$

Cl

+

$Me_4N^+F^-$

55%

Scheme 7.76

O_3, MeOH

+

O

Me_3Si

HOO

CHO

O

O

O

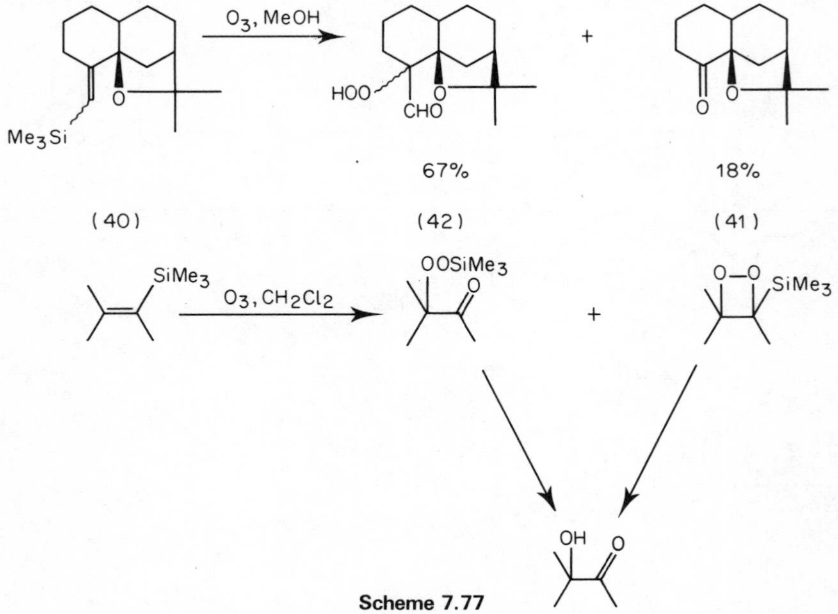

67%

18%

(40)

(42)

(41)

$SiMe_3$

O_3, CH_2Cl_2

$OOSiMe_3$

O

+

O—O

$SiMe_3$

OH

O

Scheme 7.77

Ozonolysis of the vinylsilane (40) appeared to offer a viable[89] degradative route to the norketone (41), a synthetic precursor of α-agarofuran. In fact, ozonolysis in methanol, followed by hydrolysis, afforded the α-hydroperoxy aldehyde (42) as major product. A closer investigation of this reaction with a variety of vinylsilanes showed the immediate identifiable products to be dioxetanes and α-silylperoxy carbonyl species; it is unlikely that these species are formed by single-step processes (*Scheme 7.77*).

7.5 Addendum

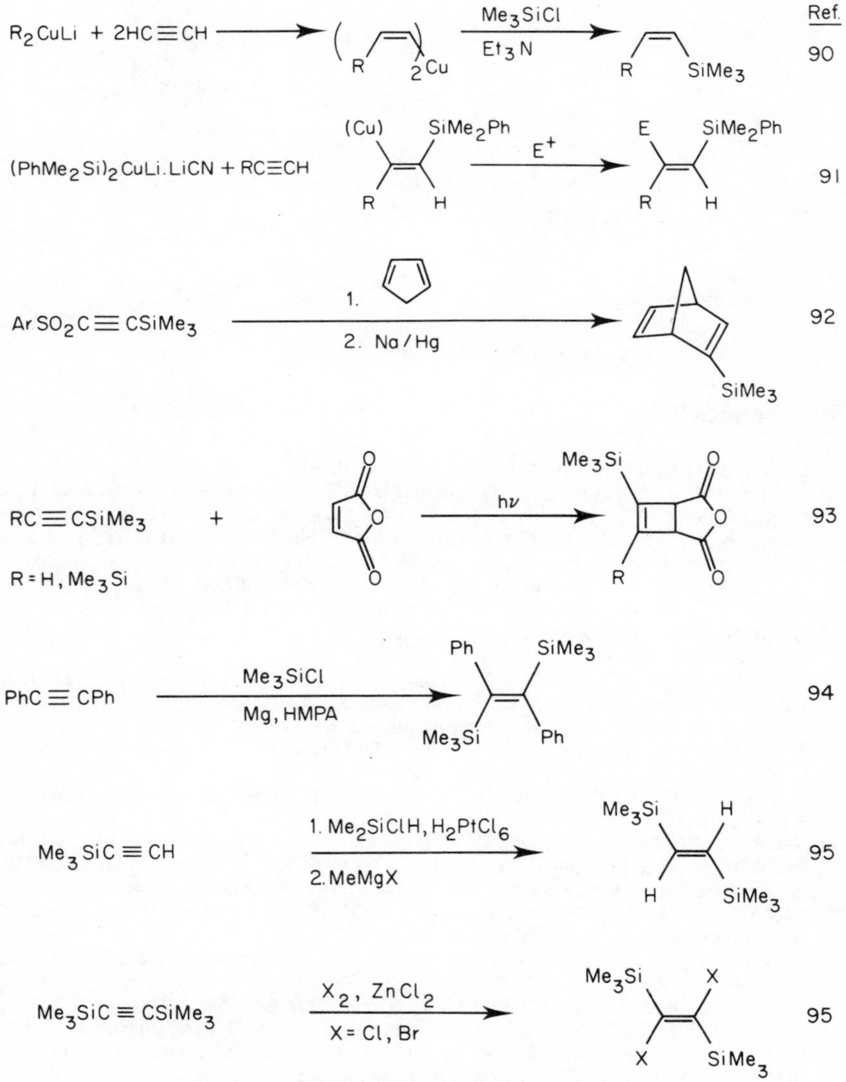

Ref.

$R_2CuLi + 2HC{\equiv}CH \rightarrow$ (R–CH=CH)$_2$Cu $\xrightarrow[Et_3N]{Me_3SiCl}$ R–CH=CH–$SiMe_3$ — 90

$(PhMe_2Si)_2CuLi.LiCN + RC{\equiv}CH \rightarrow$ R(Cu)C=CH($SiMe_2Ph$) $\xrightarrow{E^+}$ R(E)C=CH($SiMe_2Ph$) — 91

$ArSO_2C{\equiv}CSiMe_3 \xrightarrow[2.\ Na/Hg]{1.\ \text{cyclopentadiene}}$ 2-($SiMe_3$)norbornadiene — 92

$RC{\equiv}CSiMe_3$ + maleic anhydride $\xrightarrow{h\nu}$ cyclobutene adduct (Me_3Si, R) — 93

R = H, Me_3Si

$PhC{\equiv}CPh \xrightarrow[Mg,\ HMPA]{Me_3SiCl}$ Ph(Me_3Si)C=C(Ph)$SiMe_3$ — 94

$Me_3SiC{\equiv}CH \xrightarrow[2.\ MeMgX]{1.\ Me_2SiClH,\ H_2PtCl_6}$ Me_3SiCH=CH$SiMe_3$ — 95

$Me_3SiC{\equiv}CSiMe_3 \xrightarrow[X = Cl,\ Br]{X_2,\ ZnCl_2}$ Me_3Si(X)C=C(X)$SiMe_3$ — 95

Scheme 7.78 Additional routes to vinylsilanes

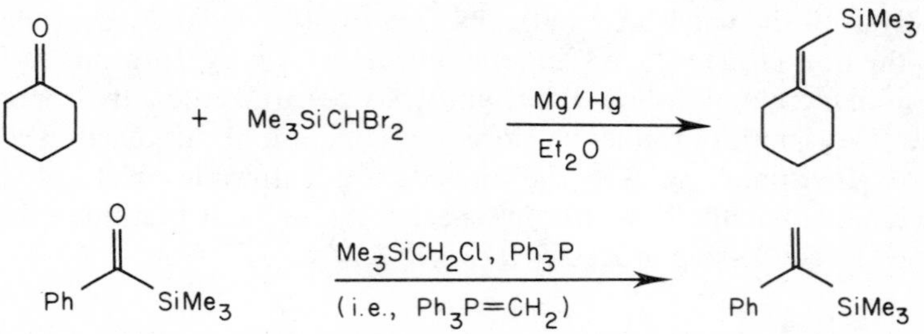

Ref.

$+ \ Me_3SiCHBr_2 \xrightarrow[Et_2O]{Mg/Hg}$ SiMe$_3$ — 96

Ph(C=O)SiMe$_3$ $\xrightarrow[\text{(i.e., } Ph_3P{=}CH_2)]{Me_3SiCH_2Cl,\ Ph_3P}$ Ph(C=CH$_2$)SiMe$_3$ — 97

Scheme 7.78 Additional routes to vinylsilanes *(continued)*

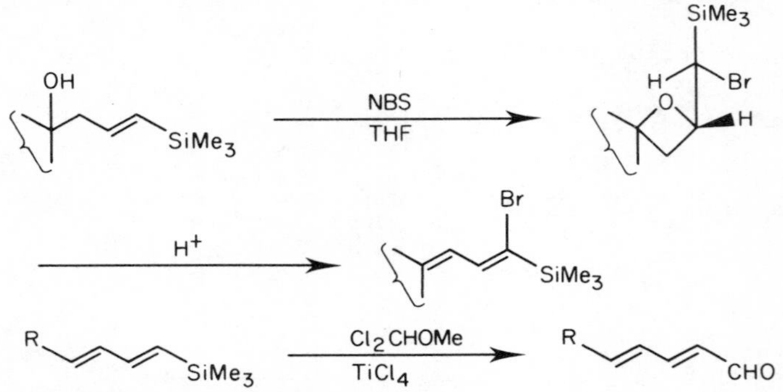

OH ... SiMe$_3$ $\xrightarrow[THF]{NBS}$ SiMe$_3$, H, Br, O, H — 98

$\xrightarrow{H^+}$ Br, SiMe$_3$

R ... SiMe$_3$ $\xrightarrow[TiCl_4]{Cl_2CHOMe}$ R ... CHO — 99

Scheme 7.79 Reactions with electrophiles

References

1 CHAN, T. H. and FLEMING, I., *Synthesis* 761 (1979)
2 SEYFERTH, D., VAUGHAN, L. G. and SUZUKI, R., *J. organometal. Chem.* **1,** 437 (1964)
3 BENKESER, R. A., BURROUS, M. L., NELSON, L. E. and SWISHER, J. V., *J. Am. chem. Soc.* **83,** 4385 (1961); BENKESER, R. A., CUNICO, R. F., DUNNY, S., JONES, P. R. and NERLEKAR, P. G., *J. org. Chem.* **32,** 2634 (1967); for discussion of mechanism, *see* CLARK, A. J. and HARROD, J. F., *J. Am. chem. Soc.* **87,** 16 (1965)
4 STORK, G. and COLVIN, E., *J. Am. chem. Soc.* **93,** 2080 (1971)
5 DUNOGUÈS, J., BOURGEOIS, P., PILLOT, J.-P., MERAULT, G. and CALAS, R., *J. organometal. Chem.* **87,** 169 (1975)
6 HUDRLIK, P. F., SCHWARTZ, R. H. and HOGAN, J. C., *J. org. Chem.* **44,** 155 (1979)
7 RYAN, J. W. and SPEIER, J. L., *J. org. Chem.* **31,** 2698 (1966)
8 BATT, D. G. and GANEM, B., *Tetrahedron Lett.* 3323 (1978)
9 TAMAO, K., MIYAKE, N., KISO, Y. and KUMADA, M., *J. Am. chem. Soc.* **97,** 5603 (1975)
10 MATSUMOTO, H., NAGASHIMA, S., KATO, T. and NAGAI, Y., *Angew. Chem. int. Edn* **17,** 279 (1978)
11 STORK, G., JUNG, M. E., COLVIN, E. and NOEL, Y., *J. Am. chem. Soc.* **96,** 3684 (1974)
12 PETROV, A. D., MIRONOV, B. F., PONOMARENKO, V. A. and CHERNYSHEV, E. A., *'Synthesis of Organosilicon Monomers',* Heywood, London (1964)
13 BAZANT, V., CHVALOVSKY, V. and RATHOUSKY, J., *'Organosilicon Compounds',* Academic Press, New York and London (1965)
14 MINH, L. Q., BILLIOTTE, J. C. and CADIOT, P., *C.r. hebd. Séanc Acad. Sci., Paris* **251,** 730 (1960)
15 MILLER, R. B. and REICHENBACH, T., *Tetrahedron Lett.* 543 (1974)
16 (*a*) EISCH, J. J. and FOXTON, M. W., *J. org. Chem.* **36,** 3520 (1971); (*b*) EISCH, J. J. and RHEE, S. G., *J. Am. chem. Soc.* **97,** 4673 (1975); (*c*) *see also* ALTNAU, G., RÖSCH, L., BOHLMANN, F. and LONITZ, M., *Tetrahedron Lett.* 4069 (1980)

17 EISCH, J. J. and DAMASEVITZ, G. A., *J. org. Chem.* **41,** 2214 (1976)
18 UCHIDA, K., UTIMOTO, K. and NOZAKI, H., *J. org. Chem.* **41,** 2215 (1976)
19 EISCH, J. J., MANFRE, R. J. and KOMAR, D. A., *J. organometal. Chem.* **159,** C13 (1978); *see also* Van HORN, D. E. and NEGISHI, E., *J. Am. chem. Soc.* **100,** 2252 (1978)
20 STORK, G. and JUNG, M. E., *J. Am. chem. Soc.* **96,** 3682 (1974)
21 (*a*) ZWEIFEL, G. and LEWIS, W., *J. org. Chem.* **43,** 2739 (1978); (*b*) MILLER, R. B. and McGARVEY, G., *J. org. Chem.* **43,** 4424 (1978)
22 UCHIDA, K., UTIMOTO, K. and NOZAKI, H., *J. org. Chem.* **41,** 2941 (1976); *Tetrahedron* **33,** 2987 (1977)
23 OBAYASHI, M., UTIMOTO, K. and NOZAKI, H., *Tetrahedron Lett.* 1805 (1977); WESTMIJZE, H., MEIJER, J. and VERMEER, P., *Tetrahedron Lett.* 1823 (1977)
24 SNIDER, B. B., KARRAS, M. and CONN, R. S. E., *J. Am. chem. Soc.* **100,** 4624 (1978); SNIDER, B. B., CONN, R. S. E. and KARRAS, M., *Tetrahedron Lett.* 1679 (1979)
25 KÖSTER, R. and HAGELEE, L. A., *Synthesis* 118 (1976)
26 (*a*) UTIMOTO, K., KITAI, M. and NOZAKI, H., *Tetrahedron Lett.* 2825 (1975); (*b*) UTIMOTO, K., KITAI, M., NARUSE, M. and NOZAKI, H., *Tetrahedron Lett.* 4233 (1975)
27 CUNICO, R. F. and CLAYTON, F. J., *J. org. Chem.* **41,** 1480 (1976)
28 SAKURAI, H., NISHIWAKI, K. and KURA, M., *Tetrahedron Lett.* 4193 (1973)
29 GRÖBEL, B.-Th. and SEEBACH, D., *Chem. Ber.* **110,** 852 (1977)
30 Van ENDE, D., DUMONT, W. and KRIEF, A., *J. organometal. Chem.* 149, C10 (1978); DUMONT, W., Van ENDE, D. and KRIEF, A., *Tetrahedron Lett.* 485 (1979)
31 SACHDEV, K., *Tetrahedron Lett.* 4041 (1976)
32 HARTZELL, S. L. and RATHKE, M. W., *Tetrahedron Lett.* 2737 (1976)
33 CAREY, F. A. and TOLER, J. R., *J. org. Chem.* **41,** 1966 (1976)
34 OTTOLENGHI, A., FRIDKIN, M. and ZILKHA, A., *Can. J. Chem.* **41,** 2977 (1963)
35 GRÖBEL, B.-Th. and SEEBACH, D., *Chem. Ber.* **110,** 867 (1977)
36 MYCHAJLOWSKIJ, W. and CHAN, T. H., *Tetrahedron Lett.* 4439 (1976); CHAN, T. H., MYCHAJLOWSKIJ, W., ONG, B. S. and HARPP, D. N., *J. organometal. Chem.,* **107,** C1 (1976); *J. org. Chem.* **43,** 1526 (1978)
37 AMOROUX, R. and CHAN, T. H., *Tetrahedron Lett.* 4453 (1978)
38 TAKIMOTO, S., INANAGA, J., KATSUKI, T. and YAMAGUCHI, M., *Bull. chem. Soc. Japan* **49,** 2335 (1976)
39 STORK, G. and GANEM, B., *J. Am. chem. Soc.* **95,** 6152 (1973)
40 BROOK, A. G. and DUFF, J. M., *Can. J. Chem.* **51,** 2024 (1973); BROOK, A. G., DUFF, J. M. and ANDERSON, D. G., *Can. J. Chem.* **48,** 561 (1970)
41 SEYFERTH, D., LEFFERTS, J. L. and LAMBERT Jr., R. L., *J. organometal. Chem.* **142,** 39 (1977)
42 SHAPIRO, R. H., *Org. Reactions* **23,** 405 (1976)
43 CHAN, T. H., BALDASSARRE, A. and MASSUDA, D., *Synthesis* 801 (1976); TAYLOR, R. T., DEGENHARDT, C. R., MELEGA, W. P. and PAQUETTE, L. A., *Tetrahedron Lett.* 159 (1977); CHAMBERLIN, A. R., STEMKE, J. F. and BOND, F. T., *J. org. Chem.* **43,** 147 (1978)
44 MIRONOV, U. F., MAKSIMOVA, N. G. and NEPOMNIVA, U. V., *Bull. Acad. Sci. USSR* 13 (1967); *Chem. Abstr.* **67,** 32719 (1967)
45 PETROV, A. D., MIRONOV, U. F. and GLUKOVTSEV, V. G., *Zh. obshch. Khim.* **27,** 1535 (1957); *J. gen. Chem. U.S.S.R.* **27,** 1609 (1957); *Chem. Abstr.* **52,** 3668 (1958)
46 NEUMANN, H. and SEEBACH, D., *Tetrahedron Lett.* 4839 (1976); *Chem. Ber.* **111,** 2785 (1978)
47 YAROSH, O. G., VORONOF, V. K. and KOMAROV, N. V., *Izv. Akad. Nauk SSSR, Ser. khim.* 875 (1971); KOMAROV, N. V. and YAROSH, O. G., *Izv. Akad. Nauk. SSSR, Ser. khim.* 1573 (1971)
48 HUNYH, C. and LINSTRUMELLE, G., *Tetrahedron Lett.* 1073 (1979)
49 CHAN, T. H., MYCHAJLOWSKIJ, W. and AMOROUX, R., *Tetrahedron Lett.* 1605 (1977)
50 KUWAJIMA, I., KATO, M. and SATO, T., *J. chem. Soc. chem. Communs* 478 (1978); SATO, T., ABE, T. and KUWAJIMA, I., *Tetrahedron Lett.* 259, 1383 (1978); MINAMI, N., ABE, T. and KUWAJIMA, I., *J. organometal. Chem.* 145, C1 (1978)
51 SOMMER, L. H., BAILEY, D. L., GOLDBERG, G. M., BUCK, C. E., BYE, T. S., EVANS, F. J. and WHITMORE, F. C., *J. Am. chem. Soc.* **76,** 1613 (1954)
52 KOENIG, K. E. and WEBER, W. P., *J. Am. chem. Soc.* **95,** 3416 (1973); KOENIG, K. E. and WEBER, W. P., *Tetrahedron Lett.* 2523 (1973)

53 *See also* BROOK, A. G., DUFF, J. M. and REYNOLDS, W. F., *J. organometal. Chem.* **121,** 293 (1976)
54 BÜCHI, G. and WÜEST, H., *Tetrahedron Lett.* 4305 (1977)
55 BOECKMAN, R. K. and BLUM, D. M., *J. org. Chem.* **39,** 3307 (1974)
56 JARVIE, A. W. P., HOLT, A. and THOMPSON, J., *J. chem. Soc. (B)* 852 (1969)
57 MILLER, R. B. and McGARVEY, G., *Synthetic Communs* **7,** 475 (1977)
58 MILLER, R. B. and McGARVEY, G., *Synthetic Communs* **8,** 291 (1978)
59 THOMAS, E. J. and WHITHAM, G. H., *J. chem. Soc. chem Communs* 212 (1979)
60 PILLOT, J.-P., DUNOGUÈS, J. and CALAS, R., *Bull. Soc. chim. Fr.* 2143 (1975); *C.r. hebd. Séanc Acad. Sci., Paris* **278,** 787, 789 (1974)
61 FLEMING, I. and PEARCE, A., *J. chem. Soc. chem. Communs* 633 (1975)
62 FRISTAD, W. E., DIME, D. S., BAILEY, T. R. and PAQUETTE, L. A., *Tetrahedron Lett.* 1999 (1979)
63 COOKE, F., SCHWINDEMAN, J. and MAGNUS, P., *Tetrahedron Lett.* 1995 (1979); COOKE, F., MOERCK, R., SCHWINDEMAN, J. and MAGNUS, P., *J. org. Chem.* **45,** 1046 (1980)
64 BARTON, T. J. and ROGIDO, R. J., *J. org. Chem.* **40,** 582 (1975)
65 YAMAMOTO, K., NUNOKAWA, O. and TSUJI, J., *Synthesis* 721 (1977); YAMAMOTO, K., YOSHITAKE, J., QUI, N. T. and TSUJI, J., *Chemy Lett.* 859 (1978)
66 PILLOT, J.-P., DUNOGUÈS, J. and CALAS, R., *Synthesis* 479 (1977)
67 JOLIBOIS, H., DOUCET, A. and PERROT, R., *Helv. chim. Acta* **58,** 1801 (1975); *see also* COREY, E. J. and ESTREICHER, H., *Tetrahedron Lett.* 1113 (1980)
68 FRITZ, G. and GROBE, J., *Z. anorg. allg. Chem.* **309,** 98 (1961)
69 SHIKHMAMEDBEKOVA, A. Z. and SULTANOV, R. A., *J. gen. Chem. U.S.S.R.* **40,** 72 (1970)
70 DÉLÉRIS, G., DUNOGUÈS, J. and CALAS, R., *J. organometal. Chem.* **93,** 43 (1975)
71 PILLOT, J.-P., DUNOGUÈS, J. and CALAS, R., *Synth. Communs* **9,** 395 (1979)
72 CHAN, T. H., LAU, P. W. K. and MYCHAJLOWSKIJ, W., *Tetrahedron Lett.* 3317 (1977)
73 MILLER, R. B. and McGARVEY, G., *Synth. Communs* **9,** 831 (1979); MILLER, R. B., personal communication
74 WEBER, W. P., FELIX, R. A., WILLARD, A. K. and KOENIG, K. E., *Tetrahedron Lett.* 4701 (1971)
75 MANSUY, D., PUSSET, J. and CHOTTARD, J. C., *J. organometal. Chem.* **110,** 139 (1976)
76 YOSHIDA, J., TAMAO, K., TAKAHASHI, M. and KUMADA, M., *Tetrahedron Lett.* 2161 (1978)
77 TAMAO, K., MATSUMOTO, H., KAKUI, T. and KUMADA, M., *Tetrahedron Lett.* 1137 (1979)
78 MUSKER, W. K. and LARSON, G. L., *Tetrahedron Lett.* 3481 (1968)
79 BROOK, A. G. and PIERCE, J. B., *J. org. Chem.* **30,** 2566 (1965)
80 SEYFERTH, D., *J. inorg. nucl. Chem.* **7,** 152 (1958)
81 SEYFERTH, D. and KAHLEN, N., *Z. naturf.* **14**b, 137 (1959)
82 STORK, G. and SINGH, J., *J. Am. chem. Soc.* **96,** 6181 (1974)
83 BOECKMAN, R. K., *J. Am. chem. Soc.* **95,** 6867 (1973); **96,** 6179 (1974); *J. org. Chem.* **38,** 4450 (1973)
84 *Org. Synth.* **58,** 152, 158 (1978)
85 HARTZELL, S. L. and RATHKE, M. W., *Tetrahedron Lett.* 2737 (1976)
86 SWENTON, J. S. and FRITZEN Jr., E. L., *Tetrahedron Lett.* 1951 (1979)
87 STANG, P. J., *Chem. Rev.* **78,** 383 (1978)
88 CUNICO, R. F. and HAN, Y.-K., *J. organometal. Chem.* **105,** C29 (1976)
89 BÜCHI, G. and WÜEST, H., *J. Am. chem. Soc.* **100,** 294 (1978)
90 ALEXAKIS, A., CAHIEZ, G. and NORMANT, J. F., *Synthesis* 826 (1979)
91 FLEMING, I. and ROESSLER, F., *J. chem. Soc. chem. Communs* 276 (1980)
92 DAVIS, A. P. and WHITHAM, G. H., *J. chem. Soc. chem. Communs* 639 (1980)
93 BIRKOFER, L. and EICHSTÄDT, D., *J. organometal. Chem.* **145,** C29 (1978)
94 DUNOGUÈS, J., CALAS, R., DUFFAUT, N., LAPOUYADE, P. and GERVAL, J., *J. organometal. Chem.* **20,** P20 (1969); DUNOGUÈS, J., BOURGEOIS, P., PILLOT, J.-R., MERAULT, G. and CALAS, R., *J. organometal. Chem.* **87,** 169 (1975)
95 BIRKOFER, L. and KÜHN, T., *Chem. Ber.* **111,** 3119 (1978)
96 MARTEL, B. and VARACHE, M., *J. organometal. Chem.* **40,** C53 (1972)
97 SEKIGUCHI, A. and ANDO, W., *J. org. Chem.* **44,** 413 (1979)
98 EHLINGER, E. and MAGNUS, P., *J. chem. Soc. chem. Communs* 421 (1980)
99 YAMAMOTO, K., OHTA, M. and TSUJI, J., *Chemy Lett.* 713 (1979)

Chapter 8

$\alpha\beta$-Epoxysilanes as precursors of carbonyl compounds and heteroatom-substituted alkenes

8.1 Preparation

Vinylsilanes are readily converted into $\alpha\beta$-epoxysilanes, normally by treatment[1] with *m*-chloroperbenzoic acid in dichloromethane or chloroform. Although vinylsilanes of diverse substitution types are quite readily available (Chapter 7), alternative convergent routes to substituted $\alpha\beta$-epoxysilanes have been devised. α-Chloro-α-trimethylsilyl carbanions/carbanionoids (1) react efficiently[2] with a wide range of aldehydes and ketones in a version of the Darzens reaction (*Scheme 8.1*); sterically hindered or readily enolizable carbonyl compounds react with a somewhat more modest yield.

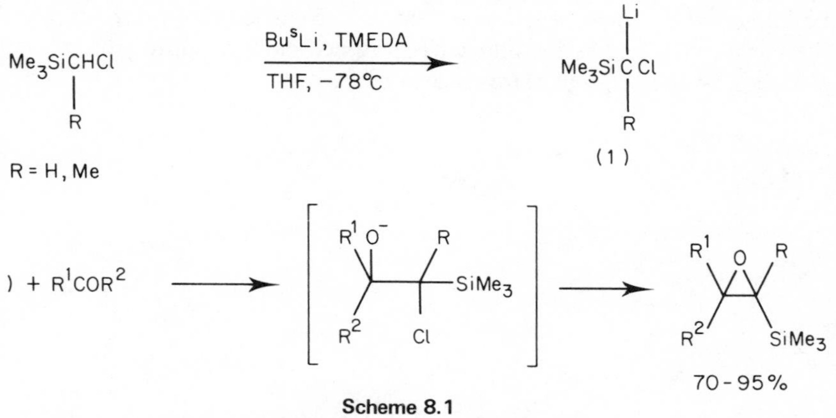

$$Me_3SiCH(R)Cl \xrightarrow[\text{THF, } -78\,°C]{Bu^sLi,\ TMEDA} Me_3SiC(Li)(R)Cl \quad (1)$$

R = H, Me

$$(1) + R^1COR^2 \longrightarrow \left[R^1R^2C(O^-)\text{–}C(R)(Cl)SiMe_3 \right] \longrightarrow \text{epoxide } R^1R^2C\text{–}O\text{–}C(R)SiMe_3$$

70–95%

Scheme 8.1

1-Diazo-1-trimethylsilyl-2-hydroxyalkanes (2), obtained from reaction of lithiotrimethylsilyldiazomethane with carbonyl compounds[3], undergo thermal decomposition to provide an alternative access to some epoxysilanes (*Scheme 8.2*); this mode of reactivity contrasts with the reaction (p. 148) of this carbanion with diaryl ketones, in which diaryl alkynes are produced in good yields.

Another non-oxidative route involves the alkylation of the lithium salts of preformed $\alpha\beta$-epoxysilanes. n-Butyl-lithium metallates the epoxide (3) exclusively at the silicon-bearing carbon. The resulting lithium species proved[4] to be relatively stable, but reacted efficiently with a variety of

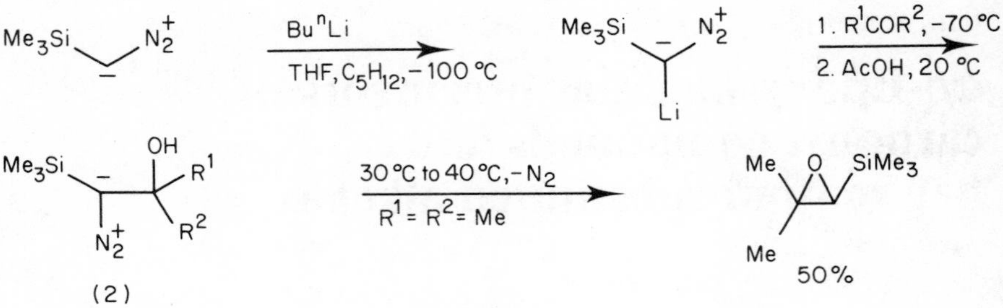

Scheme 8.2

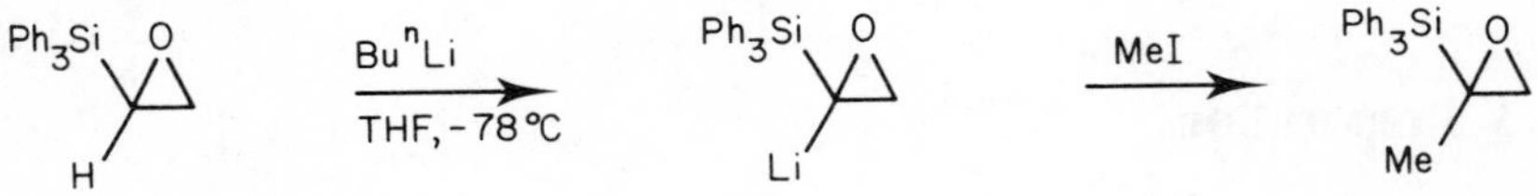

electrophiles, including methyl iodide; the full scope of this route is at present undefined.

8.2 Isomerization

αβ-Epoxysilanes are of major interest and synthetic utility, because they can be converted by simple reaction sequences into carbonyl compounds in which the carbonyl group is introduced regiospecifically at either the *α*- or the *β*-carbon of the original oxirane (*Scheme 8.3*).

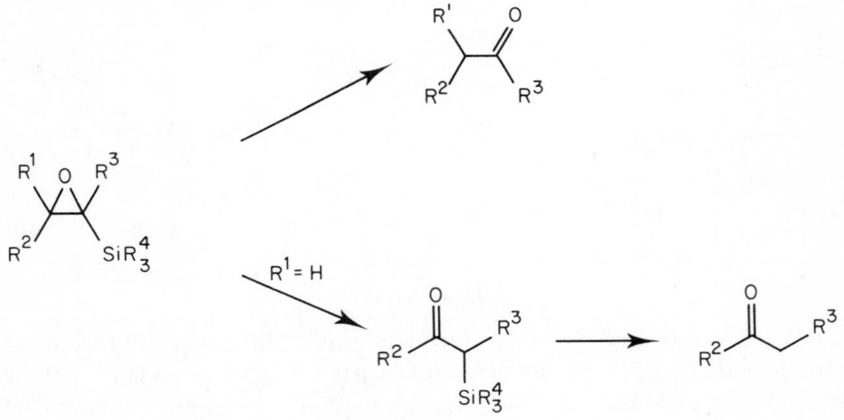

Scheme 8.3

Taking these complementary processes in turn, acyclic *αβ*-epoxysilanes undergo a smooth acid-catalysed rearrangement[5] to carbonyl compounds, where the carbonyl group appears at the carbon originally bearing silicon. Initially, only those oxiranes which could give rise to aldehydes were studied (*Scheme 8.4*).

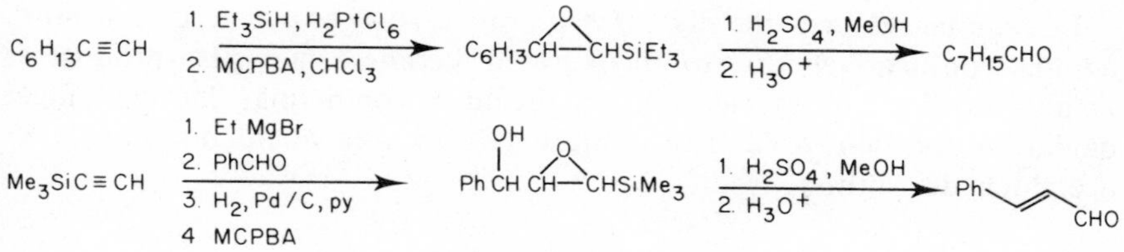

$C_6H_{13}C\equiv CH$ → (1. Et_3SiH, H_2PtCl_6; 2. MCPBA, $CHCl_3$) → $C_6H_{13}CH$–O–$CHSiEt_3$ (epoxide) → (1. H_2SO_4, MeOH; 2. H_3O^+) → $C_7H_{15}CHO$

$Me_3SiC\equiv CH$ → (1. Et MgBr; 2. PhCHO; 3. H_2, Pd/C, py; 4 MCPBA) → PhCH(OH)CH–O–$CHSiMe_3$ (epoxide) → (1. H_2SO_4, MeOH; 2. H_3O^+) → Ph–CH=CH–CHO

Scheme 8.4

Subsequently, the scope of this reaction was extended to the generation of ketones[6], providing a regiospecific alkylative alternative[7] to the Robinson annelation sequence (*Scheme 8.5*), using allyl halides/vinylsilanes (p. 50) such as (4); interestingly, displacement of the silyl group in such cases is easier than in those involving simple $\alpha\beta$-epoxysilanes, possibly because of participation by the proximate carbonyl group.

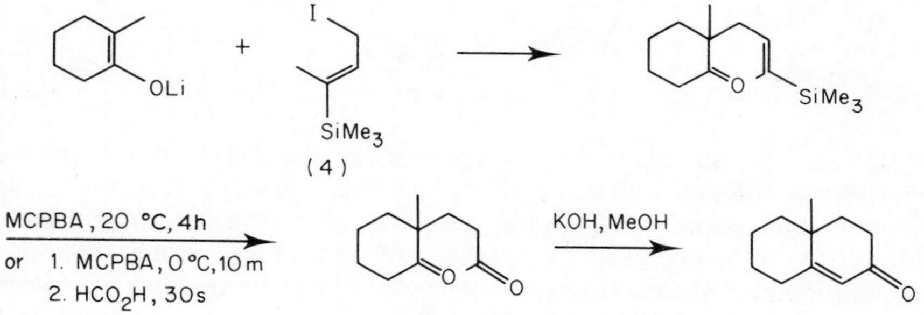

OLi + I, $SiMe_3$ (4) → O, $SiMe_3$

MCPBA, 20 °C, 4h or 1. MCPBA, 0 °C, 10 m; 2. HCO_2H, 30s → O, O → KOH, MeOH → O

Scheme 8.5

Evidence supporting such participation can be deduced from the observation that the $\alpha\beta$-epoxysilane (5) required the previous methanol/sulphuric acid conditions to rearrange, and that, in the case of five-ring annelation using the silane (6), as shown in *Scheme 8.6*, the cyclic ketal (7) was formed in quantitative yield on treatment of the vinylsilane with per-acid.

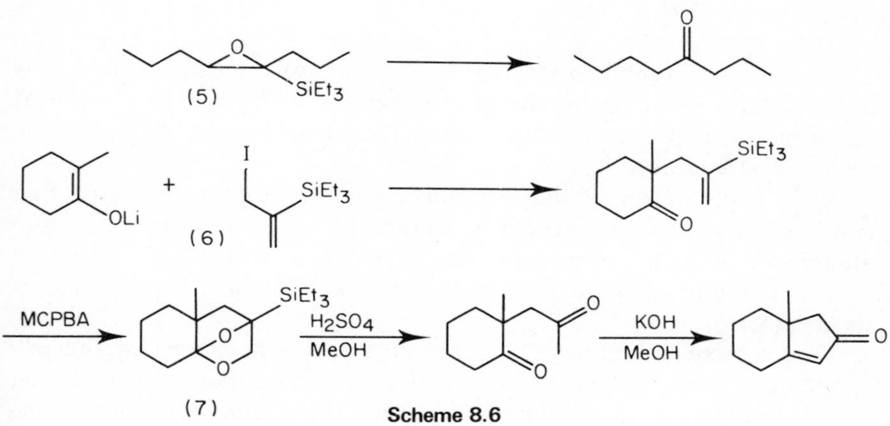

(5) O, $SiEt_3$ → O

OLi + I, $SiEt_3$ (6) → $SiEt_3$, O

MCPBA → (7) O, O, $SiEt_3$ → H_2SO_4 / MeOH → O, O → KOH / MeOH → O

Scheme 8.6

In a similar manner, the silyl vinyl cuprate (8) acts as a d^1 acetyl synthon[8], adding conjugatively[9] to *αβ*-unsaturated ketones; here the product is ketalized prior to epoxidation to preclude competing Baeyer–Villiger oxidation (*Scheme 8.7*). The cuprate (9) behaves analogously as a d^2 acetaldehyde synthon.

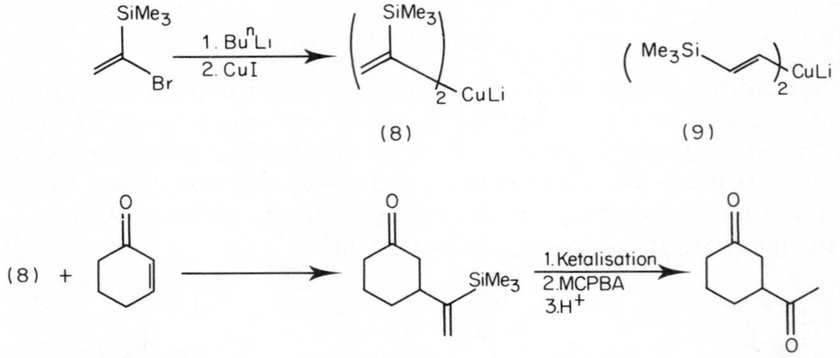

SiMe3 Br 1. Bu^{n}Li 2. CuI (SiMe3)2 CuLi (8) (Me3Si)2CuLi (9)

(8) + 1. Ketalisation 2. MCPBA 3. H$^+$ SiMe3

Scheme 8.7

Seebach[10] has applied a variety of routes to the preparation of vinylsilanes, and thence *αβ*-epoxysilanes and the derived carbonyl compounds, as illustrated in *Scheme 8.8*. However, probably the most generally applicable route to *αβ*-epoxysilanes is that of Magnus[2], using *α*-lithio-*α*-chloro(trimethyl)-methane and -ethane (p. 83); these reagents have been used by him and his group[11] in neat syntheses of 3-*R*-(+)-frontalin (10) and *Latia* luciferin (11).

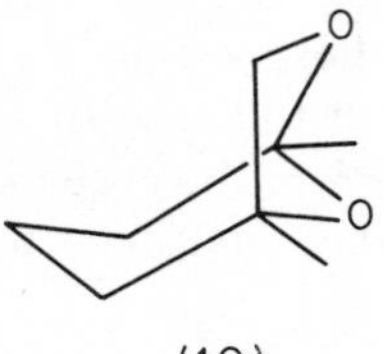

(10) 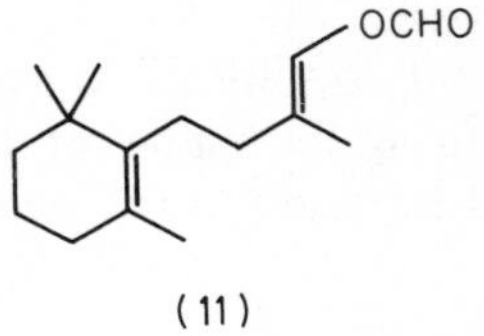 (11) OCHO

Magnus[12] has also reported that allyltrimethylsilane can function as a d^3-propanal synthon; as discussed elsewhere (p. 118), the lithio-anion of allyltrimethylsilane reacts with carbonyl electrophiles at the *γ*-position exclusively (*Scheme 8.9*).

Prior to discussion of the mechanism of this reaction, we must first consider general ring opening reactions of *αβ*-epoxysilanes. Lithium aluminium hydride effects opening[13] of such oxiranes to produce 2-silylethanols by apparent direct hydride attack at the carbon bearing silicon (*Scheme 8.10*).

It was subsequently shown[14] that this was a general phenomenon, that *αβ*-epoxysilanes underwent nucleophilic and electrophile-catalysed ring opening to give products of predominant *α*-cleavage. This result is, at first sight,

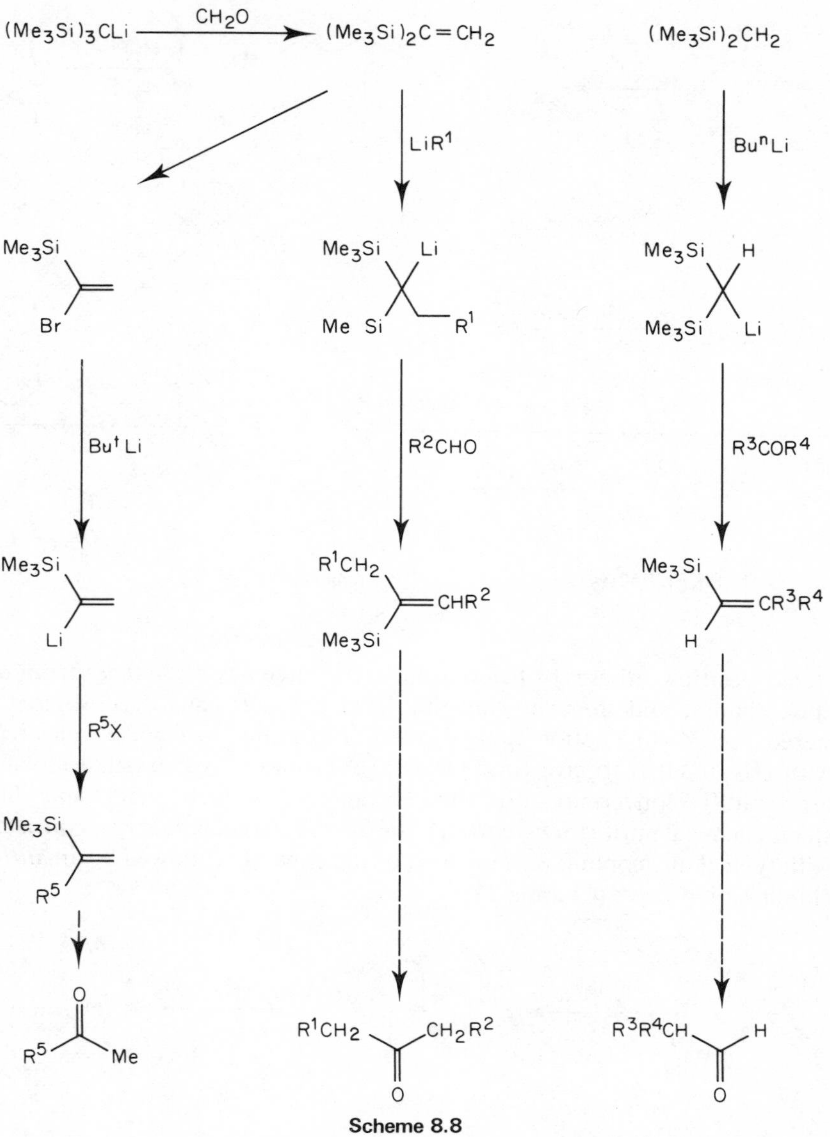

$(Me_3Si)_3CLi$ —CH_2O→ $(Me_3Si)_2C{=}CH_2$ $(Me_3Si)_2CH_2$

LiR^1

Bu^nLi

Me_3Si Br

Me_3Si Li Me Si R^1

Me_3Si H Me_3Si Li

Bu^tLi

R^2CHO

R^3COR^4

Me_3Si Li

R^1CH_2 $=CHR^2$ Me_3Si

Me_3Si $=CR^3R^4$ H

R^5X

Me_3Si R^5

O R^5 Me

R^1CH_2 CH_2R^2 O

R^3R^4CH H O

Scheme 8.8

rather unexpected, as, although a fully developed carbonium ion may not be involved in electrophile-catalysed opening, one would still expect β-cleavage to predominate, in view of the well-documented stability of cations β to silicon (Chapter 3). However, the relative orientations of the C–Si and the β-C–O bonds deviate markedly from the coplanar alignment favourable for stabilization of a developing positive charge by the C–Si bond. Indeed, the preference for α-opening in these reactions suggests that the silyl group may actually facilitate[15] bimolecular nucleophilic displacement α to silicon. Such high regioselectivity has been put to good use by Hudrlik[16], in developing

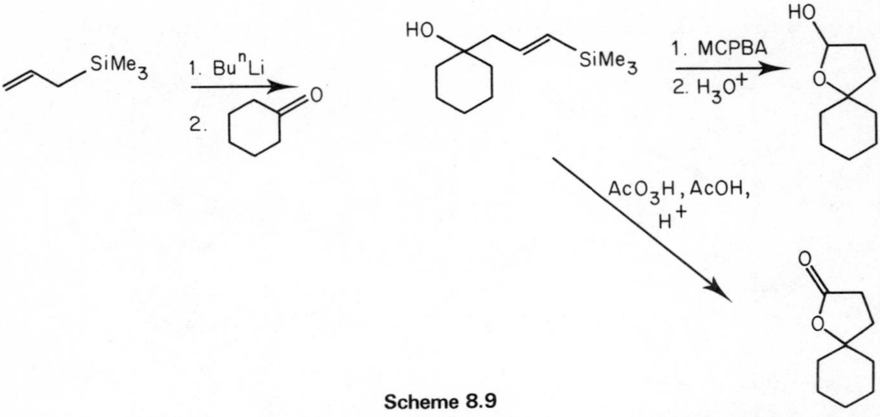

Scheme 8.9

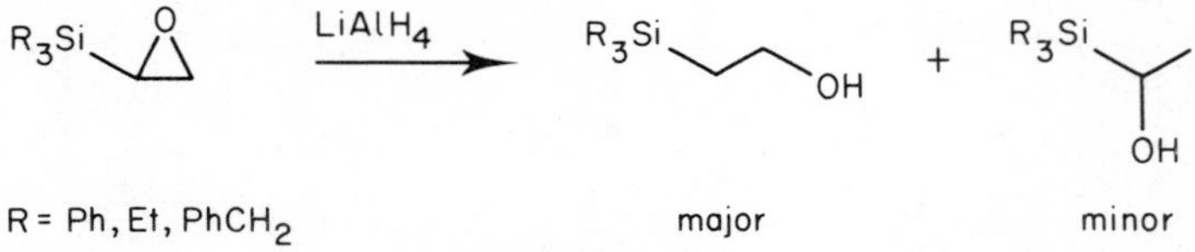

Scheme 8.10

stereospecific syntheses of heteroatom-substituted acyclic alkenes from *αβ*-epoxysilanes, and thus showing that such oxiranes can be considered as stereospecific vinyl cation equivalents. For example, *αβ*-epoxysilanes react with HBr in ether to give good yields of *α*-bromo-*β*-hydroxysilanes, which are readily converted into vinyl bromides in high yield and high stereochemical purity (*Scheme 8.11*); the overall stereochemistry is consistent with typical nucleophilic *anti* opening of the epoxide, followed by an *anti-β*-elimination process (Chapter 12).

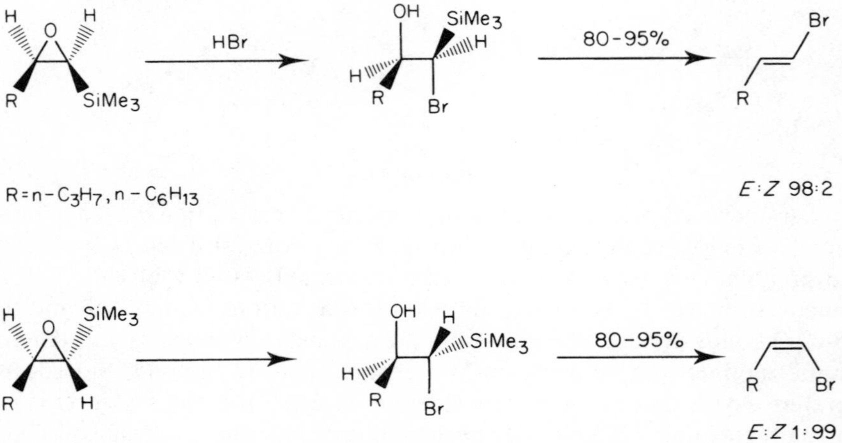

Scheme 8.11

Enol acetates were prepared in similar stereochemical purity by treatment of $\alpha\beta$-epoxysilanes with acetic acid/acetic anhydride and a catalytic amount of boron trifluoride etherate. Methanol/trifluoroacetic acid or methanol/ boron trifluoride etherate treatment of the oxiranes gave methoxy-alcohols. Although further treatment with acid, to effect *anti*-elimination, was not successful, treatment with potassium hydride effected *syn*-elimination and produced enol ethers stereoselectively (*Scheme 8.12*).

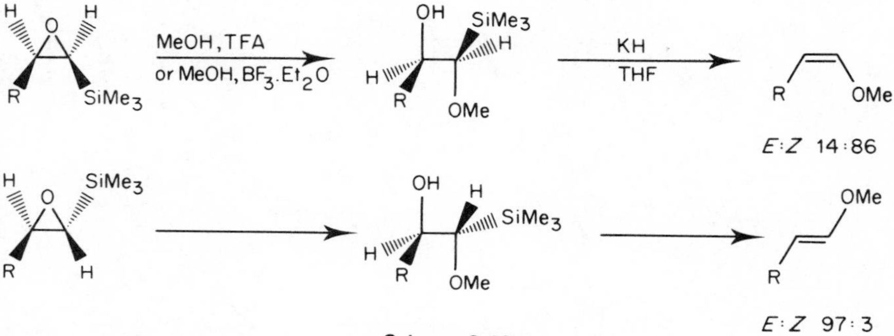

Scheme 8.12

$\alpha\beta$-Epoxysilanes undergo ring opening on treatment with lithium dialkylcuprates, regiospecifically and stereoselectively yielding β-hydroxysilanes; since these can stereoselectively lead to alkenes, this process is discussed in Chapter 12.

The high regiospecificity of such ring opening reactions is illustrated dramatically by the reactions of the two epoxides (12) and (13) with HBr and methanol respectively. With epoxide (12), one might expect that a cationic pathway would strongly favour β-opening, since the resulting cation would be both tertiary and β to silicon. With epoxide (13), nucleophilic attack should be preferred at the β-position owing to steric hindrance. However, both epoxides yield only products of α-opening under acidic conditions.

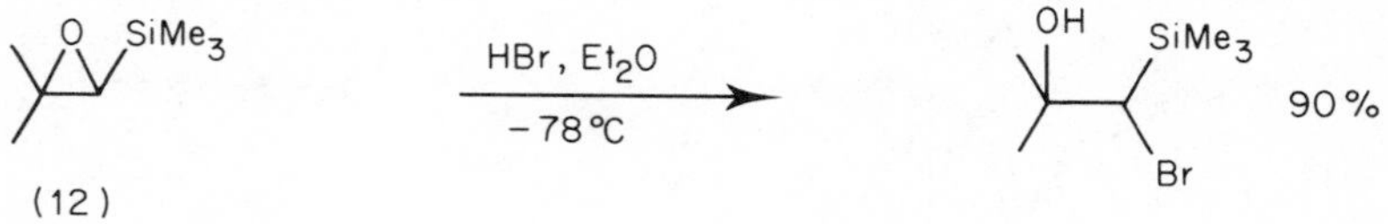

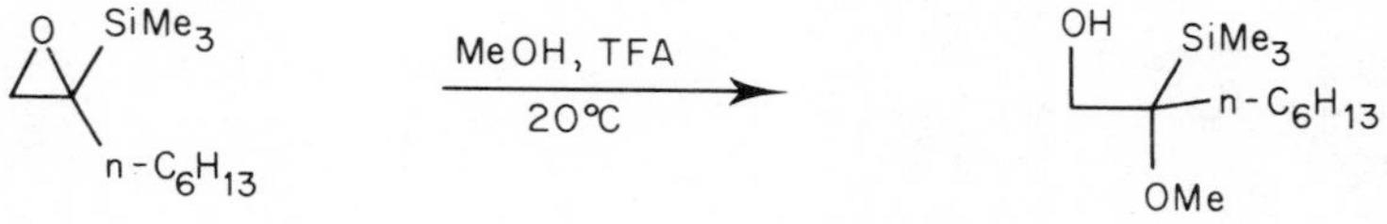

This electronic directing effect of silicon is inadequate to overcome completely the normal kinetic bias for *trans*-diaxial ring opening in conformationally rigid systems (p. 93).

On the basis of such evidence, it would therefore appear that the conversion of $\alpha\beta$-epoxysilanes into carbonyl compounds proceeds by initial *anti*-solvolysis to *αβ*-dihydroxysilanes, followed by acid-catalysed *anti*-elimination (*Scheme 8.13*); isolation[17] of the glycol (14) lends credence to this postulate, as here the trimethylsilyl group and *β*-hydroxyl group cannot achieve the preferred *anti*-periplanar geometry for acid-catalysed elimination.

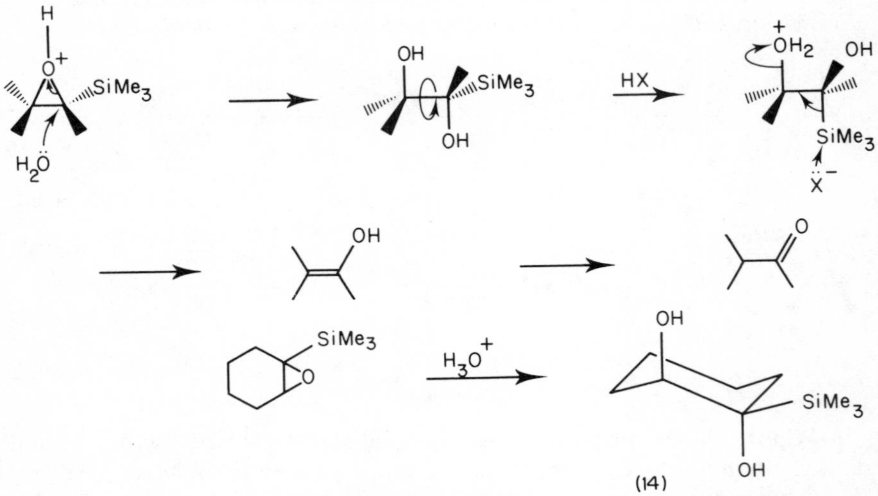

Scheme 8.13

Such stability to acid suggests[18] that the standard acidic conditions can be used only for those cases leading to acyclic carbonyl compounds. In principle, base-induced elimination, with its *syn*-stereochemical requirement, could be used for cyclic cases. However, it has recently been shown[19] that *αβ*-dihydroxysilanes undergo competitive, stereospecific elimination on treatment with potassium hydride to give isomeric mixtures of silyl enol ethers. Such base-induced elimination reactions take place via the

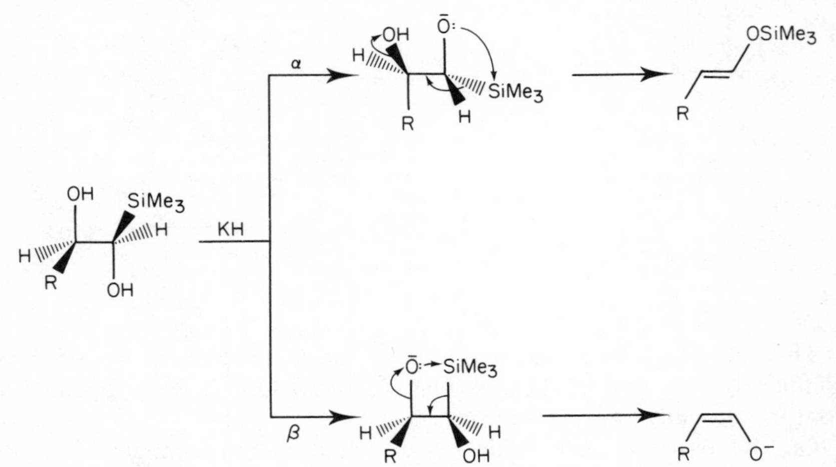

Scheme 8.14

α-oxidosilane pathway, in competition with the β-oxidosilane (*syn*) pathway. Elimination by the α-oxidosilane pathway appears to proceed with predominant *anti*-stereochemistry, with an initial Brook rearrangement (p. 30) being implicated (*Scheme 8.14*). If the silicon and the β-hydroxyl group cannot be *anti*, protiodesilylation predominates.

The alternative process, that of converting $\alpha\beta$-epoxysilanes into carbonyl compounds where the carbonyl group appears at the β-carbon, can be achieved in a number of ways. In 1963 Eisch[13] reported that triphenylsilyloxirane underwent rearrangement when treated with magnesium bromide etherate to yield triphenylsilylacetaldehyde (*Scheme 8.15*), a result confirmed and utilized by later workers[20].

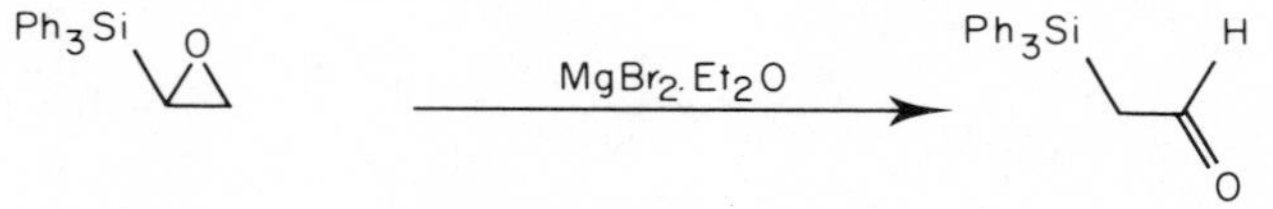

Ph_3Si O — $MgBr_2.Et_2O$ → Ph_3Si H O

Scheme 8.15

In a more general study of this rearrangement, Hudrlik[21] reported that, depending on the $\alpha\beta$-epoxysilane substrate and on reaction conditions, bromohydrins (from α-cleavage, as predicted), β-ketosilanes, or silyl enol ethers could be obtained, often in very respectable yields (*Scheme 8.16*). These products are no doubt formed sequentially, there being ample analogy for rearrangements of bromohydrins to ketones and of β-ketosilanes to silyl enol ethers[20,22].

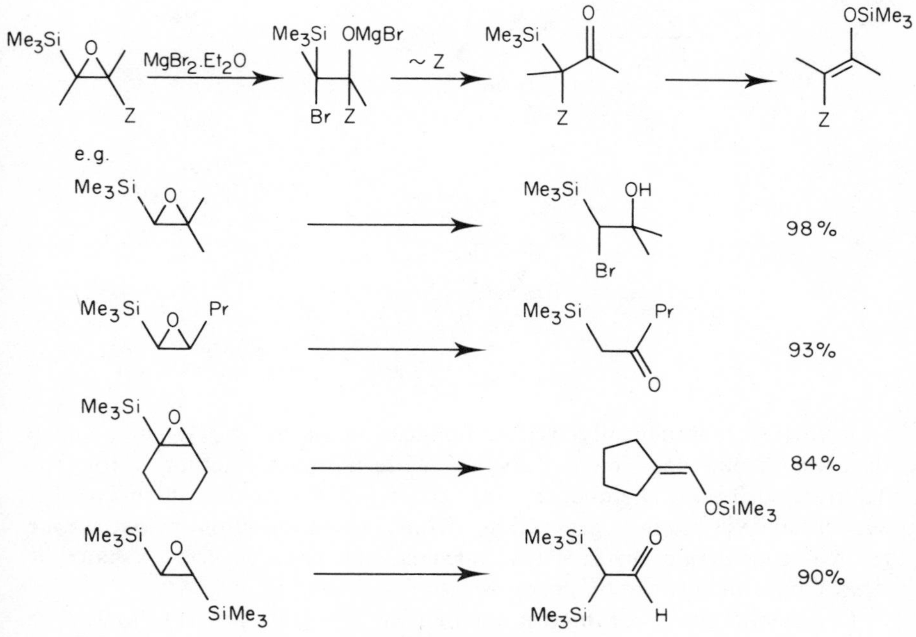

Me_3Si O Z — $MgBr_2.Et_2O$ → Me_3Si OMgBr Br Z — ~ Z → Me_3Si O Z → $OSiMe_3$ Z

e.g.

Me_3Si O → Me_3Si OH Br 98%

Me_3Si O Pr → Me_3Si Pr O 93%

Me_3Si O → $OSiMe_3$ 84%

Me_3Si O $SiMe_3$ → Me_3Si O Me_3Si H 90%

Scheme 8.16

This technique has been extended and the product β-ketosilanes employed as substrates in the silyl-Wittig olefination process (Chapter 12). Interestingly, it was also observed that whereas αβ-epoxysilanes of the general type (15) rearranged cleanly to β-ketosilanes (16), the isomeric oxiranes (17) produced significant amounts of silyl enol ethers, indicating a tendency for migration of the group *syn* to silicon.

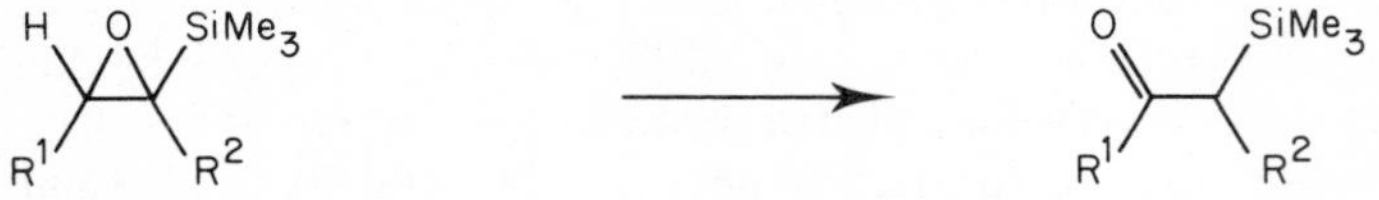

(15) → (16)

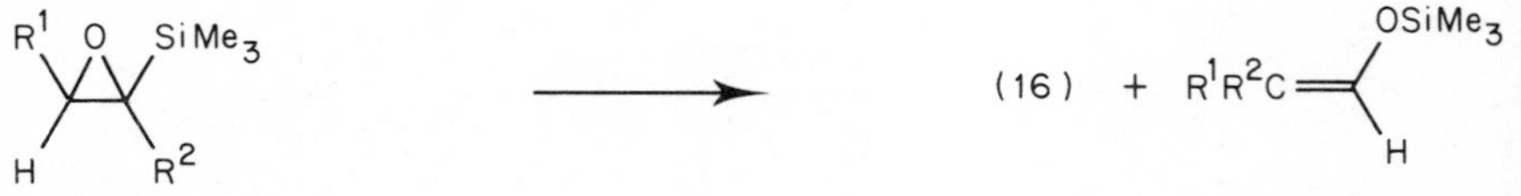

(17) → (16) + $R^1R^2C{=}CH(OSiMe_3)$

The precise reasons for such differential behaviour are not immediately apparent; some suggestions have been made, but these are based on too many assumptions to be of much significance. The synthetic problem resolved itself when it was shown[23] that either oxirane could be cleanly converted into the same β-ketosilane by treatment with hydriodic acid to give the corresponding iodohydrins and thence the β-ketosilanes by exclusive migration of hydrogen in each diastereoisomeric iodohydrin (*Scheme 8.17*).

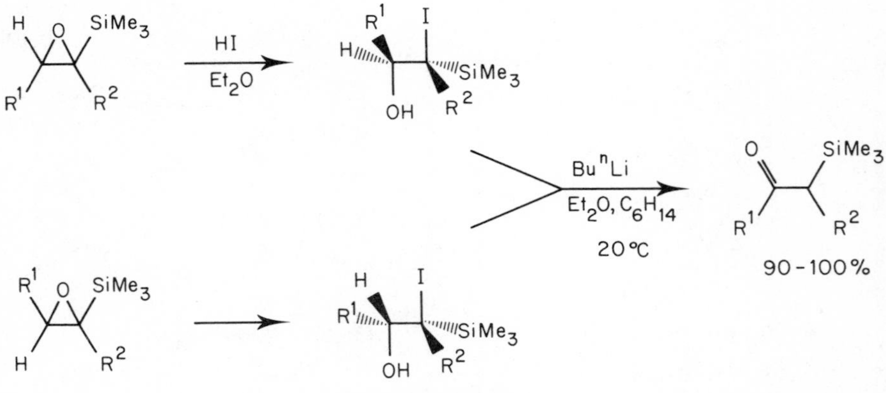

HI, Et_2O; Bu^nLi, Et_2O, C_6H_{14}, 20 °C; 90–100%

Scheme 8.17

A further, synthetically useful, application of αβ-epoxysilanes which devolves from specific α-cleavage is seen in a method[24] for the 1,2-transposition of ketone carbonyl groups. The scheme, which involves sequential vinylsilane generation from an arenesulphonylhydrazone (p. 59), epoxidation, hydride ring opening, and chromic acid oxidation is both simple and efficient (*Scheme 8.18*).

Two further points of interest arising from this work are as follows. The vinylsilane (18) gave rise to two conformationally rigid αβ-epoxysilanes; with

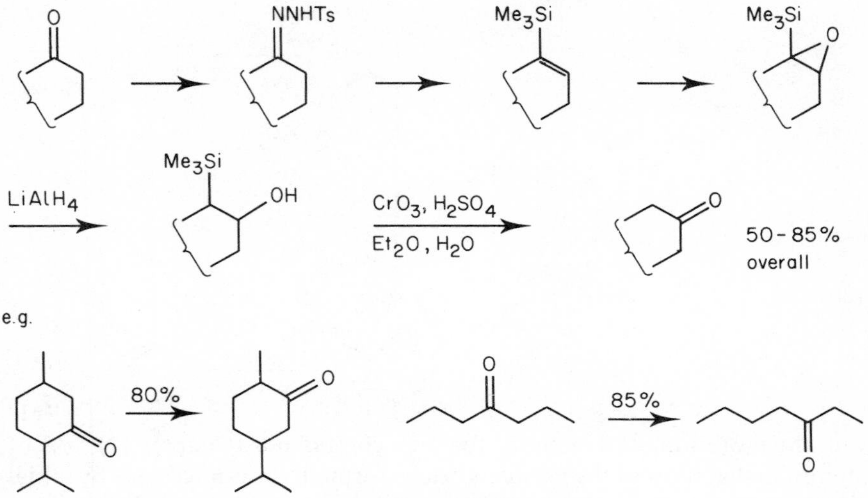

Scheme 8.18

$LiAlH_4$ as reducing agent, the electronic directing effect of silicon did not prove sufficiently strong to overcome the normal kinetic bias for *trans* diaxial ring opening, and mixtures arising from both α- and β-bond scission resulted. However, use of AlH_2Cl overcame this problem, and restored the desired regioselectivity.

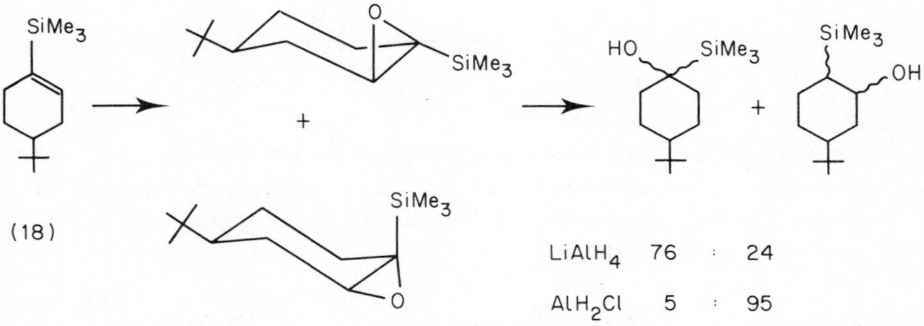

$LiAlH_4$	76	24
AlH_2Cl	5	95

The second point of interest is that activated silanes such as (19) are converted directly into the carbonyl compound by treatment with buffered *m*-chloroperbenzoic acid in dichloromethane, presumably by acid-catalysed rearrangement of the intermediate oxirane in a manner similar to that just discussed.

$\alpha\beta$-Epoxysilanes undergo pyrolytic rearrangement to silyl enol ethers and other products. Although this route to silyl enol ethers is unlikely to compete

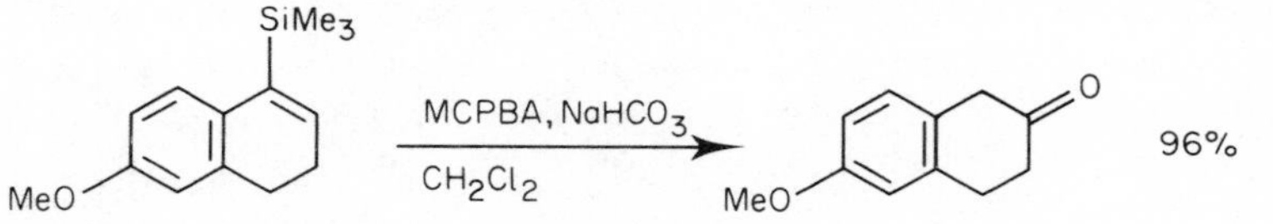

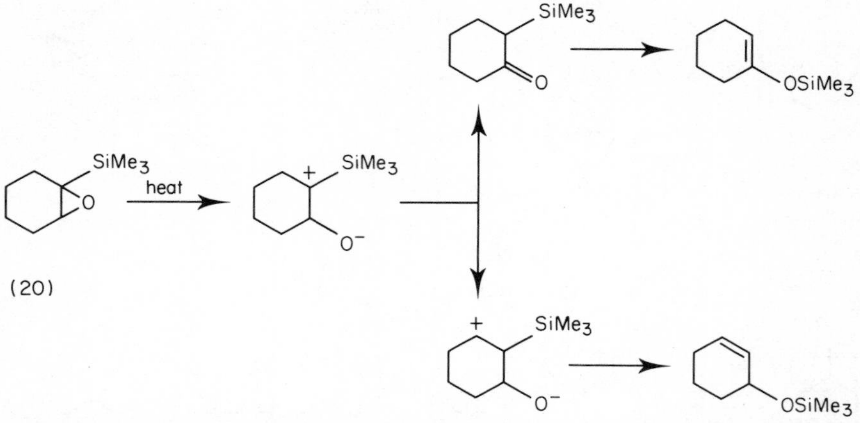

$SiMe_3$ → $OSiMe_3$

$SiMe_3$ heat → $SiMe_3$ O^-

(20)

+ $SiMe_3$ O^- → $OSiMe_3$

with the more standard methods for such compounds (Chapter 17), it is of interest in that most of the products can be formally accounted for[25] by initial α-cleavage followed by 1,2-hydride migration, as shown for oxirane (20).

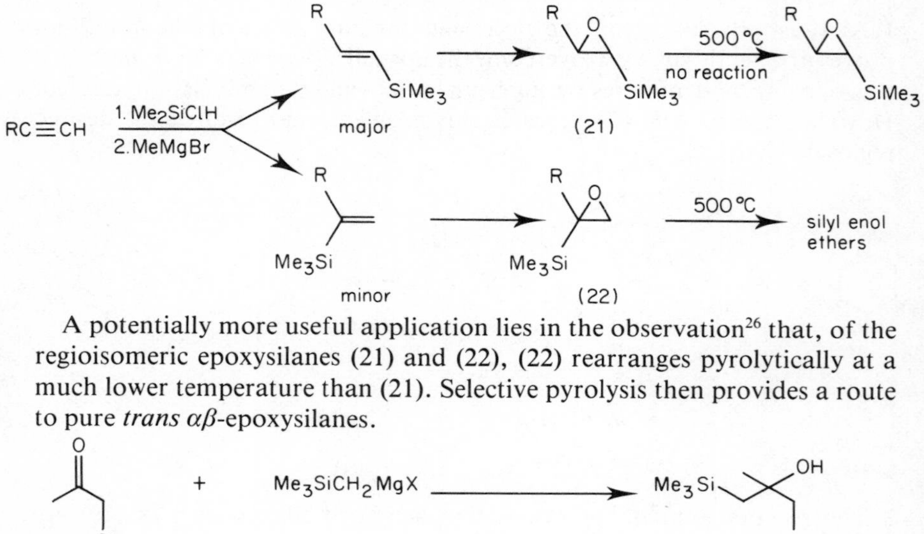

RC≡CH 1. Me_2SiClH 2. MeMgBr

R $SiMe_3$ major → R O $SiMe_3$ (21) 500 °C no reaction → R O $SiMe_3$

R Me_3Si minor → R O Me_3Si (22) 500 °C → silyl enol ethers

A potentially more useful application lies in the observation[26] that, of the regioisomeric epoxysilanes (21) and (22), (22) rearranges pyrolytically at a much lower temperature than (21). Selective pyrolysis then provides a route to pure *trans* αβ-epoxysilanes.

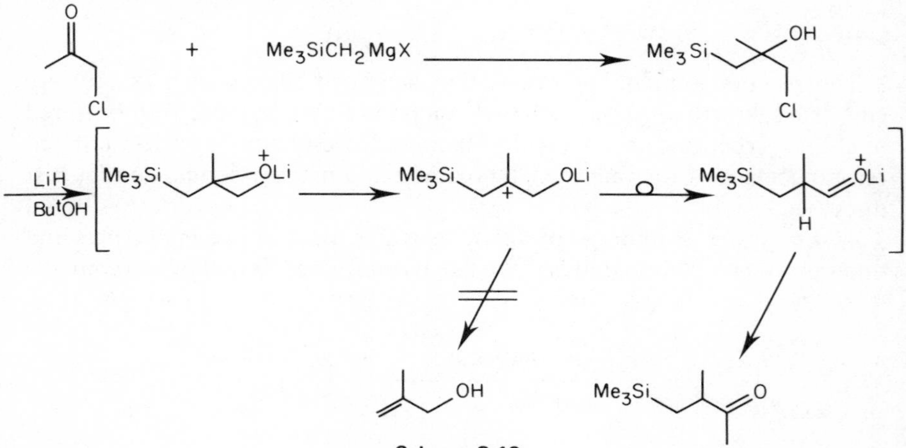

O Cl + Me_3SiCH_2MgX → Me_3Si OH Cl

LiH Bu^tOH → [Me_3Si $\overset{+}{O}Li$ → Me_3Si + OLi → Me_3Si H $\overset{+}{O}Li$]

OH Me_3Si O H

Scheme 8.19

Finally, $\beta\gamma$-epoxysilanes have no geometric constraint on C–Si stabilization of an adjacent developing carbonium ion, and are accordingly very labile. In acyclic cases[27] they have been prepared so far only *in situ*; the carbonyl products arising from their generation in basic medium (*Scheme 8.19*) can be explained by β-cleavage, followed by hydrogen migration.

Cyclic $\beta\gamma$-epoxysilanes, though also very labile, have been isolated and characterized[28]. In acidic unbuffered conditions they open cleanly to the expected products, allylic alcohols (*Scheme 8.20*).

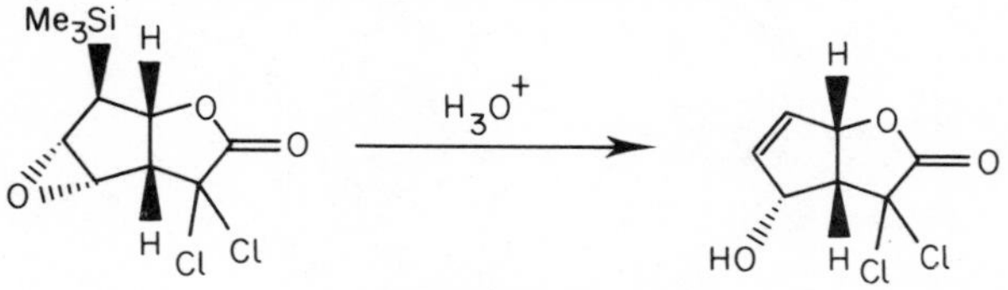

H_3O^+

Scheme 8.20

References

1 EISCH, J. J. and TRAINOR, J. T., *J. org. Chem.* **28,** 487 (1963); for an alternative oxidant *see* EHLINGER, E. and MAGNUS, P., *Tetrahedron Lett.* 11 (1980)
2 BURFORD, C., COOKE, F., EHLINGER, E. and MAGNUS, P., *J. Am. chem. Soc.* **99,** 4536 (1977); COOKE, F. and MAGNUS, P., *J. chem. Soc. chem. Communs* 513 (1977)
3 SCHÖLLKOPF, U. and SCHOLTZ, H.-U., *Synthesis* 271 (1976)
4 EISCH, J. J. and GALLE, J. E., *J. Am. chem. Soc.* **98,** 4646 (1976)
5 STORK, G. and COLVIN, E., *J. Am. chem. Soc.* **93,** 2080 (1971)
6 STORK, G. and COLVIN, E., unpublished work
7 STORK, G. and JUNG, M. E., *J. Am. chem. Soc.* **96,** 3682 (1974); STORK, G., JUNG, M., COLVIN, E. and NOEL, Y., *J. Am. chem. Soc.* **96,** 3684 (1974)
8 SEEBACH, D., *Angew. Chem. int. Edn* **18,** 239 (1979)
9 BOECKMAN, R. K. and BRUZA, K. J., *Tetrahedron Lett.* 3365 (1974)
10 GRÖBEL, B.-Th. and SEEBACH, D., *Angew. Chem. int. Edn* **13,** 83 (1974)
11 MAGNUS, P. and ROY, G., *J. chem. Soc. chem. Communs* 297 (1978)
12 AYALON-CHASS, D., EHLINGER, E. and MAGNUS, P., *J. chem. Soc. chem. Communs* 772 (1977); EHLINGER, E. and MAGNUS, P., *Tetrahedron Lett.* 11 (1980); *J. Am. chem. Soc.* **102,** 5004 (1980)
13 EISCH, J. J. and TRAINOR, J. T., *J. org. Chem.* **28,** 2870 (1963)
14 EISCH, J. J. and GALLE, J. E., *J. org. Chem.* **41,** 2615 (1976)
15 EABORN, C. and JEFFREY, J. C., *J. chem. Soc.* 4266 (1954)
16 HUDRLIK, P. F., HUDRLIK, A. M., RONA, R. J., MISRA, R. N. and WITHERS, G. P., *J. Am. chem. Soc.* **99,** 1993 (1977)
17 ROBBINS, C. M. and WHITHAM, G. H., *J. chem. Soc. chem. Communs* 697 (1976)
18 HUDRLIK, P. F., ARCOLEO, J. P., SCHWARTZ, R. H., MISRA, R. N. and RONA, R. J., *Tetrahedron Lett.* 591 (1977)
19 HUDRLIK, P. F., SCHWARTZ, R. H. and KULKARNI, A. K., *Tetrahedron Lett.* 2233 (1979); HUDRLIK, P. F., NAGENDRAPPA, G., KULKARNI, A. K. and HUDRLIK, A. M., *Tetrahedron Lett.* 2237 (1979)
20 BROOK, A. G., MacRAE, D. and BASSINDALE, A. R., *J. organometal. Chem.* **86,** 185 (1975); WILT, J. W., KOLEWE, O. and KRAEMER, J. F., *J. Am. chem. Soc.* **91,** 2624 (1969)
21 HUDRLIK, P. F., MISRA, R. N., WITHERS, G. P., HUDRLIK, A. M., RONA, R. J. and ARCOLEO, J. P., *Tetrahedron Lett.* 1453 (1976)
22 LUTSENKO, I. F., BAUKOV, Yu. I., DUDUKINA, O. V. and KRAMAROVA, E. N., *J. organometal. Chem.* **11,** 35 (1968); LARSON, G. L. and FERNANDEZ, Y. V., *J. organometal. Chem.* **86,** 193 (1975)
23 OBAYASHI, M., UTIMOTO, K. and NOZAKI, H., *Tetrahedron Lett.* 1383 (1978); *Bull. chem. Soc. Japan* **52,** 2646 (1979)

24 FRISTAD, W. E., BAILEY, T. R. and PAQUETTE, L. A., *J. org. Chem.* **43,** 1620 (1978); *Nachr. Chem. Tech. Lab.* **26,** 520 (1978)
25 HUDRLIK, P. F., WAN, C.-N. and WITHERS, G. P., *Tetrahedron Lett.* 1449 (1976); BASSINDALE, A. R., BROOK, A. G., CHEN, P. and LENNON, J., *J. organometal. Chem.* **94,** C21 (1975)
26 HUDRLIK, P. F. and WAN, C.-N., *Synth. Communs* **9,** 333 (1979)
27 HUDRLIK, P. F. and WITHERS, G. P., *Tetrahedron Lett.* 29 (1976)
28 AU-YEUNG, B.-W. and FLEMING, I., *J. chem. Soc. chem. Communs* 79 (1977)

Chapter 9
Allylsilanes

Allylsilanes, being homologues of vinylsilanes, undergo a similar regio-controlled electrophilic attack[1], the electrophile attacking the γ-carbon to induce positive charge development at the β-carbon; the intermediate or incipient β-silyl carbonium ion frequently undergoes rapid loss of the silyl group, resulting in a product of substitution with a net shift of the double bond[2]. In certain reactions, particularly [2 + 2]-cycloadditions, silyl-substituted products can be obtained and modified further. Such behaviour can be summarized as shown in *Scheme 9.1*.

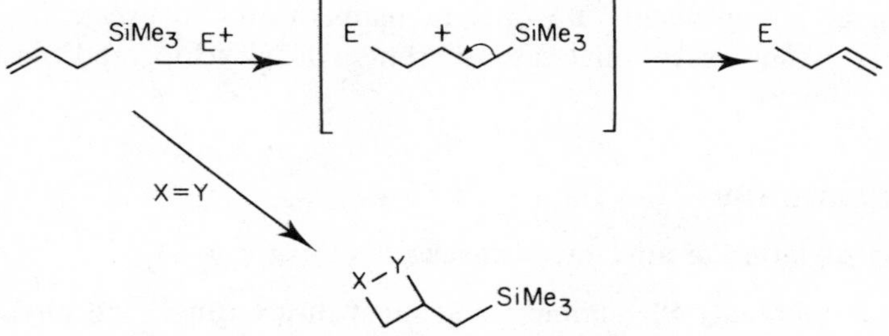

SiMe3 E+ E + SiMe3 E

X=Y

X–Y SiMe3

Scheme 9.1

No quantitative study has been reported so far on the relative reactivities of allyl-and vinyl-silanes towards electrophilic attack. However, with silanes which are simultaneously vinylic and allylic, allylsilyl reactivity takes precedence over vinylsilyl. For example, the disilylpropene (1) undergoes

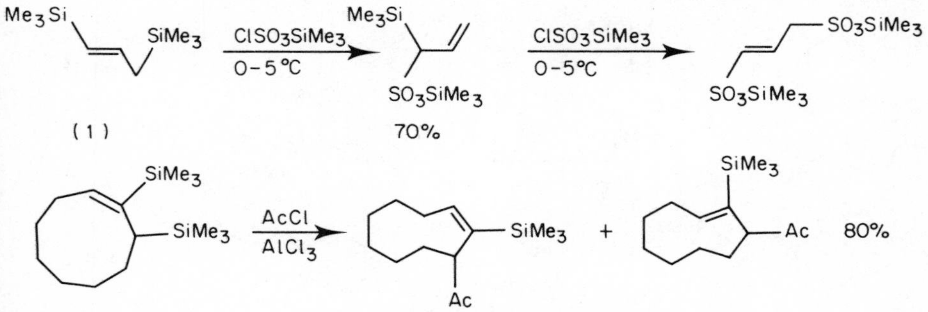

Me3Si SiMe3 ClSO3SiMe3 0–5 °C Me3Si SO3SiMe3 70% ClSO3SiMe3 0–5 °C SO3SiMe3 SO3SiMe3

(1)

SiMe3 SiMe3 AcCl AlCl3 SiMe3 Ac + SiMe3 Ac 80%

(2)

Scheme 9.2 60 : 40

stepwise sulphonation[3], and the cyclononene (2) undergoes electrophilic acylation with exclusive replacement[4] of the allylic silyl group (*Scheme 9.2*).

Other pieces of evidence are in accord with these experimental findings. If we consider again the reaction pathway followed by each substrate on electrophilic attack, i.e., β-carbonium ion development, such a pathway will have a lower activation energy in the case of allylsilanes. Here, stabilization of the developing charge can be continuous, as the β-C–Si bond can overlap with the π-system[5] with no geometric constraints (such orbital overlap should also raise the energy of the HOMO of the π-system, thus increasing its reactivity). With vinylsilanes, on the other hand, maximum stabilization requires rotation of the originally coplanar C–Si bond through 90 degrees (Chapter 7). Indeed, vinylsilanes are *ca.* 8 kJ mol^{-1} *more* stable than their allyl isomers; heating allyltriphenylsilane in quinoline at reflux results[6] in its isomerization to a mixture of (*Z*)- and (*E*)-propenyltriphenylsilane (*Scheme 9.3*).

Ph_3Si quinoline, heat Ph_3Si + Ph_3Si

Scheme 9.3

Finally, it must be appreciated that unlike many other metal-allyl systems, allylsilanes are regio-stable[6] at normal temperatures, 1,3-sigmatropic shifts occuring at a significant rate only at temperatures in excess of 500 °C (Chapter 5). The exceptional case of trimethylsilylcyclopentadiene will be discussed shortly.

9.1 Preparation

9.1.1 By silylation of allyl–metal species

This is undoubtedly the simplest and most direct route[7-9] to allylsilanes, especially when they are symmetrical or, as in the second two cases shown in *Scheme 9.4*, heavily biased sterically and/or electronically. For other examples, the review[1] by Chan and Fleming is highly recommended.

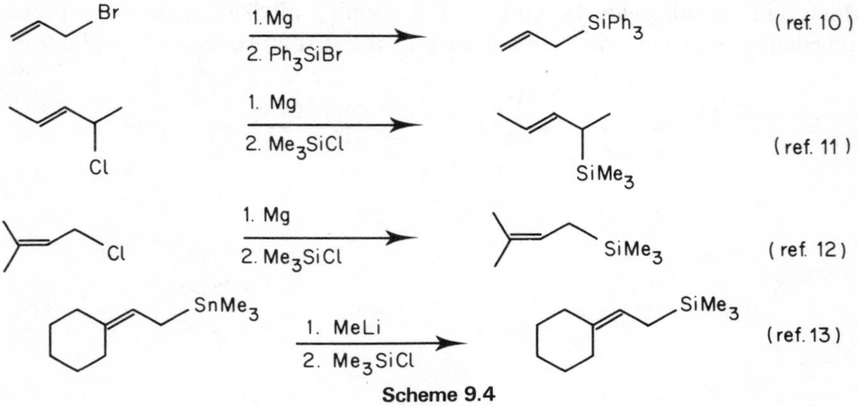

Br 1. Mg 2. Ph_3SiBr $SiPh_3$ (ref. 10)

Cl 1. Mg 2. Me_3SiCl $SiMe_3$ (ref. 11)

Cl 1. Mg 2. Me_3SiCl $SiMe_3$ (ref. 12)

$SnMe_3$ 1. MeLi 2. Me_3SiCl $SiMe_3$ (ref. 13)

Scheme 9.4

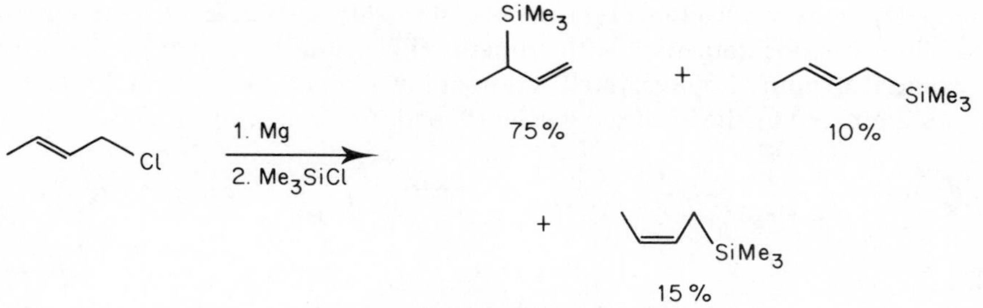

Scheme 9.5

In some cases, regioselectivity can be a problem[6] (*Scheme 9.5*), and separation of the required isomer can be tedious. However, if the desired isomer is also the thermodynamically favoured one, then fluoride ion-promoted isomerization[14] can prove to be of great value. For example, in the case cited above, the terminal silylated isomer is the more stable, and the mixture can be converted quantitatively into one regioisomer by heating in the presence of a catalytic amount of fluoride ion (*Scheme 9.6*); the allylsilane (3) undergoes a similar isomerization.

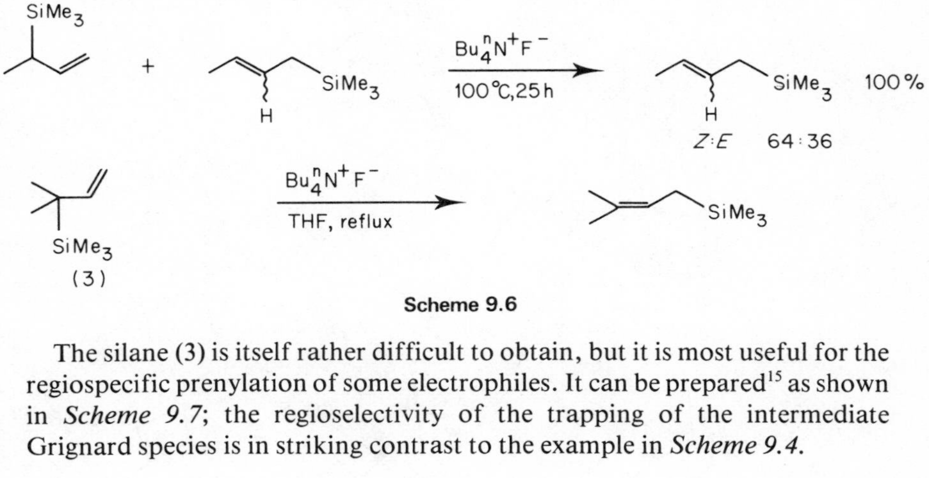

Scheme 9.6

The silane (3) is itself rather difficult to obtain, but it is most useful for the regiospecific prenylation of some electrophiles. It can be prepared[15] as shown in *Scheme 9.7*; the regioselectivity of the trapping of the intermediate Grignard species is in striking contrast to the example in *Scheme 9.4*.

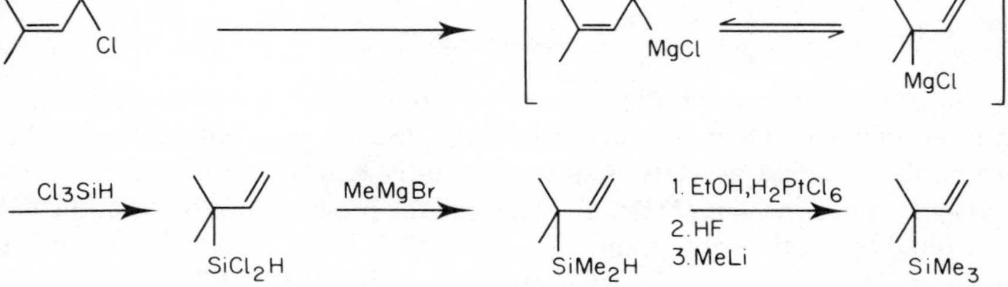

Scheme 9.7

5-Trimethylsilylcyclopentadiene (4) is readily available by reaction of sodium cyclopentadienide with trimethylsilyl chloride. It can be obtained isomerically pure[16] by recrystallization at low temperature, but at 30 °C it is contaminated by the isomeric species (5) and (6), in the ratio 90:7:3.

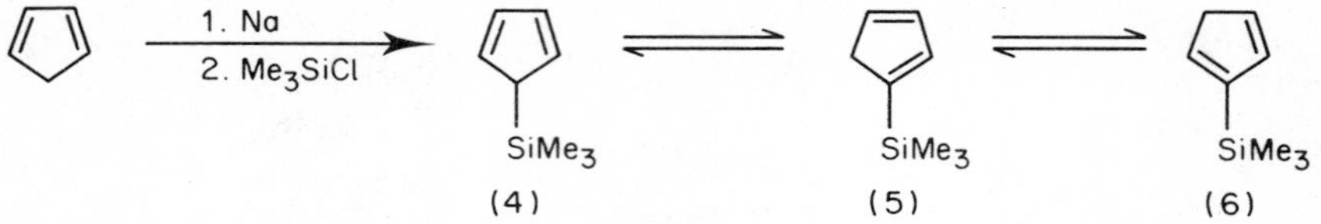

1. Na
2. Me_3SiCl
$SiMe_3$ (4) $SiMe_3$ (5) $SiMe_3$ (6)

At low temperature, and with reactive dienophiles, the rate of interconversion, by [1,5]-hydrogen shifts[17], is much slower than the rate of reaction, and adducts derived from (4) can be isolated readily (*Scheme 9.8*), although isomer (6) is the most reactive of the three. At higher temperatures, the rate of interconversion becomes more rapid, and the preferential formation of adducts from isomer (6) is then observed.

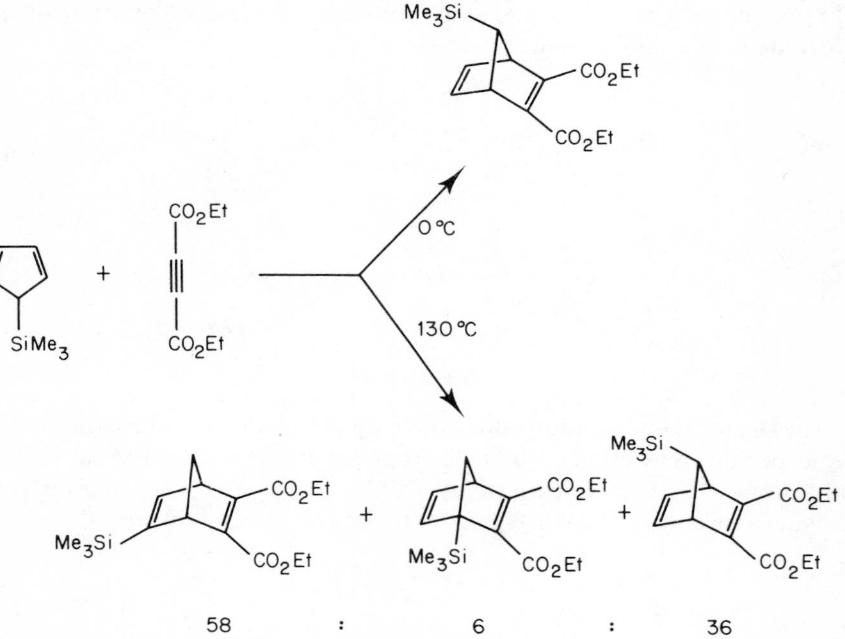

Me_3Si CO_2Et CO_2Et
0 °C
$SiMe_3$ + CO_2Et CO_2Et
130 °C
Me_3Si CO_2Et CO_2Et + Me_3Si CO_2Et CO_2Et + Me_3Si CO_2Et CO_2Et
58 : 6 : 36

Scheme 9.8

Satisfactory use of this diene therefore requires highly reactive dienophiles (or enophiles), which in turn allow the use of low temperatures. For example[18], methyl acrylate is not sufficiently reactive to give directly the 7-silylated norbornene (7) but gives instead the products (8), by reaction with the relatively reactive but minor component (6). Lewis acid activation of the dienophile with boron trifluoride etherate, on the other hand, gives as major product the desired 7-substituted isomer (*Scheme 9.9*).

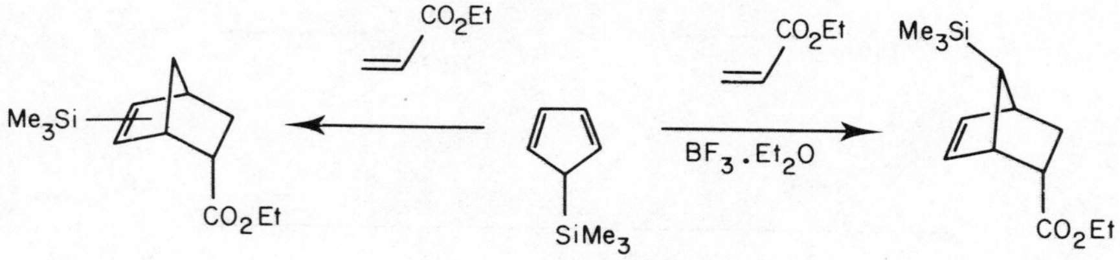

Scheme 9.9

The further use of such products, and of those formed by reaction of (4) with enophiles such as dichloroketene[19] (*Scheme 9.10*), will be discussed shortly.

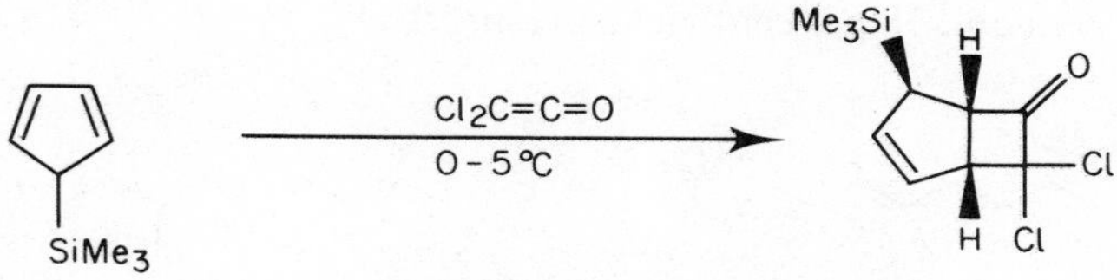

Scheme 9.10

9.1.2 By Wittig condensation/olefination

Seyferth[13] has described a flexible, relatively straightforward route to a wide variety of allylsilanes, using the homologous Wittig reagents (9) and (10) (*Scheme 9.11* and *Table 9.1*).

Table 9.1 Seyferth–Wittig route to allylsilanes

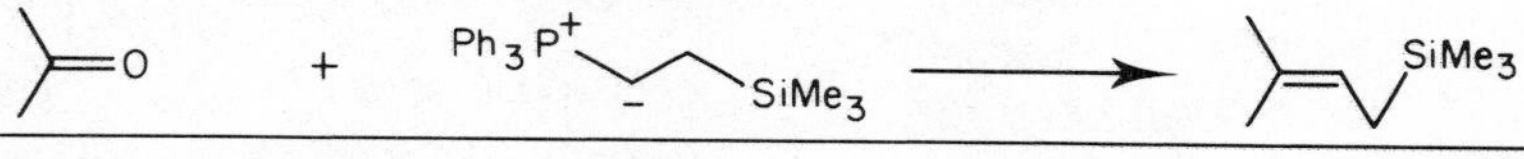

Ylide	*Carbonyl compound*	*Allylsilane*	*Yield per cent*	*E:Z ratio*
(9)	n-$C_6H_{13}CHO$	n-$C_6H_{13}CH=CHCH_2SiMe_3$	71	75:25
	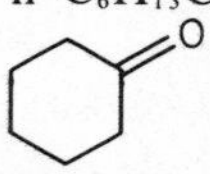	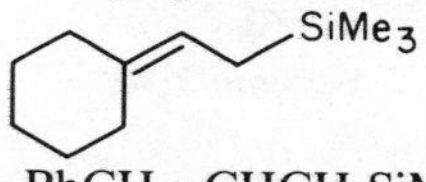	85	
	PhCHO	$PhCH=CHCH_2SiMe_3$	63	64:36
	Et_2CO	$Et_2C=CHCH_2SiMe_3$	38	
	$(CF_3)_2CO$	$(CF_3)_2C=CHCH_2SiMe_3$	43	
(10)	EtCHO	$EtCH=C(Me)CH_2SiMe_3$	74	50:50
	PhCHO	$PhCH=C(Me)CH_2SiMe_3$	72	50:50

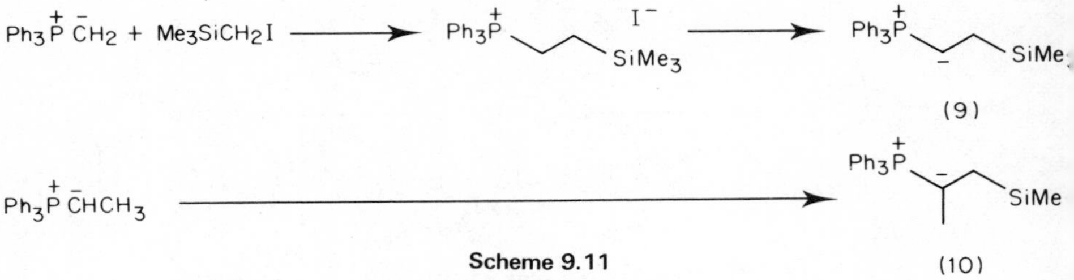

$Ph_3\overset{+}{P}\overset{-}{C}H_2 + Me_3SiCH_2I \longrightarrow Ph_3\overset{+}{P}CH_2CH_2CH_2SiMe_3\ I^- \longrightarrow$ (9)

$Ph_3\overset{+}{P}\overset{-}{C}HCH_3 \longrightarrow$ (10)

Scheme 9.11

The more substituted ylide (10) did not condense with cyclohexanone; instead it acted as a base, deprotonating the ketone and regenerating a phosphonium salt. Fleming[20] has added more experimental detail to this process, with some improvements, and has made the allylsilanes (11) and (12) in 60 per cent and 78 per cent yields, respectively.

(11) (12)

9.1.3 By catalytic 1,4-hydrosilylation of 1,3-dienes

This method has been used successfully (*Scheme 9.12*) for the preparation[21] of 3-trimethylsilylcyclopentene (13), but is of less utility with unsymmetrical dienes, as shown by the hydrosilylation of isoprene[22].

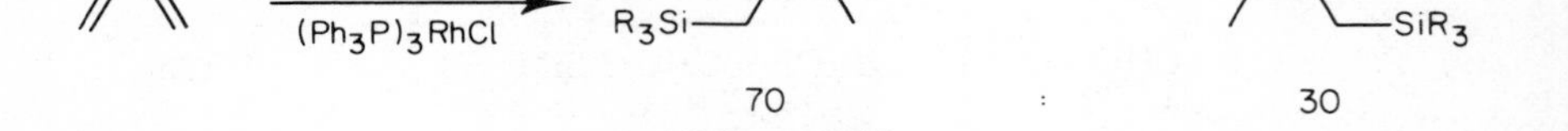

$+ X_2YSiH \xrightarrow[\text{catalyst}]{(PhCN)_2PdCl_2.2Ph_3P}$ —SiX_2Y $\xrightarrow{MeMgBr}$ —$SiMe_3$ (13)

X = Cl, Y = Me 90%

X = Y = Cl 83%

$\xrightarrow[(Ph_3P)_3RhCl]{R_3SiH}$ R_3Si— + —SiR_3

70 : 30

Scheme 9.12

9.1.4 By Diels–Alder cycloaddition

Several terminally silylated 1,3-butadienes have been studied as Diels–Alder dienes: reaction will, of course, produce cyclic allylsilanes. The silyl group exerts little directing influence on the orientation of cycloaddition[23], and product regioselectivity with unsymmetrical dienophiles is controlled by other substituents, if any, on the diene (*Scheme 9.13*). Such lack of regioselectivity is not without value. Since the trimethylsilyl group exerts such a weak directing effect, any other substituent on the diene should exert its directing effect more or less unimpeded, as illustrated with the diene (14).

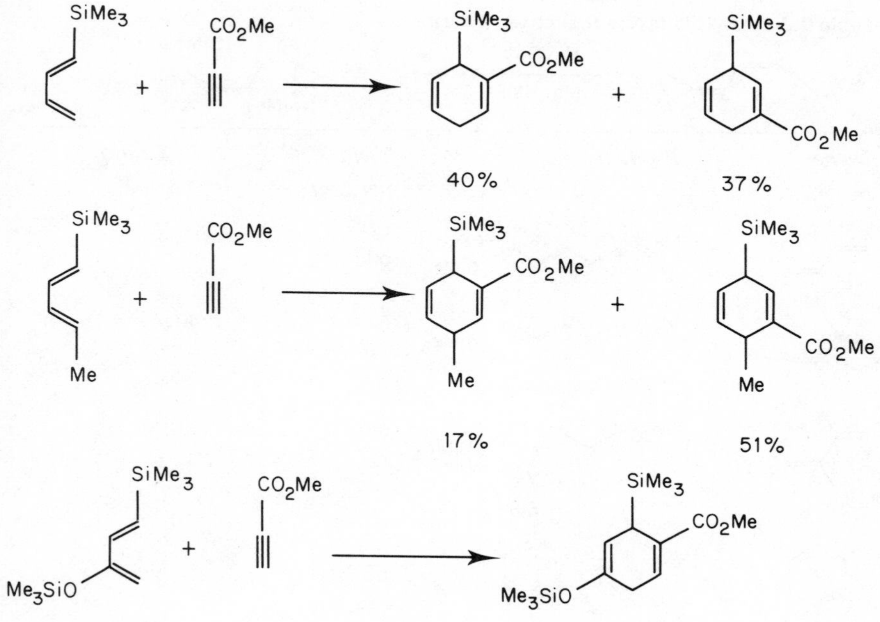

Scheme 9.13

9.1.5 By reductive silylation

Calas, Dunoguès and their collaborators have reported[24,25] many examples of reductive silylations using chlorosilane/metal systems, some of which are illustrated in *Table 9.2*; this is discussed also in Chapter 21. Aromatic ring systems can undergo a similar process[26], as can allenes[27], as shown in *Scheme 9.14*.

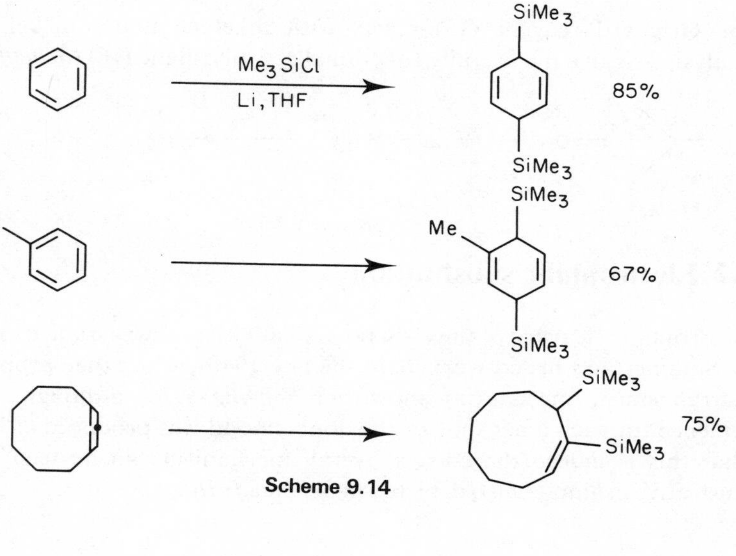

Scheme 9.14

Table 9.2 Allylsilanes by reductive silylation

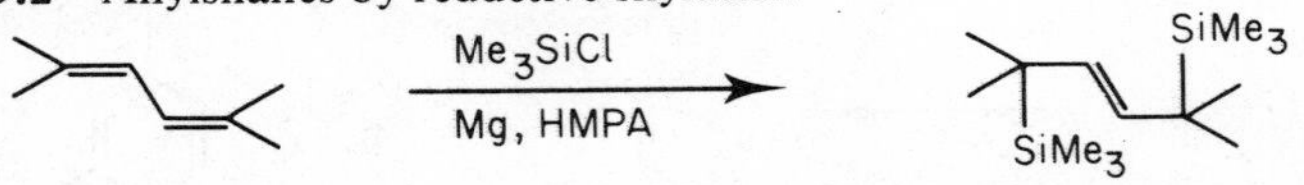

Me_3SiCl / Mg, HMPA

Diene	Products	Yield per cent	E:Z ratio
 butadiene	 $Me_3SiCH_2CH=CHCH_2SiMe_3$	60	
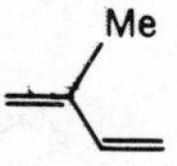 isoprene	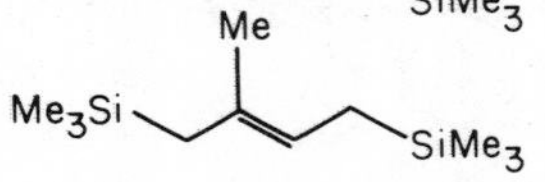 $Me_3SiCH_2C(Me)=CHCH_2SiMe_3$	64	40:60
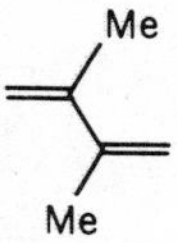 2,3-dimethylbutadiene	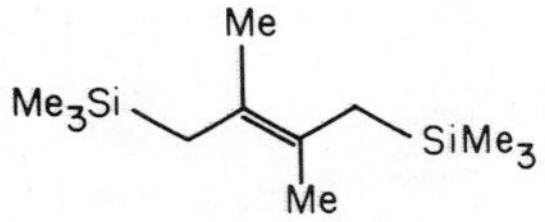 $Me_3SiCH_2C(Me)=C(Me)CH_2SiMe_3$	66	65:35
 1,3-cyclohexadiene	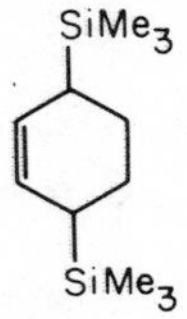 3,6-bis(trimethylsilyl)cyclohexene	65	
 1,3-cyclooctadiene	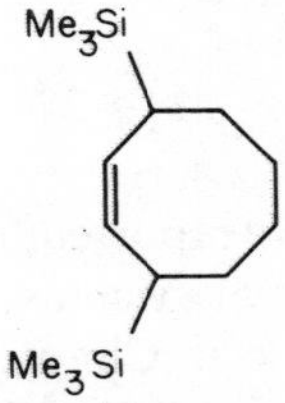 3,8-bis(trimethylsilyl)cyclooctene	68	

9.1.6 Miscellaneous

The Grignard reagent (15) reacts with diketene under nickel chloride catalysis[28] to give the usefully functionalized allylsilane (16) (*Scheme 9.15*).

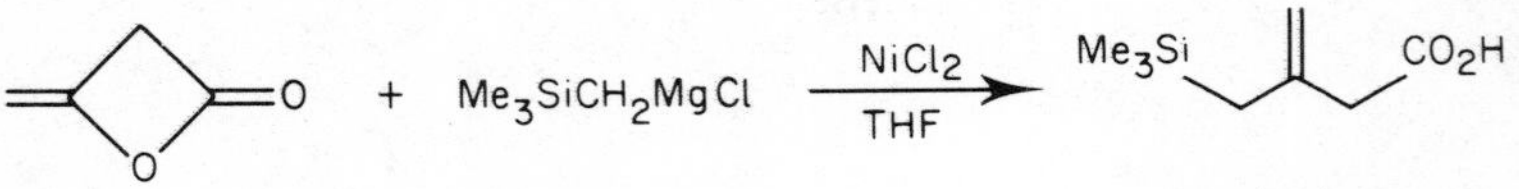

Me_3SiCH_2MgCl (15) + diketene $\xrightarrow{NiCl_2,\ THF}$ $Me_3SiCH_2C(=CH_2)CH_2CO_2H$ (16)

Scheme 9.15

9.2 Electrophilic substitution

Electrophilic cleavage of the C–Si bond of allylsilanes was studied extensively by Sommer and his co-workers in the late 1940s, when they proposed[2] the rearrangement mechanism shown in *Scheme 9.16*, although no direct evidence for such a net shift of the double bond was produced at that time. That this is indeed the correct overall mechanism can be seen from the illustrative examples also depicted in *Scheme 9.16*.

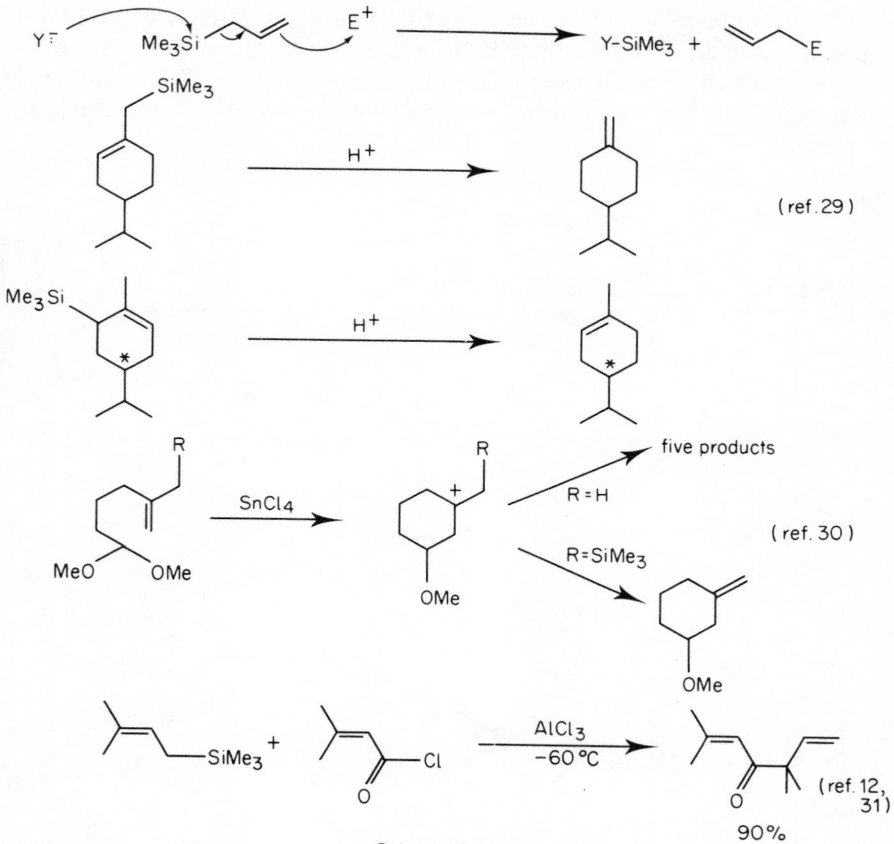

Scheme 9.16

Many other electrophiles can, of course, be used, and the scope of this general and fundamental transposition can be seen to advantage in the elegant studies of Fleming[1]. Diels–Alder cycloaddition between 1-trimethylsilylbutadiene[32] and maleic anhydride gives[33] the adduct (17), the relative stereochemistry being that derived by assumption of operation of the *endo*-rule. The diene itself was best prepared (*Scheme 9.17*) from 3-trimethylsilylprop-2-en-1-ol[34] by a silyl-Wittig reaction (Chapter 12) on the derived aldehyde. Protiodesilylation of the anhydride (17) gave the new anhydride (18) with net shift of the double bond. Treatment of the diacid (19) with unbuffered peracetic acid gave the allyl alcohol (21), presumably via the corresponding epoxide (20). Two points should be made here: first, the stereochemistry of the allylic alcohol produced is probably due to initial epoxidation having occurred from the less-hindered face of the cyclohexene ring; secondly, (acidic) unbuffered conditions are essential for success, Hudrlik[35] having shown (Chapter 8) that $\beta\gamma$-epoxysilanes undergo opening under *basic* conditions to give carbonyl compounds, a hydride shift being implicated.

Phenylsulphenylation of the diester (22) gave the allyl sulphide (23) as major isomer, steric congestion again resulting in the observed stereoselectivity; this allyl sulphide could be converted, via [2,3]-sigmatropic rearrangement[36] of the corresponding sulphoxide, into the allyl alcohol (24).

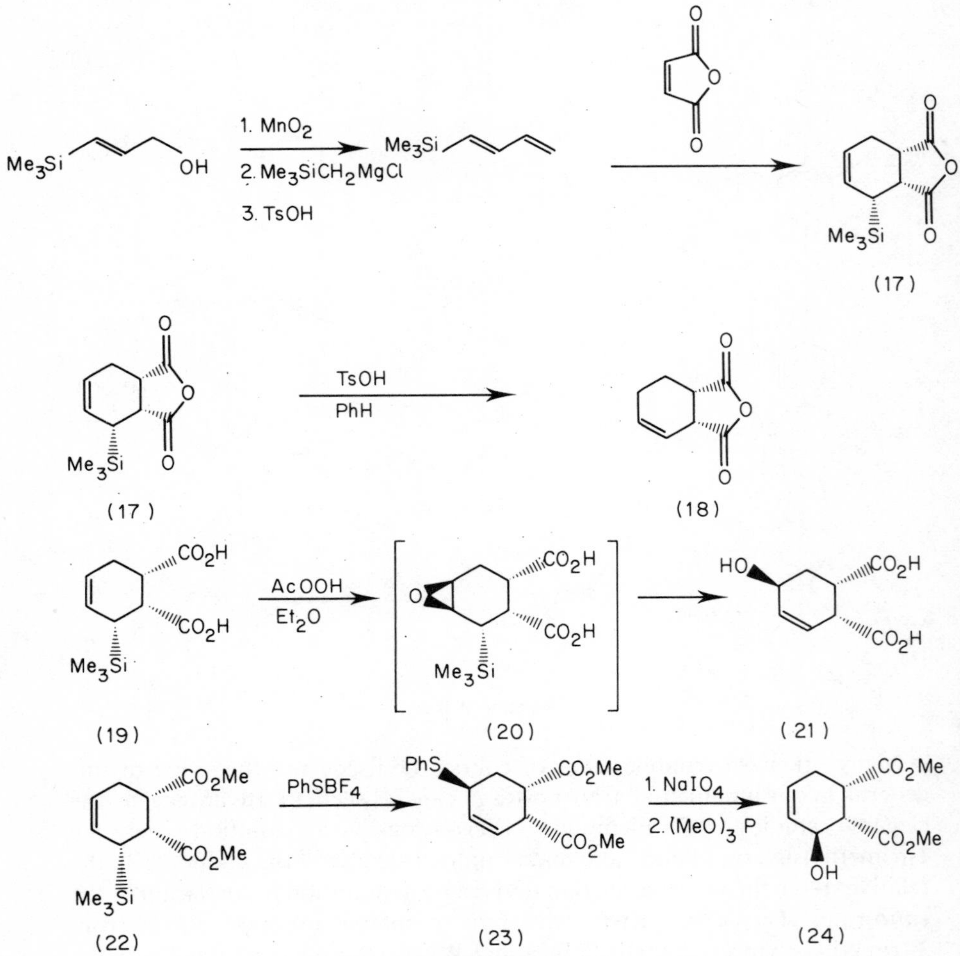

Scheme 9.17

Attempts to extend this methodology to less carbon-symmetrical dienes were largely unsuccessful, little regioselectivity being observed (p. 102). Heterodienes, on the other hand, should prove much more useful, the hetero substituent easily overriding the very weak directing effect of the silyl substituent. Other dienes have been used with more success. Dichloroketene is sufficiently reactive an enophile to react[19] with 5-trimethylsilylcyclopentadiene at a temperature low enough to preclude [1,5]-hydrogen shifts taking place in the diene (p. 100). The [2 + 2]-cycloaddition proceeds regiospecifically, to give the adduct (25), which on reaction with a range of

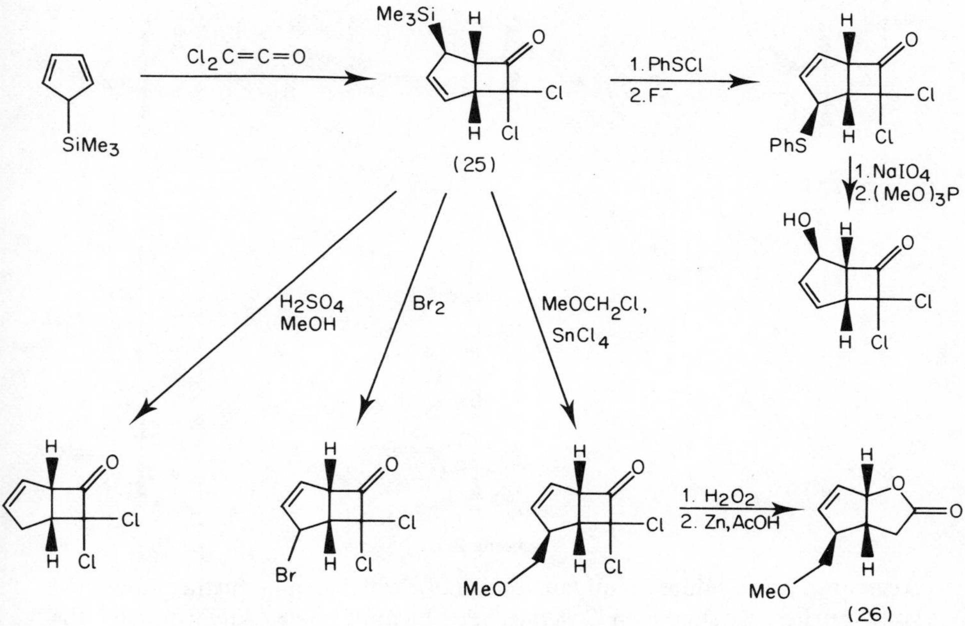

Scheme 9.18

electrophiles gives the results shown in *Scheme 9.18.* The lactone (26) is a key intermediate in prostaglandin synthesis.

One important point must be made at this juncture. In the previous six-membered ring series, electrophilic displacement took place with overall *anti*-stereochemistry, whereas here *syn*-displacement is reported to occur. As a further complicating factor, the lactone (27) apparently undergoes *endo*-epoxidation, a prediction made by analogy with the corresponding deschloro, des-silyl lactone; it also undergoes sulphenylation[37a] with *anti*-stereoselectivity (*Scheme 9.19*). On the other hand, an *acyclic* allylsilane has been shown to undergo acylation with a high degree of *syn*-stereoselectivity[37b]. The only safe conclusion which can be drawn from such results is that the stereochemical relationship of the silyl group may play little if any part in directing the orientation of attack, and that any observed steric orientation is a feature of the overall molecular geometry. To be forced to such a conclusion is rather perplexing, but can be justified by arguing that in configurationally biased and relatively rigid systems, initial electrophilic attack will be *sterically* governed; certainly, subsequent collapse to product must then involve substantial carbonium ion development adjacent to the C–Si bond to allow adequate overlap for such collapse.

The lactone (27) has found further utility in a synthesis of loganin. Here, the critical step was reaction of the allylsilane (28) with chlorosulphonyl isocyanate. It had already been shown[38] that simple allylsilanes react with this reagent to produce *β*-lactams; these *β*-lactams are thermally unstable, and

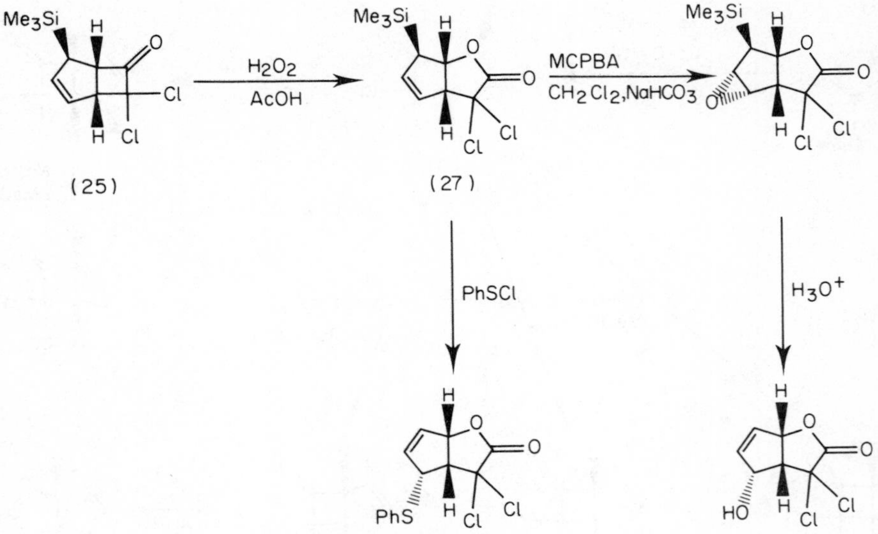

Scheme 9.19

rearrange on standing to imidate esters (29), which can be further converted into nitriles, as shown in *Scheme 9.20.* Fleming instead hydrolysed[19] the intermediate imidate ester under carefully controlled conditions to the corresponding acid, which was then transformed into the racemic aglucone acetate (30) of loganin.

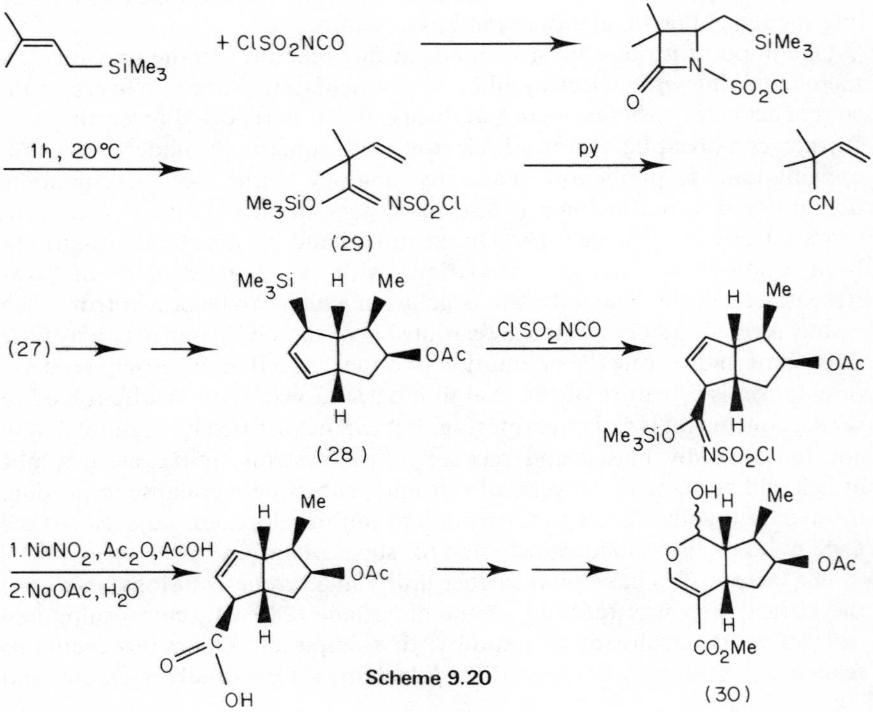

Scheme 9.20

This characteristic reactivity of allylsilanes has provided a method[20] for the replacement of carbonyl oxygen by both a vinyl group and a variable alkyl substituent (*Scheme 9.21*). The first step of this two-stage sequence employs the invaluable method of Seyferth[13] and his co-workers (p. 101) to convert the carbonyl compound into the corresponding allylsilane, which is then exposed to a variety of electrophiles, normally in the presence of a Lewis acid to enhance electrophilicity. Halides which can give rise to good equilibrium

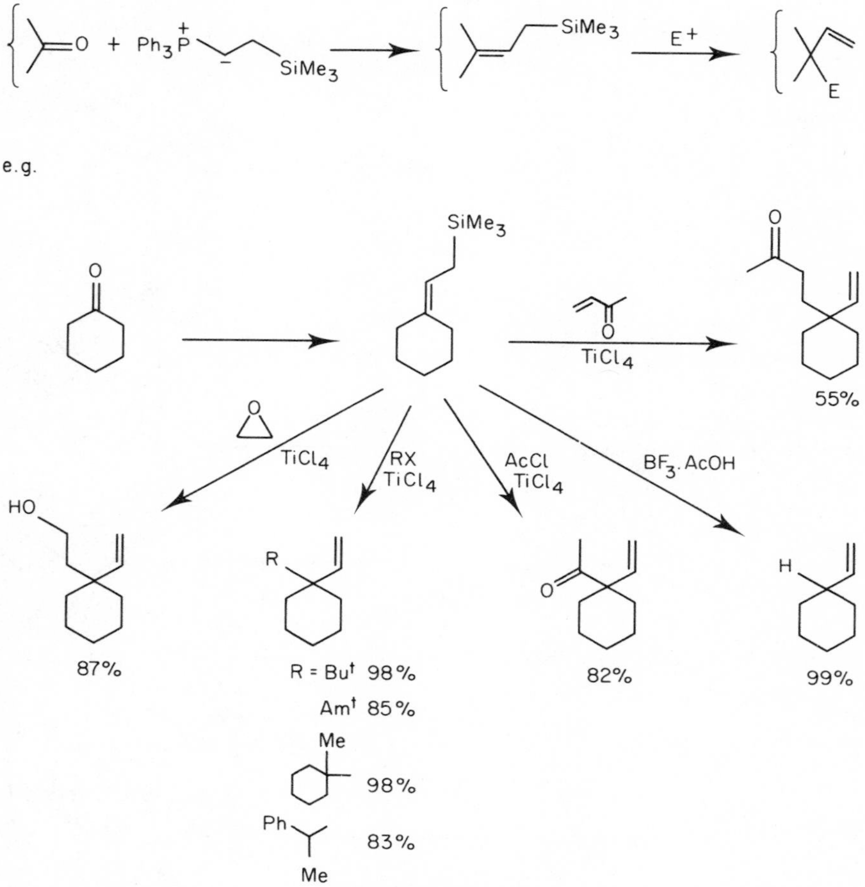

O + $Ph_3\overset{+}{P}$ $\overset{-}{}$ $SiMe_3$ → $SiMe_3$ —E^+→ E

e.g.

$SiMe_3$ O $TiCl_4$ 55%

$TiCl_4$ RX $TiCl_4$ AcCl $TiCl_4$ BF_3.AcOH

HO   R   O   H

87%   R = Bu^t 98%   82%   99%

Am^t 85%

Me 98%

Ph Me 83%

Scheme 9.21

cation concentrations, such as t-butyl chloride, t-amyl chloride, 1-methylcyclohexyl chloride and α-methylbenzyl bromide, all react smoothly in the presence of titanium(IV) chloride; less reactive halides such as isopropyl chloride and, surprisingly, benzyl chloride, fail to react. Boron trifluoride–acetic acid is the reagent system of choice for the protiodesilylation of such allylsilanes. In general, allylsilanes behave in a remarkably similar manner to silyl enol ethers (Chapter 17); electrophiles which react well with one will normally react equally well with the other.

Titanium(IV) chloride has played a major role[39] as a Lewis acid in enhancing the reactivity of weak electrophiles such as carbonyl compounds and their acetals, some examples of which are given in *Scheme 9.22*.

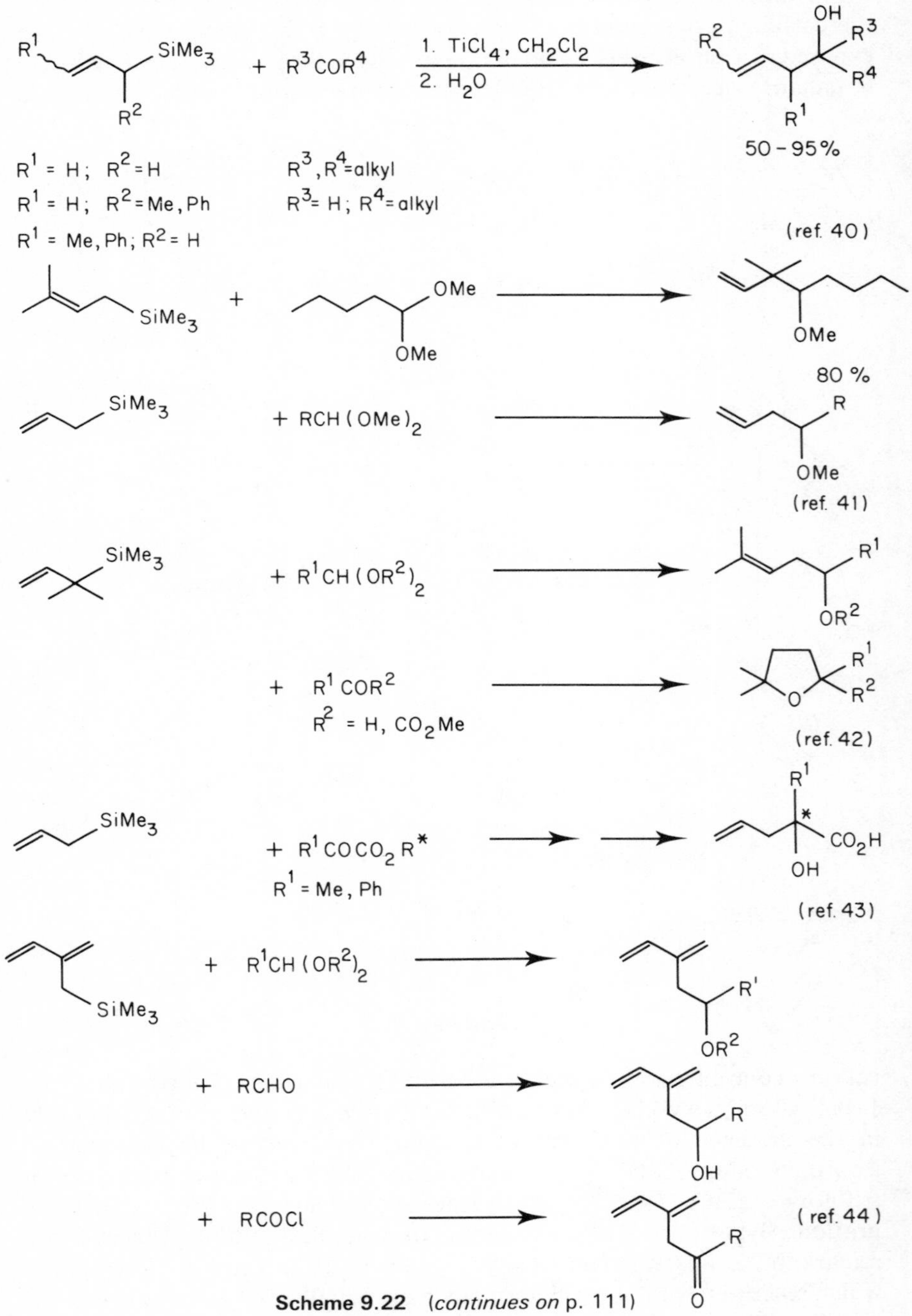

R^1 / $SiMe_3$ / R^2 + R^3COR^4 → 1. $TiCl_4$, CH_2Cl_2 2. H_2O → R^2 / OH / R^3 / R^4 / R^1

50 – 95%

$R^1 = H$; $R^2 = H$ — R^3, R^4 = alkyl

$R^1 = H$; R^2 = Me, Ph — R^3 = H; R^4 = alkyl

R^1 = Me, Ph; R^2 = H

(ref. 40)

$SiMe_3$ + OMe / OMe → OMe

80 %

$SiMe_3$ + $RCH(OMe)_2$ → R / OMe

(ref. 41)

$SiMe_3$ + $R^1CH(OR^2)_2$ → R^1 / OR^2

+ R^1COR^2, R^2 = H, CO_2Me → R^1 / O / R^2

(ref. 42)

$SiMe_3$ + $R^1COCO_2R^*$, R^1 = Me, Ph → → R^1 / * / CO_2H / OH

(ref. 43)

$SiMe_3$ + $R^1CH(OR^2)_2$ → R' / OR^2

+ RCHO → R / OH

+ RCOCl → R / O (ref. 44)

Scheme 9.22 (*continues on* p. 111)

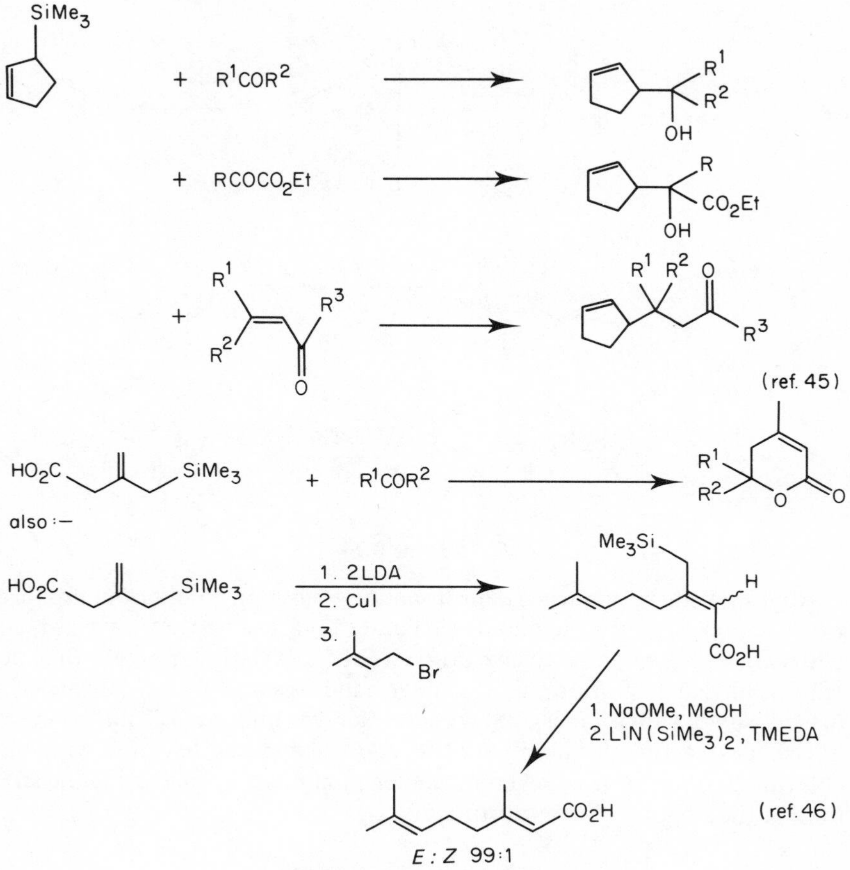

$SiMe_3$ + R^1COR^2 → OH, R^1, R^2

+ $RCOCO_2Et$ → R, CO_2Et, OH

+ R^1, R^2, R^3, O → R^1 R^2 O, R^3

(ref. 45)

HO_2C … $SiMe_3$ + R^1COR^2 → R^1, R^2, O, O

also :—

HO_2C … $SiMe_3$ → 1. 2LDA 2. CuI 3. Br → Me_3Si, H, CO_2H

1. NaOMe, MeOH
2. $LiN(SiMe_3)_2$, TMEDA

CO_2H (ref. 46)

E : *Z* 99:1

Scheme 9.22 (*continued*)

With enones as substrates, 1,4-addition[47] of the allyl unit is normally observed (*Scheme 9.23*) and such methodology has been used, *inter alia*, in a synthesis[48] of (+)-nootkatone from (−)-β-pinene.

Quinones, on the other hand, undergo 1,2-addition[49] (*Scheme 9.24*); the intermediate products suffer a rapid [1,2]-allylic shift to give the aromatic hydroquinones. 2,6-Dialkylbenzoquinones cannot undergo this rearrangement, and the initial adducts are therefore isolable.

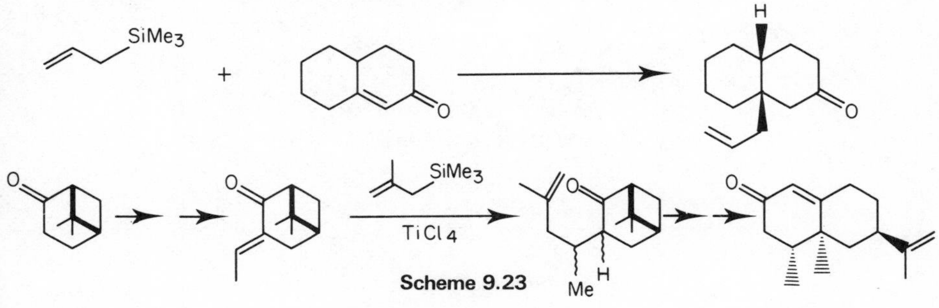

$SiMe_3$ + O → H, O

O → → O → $SiMe_3$ / $TiCl_4$ → O, H, Me → → O

Scheme 9.23

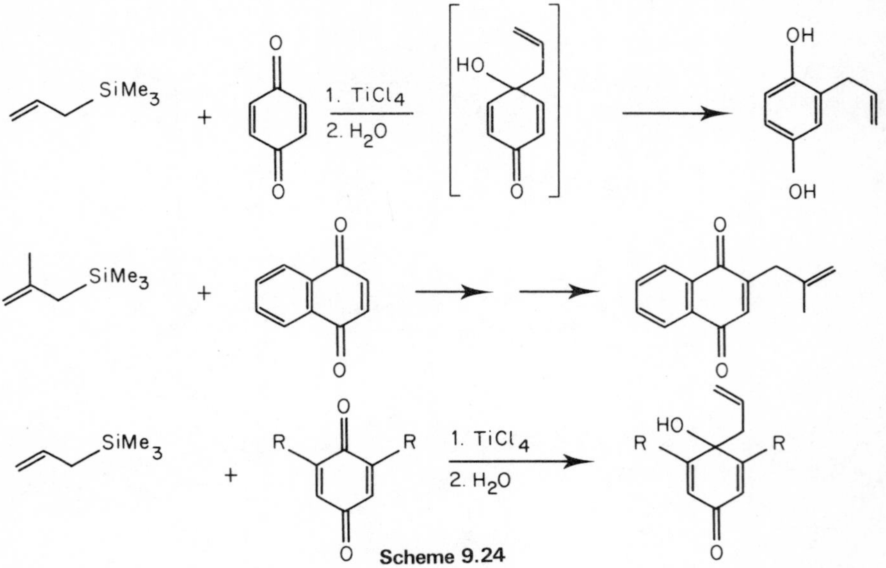

SiMe3 + O O 1. TiCl4 2. H2O [HO O] → OH OH

SiMe3 + O O → → O O

SiMe3 + R O R O 1. TiCl4 2. H2O → HO R R O

Scheme 9.24

Allyl silyl ethers[50] undergo deprotonation to give an equilibrium mixture[51] (*see* Chapter 5) of the metalloids (31) and (32) (*Scheme 9.25*), which reacts with carbon electrophiles to give products of *C*-alkylation (p. 209). Reaction with trialkylsilyl chlorides, on the other hand, results in[31,52] *O*-silylation, to afford silyloxyallylsilanes (33). These can be utilized as functionalized allylsilanes, acting as homo-enolate equivalents[31] in reaction with acid chlorides; competitive *O*-acylation can be minimized by increasing the steric bulk of the silyloxy silicon substituents.

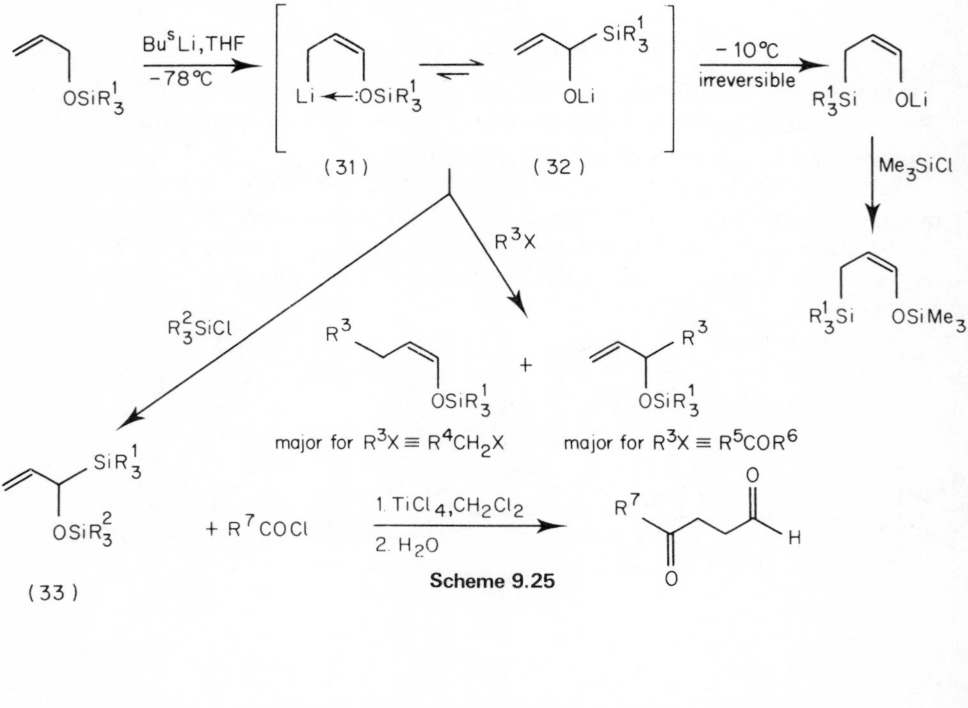

$OSiR^1_3$ —Bu^sLi, THF, −78°C→ [Li ←:$OSiR^1_3$ ⇌ SiR^1_3 OLi] —−10°C irreversible→ R^1_3Si OLi

(31) (32)

Me_3SiCl ↓ R^1_3Si $OSiMe_3$

R^3X → R^3 $OSiR^1_3$ + R^3 $OSiR^1_3$

major for $R^3X \equiv R^4CH_2X$ major for $R^3X \equiv R^5COR^6$

R^2_3SiCl → SiR^1_3 $OSiR^2_3$ (33) + R^7COCl —1. $TiCl_4$, CH_2Cl_2 2. H_2O→ R^7 O O H

Scheme 9.25

Many Lewis acids other than titanium(IV) chloride can be advantageously employed. Dunogùes, Calas, and their co-workers have described[53] the reaction of allyltrimethylsilane with electrophiles such as chloral and chloroacetone in the presence of gallium(III) chloride, aluminium(III) chloride, or indium(III) chloride (*Scheme 9.26*).

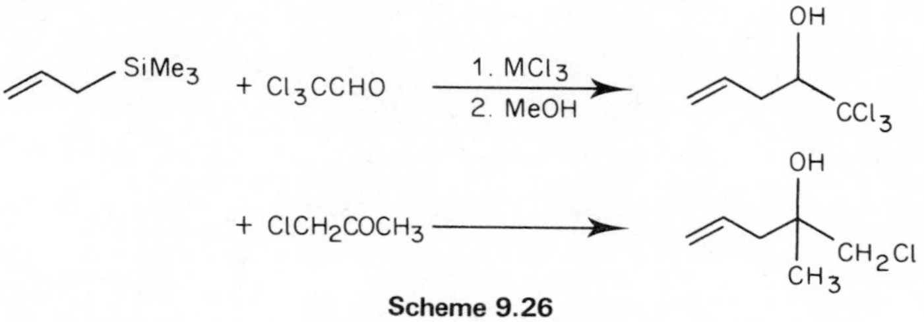

SiMe3 + Cl3CCHO —(1. MCl3; 2. MeOH)→ OH, CCl3

+ ClCH2COCH3 ——→ OH, CH2Cl, CH3

Scheme 9.26

A similar range of Lewis acids induces the reaction between acid chlorides and allyltrimethylsilane; in simple cases, this provides an exceptionally easy route[54] to allyl ketones or, if desired, to (*E*)-propenyl ketones (*Scheme 9.27*). More complex disilanes were also studied; in unsymmetrical cases, steric hindrance seems to control the regioselectivity of reaction.

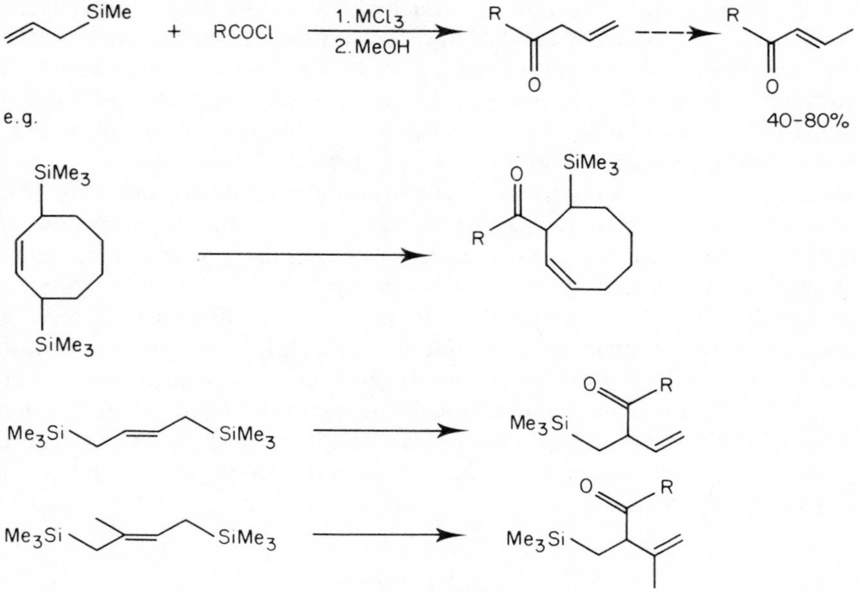

SiMe + RCOCl —(1. MCl3; 2. MeOH)→ R, O ---→ R, O 40–80%

e.g.

SiMe3, SiMe3 ——→ O, SiMe3, R

Me3Si, SiMe3 ——→ O, R, Me3Si

Me3Si, SiMe3 ——→ O, R, Me3Si

Scheme 9.27

An interesting chemoselectivity of Lewis acid is seen[55] in the reactions of α-keto- and β-keto-acetals. Regardless of whether aluminium(III) chloride or titanium(IV) chloride is employed, β-keto-acetals react selectively at the acetal carbon; with α-keto-acetals, on the other hand, aluminium(III) chloride induces attack at the carbonyl carbon, whereas titanium(IV) chloride activates both sites to the same extent (*Scheme 9.28*).

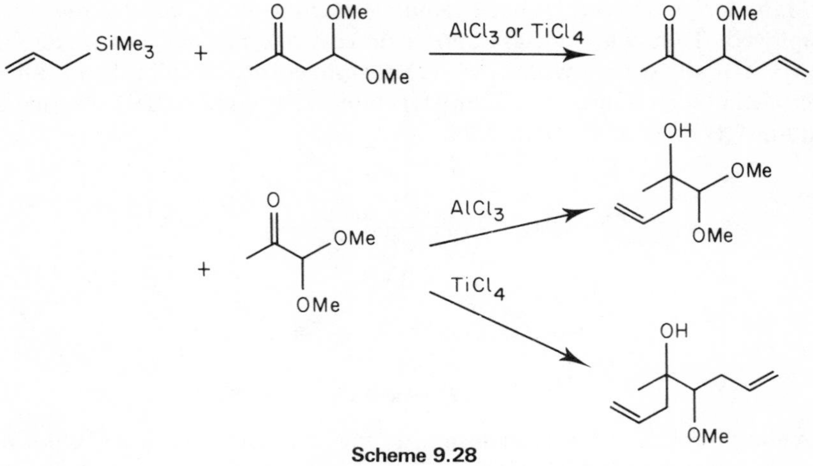

Scheme 9.28

In the above examples, the Lewis acid is used normally in stoichiometric amounts. In contrast, *catalytic* amounts of trimethylsilyl trifluoromethanesulphonate[56] bring about reaction between allyltrimethylsilane and a range of acetals; under such mild conditions, aldehydes and ketones are unreactive.

All examples discussed so far have used Lewis acids to enhance the electrophilicity of the organic reagent, be it alkyl halide (including adamantyl chloride[57]), carbonyl compound, acid chloride or acetal. An alternative methodology, which involves generation[58] of an allyl anion or its equivalent, is to treat the allylsilane with fluoride ion: this constitutes a most effective, relatively non-basic route to the allyl anion (the pK_a of propene is *ca.* 36). Allyl anions generated in such a manner show considerable chemoselectivity, aldehydes being attacked in preference to ketones, with esters being comparatively unreactive; $\alpha\beta$-unsaturated carbonyl compounds undergo competitive 1,2- and 1,4-addition. In contrast to all earlier cases, the mechanistic requirement for a net double-bond shift in the product is now absent; substitution now occurs regioselectively at the *less* substituted end of the allyl chain (*Scheme 9.29*), although the degree of regioselection is not great. Allyl anions as such may not be involved in this reaction. An allylically labile, hypervalent anionic silicon intermediate (*see also* Chapters 11 and 17) may well be the reacting species.

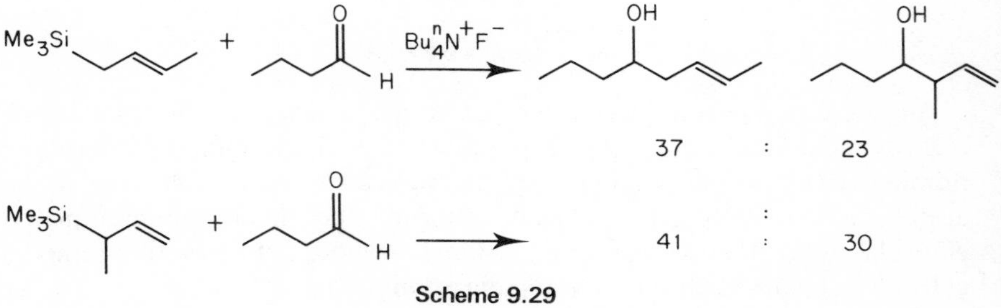

Scheme 9.29

Similarly, treatment of the allylsilane (34) with fluoride ion[59] resulted in smooth closure (*Scheme 9.30*), by intramolecular attack of the allyl anion (or its equivalent) on the proximal aldehyde; this is the first reported case of such a 'push' intramolecular attack on a free carbonyl group, the earlier example[30] of Fleming and his co-workers (p. 105) being a 'pull' process of Lewis acid-catalysed attack on an acetal. Preparation of the requisite allylsilane by anion silylation required favourable alkoxide coordination of the allyl-lithium for success, with failure being experienced with homologous, longer chain, primary alcohols.

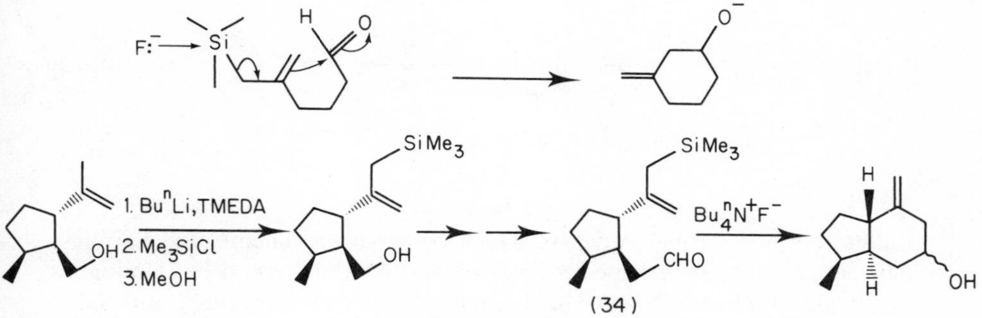

F:⁻ Si H O O⁻

SiMe$_3$ SiMe$_3$ H

1. Bu^{n}Li,TMEDA 2. Me$_3$SiCl 3. MeOH OH OH CHO Bu$^n_4$N$^+$F$^-$ H OH

(34)

Scheme 9.30

Trost[60] has ingeniously utilized allylsilanes to provide a route to palladium(0)-complexed trimethylenemethane (35). Based on the hypothesis that a silylmethylallyl cation might decompose as shown in *Scheme 9.31*, he was able to extend his elegant studies[61] on π-allylpalladium complexes. The complex (35) underwent cycloaddition with electron-deficient alkenes, leading to functionalized methylenecyclopentanes.

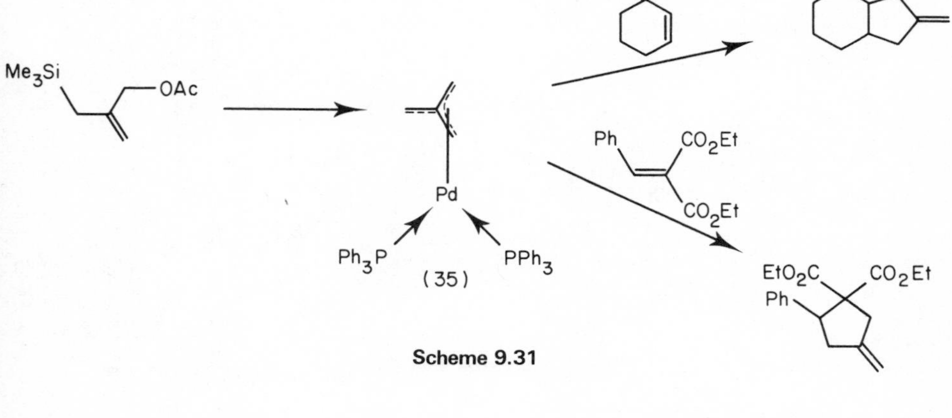

Me$_3$Si +

O O

Me$_3$Si OAc

Ph CO$_2$Et CO$_2$Et

Pd

Ph$_3$P PPh$_3$ (35)

EtO$_2$C CO$_2$Et Ph

Scheme 9.31

An interesting case where the focus of attention is on the silicon-containing product is seen in a method[62] for the *in situ* generation of the highly useful (Chapter 18) iodotrimethylsilane. Based on an earlier report[63] on the reaction of allyltrimethylsilane with iodine, in which only allyl iodide was isolated, Jung has re-investigated this reaction as a source of iodotrimethylsilane (*Scheme 9.32*). This route has the disadvantage of co-producing reactive allyl

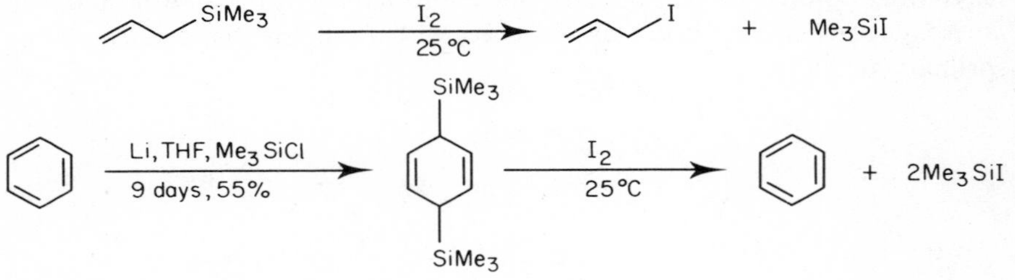

Scheme 9.32

iodide. An alternative sequence, which co-produces (chemically) harmless benzene, uses the disilyl species (36) prepared by the method[64] of Dunoguès and Calas (*Scheme 9.14*). The French school has subjected[64] this same compound (36) to a different nucleophile, hydroxide ion, in a convenient route to 1,4-cyclohexadiene (*Scheme 9.33*).

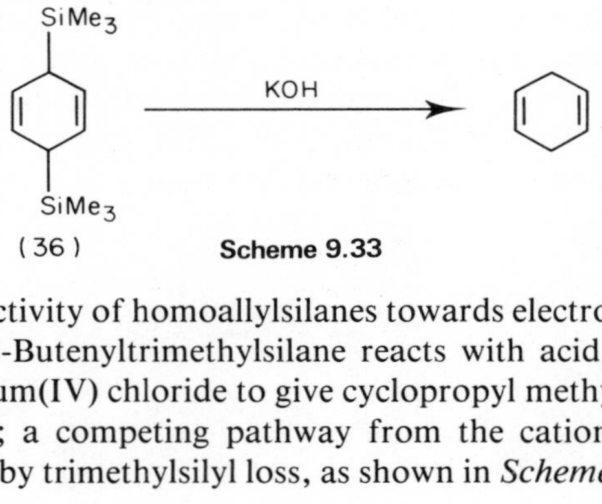

Scheme 9.33

Finally, the reactivity of homoallylsilanes towards electrophilic attack has been described. 3-Butenyltrimethylsilane reacts with acid chlorides in the presence of titanium(IV) chloride to give cyclopropyl methyl ketones (38) in moderate yields[65]; a competing pathway from the cation (37) is hydride transfer followed by trimethylsilyl loss, as shown in *Scheme 9.34*.

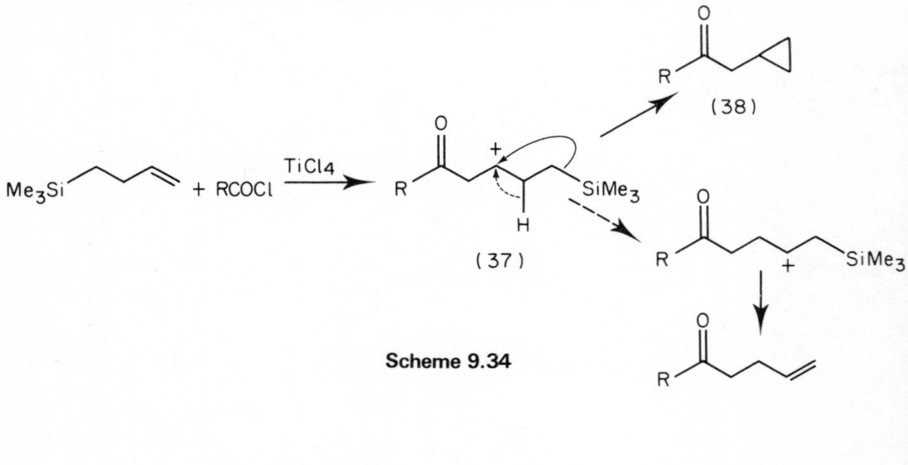

Scheme 9.34

Cyclopropyl ketones can themselves be prepared[66] in good yield and in wide range (*Scheme 9.35*) by Lewis acid-catalysed attack of an acid chloride on cyclopropyltrimethylsilane.

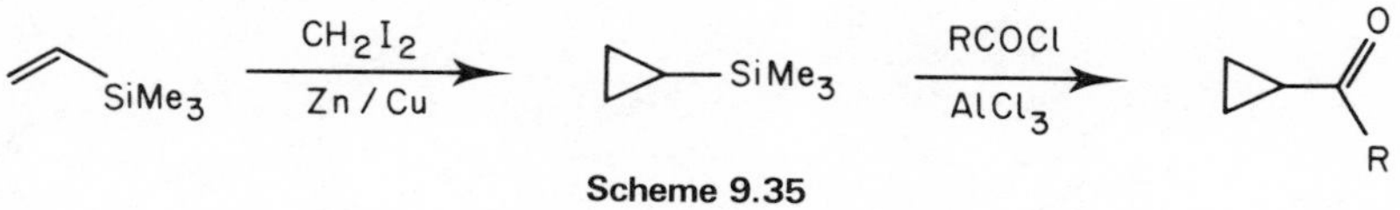

CH_2I_2, Zn / Cu; RCOCl, $AlCl_3$

Scheme 9.35

9.3 Other selected examples of silyl control of carbonium ion formation and collapse

Allylphosphine oxides[67] and allyl sulphides[36] are synthetically versatile intermediates, but their utility is directly related to their availability. A general scheme for their preparation is shown in *Scheme 9.36*.

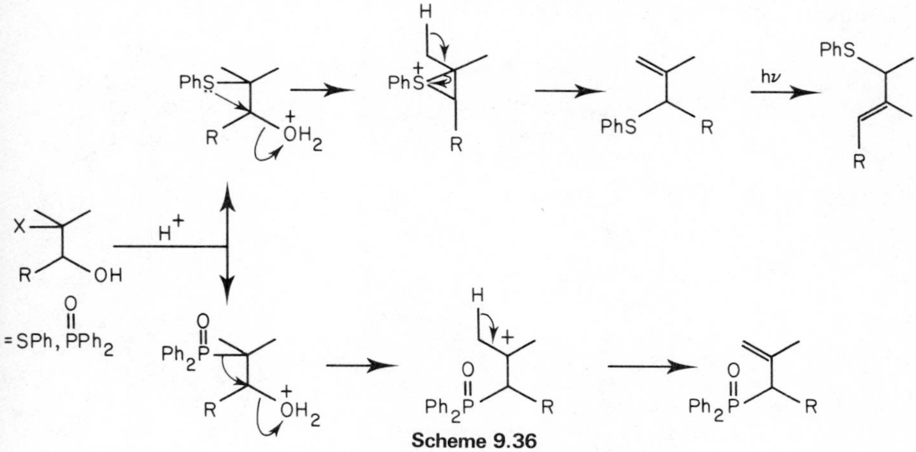

X = SPh, PPh_2

Scheme 9.36

These routes are limited by the intermediacy and subsequent collapse of carbonium or episulphonium ions, when thermodynamic control of product formation is normally observed. For example, the carbinol (39) gives only the more substituted alkene (41). If, however, the carbinol (40) is used instead, *only* the alkene (42) is formed, and the rate of rearrangement is enhanced[30,68] (*Scheme 9.37*).

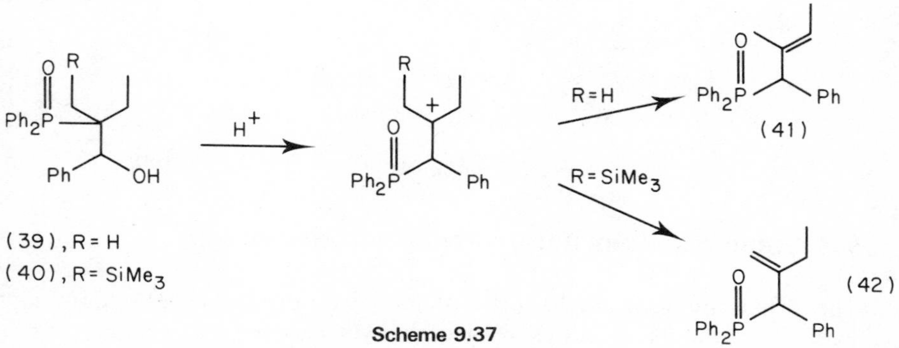

(39), R = H
(40), R = $SiMe_3$

Scheme 9.37

Similarly, in the case of β-hydroxyalkyl phenyl sulphides, silicon mediates migration[69] from a *secondary* carbon atom to a *tertiary* cationic site (*Scheme 9.38*).

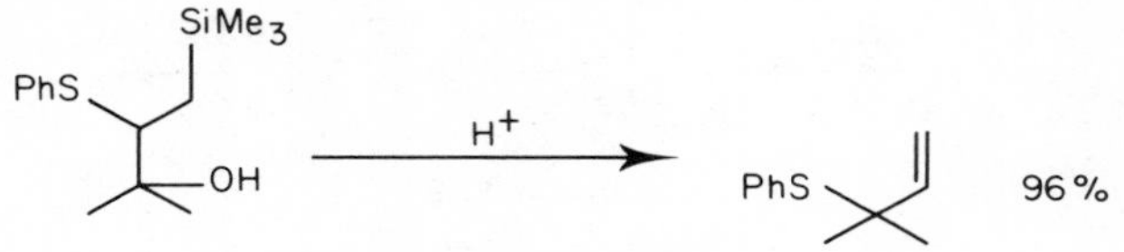

Scheme 9.38

As previously discussed (p. 100), 5-trimethylsilylcyclopentadiene reacts with suitably reactive (or activated) dienophiles to give 7-silylnorbornenes. Fleming has described[18] how the presence of the C–Si bond can encourage and direct carbonium ion rearrangements in such a framework to produce 7-substituted norbornenes (*Scheme 9.39*).

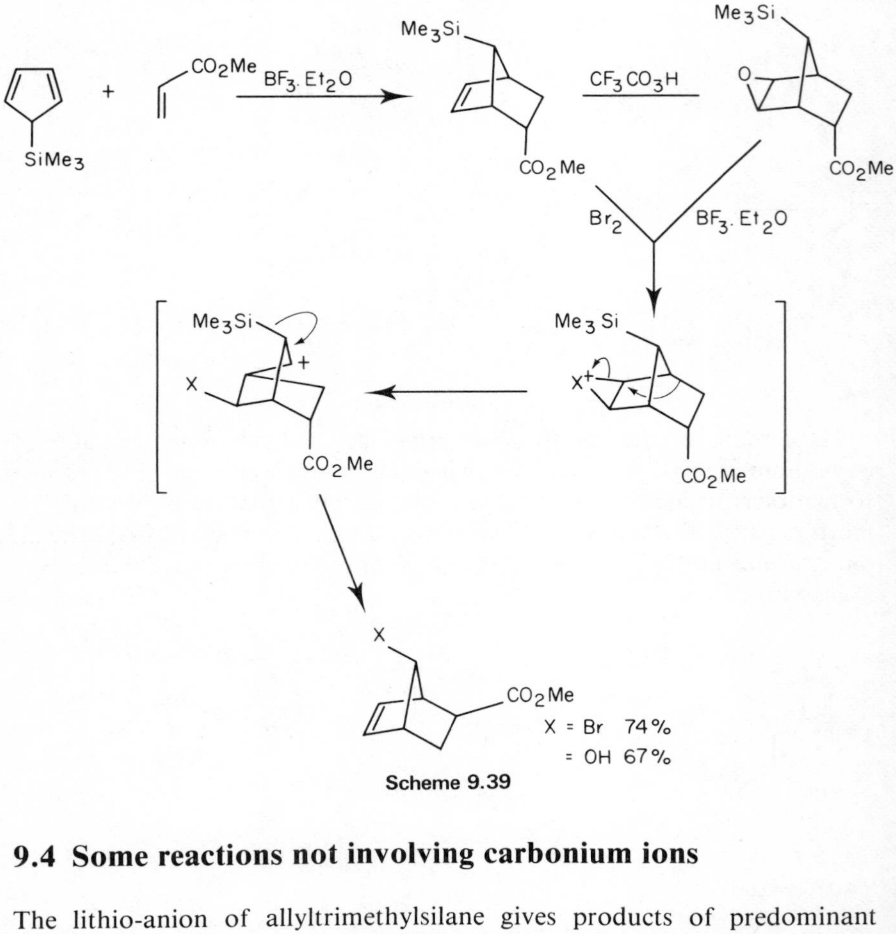

Scheme 9.39

9.4 Some reactions not involving carbonium ions

The lithio-anion of allyltrimethylsilane gives products of predominant γ-attack[70–74] in its reactions with carbonyl electrophiles. Change of the

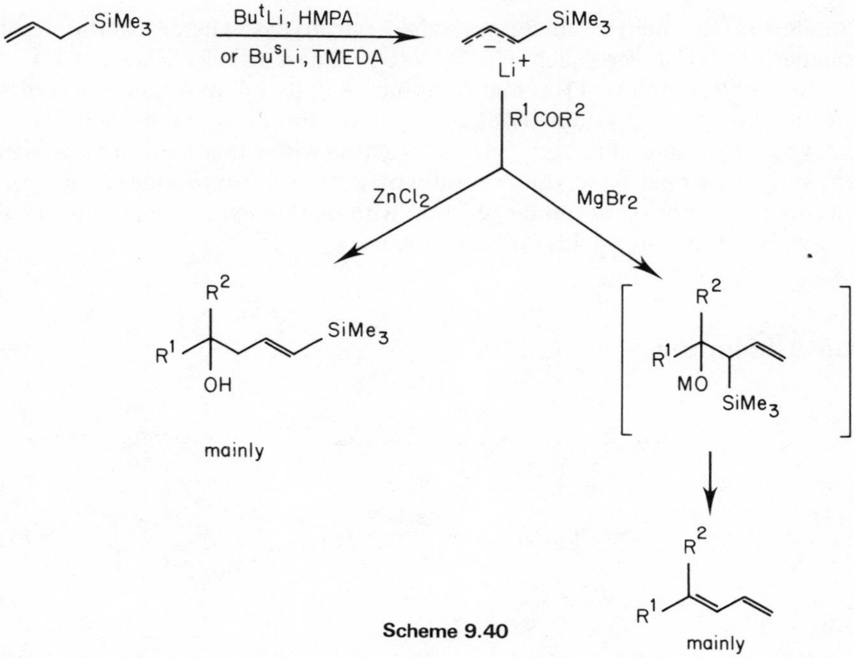

SiMe$_3$ — Bu^{t}Li, HMPA or Bu^{s}Li, TMEDA → SiMe$_3$, Li$^+$

R^1COR^2

ZnCl$_2$ — MgBr$_2$

R^2, R^1, OH, SiMe$_3$

mainly

R^2, R^1, MO, SiMe$_3$

R^2, R^1

mainly

Scheme 9.40

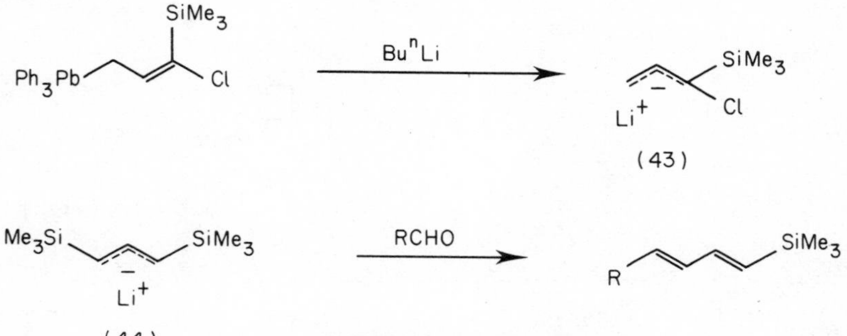

SiMe$_3$, Ph$_3$Pb, Cl — Bu^{n}Li → SiMe$_3$, Cl, Li$^+$ (43)

Me$_3$Si, SiMe$_3$, Li$^+$ (44) — RCHO → R, SiMe$_3$

Scheme 9.41

SiMe$_3$ — O$_2$ or Bu^{t}OOH, (Ph$_3$P)$_3$ RhCl → O, SiMe$_3$

SiMe$_3$ + O=C(CO$_2$Et)$_2$ → HO, CO$_2$Et, CO$_2$Et, SiMe$_3$ → → CO$_2$H, SiMe$_3$

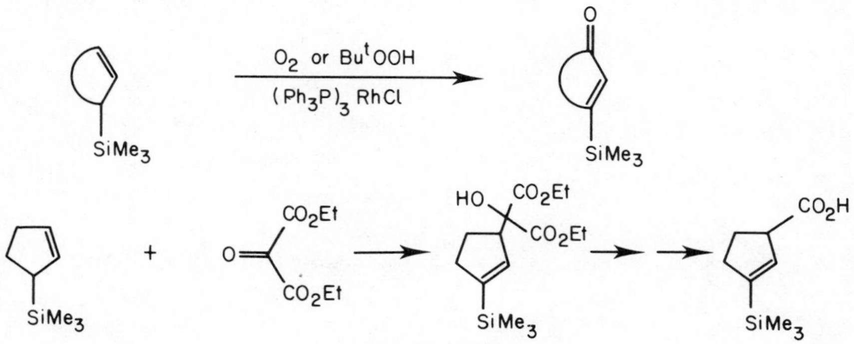

Scheme 9.42

counterion to zinc(II)[75] increases this γ-selectivity, whereas addition of magnesium(II)[76] inverts such selectivity to favour α-attack (*Scheme 9.40*).

The α-chloro-anion (43) is highly ambident in its site of attack[77], whereas the anion (44), being symmetrical, gives unique products[33] (*Scheme 9.41*).

Cyclic allylsilanes undergo[78] an 'ene' reaction with either molecular oxygen or t-butyl hydroperoxide, under rhodium(I) catalysis, to produce β-silyl $\alpha\beta$-unsaturated ketones. A similar reaction with diethyl oxomalonate[79] leads to γ-silyl $\beta\gamma$-unsaturated acids (*Scheme 9.42*).

9.5 Addendum

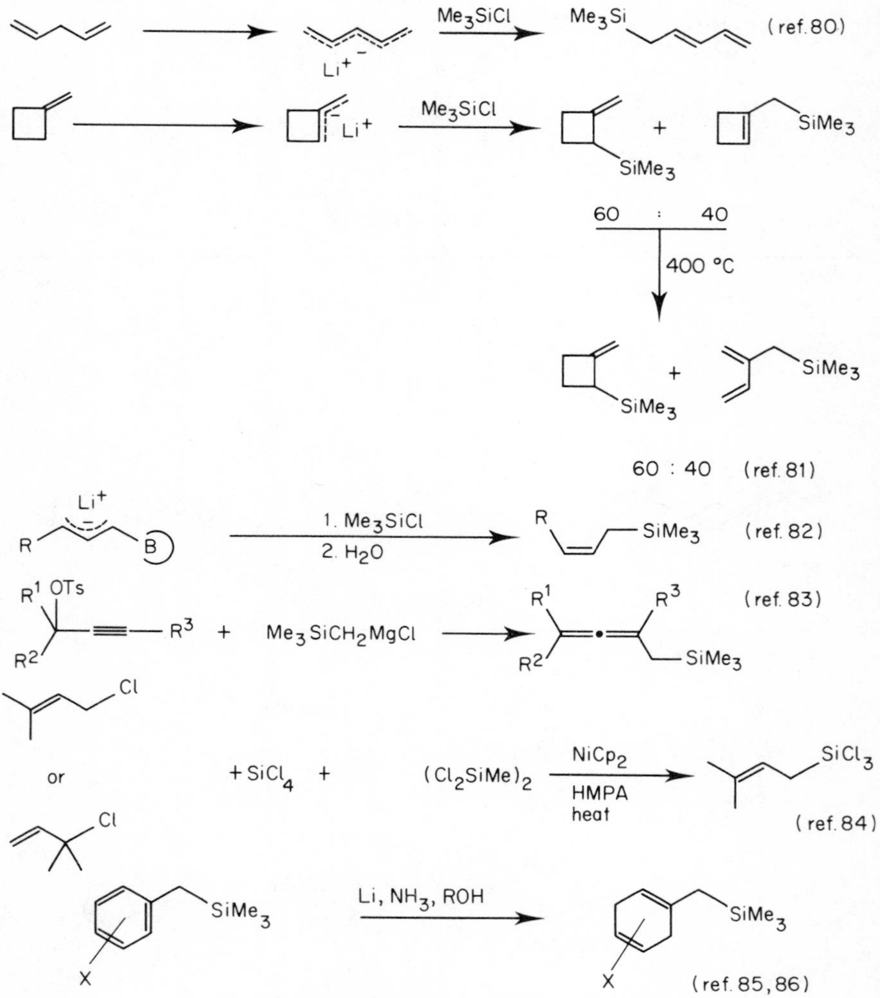

Scheme 9.43 Additional routes to allylsilanes

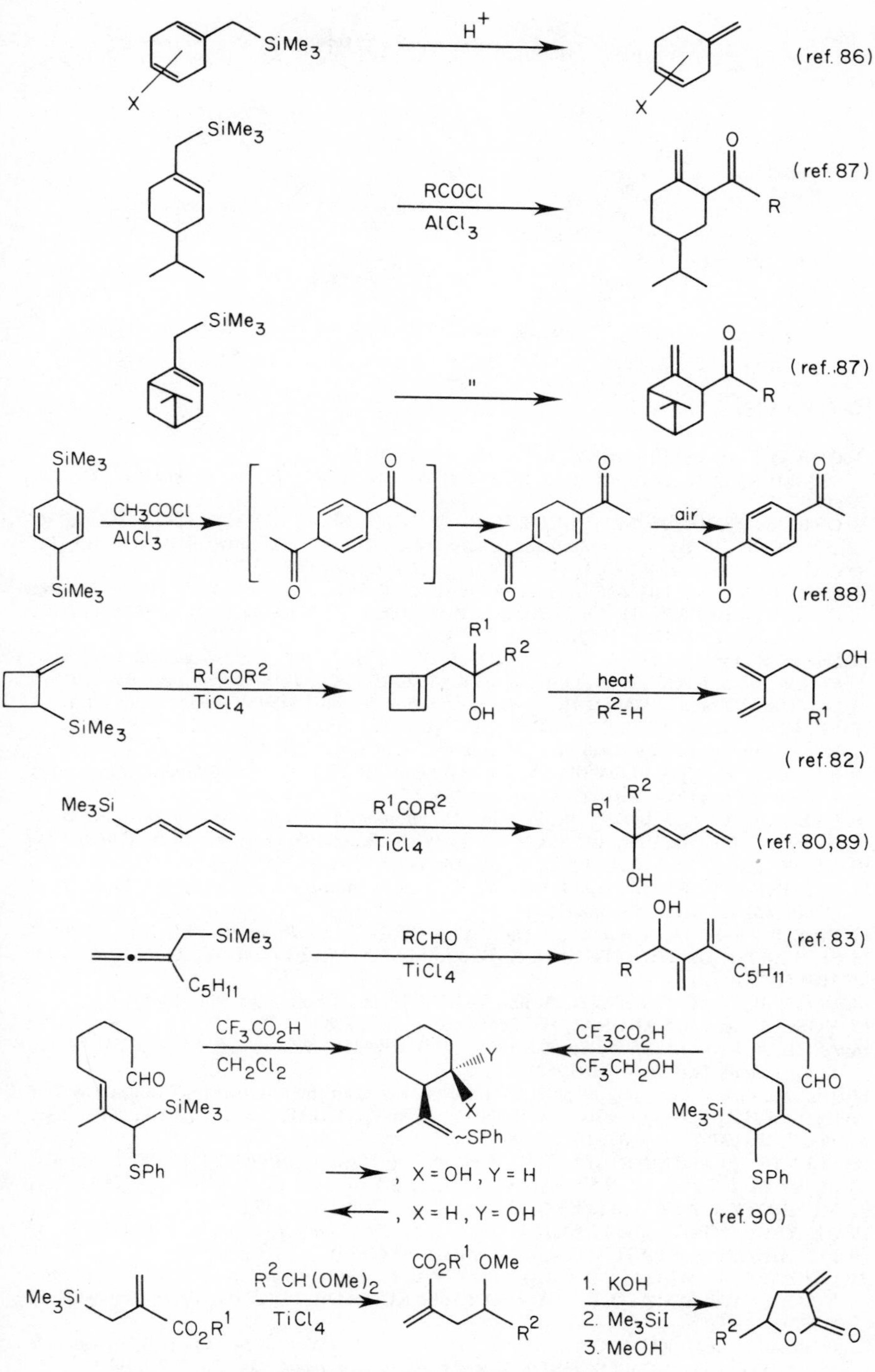

H^+ (ref. 86)

RCOCl, $AlCl_3$ (ref. 87)

" (ref. 87)

CH_3COCl, $AlCl_3$; air (ref. 88)

R^1COR^2, $TiCl_4$; heat, R^2 = H (ref. 82)

R^1COR^2, $TiCl_4$ (ref. 80,89)

RCHO, $TiCl_4$ (ref. 83)

CF_3CO_2H, CH_2Cl_2; CF_3CO_2H, CF_3CH_2OH

→ , X = OH, Y = H

← , X = H, Y = OH

(ref. 90)

$R^2CH(OMe)_2$, $TiCl_4$; 1. KOH 2. Me_3SiI 3. MeOH

Scheme 9.44 Reactions with electrophiles (ref. 91)

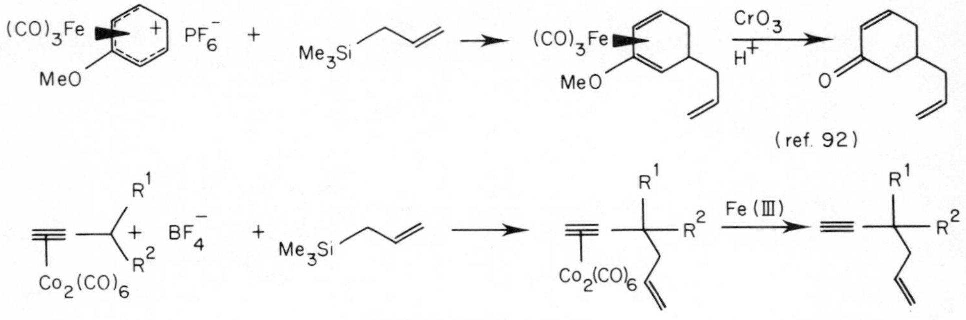

(ref. 92)

(ref. 92)

Scheme 9.44 (*continued*)

References

1 CHAN, T. H. and FLEMING, I., *Synthesis* 761 (1979)
2 SOMMER, L. H., TYLER, L. J. and WHITMORE, F. C., *J. Am. chem. Soc.* **70,** 2872 (1948)
3 GRIGNON-DUBOIS, M., PILLOT, J.-P., DUNOGUÈS, J., DUFFAUT, N., CALAS, R. and HENNER, B., *J. organometal. Chem.* **127,** 135 (1977); *see also* BOURGEOIS, P., CALAS, R. and MERAULT, G., *J. organometal. Chem.* **141,** 23 (1977)
4 LAGUERRE, M., DUNOGUÈS, J. and CALAS, R., *Tetrahedron Lett.* 57 (1978); *see also* PANDY-SZEKERES, D., DÉLÉRIS, G., PICARD, J.-P., PILLOT, J.-P. and CALAS, R., *Tetrahedron Lett.* 4267 (1980)
5 *See,* for example, SCHWEIG, A., WEIDER, U. and MANUEL, G., *J. organometal. Chem.* **67,** C4 (1974); HARTMAN, G. D. and TRAYLOR, T. G., *Tetrahedron Lett.* 939 (1975)
6 SLUTSKY, J. and KWART, H., *J. Am. chem. Soc.* **95,** 8678 (1973); *see* footnote 28(a)
7 PETROV, A. D., MIRONOV, B. F., PONOMARENKO, V. A. and CHERNYSHEV, E. A., *'Synthesis of Organosilicon Monomers',* Heywood, London (1964)
8 BAZANT, V., CHVALOVSKY, V. and RATHOUSKY, J., *'Organosilicon Compounds',* Academic Press, New York and London (1965)
9 EABORN, C. and BOTT, R. W., in *'Organometallic Compounds of the Group IV Elements',* Ed. MacDiarmid, A. G., vol. 1, part 1, Marcel Dekker, New York (1968)
10 HENRY, M. C. and NOLTES, J. G., *J. Am. chem. Soc.* **82,** 555 (1960)
11 MIRONOV, V. F. and NEPOMNINA, V. V., *Izv. Akad. Nauk SSSR, Ser. khim.* 1419 (1960); *Chem. Abstr.* **55,** 358 (1961)
12 PILLOT, J.-P., DUNOGUÈS, J. and CALAS, R., *Tetrahedron Lett.* 1871 (1976)
13 SEYFERTH, D., WURSTHORN, K. R. and MAMMARELLA, R. E., *J. org. Chem.* **42,** 3104 (1977)
14 HOSOMI, A., SHIRAHATA, A. and SAKURAI, H., *Chem. Lett.* 901 (1978)
15 HOSOMI, A. and SAKURAI, H., *Tetrahedron Lett.* 2589 (1978)
16 ASHE, A. J., *J. Am. chem. Soc.* **92,** 1233 (1970); KRAIHANZEL, C. S. and LOSEE, M. L., *J. Am. chem. Soc.* **90,** 4701 (1968)
17 The rate of hydrogen migration is 10^6 times slower than the rate of Me_3Si migration. For [1,2]- or [1,5]-silatropic shifts, *see* BONNY, A. and STOBART, S. R., *J. Am. chem. Soc.* **101,** 2247 (1979); SPANGLER, C. W., *Chem. Rev.* **76,** 187 (1976)
18 FLEMING, I. and MICHAEL, J. P., *J. chem. Soc. chem. Communs* 245 (1978); for similar use of nitroethene *see* RANGANATHAN, D., RAO, C. B., RANGANATHAN, S., MEHROTRA, A. K. and IYENGAR, R., *J. org. Chem.* **45,** 1185 (1980)
19 AU-YEUNG, B.-W. and FLEMING, I., *J. chem. Soc. chem. Communs* 79, 81 (1977)
20 FLEMING, I. and PATERSON, I., *Synthesis* 446 (1979)
21 OJIMA, I., KUMAGAI, M. and MIYAZAWA, Y., *Tetrahedron Lett.* 1385 (1977); *see also* KISO, Y., YAMAMOTO, K., TAMAO, K. and KUMADA, M., *J. Am. chem. Soc.* **94,** 4373 (1972)
22 OJIMA, I. and KUMAGAI, M., *J. organometal. Chem.* **134,** C6 (1977); *see also* WRIGHTON, M. S. and SCHROEDER, M. A., *J. Am. chem. Soc.* **96,** 6235 (1974)
23 FLEMING, I. and PERCIVAL, A., *J. chem. Soc. chem. Communs* 681 (1976); 178 (1978)

24 CALAS, R. and DUNOGUÈS, J., *Organometal. Chem. Rev.* **2,** 277 (1976)
25 DUNOGUÈS, J., CALAS, R., DÉDIER, J. and PISCIOTTI, F., *J. organometal. Chem.* **25,** 51 (1970)
26 LAGUERRE, M., DUNOGUÈS, J., CALAS, R. and DUFFAUT, N., *J. organometal. Chem.* **112,** 49 (1976); *see also* BIRKOFER, L. and RAMADAN, N., *Chem. Ber.* **104,** 138 (1971)
27 LAGUERRE, M., DUNOGUÈS, J. and CALAS, R., *Tetrahedron Lett.* 57 (1978)
28 ITOH, K., FUKUI, M. and KURACHI, Y., *J. chem. Soc. chem. Communs* 500 (1977)
29 FRAINNET, E., *Bull. Soc. chim. Fr.* 1441 (1959); FRAINNET, E. and CALAS, R., *C.r. hebd. Séanc Acad. Sci., Paris* **240,** 203 (1955)
30 FLEMING, I., PEARCE, A. and SNOWDEN, R. L., *J. chem. Soc. chem. Communs* 182 (1976); *see also* HUGHES, L. R., SCHMIDT, R. and JOHNSON, W. S., *Bio-org. Chem.* **8,** 513 (1979)
31 HOSOMI, A., HASHIMOTO, H. and SAKURAI, H., *J. org. Chem.* **43,** 2551 (1978)
32 SADYKH-ZADE, S. I. and PETROV, A. D., *J. gen. Chem. U.S.S.R.* **28,** 1591 (1958)
33 CARTER, M. J. and FLEMING, I., *J. chem. Soc. chem. Communs* 679 (1976)
34 STORK, G., JUNG, M., COLVIN, E. and NOEL, Y., *J. Am. chem. Soc.* **96,** 3684 (1974)
35 HUDRLIK, P. F. and WITHERS, G. P., *Tetrahedron Lett.* 29 (1976)
36 EVANS, D. A. and ANDREWS, G. C., *Accts chem. Res.* **7,** 147 (1974)
37 (*a*) FLEMING, I. and AU-YEUNG, B.-W., *Tetrahedron* **37** (1980); (*b*) WETTER, H., SCHERER, P. and SCHWEIZER, W. B., *Helv. chim. Acta* **62,** 1985 (1979)
38 DÉLÉRIS, G., DUNOGUÈS, J. and CALAS, R., *J. organometal. Chem.* **116,** C45 (1976)
39 MUKAIYAMA, T., *Angew Chem. int. Edn* **16,** 817 (1977)
40 HOSOMI, A. and SAKURAI, H., *Tetrahedron Lett.* 1295 (1976)
41 HOSOMI, A., ENDO, M. and SAKURAI, H., *Chemy Lett.* 941 (1976)
42 HOSOMI, A. and SAKURAI, H., *Tetrahedron Lett.* 2589 (1978)
43 OJIMA, I., MIYAZAWA, Y. and KUMAGAI, M., *J. chem. Soc. chem. Communs* 927 (1976)
44 HOSOMI, A., SAITO, M. and SAKURAI, H., *Tetrahedron Lett.* 429 (1979); for use as a regioselective Diels–Alder diene, *see* HOSOMI, A., SAITO, M. and SAKURAI, H., *Tetrahedron Lett.* 355 (1980)
45 OJIMA, I., KUMAGAI, M. and MIYAZAWA, Y., *Tetrahedron Lett.* 1385 (1977)
46 ITOH, K., FUKUI, M. and KURACHI, Y., *J. chem. Soc. chem. Communs* 500 (1977); ITOH, K., YOGO, T. and ISHII, Y., *Chemy Lett.* 103 (1977)
47 HOSOMI, A. and SAKURAI, H., *J. Am. chem. Soc.* **99,** 1673 (1977)
48 YANAMI, T., MIYASHITA, M. and YOSHIKOSHI, A., *J. chem. Soc. chem. Communs* 525 (1979); *J. org. Chem.* **45,** 607 (1980); *see also* HOSOMI, A., KOBAYASHI, H. and SAKURAI, H., *Tetrahedron Lett.* 955 (1980)
49 HOSOMI, A. and SAKURAI, H., *Tetrahedron Lett.* 4041 (1977)
50 STILL, W. C. and MACDONALD, T. L., *J. Am. chem. Soc.* **96,** 5561 (1974); *see also* WRIGHT, A., LING, D., BOUDJOUK, P. and WEST, R., *J. Am. chem. Soc.* **94,** 4784 (1972)
51 BROOK, A. G., *Accts chem. Res.* **7,** 77 (1974)
52 STILL, W. C., *J. org. Chem.* **41,** 3063 (1976)
53 DÉLÉRIS, G., DUNOGUÈS, J. and CALAS, R., *J. organometal. Chem.* **93,** 43 (1975)
54 CALAS, R., DUNOGUÈS, J., BIRAN, C., PISCIOTTI, F. and ARREGUY, B., *J. organometal. Chem.* **85,** 149 (1975)
55 OJIMA, I. and KUMAGAI, M., *Chemy Lett.* 575 (1978)
56 TSUNODA, T., SUZUKI, M. and NOYORI, R., *Tetrahedron Lett.* 71 (1980)
57 SASAKI, T., USUKI, A. and OHNO, M., *Tetrahedron Lett.* 4925 (1978)
58 HOSOMI, A., SHIRATA, A. and SAKURAI, H., *Tetrahedron Lett.* 3043 (1978)
59 SARKAR, T. K. and ANDERSEN, N. H., *Tetrahedron Lett.* 3513 (1978); *see also* TROST, B. M. and VINCENT, J. E., *J. Am. chem. Soc.* **102,** 5680 (1980)
60 TROST, B. M. and CHAN, D. M. T., *J. Am. chem. Soc.* **101,** 6432 (1979); **102,** 6359 (1980); *see also* KNAPP, S., O'CONNOR, U. and MOBILIO, D., *Tetrahedron Lett.* 4557 (1980)
61 TROST, B. M., STREGE, P. L., WEBER, L., FULLERTON, T. J. and DIETSCHE, T. J., *J. Am. chem. Soc.* **100,** 3407 (1978); TROST, B. M., WEBER, L., STREGE, P. L., FULLERTON, T. J. and DIETSCHE, T. J., *J. Am. chem. Soc.* **100,** 3416, 3426 (1978); TROST, B. M. and Ver HOEVEN, T. R., *J. Am. chem. Soc.* **100,** 3435 (1978)
62 JUNG, M. E. and BLUMENKOPF, T. A., *Tetrahedron Lett.* 3657 (1978)
63 GRAFSTEIN, D., *J. Am. chem. Soc.* **77,** 6650 (1955)

64 DUNOGUÈS, J., CALAS, R. and ARDOIN, N., *J. organometal. Chem.* **43,** 127 (1972)
65 SAKURAI, H., IMAI, T. and HOSOMI, A., *Tetrahedron Lett.* 4045 (1977)
66 GRIGNON-DUBOIS, M., DUNOGUÈS, J. and CALAS, R., *Synthesis* 737 (1976)
67 WARREN, S., *Accts chem. Res.* **11,** 401 (1978)
68 DAVIDSON, A. H., EARNSHAW, C., GRAYSON, J. I. and WARREN, S., *J. chem. Soc. Perkin I* 1452 (1977); DAVIDSON, A. H., FLEMING, I., GRAYSON, J. I., PEARCE, A., SNOWDEN, R. L. and WARREN, S., *J. chem. Soc. Perkin I* 550 (1977)
69 BROWNBRIDGE, P., FLEMING, J., PEARCE, A. and WARREN, S., *J. chem. Soc. chem. Communs* 751 (1976); BROWNBRIDGE, P. and WARREN, S., *J. chem. Soc. Perkin I* 1131 (1977)
70 STORK, G. and COLVIN, E., unpublished observations
71 CORRIU, R. J. P., LANNEAU, G. F., LECLERQ, D. and SAMATE, D., *J. organometal. Chem.* **144,** 155 (1978)
72 AYALON-CHASS, D., EHLINGER, E. and MAGNUS, P., *J. chem. Soc. chem. Communs* 772 (1977)
73 EHLINGER, E. and MAGNUS, P., *J. chem. Soc. chem. Communs* 421 (1979)
74 YAMAMOTO, K., OHTA, M. and TSUJI, J., *Chemy Lett.* 713 (1979)
75 EHLINGER, E. and MAGNUS, P., *Tetrahedron Lett.* 11 (1980); *J. Am. chem. Soc.* **102,** 5004 (1980)
76 LAU, P. W. K. and CHAN, T. H., *Tetrahedron Lett.* 2383 (1978)
77 MAMMARELLA, R. E. and SEYFERTH, D., *J. organometal. Chem.* **156,** 279 (1978)
78 REUTER, J. M., SINHA, A. and SALOMON, R. G., *J. org. Chem.* **43,** 2438 (1978)
79 SALOMON, M. F., PARDO, S. N. and SALOMON, R. G., *J. Am. chem. Soc.* **102,** 2473 (1980); *see also* GOPALAN, A., MOERCK, R. and MAGNUS, P., *J. chem. Soc. chem. Communs* 548 (1979)
80 SEYFERTH, D. and PORNET, J., *J. org. Chem.* **45,** 1721 (1980)
81 YATAGAI, H., YAMAMOTO, Y. and MURUYAMA, K., *J. Am. chem. Soc.* **102,** 4548 (1980)
82 WILSON, S. R., PHILLIPS, L. R. and NATALIE, K. J., *J. Am. chem. Soc.* **101,** 3340 (1979)
83 MONTURY, M., PSAUME, B. and GORÉ, J., *Tetrahedron Lett.* 163 (1980)
84 LEFORT, M., SIMMONET, C., BIROT, M., DÉLÉRIS, G., DUNOGUÈS, J. and CALAS, R., *Tetrahedron Lett.* 1857 (1980)
85 EABORN, C., JACKSON, R. A. and PEARCE, R., *J. chem. Soc. Perkin (I)* 470 (1975)
86 COUGHLIN, D. J. and SALOMON, R. G., *J. org. Chem.* **44,** 3784 (1979)
87 PILLOT, J.-P., DÉLÉRIS, G., DUNOGUÈS, J. and CALAS, R., *J. org. Chem.* **44,** 3397 (1979)
88 LAGUERRE, M., DUNOGUÈS, J. and CALAS, R., *Tetrahedron Lett.* 831 (1980)
89 PORNET, J., *Tetrahedron Lett.* 2049 (1980); HOSOMI, A., SAITO, M. and SAKURAI, H., *Tetrahedron Lett.* 3783 (1980)
90 ITOH, A., OSHIMA, K. and NOZAKI, H., *Tetrahedron Lett.* 1783 (1979)
91 HOSOMI, A., HASHIMOTO, H. and SAKURAI, H., *Tetrahedron Lett.* 951 (1980)
92 KELLY, L. F., NARULA, A. S. and BIRCH, A. J., *Tetrahedron Lett.* 871, 2455 (1980)
93 O'BOYLE, J. E. and NICHOLAS, K. M., *Tetrahedron Lett.* 1595 (1980)

Chapter 10

Arylsilanes

10.1 Preparation

The preparation of aryl- and heteroaryl-trimethylsilanes has been comprehensively surveyed[1]. In practice, such silanes are obtained either by transmetallation[2-5] of an aryl organometallic with chlorotrimethylsilane, or by cycloaddition[6] of diynes with alkynyl silanes. A subsidiary approach is nuclear modification of an arylsilane obtained by one of the above techniques. Since the area has been well reviewed, it would be inappropriate to go into extensive detail here. Instead, the various general approaches will be outlined only; for particular information, the sources already cited are recommended.

The most generally useful route to arylsilanes involves reaction of a preformed aryl organometallic species with chlorotrimethylsilane. The requisite organometallics can be obtained in a variety of ways; with but one exception all require aryl halides as precursors. The exception, also shown in *Scheme 10.1*, is generation by proton abstraction from a suitably activated, e.g., methoxy-substituted, arene.

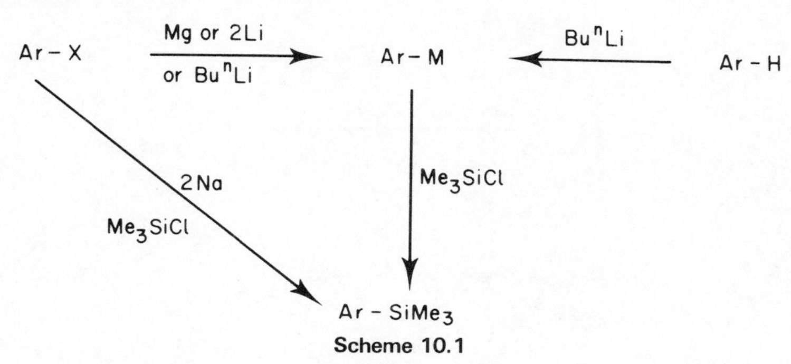

Scheme 10.1

All of these routes proceed with complete retention of orientation of the original substitution, or of regioselectivity of proton removal. Their synthetic scope is limited only by the availability of the requisite aryl halide or activated arene. The subsidiary approach, that of modifying an already silylated arene, can give mixtures of substituted and *ipso*-desilylated products

(*Scheme 10.2*), and is accordingly much less satisfactory for such purposes. The reasons for *ipso*-desilylation occurring on electrophilic attack will be discussed shortly, as will its great potential for the synthesis of regiospecifically substituted arenes.

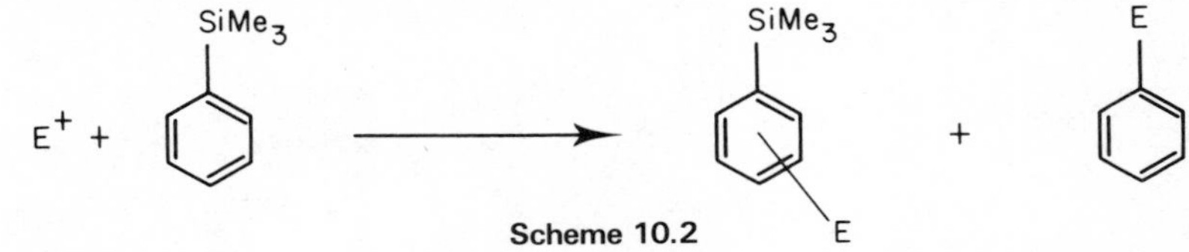

Scheme 10.2

A reductive addition/oxidative rearomatization reaction of somewhat limited scope, but of commendable simplicity, leads[7,8] directly from benzene to 1,4-bis(trimethylsilyl)benzene (*Scheme 10.3*).

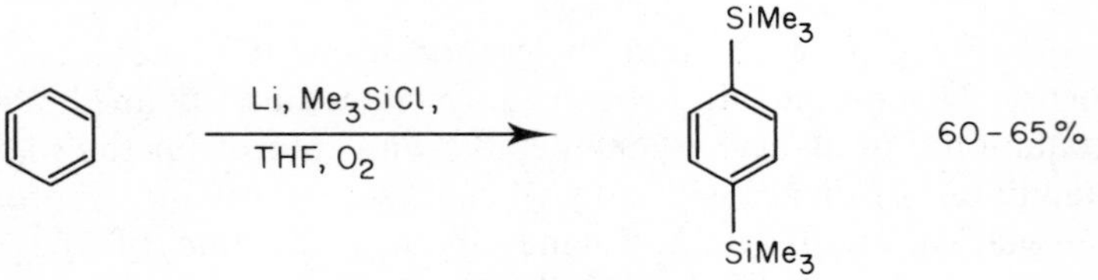

Scheme 10.3

The other major mode of construction of arylsilanes utilizes cycloaddition processes to create the aromatic system. The most dramatic examples of this technique are to be seen in the work of Vollhardt[6,9] and his collaborators, who have studied the cobalt-catalysed co-oligomerization of $\alpha\omega$-diynes with bis(trimethylsilyl)ethyne and other related monoalkynes (*Scheme 10.4*). Owing to steric congestion, the heavily substituted monoalkynes employed are most reluctant to autocyclize, although when relatively large amounts of cobalt catalyst are employed, a variety of autocyclization products can be obtained[10].

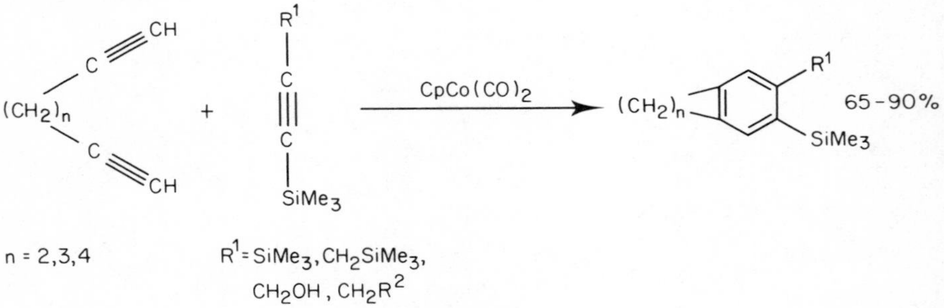

Scheme 10.4

10.2 Electrophile-induced desilylation

Aryl carbon–silicon bonds are readily cleaved[11] by a variety of electrophilic reagents, many examples existing in the early organosilicon literature[12]. Two such examples are shown in *Scheme 10.5*.

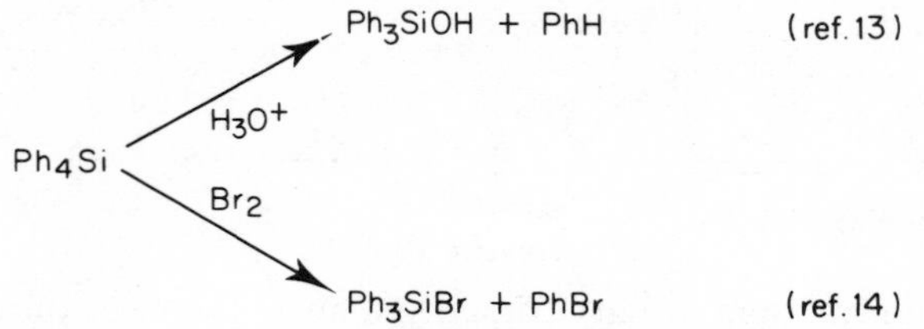

$Ph_4Si \xrightarrow{H_3O^+} Ph_3SiOH + PhH$ (ref. 13)

$Ph_4Si \xrightarrow{Br_2} Ph_3SiBr + PhBr$ (ref. 14)

Scheme 10.5

Extensive studies by Eaborn[15] and others have led to the conclusion that such cleavages occur by the same mechanism as that of electrophilic aromatic substitution, with the difference that in the second step a carbon–silicon bond is broken in the sense C^-Si^+, rather than a carbon–hydrogen bond in the direction C^-H^+. Just as aromatic substitutions involve intermediate delocalized cations such as (1), arylsilanes can react via delocalized cations such as (2). Electrophilic attack occurs at the ring carbon carrying the silyl group, i.e., at the *ipso* position, because of the stabilization offered to an adjacent carbonium ion by the carbon–silicon bond (the β-effect, Chapter 3). A β-carbonium ion would certainly be generated by electrophilic attack at the *meta* position, but the stabilizing coplanarity of the carbon–silicon bond and the vacant p (or π) orbital cannot be achieved in such a species (3), the orbitals involved being orthogonal to one another. The overall result of such regio-controlled attack is one of *ipso*-substitution (*Scheme 10.6*).

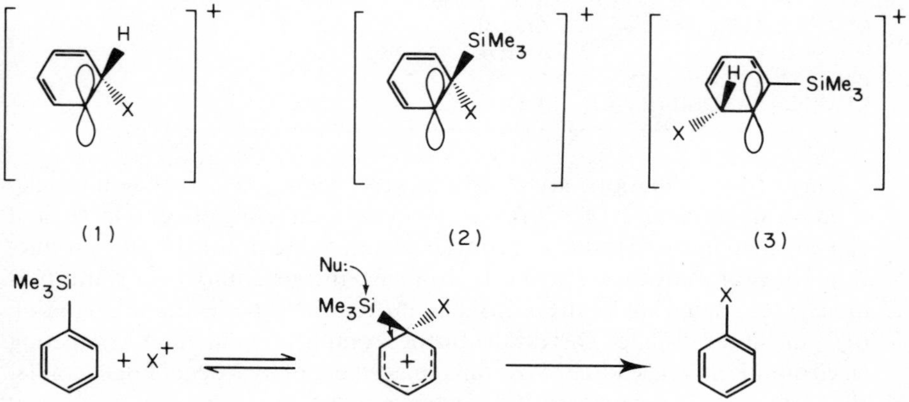

(1) (2) (3)

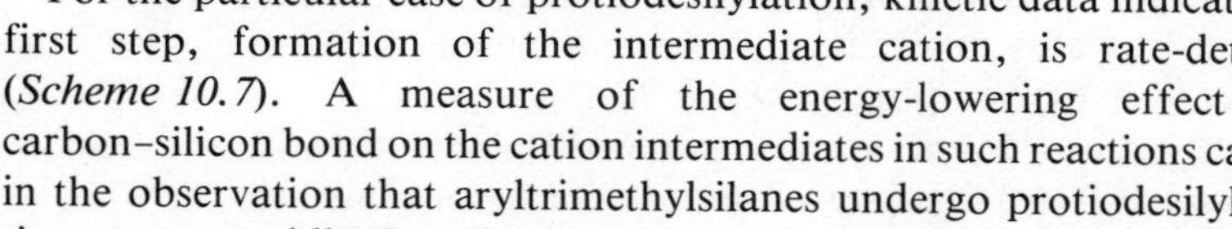

Scheme 10.6

For the particular case of protiodesilylation, kinetic data indicate that the first step, formation of the intermediate cation, is rate-determining (*Scheme 10.7*). A measure of the energy-lowering effect of the carbon–silicon bond on the cation intermediates in such reactions can be seen in the observation that aryltrimethylsilanes undergo protiodesilylation 10^4 times more rapidly than do the parent arenes undergo hydrogen exchange. Similarly, phenyltriethylsilane is hydrolysed by dilute aqueous acid to benzene 10^4 times more rapidly than the all-carbon analogue, t-heptylbenzene, is hydrolysed. As one might expect, aryltriphenylsilanes are cleaved

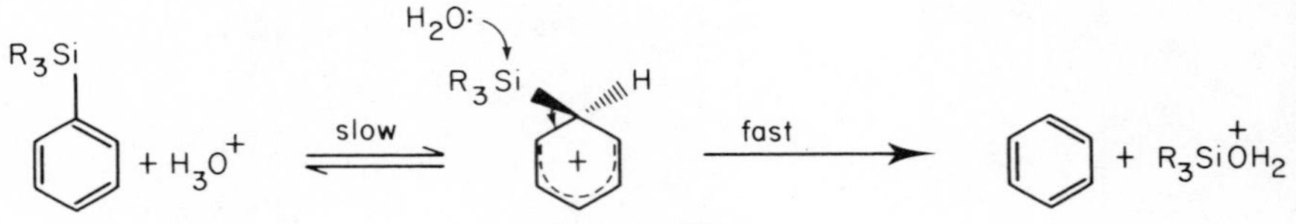

Scheme 10.7

much less readily[16] than are aryltrimethylsilanes, the electron-withdrawing effect of the phenyl groups reducing the ability of the carbon–silicon bond to stabilize a developing β-cation.

With most electrophilic reagents, the electrophile effects overall replacement of silicon more rapidly than it does of hydrogen, with resulting *ipso*-substitution. A selection of such electrophiles is given in *Table 10.1*.

Table 10.1 Aromatic electrophilic *ipso*-desilylation

$$Ar-SiR_3 + E^+ \longrightarrow Ar-E$$

Electrophilic system	*Product type*	*References*
H^+	Ar–H	17
Br_2	Ar–Br	18
I_2	Ar–I	18, 19, 21
ICl	Ar–I	20, 21
CNCl, $AlCl_3$	Ar–CN	22
Cl_2, Fe	Ar–Cl	23
SO_3	$ArSO_3SiMe_3$	24
R^1SO_2Cl, $AlCl_3$	$ArSO_2R^1$	25
R^1COCl, $AlCl_3$	$ArCOR^1$	26
$Pb(OCOCF_3)_4$	ArOH	27
KF, MCPBA, DMF(R = Cl)	ArOH	28

These transformations are clearly of great value, in that they normally produce single isomers. The strength of the *ipso*-directing effect is large, and can direct addition to rather unpromising sites, as exemplified by the production[29] of *o*-bromobenzoic acid (4). However, the generality of this utility is moderated somewhat by the occasional difficulties experienced in preparing the requisite arylsilane. There is the further complicating factor of competing electrophilic attack leading to overall replacement of hydrogen. For example, the MeO grouping is apparently[19] a stronger *ortho–para*-directing substituent in bromination than is the Me_3Si grouping an *ipso*-director and yet, in acylation[26], clean *ipso*-substitution is reported to take place (*Scheme 10.8*).

Nitration can also take an unpredictable course, as illustrated by the contrasting pairs of examples in *Scheme 10.9*.

However, such drawbacks are minor when compared with the demonstrably broad utility of this *ipso*-substitution technique. *Scheme 10.10* shows one representative example of this utility: the Calas and Dunoguès group[34] have synthesized regiospecifically a wide range of disubstituted benzene derivatives by stepwise *ipso*-replacement of the trimethylsilyl groups of *o*-, *m*-, and *p*-bis(trimethylsilyl)benzene.

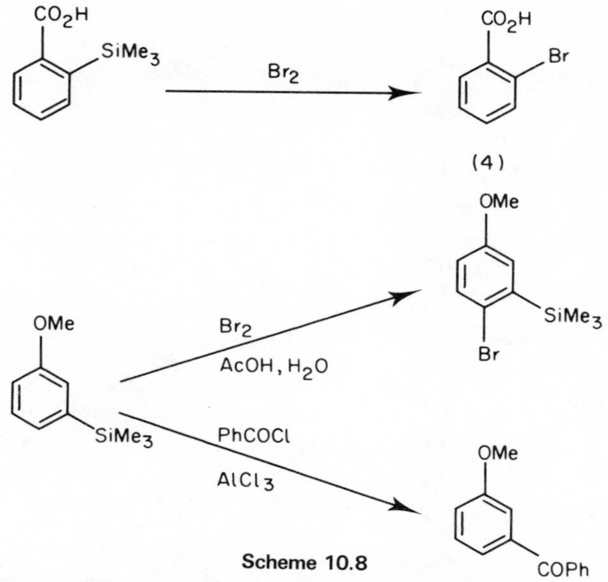

(4)

Scheme 10.8

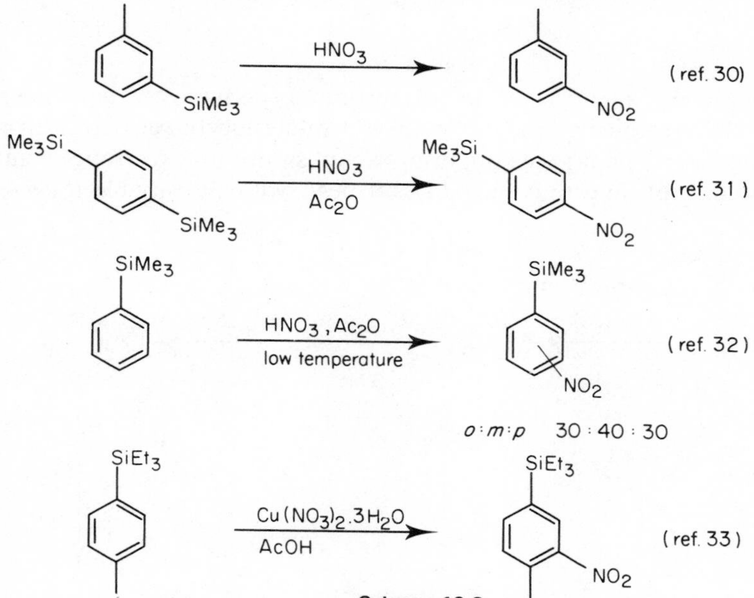

(ref. 30)

(ref. 31)

(ref. 32)

o:*m*:*p* 30 : 40 : 30

(ref. 33)

Scheme 10.9

As discussed earlier, the cobalt-catalysed co-oligomerization of $\alpha\omega$-diynes with bis(trimethylsilyl)ethyne and related species is a most fruitful route to arylsilanes. A particular example[35] is shown in *Scheme 10.11*. Kinetic data on the deuteriodesilylation of the benzocyclobutene (5) so produced indicate that, owing to steric acceleration, the first displacement of a silyl group occurs 36–42 times more rapidly than does the second. Selective stepwise

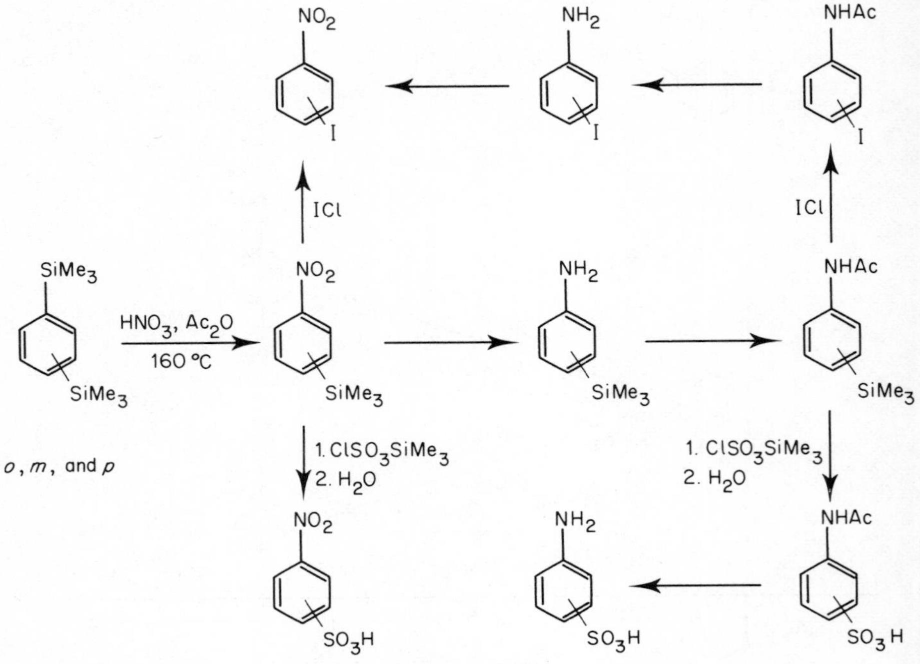

Scheme 10.10

ipso-replacement then leads to disubstituted derivatives, as shown. Additionally, treatment of the species (5) with dilute aqueous acid, or with an electrophile with a poor gegenion, brings about substitutive rearrangement[37] to compound (6), and thus introduces a new synthetic variable (*Scheme 10.11*).

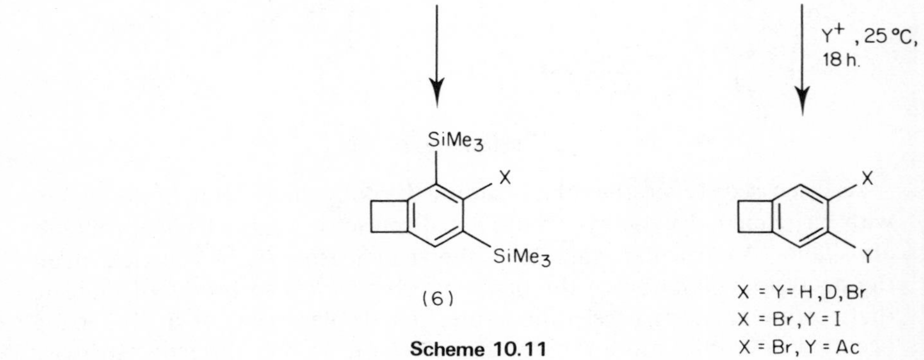

Scheme 10.11

The same methodology has been used for the production of 3,4-bis(trimethylsilyl)benzocyclobutene[36] (7), of the air-sensitive benzocyclobutadiene[38] (8), and of the highly strained tricyclic system[39] (9) (*Scheme 10.12*).

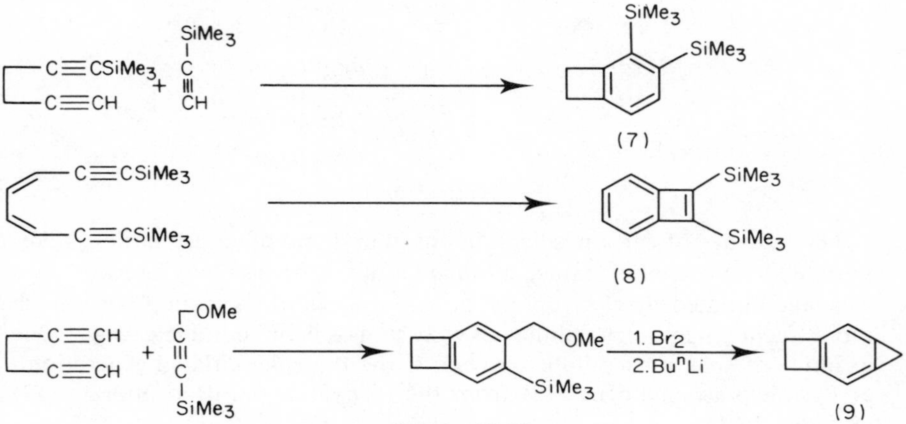

Scheme 10.12

The general technique reached its current peak in a total synthesis[40] of (±)-oestrone (10) (*Scheme 10.13*). Treatment of the bis-silylated steroid (11) with trifluoroacetic acid regioselectively produces the 2-protio-3-trimethylsilyl aromatic nucleus in 80 per cent yield. Oxidative aryl–oxygen fission then occurs almost quantitatively on reaction with lead(IV) trifluoroacetate[27] to yield (±)-oestrone by the shortest racemic synthesis of this compound reported to date.

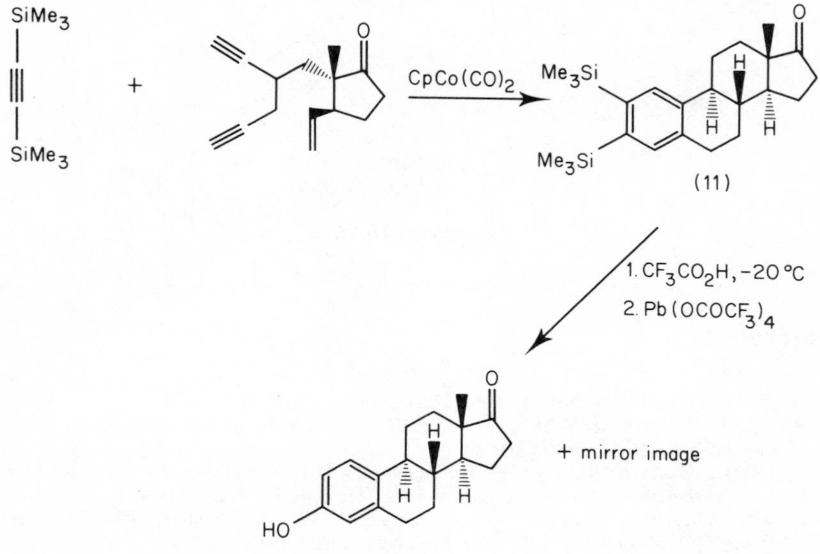

Scheme 10.13

The electrophile-induced desilylations discussed so far have all concentrated on the desilylated products. In contrast, a method[41] for the preparation of trimethylsilyl esters of strong acids focuses attention on the silicon-containing fragment (*Scheme 10.14*).

$$Ar{-}SiMe_3 + HX \longrightarrow Me_3SiX + ArH$$

$$X = C_nF_{2n-1}SO_3$$
$$= C_nF_{2n-1}CO_2$$
$$= F_2P(O)O$$

Scheme 10.14

The cleavage of carbon–silicon bonds in aryl and other aromatic systems initiated by nucleophilic attack at silicon is not nearly as ready a process as is cleavage initiated by electrophilic attack at carbon[4]. It is, however, much more facile than nucleophilic alkyl carbon–silicon bond cleavage. One example of such nucleophilic cleavage is the fluoride-initiated generation[42] of cycloheptatrienylidene (12) from the tropylium tetrafluoroborate (13) (*Scheme 10.15*).

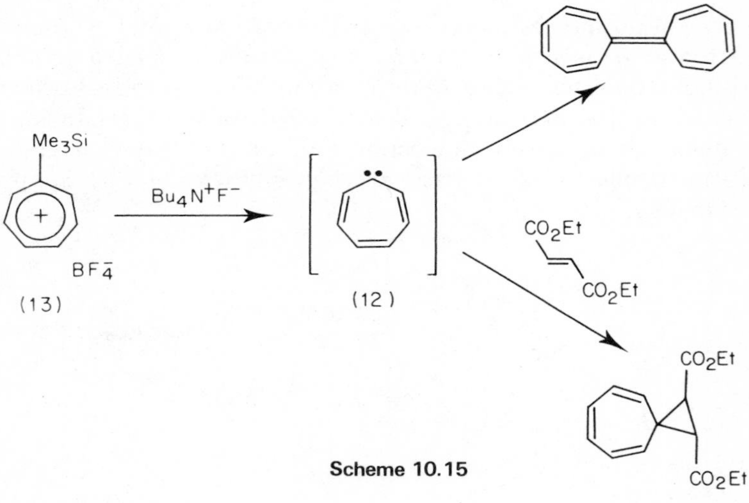

Scheme 10.15

References

1 HÄBICH, D. and EFFENBERGER, F., *Synthesis* 841 (1979)
2 BAZANT, V., CHVALOVSKY, V. and RATHOUSKY, J., *'Organosilicon Compounds'*, Academic Press, New York (1965)
3 PETROV, A. D., MIRONOV, B. F., PONOMARENKO, V. A. and CHERNYSHEV, E. A., *'Synthesis of Organosilicon Monomers'*, Heywood, London (1964)
4 EABORN, C. and BOTT, R. W., in *'Organometallic Compounds of the Group IV Elements'*, Ed. MacDiarmid, A. G., vol. 1, part 1, Marcel Dekker, New York (1968)
5 CALAS, R. and DUNOGUÈS, J., *Organometal. Chem. Rev., Organosilicon Rev.* **2**, 277 (1976)

6 VOLLHARDT, K. P. C., *Accts chem. Res.* **10,** 1 (1977); FUNK, R. L. and VOLLHARDT, K. P. C., *Chem. Soc. Rev.* **9,** 41 (1980)
7 NORMANT, H. and CUVIGNY, T., *Organometal. chem. Synth.* **1,** 223 (1971)
8 DUNOGUÈS, J., CALAS, R. and ARDOIN, N., *J. organometal. Chem.* **43,** 127 (1972)
9 GESING, E. R. F., SINCLAIR, J. A. and VOLLHARDT, K. P. C., *J. chem. Soc. chem. Communs* 286 (1980)
10 FRITCH, J. R., VOLLHARDT, K. P. C., THOMPSON, M. R. and DAY, V. W., *J. Am. chem. Soc.* **101,** 2768 (1979)
11 CHAN, T. H. and FLEMING, I., *Synthesis* 761 (1979)
12 KIPPING, F. S., *Proc. R. Soc. A.* **159,** 139 (1937)
13 KIPPING, F. S. and LLOYD, L. L., *Trans. chem. Soc.* **79,** 449 (1901)
14 LADENBURG, A., *Chem. Ber.* **40,** 2274 (1907)
15 EABORN, C., *J. organometal. Chem.* **100,** 43 (1975)
16 GILMAN, H. and NOBIS, J. F., *J. Am. chem. Soc.* **72,** 2629 (1950)
17 EABORN, C., JENKINS, I. D. and WALTON, D. R. M., *J. chem. Soc. Perkin II* 596 (1974)
18 PRAY, B. O., SOMMER, L. H., GOLDBERG, G. M., KERR, G. T., Di GIORGIO, P. A. and WHITMORE, F. C., *J. Am. chem. Soc.* **70,** 433 (1948)
19 EABORN, C. and WEBSTER, D. E., *J. chem. Soc.* 179 (1960)
20 STOCK, L. M. and SPECTOR, A. R., *J. org. Chem.* **28,** 3272 (1963)
21 FÉLIX, G., DUNOGUÈS, J., PISCIOTTI, F. and CALAS, R., *Angew. Chem. int. Edn* **16,** 488 (1977)
22 BARTLETT, E. H., EABORN, C. and WALTON, D. R. M., *J. organometal. Chem.* **46,** 267 (1972)
23 PONOMARENKO, V. A., SNEGOVA, A. D. and EGOROV, Yu. P., *Izv. Akad. Nauk SSSR, Ser khim.* 244 (1960); *Chem. Abstr.* **54,** 20932 (1960)
24 BOTT, R. W., EABORN, C. and HASHIMOTO, T., *J. organometal. Chem.* **3,** 442 (1965); CALAS, R., BOURGEOIS, P. and DUFFAUT, N., *C.r. hebd. Séanc Acad. Sci., Paris* **C263,** 243 (1966)
25 BATTACHARYA, S. N., EABORN, C. and HASHIMOTO, T., *J. chem. Soc. (C)* 1367 (1969)
26 DEY, K., EABORN, C. and WALTON, D. R. M., *Organometal. chem. Synth.* **1,** 151 (1970/71)
27 KALMAN, J. R., PINHEY, J. T. and STERNHELL, S., *Tetrahedron Lett.* 5369 (1972); BELL, H. C., KALMAN, J. R., PINHEY, J. T. and STERNHELL, S., *Tetrahedron Lett.* 853 (1974)
28 TAMAO, K., KAKUI, T. and KUMADA, M., *J. Am. chem. Soc.* **100,** 2268 (1978)
29 HASHIMOTO, T., *J. pharm. Soc. Japan* **87,** 528 (1967)
30 CHVALOVSKÝ, V. and BAZANT, V., *Colln czech. chem. Commun* **16,** 580 (1951)
31 DEANS, F. B. and EABORN, C., *J. chem. Soc.* 498 (1957)
32 SPEIER, J. L., *J. Am. chem. Soc.* **75,** 2930 (1953)
33 BENKESER, R. A. and BRUMFIELD, P. E., *J. Am. chem. Soc.* **73,** 4770 (1951); BENKESER, R. A. and LANDESMAN, H., *J. Am. chem. Soc.* **76,** 904 (1954)
34 FÉLIX, G., DUNOGUÈS, J. and CALAS, R., *Angew. Chem. int. Edn* **18,** 402 (1979)
35 AALBERGSBERG, W. G. L., BARKOVICH, A. J., FUNK, R. L., HILLARD, R. L. and VOLLHARDT, K. P. C., *J. Am. chem. Soc.* **97,** 5600 (1975); HILLARD, R. L. and VOLLHARDT, K. P. C., *J. Am. chem. Soc.* **99,** 4058 (1977)
36 HILLARD, R. L. and VOLLHARDT, K. P. C., *Angew. Chem. int. Edn* **16,** 399 (1977)
37 Cf. SEYFERTH, D. and WHITE, D. L., *J. Am. chem. Soc.* **94,** 3132 (1972); *see also* Chapter 5
38 VOLLHARDT, K. P. C. and YEE, L. S., *J. Am. chem. Soc.* **99,** 2010 (1977)
39 SAWARD, C. J. and VOLLHARDT, K. P. C., *Tetrahedron Lett.* 4539 (1975)
40 FUNK, R. L. and VOLLHARDT, K. P. C., *J. Am. chem. Soc.* **99,** 5483 (1977); **101,** 215 (1979); **102,** 5253 (1980)
41 HÄBICH, D. and EFFENBERGER, F., *Synthesis* 755 (1978)
42 REIFFEN, M. and HOFFMANN, R. W., *Tetrahedron Lett.* 1107 (1978)

Chapter 11

Organosilyl anions

The preparation and fundamental chemistry of organosilyl metallic compounds has been well reviewed[1] up to 1970.

11.1 Preparation

The most flexible route to organosilyl anions utilizes organodisilanes, which suffer Si–Si bond fission when treated with an alkali metal, an alkali metal hydride, or a good nucleophile for silicon (*Scheme 11.1*).

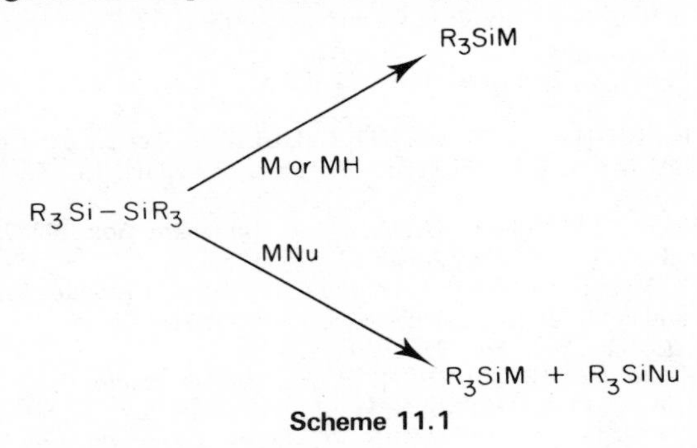

Scheme 11.1

Paradoxically, the requisite disilanes are obtained[2,3] by Wurtz coupling of an organohalogenosilane (*Scheme 11.2*).

The first cleavage process, utilizing an alkali metal, requires at least one of the R groups to be aryl; within this limitation, a reasonable range (1), (2), (3), (4), of triorganosilyl metals can be prepared[4,5], often in good yield.

$$2R_3SiCl + 2M \xrightarrow[(HMPA)]{THF\ or\ Et_2O} R_3Si-SiR_3 + 2MX$$

e.g.

$$2Me_3SiCl + 2Li \xrightarrow[reflux, 8h]{THF} Me_3Si-SiMe_3 \quad 97\%$$

Scheme 11.2

Ph_3SiNa 68% (1)

$Ph_2MeSiLi$ 74% (2)

Ph_3SiLi 79% (3)

$PhMe_2SiLi$ 47% (4)

Trialkylsilyl-metals cannot be obtained in this manner, but are available by metal–metal exchange[6] of disilylmercurials with lithium (*Scheme 11.3*). The dangers inherent in handling volatile mercury compounds make this route somewhat less than attractive.

$$(R_3Si)_2Hg + 2Li \longrightarrow 2R_3SiLi + Hg$$

R = Me, Et

Scheme 11.3

The second cleavage process employs[7] either sodium or potassium hydride, and is well suited to the preparation of trialkylsilyl-metals (*Scheme 11.4*). An additional advantage is that if the triorganosilyl hydride is readily available, then cleavage of the Si–H bond occurs under similar conditions. Either way, solutions of triorganosilyl-metals are formed in good yield and free from any by-products.

$$Me_3Si-SiMe_3 + 2MH \xrightarrow[\text{HMPA}]{\text{DME or}} Me_3SiM + [Me_3SiH] \longrightarrow 2\,Me_3SiM$$

M = Na, K

$$R_3SiH + KH \xrightarrow[\text{HMPA}]{\text{DME or}} R_3SiK$$

R = Et, Ph

Scheme 11.4

The third process, that of cleaving a disilane with a good nucleophile for silicon, also gives access to a good range of trialkylsilyl-metals (*Scheme 11.5*) and other more highly functionalized derivatives[8], when the presence of HMPA is not required.

$$Me_3Si-SiMe_3 + MOMe \xrightarrow{\text{HMPA}} Me_3SiM + Me_3SiOMe$$

M = Na, K

$$(MeO)_2Si(Me)-Si(Me)(OMe)_2 + NaOMe \xrightarrow{\text{THF}} (MeO)_2Si(Me)Na + (MeO)_3SiMe$$

Scheme 11.5

Trimethylsilyl-lithium itself is obtained[9] by cleavage of hexamethyldisilane with methyl-lithium in HMPA, when a deep red solution of the reagent results (*Scheme 11.6*).

$$Me_3Si-SiMe_3 + MeLi \xrightarrow[\text{0 °C, 5 min}]{\text{HMPA}} Me_3SiLi + Me_4Si$$

Scheme 11.6

A rather more specialized situation is seen in the cases of organopentafluorosilicates. Catalysed addition[10] of trichlorosilane to terminal alkenes and alkynes, followed by the addition of an aqueous solution of potassium fluoride, produces[11] highly reactive organopentafluorosilicates (*Scheme 11.7*), discussed further on p. 138.

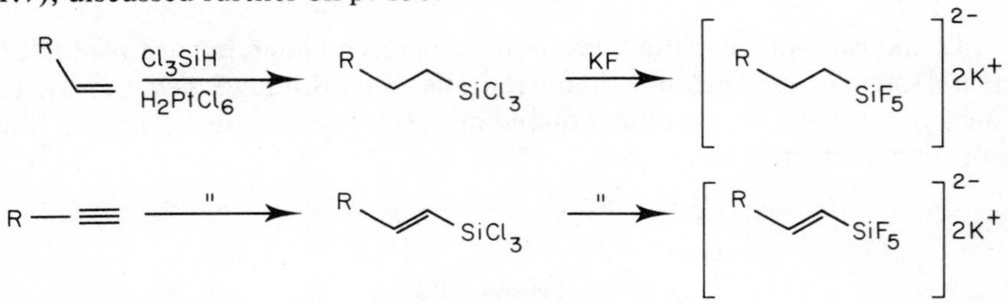

$$RCH=CH_2 \xrightarrow[H_2PtCl_6]{Cl_3SiH} RCH_2CH_2SiCl_3 \xrightarrow{KF} [RCH_2CH_2SiF_5]^{2-}\ 2K^+$$

$$RC\equiv CH \xrightarrow{''} RCH=CHSiCl_3 \xrightarrow{''} [RCH=CHSiF_5]^{2-}\ 2K^+$$

Scheme 11.7

11.2 Reactions

Organosilyl anions are strong bases and good nucleophiles, and must be generated and reacted in aprotic media.

11.2.1 Alkylation

This is generally successful with primary and secondary alkyl chlorides. Bromides and iodides give very poor yields, owing to competing electron transfer processes, the major product frequently being the coupled disilane when HMPA is solvent[12]. In DME or THF plus 18-crown-6, the dominant process becomes one of bimolecular nucleophilic substitution (*Scheme 11.8*), and tetrasubstituted silanes can be obtained in excellent yields[7].

$$Ph_3SiK + PhCH_2Cl \xrightarrow{DME} Ph_3SiCH_2Ph \quad 70\%$$

$$Me_3SiNa + n\text{-}C_{12}H_{25}Br \xrightarrow[18\text{-crown-}6]{THF} Me_3Si(CH_2)_{11}CH_3 \quad 75\%$$

Scheme 11.8

11.2.2 Reaction with carbonyl compounds

This is a good route to silyl carboxylic acids, by reaction with CO_2, and to silylmethanols by reaction with aliphatic aldehydes or ketones[13] (*Scheme 11.9*).

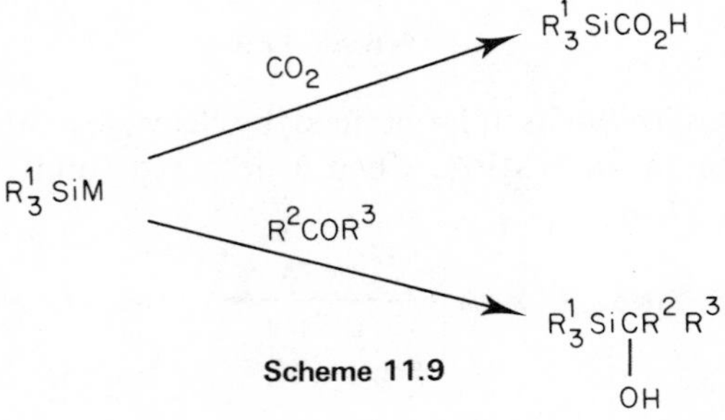

Scheme 11.9

With aromatic aldehydes and ketones, the initial adducts undergo[14,15] Brook rearrangement (Chapter 5), and silyl ethers are isolated (*Scheme 11.10*).

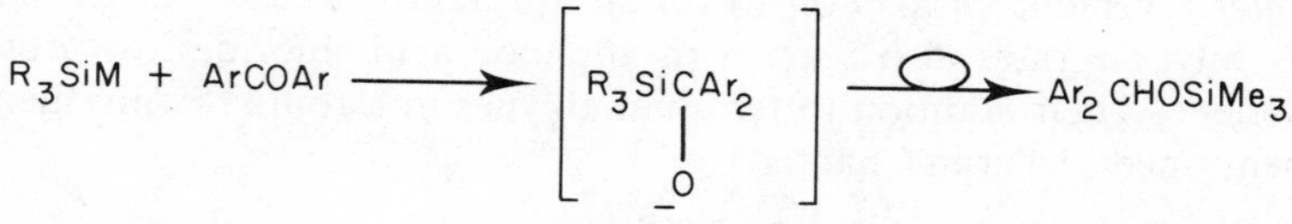

$$R_3SiM + ArCOAr \longrightarrow \left[R_3SiCAr_2(O^-) \right] \longrightarrow Ar_2CHOSiMe_3$$

Scheme 11.10

Trimethylsilyl-lithium reacts with $\alpha\beta$-unsaturated ketones to give products of exclusive 1,4-addition[9,16] (*Scheme 11.11*). That these are kinetic products was demonstrated by separate generation of a product (5) of 1,2-addition; this species (5) proved to be completely stable under the conditions used for conjugate addition.

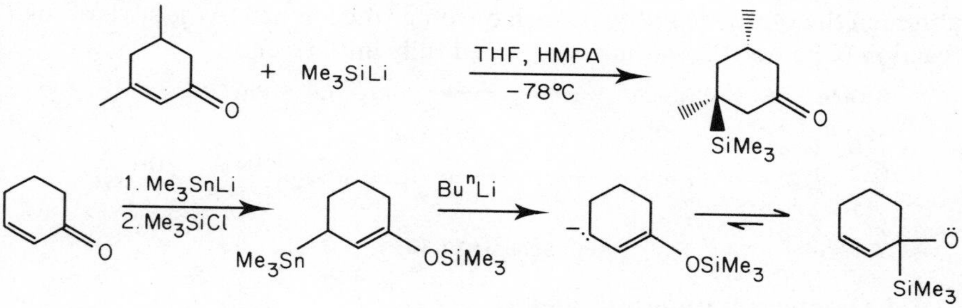

+ Me_3SiLi THF, HMPA −78°C $SiMe_3$ O

1. Me_3SnLi 2. Me_3SiCl Me_3Sn $OSiMe_3$ Bu^nLi $OSiMe_3$ Ö⁻ $SiMe_3$ (5)

Scheme 11.11

With cyclic $\alpha\beta$-unsaturated ketones, a strong preference for axial addition is observed, as is a susceptibility to steric hindrance. Addition[17] of copper(I) iodide forms a silyl cuprate species, and allows higher reaction temperatures

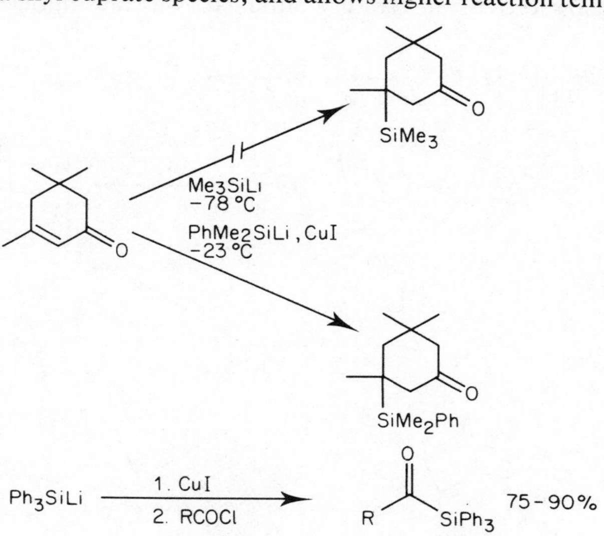

Me_3SiLi −78 °C (no reaction) → $SiMe_3$ product

$PhMe_2SiLi$, CuI −23 °C → $SiMe_2Ph$ product

$$Ph_3SiLi \xrightarrow[2.\ RCOCl]{1.\ CuI} RCOSiPh_3 \quad 75–90\%$$

Scheme 11.12 R = Me, Et, Bu^t

to be employed. This minimizes the effects of steric hindrance, and a wide range of $\alpha\beta$-unsaturated substrates can be smoothly converted into the corresponding γ-ketosilanes (*Scheme 11.12*). The utility of this sequence in providing reversible protection to $\alpha\beta$-unsaturated ketones is discussed on p. 157. Silyl cuprates also react with aliphatic acid chlorides, to produce acylsilanes[18]. Their addition to terminal alkynes in a route to vinylsilanes[19] was mentioned earlier in Chapter 7.

11.2.3 Reactions with ethers

Aryl alkyl ethers are cleaved[20] by triphenylsilyl-lithium to give products of *O*-alkyl fission, albeit in rather modest yields. Cyclic ethers, on the other hand, give good to excellent yields[21] of ω-triorganosilyl-methanols (*Scheme 11.13*), with, when applicable, the regioselectivity expected from bimolecular nucleophilic attack. Epoxides[22] are themselves opened to β-hydroxysilanes, although the synthetic utility of such opening when linked to the silyl-Wittig reaction (Chapter 12) was not recognized fully until recently.

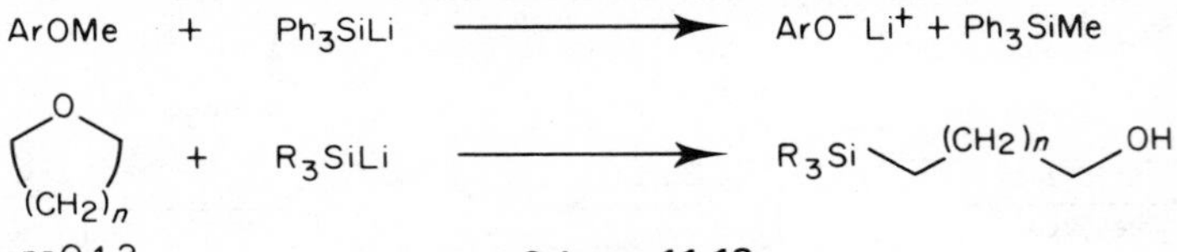

$$ArOMe + Ph_3SiLi \longrightarrow ArO^- Li^+ + Ph_3SiMe$$

$$\text{cyclic ether } (CH_2)_n + R_3SiLi \longrightarrow R_3Si\text{–}CH_2\text{–}(CH_2)_n\text{–}CH_2\text{–}OH$$

$n = 0,1,2$

Scheme 11.13

11.2.4 Organopentafluorosilicates

Organopentafluorosilicates, described earlier on p. 136, react, in some cases exothermically, with a wide range of electrophilic species such as the halogens and halogenoids[10], copper(II) halides[23], and *m*-chloroperbenzoic acid[24]. In all cases, regioselectivity is readily attained via the initial addition of

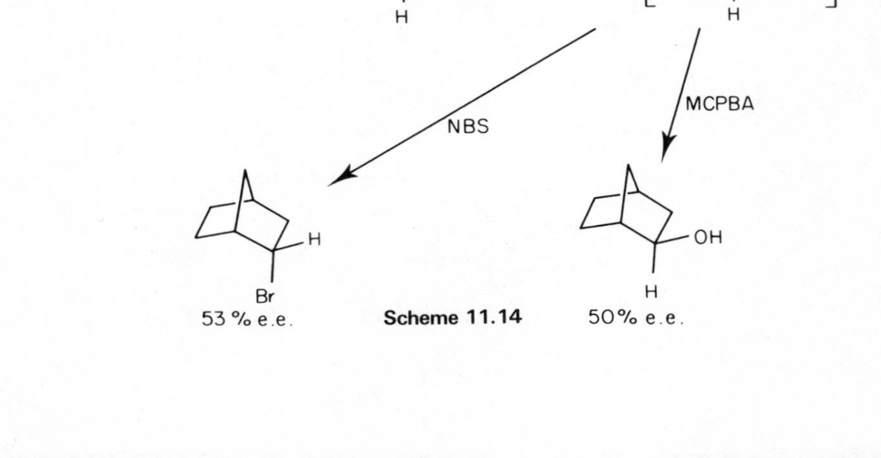

$$[R\text{–}CH_2CH_2\text{–}SiF_5]^{2-}\,2K^+ \xrightarrow{E^+} R\text{–}CH_2CH_2\text{–}E \quad 54\text{–}86\%$$

Norbornene $\xrightarrow[Pd^*]{Cl_3SiH}$ exo-2-norbornyl-$SiCl_3$ $\xrightarrow{KF}$ $[\text{norbornyl-}SiF_5]^{2-}\,2K^+$

NBS → exo-2-norbornyl bromide, 53 % e.e.

MCPBA → exo-2-norborneol, 50% e.e.

Scheme 11.14

trichlorosilane to the alkene. The cleavage reaction can show stereoselectivity[25], allowing asymmetric synthesis[26] of chiral alcohols and bromides from alkenes (*Scheme 11.14*). These sequences of reactions are unique in that they represent the first practical methods for the cleavage of an alkyl–silicon bond to give an alkyl halide or alkanol; one-electron transfer processes are implicated[27] in the cleavage step.

The species derived from alkynes can be allylated[11] to produce 1,4-dienes in reasonable yields (*Scheme 11.15*); bromination[10], methoxycarbonylation[28], and thiocyanation[29] processes have been described also.

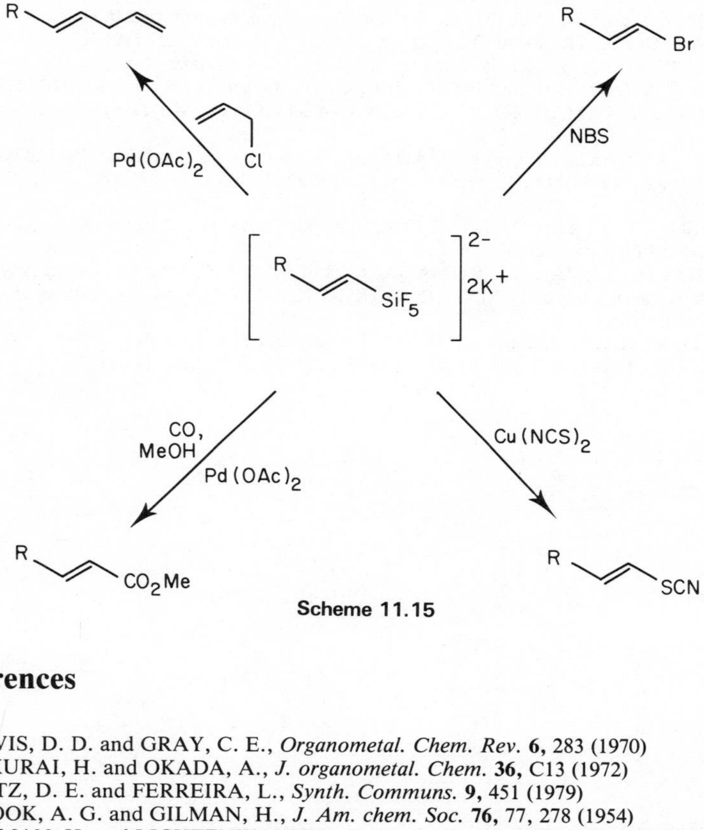

R Br NBS Pd(OAc)$_2$ Cl R SiF$_5$ 2− 2K$^+$ CO, MeOH Pd(OAc)$_2$ Cu(NCS)$_2$ R CO$_2$Me R SCN

Scheme 11.15

References

1 DAVIS, D. D. and GRAY, C. E., *Organometal. Chem. Rev.* **6,** 283 (1970)
2 SAKURAI, H. and OKADA, A., *J. organometal. Chem.* **36,** C13 (1972)
3 SEITZ, D. E. and FERREIRA, L., *Synth. Communs.* **9,** 451 (1979)
4 BROOK, A. G. and GILMAN, H., *J. Am. chem. Soc.* **76,** 77, 278 (1954)
5 GILMAN, H. and LICHTENWALTER, G. D., *J. Am. chem. Soc.* **80,** 608 (1958)
6 HENGGE, E. and HOLTSCHMIDT, N., *J. organometal. Chem.* **12,** P5 (1968); VYAZANKIN, N. S., RAZUVAEV, G. A., GLADYSHEV, E. N. and KORNEVA, S. P., *J. organometal. Chem.* **7,** 353 (1967)
7 CORRIU, R. J. P. and GUERIN, C., *J. chem. Soc. chem. Communs* 168 (1980)
8 SAKURAI, H., OKADA, A., KIRA, M. and YONEZAWA, K., *Tetrahedron Lett.* 1511 (1971); SAKURAI, H. and KONDO, F., *J. organometal. Chem.* **92,** C46 (1975); WATANABE, H., HIGUCHI, K., KOBAYASHI, M., HARA, M., KOIKE, Y., KITAHARA, T. and NAGAI, Y., *J. chem. Soc. chem. Communs* 534 (1977)
9 STILL, W. C., *J. org. Chem.* **41,** 3063 (1976); ILSLEY, W. H., SCHAAF, T. F., GLICK, M. D. and OLIVER, J. P., *J. Am. chem. Soc.* **102,** 3769 (1980)
10 TAMAO, K., YOSHIDA, J., TAKAHASHI, M., YAMAMOTO, H., KAKUI, T., MATSUMOTO, H., KURITA, A. and KUMADA, M., *J. Am. chem. Soc.* **100,** 290 (1978)

11 YOSHIDA, J., TAMAO, K., TAKAHASHI, M. and KUMADA, M., *Tetrahedron Lett.* 2161 (1978)
12 *See,* for example, SAKURAI, H., OKADA, A., UMINO, H. and KIRA, M., *J. Am. chem. Soc.* **95,** 955 (1973)
13 GILMAN, H. and WU, T. C., *J. Am. chem. Soc.* **75,** 2935 (1953); *J. Am. chem. Soc.* **76,** 2502 (1954); GILMAN, H. and LICHTENWALTER, G. D., *J. Am. chem. Soc.* **80,** 2680 (1958)
14 BROOK, A. G., *J. Am. chem. Soc.* **80,** 1886 (1958); *Accts chem. Res.* **7,** 77 (1974)
15 WRIGHT, A. and WEST, R., *J. Am. chem. Soc.* **96,** 3214 (1974)
16 STILL, W. C. and MITRA, A., *Tetrahedron Lett.* 2659 (1978)
17 AGER, D. J. and FLEMING, I., *J. chem. Soc. chem. Communs* 177 (1978)
18 DUFFAUT, N., DUNOGUÈS, J., BIRAN, C., CALAS, R. and GERVAL, J., *J. organometal. Chem.,* **161,** C23 (1978)
19 FLEMING, I. and ROESSLER, F., *J. chem. Soc. chem. Communs* 276 (1980)
20 GILMAN, H. and TREPKA, W. J., *J. organometal. Chem.* **1,** 222 (1964)
21 WITTENBERG, D., AOKI, D. and GILMAN, H., *J. Am. chem. Soc.* **80,** 5933 (1958)
22 GILMAN, H., AOKI, D. and WITTENBERG, D., *J. Am. chem. Soc.* **81,** 1107 (1959)
23 YOSHIDA, J., TAMAO, K., KURITA, A. and KUMADA, M., *Tetrahedron Lett.* 1809 (1978)
24 TAMAO, K., KAKUI, T. and KUMADA, M., *J. Am. chem. Soc.* **100,** 2268 (1978)
25 TAMAO, K., YOSHIDA, J., MURATA, M. and KUMADA, M., *J. Am. chem. Soc.* **102,** 3267 (1980)
26 HAYASHI, T., TAMAO, K., KATSURO, Y., NAKAE, I. and KUMADA, M., *Tetrahedron Lett.* 1871 (1980)
27 YOSHIDA, J., TAMAO, K., KUMADA, M. and KAWAMURA, T., *J. Am. chem. Soc.* **102,** 3269 (1980); *see also* TAMAO, K., KAKUI, T. and KUMADA, M., *Tetrahedron Lett.* 4105 (1980)
28 TAMAO, K., KAKUI, T. and KUMADA, M., *Tetrahedron Lett.* 619 (1979)
29 TAMAO, K., KAKUI, T. and KUMADA, M., *Tetrahedron Lett.* 111 (1980)

Chapter 12

Alkene synthesis by 1,2-elimination reactions of β-functional organosilanes

In 1946, Sommer and Whitmore reported[1] that 2-chloroethyltrichlorosilane was so reactive towards attack by base that it could be titrated with alkali, and that *ethene* was produced (*Scheme 12.1*). Continuing their studies, and concentrating on the high reactivity shown by organosilanes with a leaving group in the β-position (Chapter 3), they reported[2] that 2-hydroxypropyltrialkylsilanes underwent rapid conversion into propene on treatment with dilute sulphuric acid (*Scheme 12.1*). In 1948, they published[3] a full paper on their investigations of such systems, finding that β-chloroalkylsilanes could be induced to undergo 1,2-elimination by treatment with alcoholic bases, aqueous alkali, potassium acetate in acetic acid, methylmagnesium bromide, small amounts of aluminium(III) chloride, silver nitrate in methanol, or, occasionally, with water or heat alone.

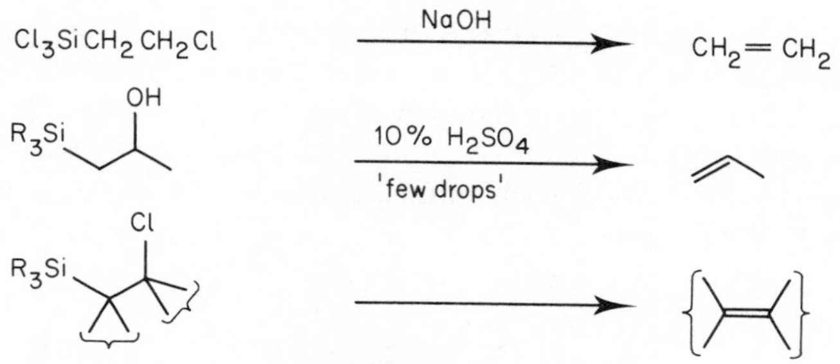

$$Cl_3SiCH_2CH_2Cl \xrightarrow{NaOH} CH_2{=}CH_2$$

R_3Si, OH — 10% H_2SO_4 'few drops' →

R_3Si, Cl →

Scheme 12.1

Much subsequent work, largely of a physical organic nature, was performed on such systems[4]. Only recently has its scope been extended to provide a general and extremely useful synthesis of alkenes. Before discussing such a synthetic strategy, it must be noted that, in 1962, Gilman and Tomasi[5] reported another example of this reaction, an example which additionally illustrated that O–Si bond formation is preferred over O–P bond formation; the silyl-substituted phosphorus ylide (1) reacted with benzophenone to produce tetraphenylallene, a plausible mechanism for which is shown in *Scheme 12.2*.

The potential of this elimination reaction then lay dormant for a further period. It is only within the last ten years or so that its full scope has been

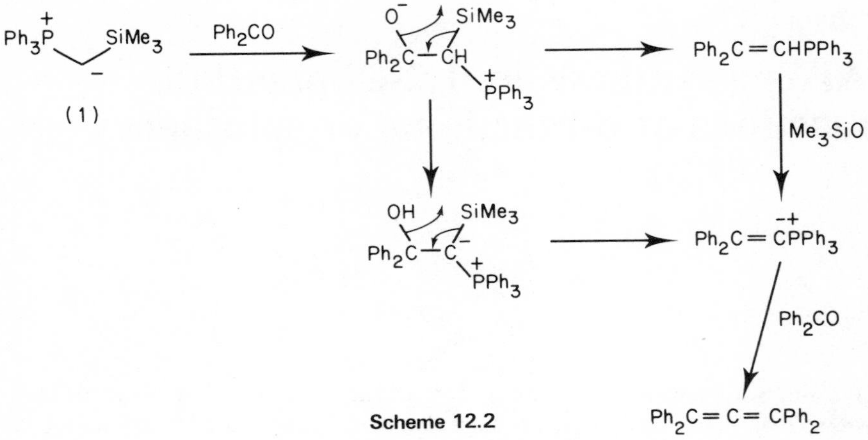

$Ph_3\overset{+}{P}\overset{-}{C}H_2SiMe_3$ (1) $\xrightarrow{Ph_2CO}$ $Ph_2C(O^-)-CH(SiMe_3)\overset{+}{P}Ph_3$ ⟶ $Ph_2C{=}CHPPh_3$

$Ph_2C(OH)-\overset{-}{C}(SiMe_3)\overset{+}{P}Ph_3$ ⟶ $Ph_2C{=}\overset{-+}{C}PPh_3$ (via Me_3SiO^-) $\xrightarrow{Ph_2CO}$ $Ph_2C{=}C{=}CPh_2$

Scheme 12.2

realized and its mechanism defined, particularly noteworthy contributions having been made and summarized by Peterson[6], Hudrlik[7], and Chan[8] and their co-workers.

For the purposes of this Chapter, and for convenience, β-hydroxyalkylsilanes and β-halogenoalkylsilanes will be treated separately, although alkenes are produced in both cases.

12.1 β-Hydroxyalkylsilanes

A major factor which hindered full utilization of this elimination reaction was a comparative lack of adequate routes to the requisite substrates. One of the most attractive routes, condensation of an α-silylcarbanion or carbanionoid with a carbonyl compound (*Scheme 12.3*) suffered from a deficiency of availability of such anions, a situation which has now changed dramatically (Chapter 4).

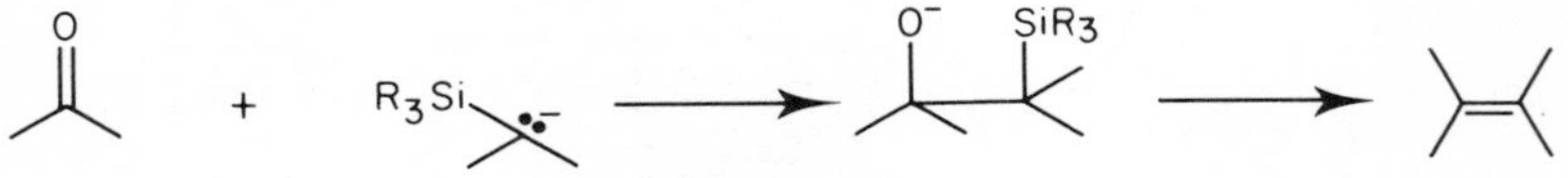

$R_2C{=}O$ + $R_3Si\overset{-}{C}R_2$ ⟶ $R_2C(O^-)-CR_2(SiR_3)$ ⟶ $R_2C{=}CR_2$

Scheme 12.3

Other routes to β-hydroxysilanes have been developed, although they are rarely of such general applicability as are those involving α-silylcarbanions. These routes include the regiospecific opening[9] of αβ-epoxysilanes (Chapter 8) with dialkylcuprates, of oxiranes with silyl anions[28,29], the reactions of β-ketosilanes with hydride reducing agents or organometallic species such as Grignard reagents and alkyl-lithiums[10–12], the reactions[13] of α-haloacylsilanes (p. 273) with certain Grignard reagents (*Scheme 12.4*).

The mechanistic parallel between this reaction and the Wittig reaction is obvious, although, as Gilman and Tomasi found, in some competitive situations this silyl-Wittig reaction is preferred. The reaction is often referred

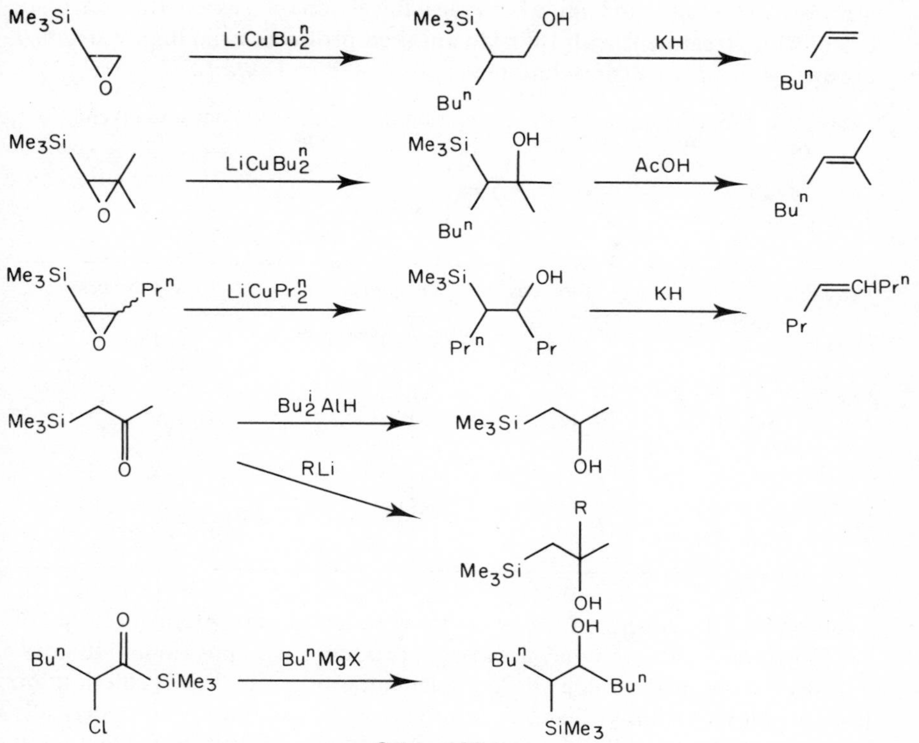

Scheme 12.4

to as Peterson Olefination, since a benchmark paper by Peterson[14] re-awakened interest in it. Before going into detail, it is appropriate to note several important facets of the overall transformation (*Scheme 12.5*).

In general, if X is electron-withdrawing and anion-stabilizing, then the alkene is isolated directly, the elimination reaction of the *β*-oxidosilane (2) being rapid. If, on the other hand, X is not anion-stabilizing, the *β*-hydroxysilane is often isolable, and can subsequently be converted into the alkene on treatment with either basic or acidic reagents. Alternatively, the

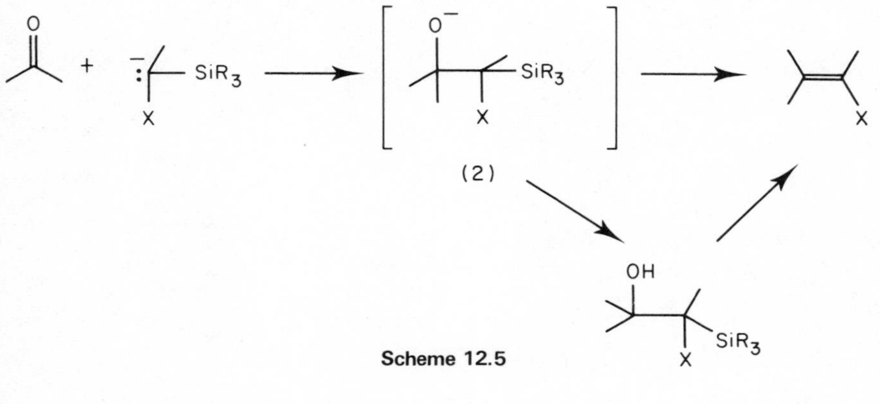

Scheme 12.5

alcohol can be converted into a better leaving group such as a trifluoroacetate or chloride; treatment with fluoride ion then produces even highly strained alkenes. A range of such techniques is illustrated in *Table 12.1*

Table 12.1 Some reagents for the conversion of β-hydroxysilanes into alkenes

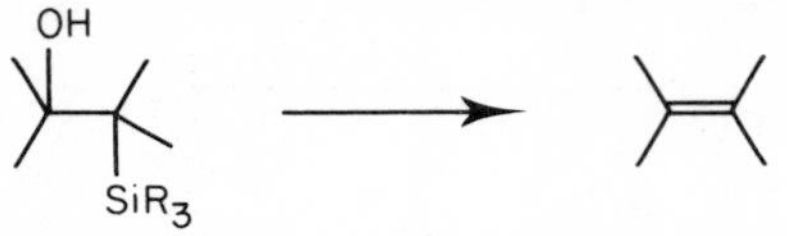

Reagents	*References*	*Reagents*	*References*
NaH	9, 14, 15, 17	$MeSO_2Cl/Et_3N$	9, 15
KH	9, 14, 15	AcCl	18
$KOBu^t$	14	SO_2Cl_2	18
NaOAc/AcOH	10, 12, 16	1. Bu^nLi 2. $MeSO_2Cl$	19
KF/AcOH	12		
$BF_3.Et_2O$	9, 15	1. AcCl 2. F^-	21
H_2SO_4	9, 14, 15		
$HClO_4$	11	1. $SOCl_2$ or $(CF_3CO)_2O$ 2. $Et_4NF/DMSO$	20

Secondly, the alkenes formed in these reactions are often mixtures of (*Z,E*)-isomers, although they are *always* produced with positional integrity. Thirdly, such methodology is complementary to the Wittig olefination process: all other things being equal, the silyl carbanionoid is *less* sterically hindered and *more* basic than the corresponding Wittig reagent. For example, the (±)-β-gorgonene precursor (3) failed to react with methylenetriphenylphosphorane at the relatively hindered carbonyl group, success being achieved[16] with trimethylsilylmethylmagnesium chloride. On the other hand, the ketone (4) reacted smoothly under Wittig conditions (*Scheme 12.6*), whereas trimethylsilylmethylmagnesium chloride acted as a base and abstracted an α-proton, with aldol products being isolated[22].

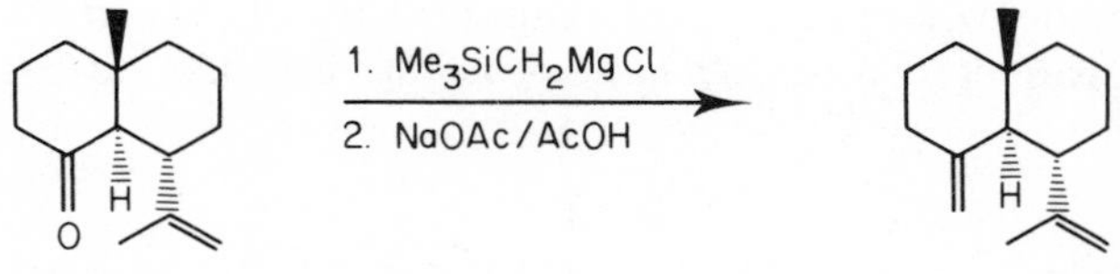

1. Me_3SiCH_2MgCl
2. NaOAc/AcOH

(3)

$Ph_3\overset{+}{P}-\overset{-}{C}H_2$

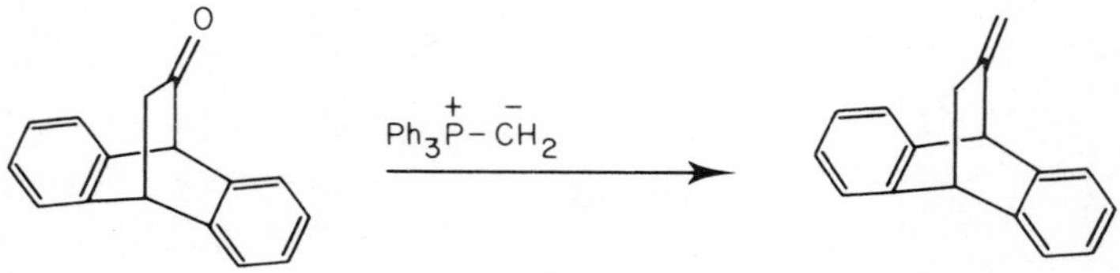

(4)

Scheme 12.6

Lastly, and of fundamental importance, is the mechanism of this elimination reaction[23]. In an elegant study, Hudrlik[15] reported that eliminative reaction of a diastereoisomerically highly enriched β-hydroxysilane led to either one or the other alkene isomer as product, depending upon whether acidic or basic conditions were employed; the elimination was stereospecific, after allowing for diastereoisomeric purity. For example, the alcohol (5) can be obtained as a 15:1 mixture of diastereoisomers by the oxidation–reduction sequence shown in *Scheme 12.7*. Exposure of this alcohol on the one hand to potassium hydride or to

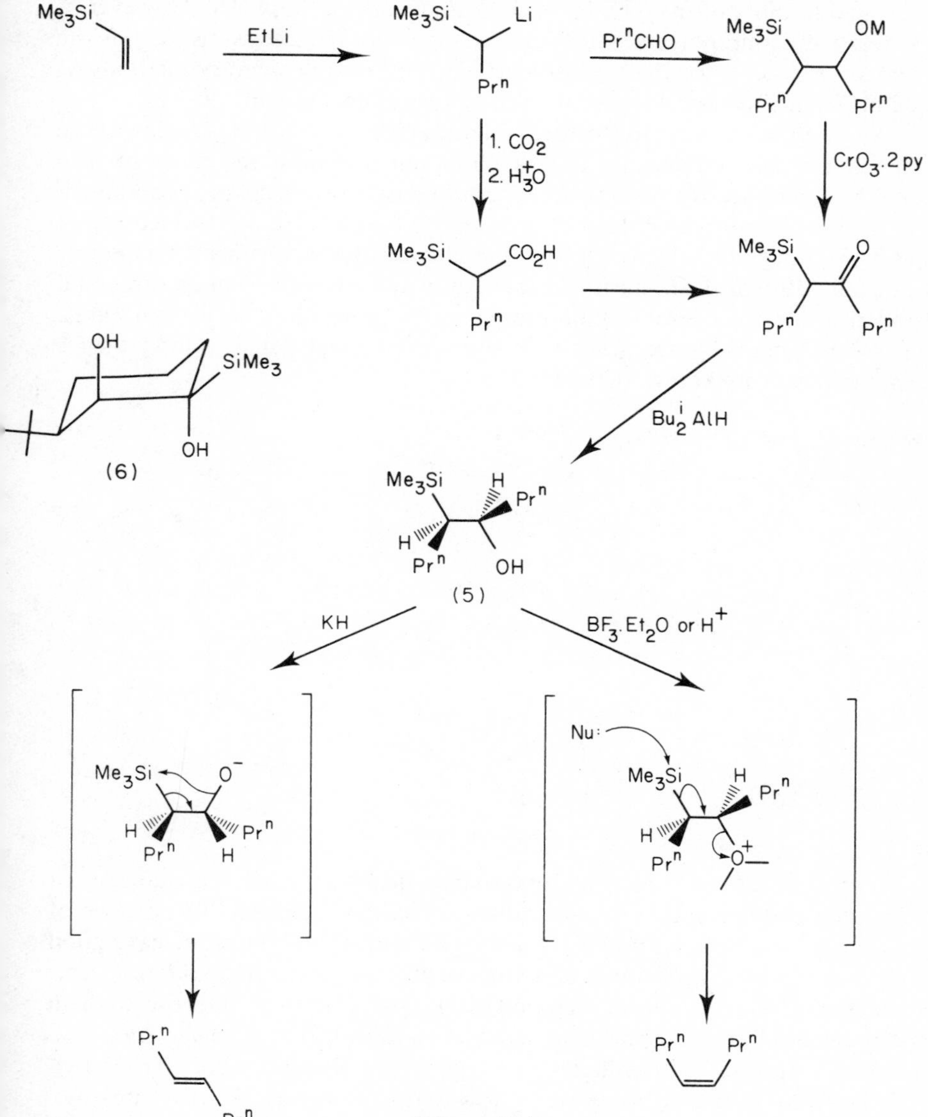

(6)

(5)

Scheme 12.7

boron trifluoride etherate or sulphuric acid on the other, led to (*E*)- or (*Z*)-alkenes, respectively. On the assumption that the more abundant diastereoisomer is the *threo* one (5), as illustrated (and as predicted by Cram's Rule[24], if trimethylsilyl is considered to be larger than n-propyl), then these different elimination pathways reflect the requirement, in the former case, for *syn*-elimination to occur in order that an Si–O bond might be formed. In the latter case, the usual stereoelectronic factors determine the antiperiplanar geometry of elimination[25]. Additional evidence for the requirement of antiperiplanar geometry in such acid-promoted elimination is provided by the acid stability of the *β*-hydroxysilane (6) (*see also* Chapter 8). However, it is relatively unimportant which diastereoisomer is in excess, as long as the excess is large, since the two stereochemically complementary methods of elimination allow production of either alkene isomer at will.

An alternative route to *β*-hydroxysilanes discussed earlier is regiospecific opening of epoxysilanes (p. 142); it turns out that these reactions are also highly stereoselective, with the alkenes ultimately produced being obtained[9] in greater than 99 per cent geometric purity (*Scheme 12.8*). The scope and utility of this stereoselective synthesis of alkenes depends critically on the ease of availability of the requisite geometrically defined epoxysilanes; this could constitute a considerable limitation on its applicability in certain cases, although many stereoselective routes to vinylsilanes (and thence epoxysilanes) are known (Chapter 7).

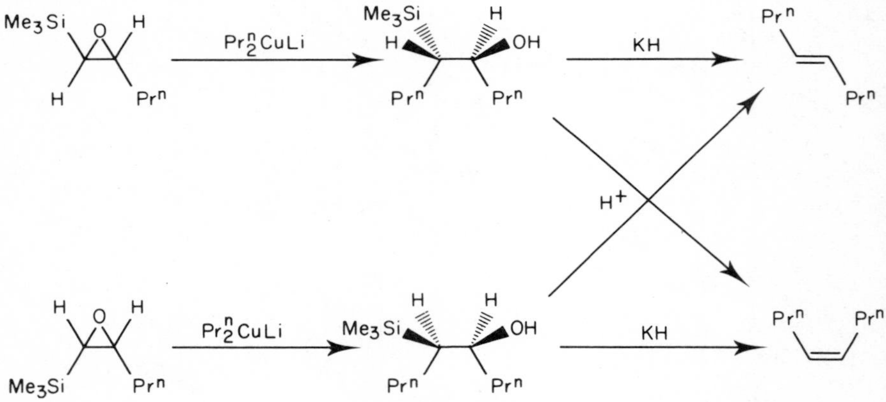

Scheme 12.8

Yet another route to *β*-hydroxysilanes discussed earlier is addition of organometallic reagents to *β*-ketosilanes. Hudrlik[10] showed that reaction of Grignard and organolithium reagents with trimethylsilylacetone gave good yields of *β*-hydroxysilanes, and thence alkenes. Extending this concept, Ruden and Gaffney[11] have prepared several *β*-ketosilanes, and reacted them with several lithiated acetic acid derivatives, as shown in *Scheme 12.9*.

The requisite *β*-ketosilanes are quite readily available. Trimethylsilylacetone[26] itself is prepared by low-temperature reaction of trimethylsilylmethylmagnesium chloride with acetic anhydride (*Scheme 12.10*).

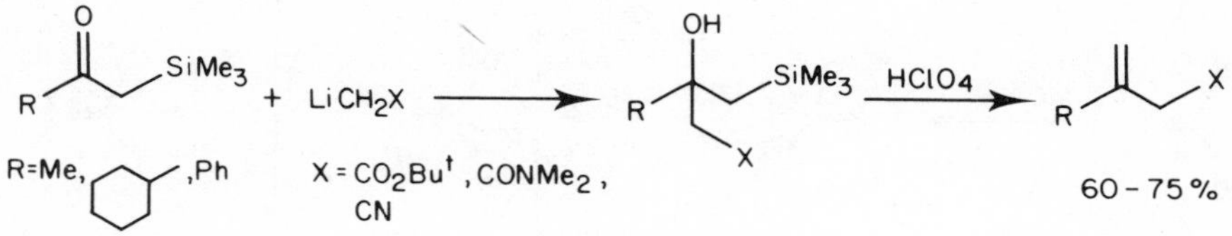

$RCOCH_2SiMe_3$ + $LiCH_2X$ → $RC(OH)(CH_2X)CH_2SiMe_3$ —$HClO_4$→ $RC(=CH_2)CH_2X$

R=Me, cyclohexyl, Ph; X = CO_2Bu^t, $CONMe_2$, CN; 60–75%

Scheme 12.9

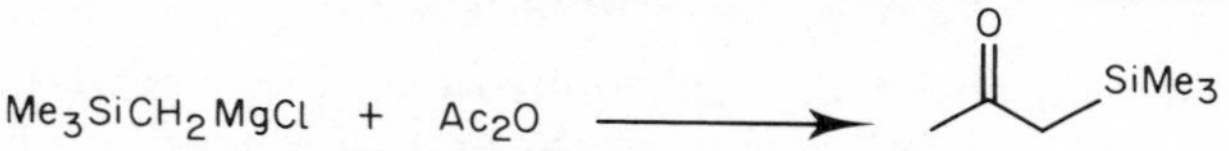

Me_3SiCH_2MgCl + Ac_2O → $MeCOCH_2SiMe_3$

Scheme 12.10

Higher *β*-ketosilanes are best prepared, somewhat paradoxically, by the oxidation of *β*-hydroxysilanes[15].

This general route can be extended even further to provide a stereoselective synthesis of trisubstituted alkenes. *β*-Ketosilanes react with alkyl-lithium reagents to afford predominantly one diastereoisomer[12,27] of the two possible *β*-oxidosilanes; acidic or basic work-up then leads to the stereoselective production of trisubstituted alkenes. The diastereoisomer shown is that one predicted on the basis of Cram's Rule, and also on the discussed mechanisms of elimination (*Scheme 12.11*).

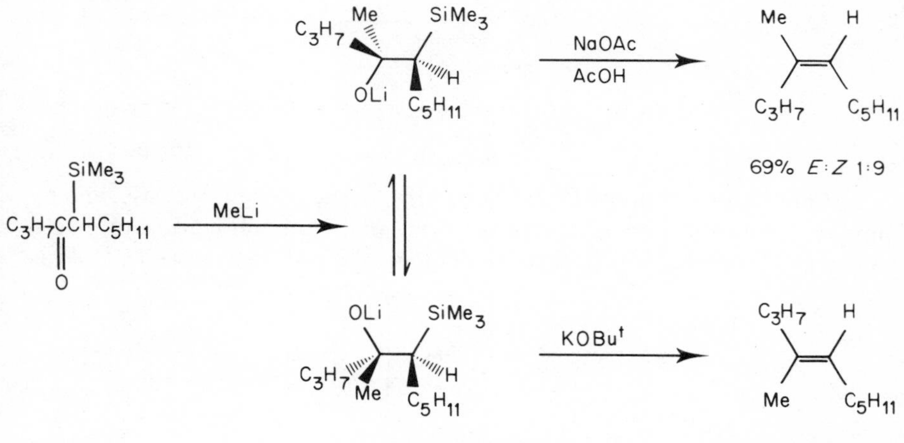

$C_3H_7COCH(SiMe_3)C_5H_{11}$ —MeLi→ $C_3H_7C(Me)(OLi)CH(SiMe_3)C_5H_{11}$ (two conformers)

—NaOAc, AcOH→ Me(C₃H₇)C=CH(C_5H_{11}): 69% *E*:*Z* 1:9

—$KOBu^t$→ (C_3H_7)MeC=CH(C_5H_{11}): 74% *E*:*Z* 9:1

Scheme 12.11

The strict geometric requirements for elimination can be put to further use, as illustrated in an elegant procedure for the geometric isomerization of alkenes. Trimethylsilylpotassium[28] and phenyldimethylsilyl-lithium[29] both effect the smooth conversion of oxiranes into alkenes, nucleophilic ring opening being followed by rotation and spontaneous *syn β*-elimination, as shown in *Scheme 12.12*; this provides an excellent alternative to earlier Wittig-based methods[30] for the controlled geometric isomerization of alkenes via their oxiranes. Several instances of organosilyl anions being used to open oxiranes had been reported[31] earlier, but the *β*-hydroxysilane products had never been put to such further use.

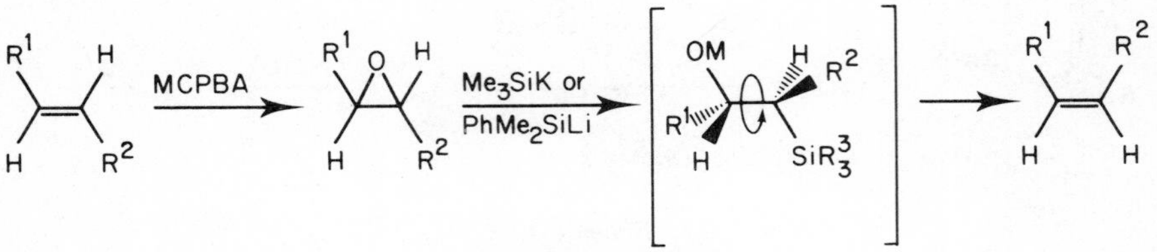

$$R^1CH{=}CHR^2 \xrightarrow{MCPBA} \text{epoxide} \xrightarrow{Me_3SiK \text{ or } PhMe_2SiLi} \left[R^1CH(OM)\text{–}CH(R^2)SiR^3_3 \right] \longrightarrow R^1CH{=}CHR^2$$

Scheme 12.12

The silicon- and phosphorus-substituted diazomethanes (7) and (8), as their metal salts, convert[32] some ketones and aldehydes into homologous alkynes (*Scheme 12.13*). Recently the scope of this reaction has been extended[33], although it remains inapplicable to those cases involving dialkyl ketones.

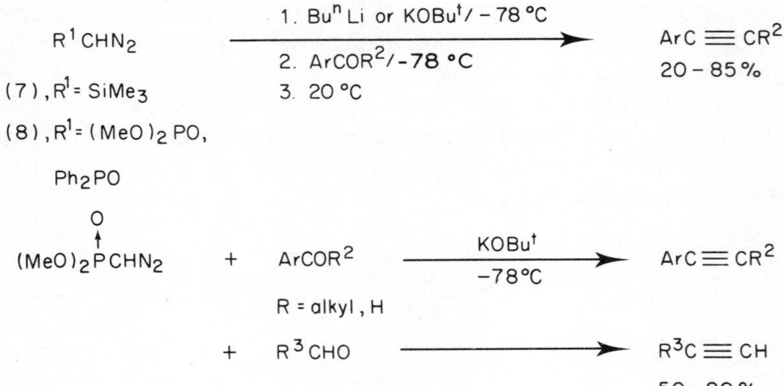

$$R^1CHN_2 \xrightarrow[\text{2. } ArCOR^2/-78\,°C \quad \text{3. } 20\,°C]{\text{1. } Bu^nLi \text{ or } KOBu^t/-78\,°C} ArC{\equiv}CR^2 \quad 20-85\%$$

(7), $R^1 = SiMe_3$

(8), $R^1 = (MeO)_2PO$, Ph_2PO

$$(MeO)_2P(\rightarrow O)CHN_2 + ArCOR^2 \xrightarrow[-78\,°C]{KOBu^t} ArC{\equiv}CR^2$$

R = alkyl, H

$$+ R^3CHO \longrightarrow R^3C{\equiv}CH \quad 50-80\%$$

Scheme 12.13

The mechanism proposed[32] to be operative in the phosphorus case, and by implication in the silicon case also, is one of initial elimination to give a diazovinyl species, which then undergoes skeletal rearrangement. This

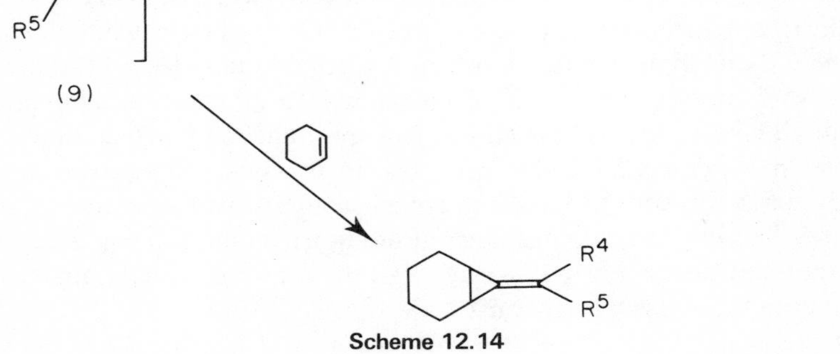

$$R^1\ddot{C}^-N_2 + R^4COR^5 \xrightarrow{-R^1O^-} \left[R^4R^5C{=}\ddot{C}^-N_2^+ \right] \xrightarrow{-N_2} \left[R^4R^5C{=}C\colon \right] \longrightarrow R^4C{\equiv}CR^5$$

(9)

(9) + cyclohexene → 7-(R⁴R⁵C=)bicyclo[4.1.0]heptane

Scheme 12.14

suggestion has been substantiated by the successful trapping[34] (*Scheme 12.14*) of dialkylvinylidenes (9), derived from dialkyl ketones (the one class of substrate which does not give rise to alkynes, presumably because skeletal rearrangement is so slow compared with other decomposition pathways).

Further examples of the silyl-Wittig olefination reaction are summarized in *Table 12.2,* as are some related processes which do not result in the ultimate production of alkenes.

Table 12.2 Some additional examples of the silyl-Wittig reaction

$R^3_3Si(M)CR^4R^5 + R^1COR^2 \longrightarrow R^1R^2C{=}CR^4R^5$

α-Metallosilane	*Products*	*Notes*	*References*
Me_3SiCH_2MgX	$R^1R^2C{=}CH_2$	1	14, 17, 18, 35
$Ph_3SiCH(Li)C_5H_{11}$	$R^1R^2C{=}CH(C_5H_{11})$ (E/Z)	2	17
$Ph_3SiCH(Li)(CH_2)_4CHMe_2$	$R^1CH{=}CH(CH_2)_4CHMe_2$	2, 3	18
$Me_3SiCH(Li)CH{=}CHSiMe_3$	$R^1CH{=}CHCH{=}CHSiMe_3$		36
$Me_3SiCH(Li)SiMe_3$	$R^1R^2C{=}CHSiMe_3$	4	37, 38
$(Me_3Si)_3CM$, M = Na, Li	$R^1R^2C{=}C(SiMe_3)_2$	4	37, 38
$(Me_3Si)_2C(Li)SR$	$R^1CH{=}C(SiMe_3)SR$	5	38
$Me_3SiC(Li)(SnMe_3)SR^6$	$R^1CH{=}C(SnMe_3)SR^6$	5	38
$Me_3SiC(Li)(SMe)_2$	$R^1R^2C{=}C(SMe)_2$	1	38

Table 12.2—*cont.*

Reagents	*References*	*Notes*	*References*
$Me_3SiC(Li)(SePh)_2$	$R^1CH{=}C(SePh)_2$	5	38
$Me_3SiC(Li)(SiMe_3)Br$	$R^1CH{=}C(Br)SiMe_3$	2, 5	39
$Me_3SiCH(Li)R$	$R^1R^2C{=}CHR$	1, 2	
R = Ph_2P, $Ph_2P(S)$, MeS, Ph			14, 17
R = PhS, $(EtO)_2P(O)$, PhS(O)			40
$Me_3SiCH(Li)CO_2Li$	$R^1R^2C{=}CHCO_2H$	1, 2	41
$Me_3SiCH(Li)CO_2R$	$R^1R^2C{=}CHCO_2R$	1, 2	
R = Me			42
= Et			43
= Bu^t			44
$Me_3SiCH(Li)CO_2Bu^t$	$R^1C(OLi){=}CHCO_2Bu^t$	6	45
3-Me_3Si-3-Li-5-R-lactone (O, O, R)	3-ethylidene-5-R-lactone (O, O, R)	7	41
$Me_3SiCH(Li)COSR$	$R^1R^2C{=}CHCOSR$	1, 8	46
$PhMe_2SiC(Li)(R)CN$	$R^1R^2C{=}C(R)CN$	1, 2	47
R = Me, Et, $PhCH_2$			
$Me_3SiCH(Li)CONMe_2$	$R^1C(X){=}CHCONMe_2$ X = NR_2, OR	9	48

Table 12.2—*cont.*

Reagents	*References*	*Notes*	*References*
	$R^1R^2C{=}CHCONMe_2$	10	49
$Me_3SiC(Cl)(Li)CO_2Bu^t$	$R^1R^2C{=}C(Cl)CO_2Bu^t$	1, 2	50
$Me_3SiC(SiMe_3)(Li)CO_2Bu^t$	$R^1CH{=}C(SiMe_3)CO_2Bu^t$	5	51
3-Li-3-Me_3Si-1-Ph-azetidin-2-one	$R^1R^2C{=}$(1-Ph-azetidin-2-one-3-ylidene)		52
$Me_3SiCH(Li)$-(benzothiazol-2-yl)	$R^1R^2C{=}CHCHO$	1, 2, 11	53
$Me_3SiC(Li)(R)CH{=}NR^6$ R = H, Me	$R^1R^2C{=}CHCHO$	1, 8, 11	54
$Ph_3SiCH(Li)N{=}CPh_2$	$R^1R^2C{=}CH{-}N{=}CPh_2$	1	55
$Me_3SiCH(Li)$-(4,4,6-trimethyl-5,6-dihydro-4H-1,3-oxazin-2-yl)	$R^1R^2C{=}CH$-(4,4,6-trimethyl-5,6-dihydro-4H-1,3-oxazin-2-yl)	1	56
$Me_3SiC(Li)(SiMe_3)$-(4,4,6-trimethyl-5,6-dihydro-4H-1,3-oxazin-2-yl)	$R^1R^2C{=}C(SiMe_3)$-(4,4,6-trimethyl-5,6-dihydro-4H-1,3-oxazin-2-yl)	5	56
2-Li-2-Me_3Si-1,3-dithiane	$R^1R^2C{=}$(1,3-dithian-2-ylidene)	1	57–60
	(1,3-dithian-2-ylidene)$=S{=}O$	12	61

Table 12.2—*cont.*

Reagents	*References*	*Notes*	*References*
Me_3Si, $\overset{+}{S}Me_2$	R^1, O, R^2	13	62
	$SiMe_3$, O, R	14	63
$SiMe_3$	R^1, R^2, OH, $SiMe_3$	15	35, 64, 65
	R^1, R^2, =CH—CH=CH_2	16	35, 64

[1]General applicability.
[2]As $E:Z$ mixtures.
[3]With $CH_3(CH_2)_9CHO$.
[4]With aldehydes and non-enolizable ketones.
[5]With aldehydes only.
[6]With acyl imidazoles, RCOIm.
[7]With CH_3CHO, only (*E*)-isomer formed.
[8]Mainly (*E*)-isomers with aldehydes.
[9]With R^1COX, $X = NR_2$ or OR.
[10]With a benzoquinone derivative.
[11]After appropriate work-up.
[12]On reaction with SO_2.
[13]Possibly by initial desilylation of ylide. Other pathways are also observed.
[14]With a range of vinyl ketones.
[15]Simple anion always shows γ-attack; *see also* p. 118.
[16]Addition of $MgBr_2$ to anion inverts regiospecificity to predominant α-attack.

12.2 β-Halogenoalkylsilanes and related species

β-Halogenosilanes have been known for a considerable time, as has their pronounced lability[4] towards β-elimination, heat alone frequently being sufficient. However, many unsaturated β-halogenosilanes are quite stable, and rather severe conditions have to be employed to promote elimination. In 1972, Cunico and Dexheimer made use of the strong affinity of silicon for fluoride ion, and generated[66] ethyne from 2-chlorovinyltrimethylsilane (*Scheme 12.15*), and, later, benzyne by a similar method[67], albeit in very modest yield in the latter case.

Chan, in an attempt to extend the silyl-Wittig reaction to the preparation of allenes (*Scheme 12.16*), found that the intermediate adducts did not eliminate spontaneously, and could be isolated[20] as the corresponding

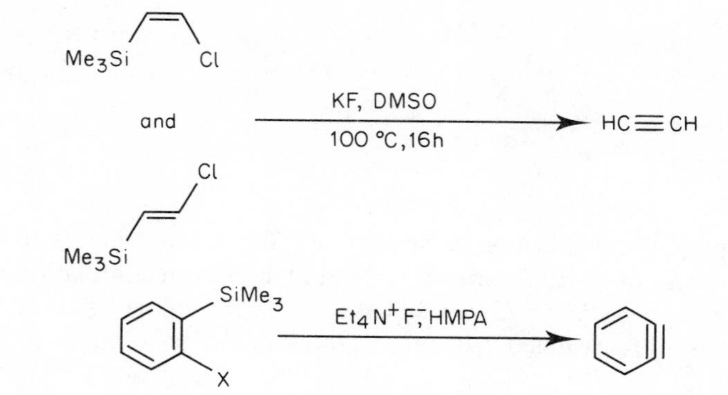

Scheme 12.15

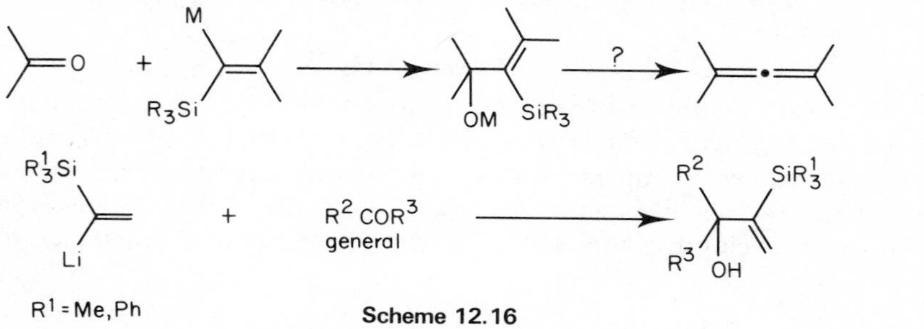

Scheme 12.16

alcohols, an observation later confirmed by Seebach[68]. When such alcohols had been obtained, it was discovered that they were most reluctant to eliminate the elements of triorganosilanol. Indeed, treatment with fluoride ion resulted in clean desilylation[69], to afford the corresponding vinyl alcohols. Further, attempted conversion of the alcohol function into a better leaving group resulted in elimination; this afforded 1,3-dienes in those cases involving aliphatic ketones such as cyclohexanone (*Scheme 12.17*).

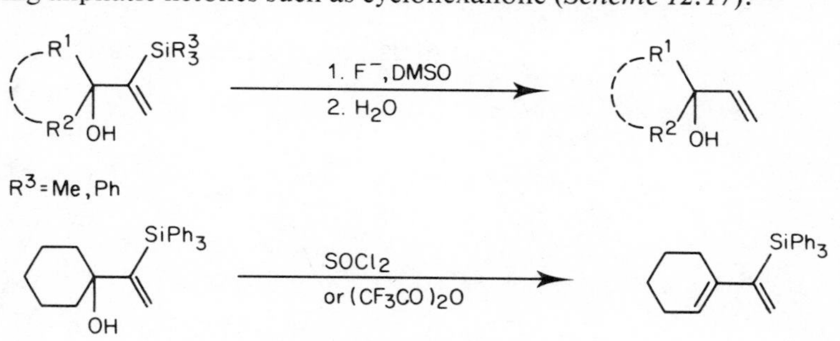

Scheme 12.17

However, the adducts derived from aldehydes or from diaryl ketones could be converted[70] into chlorides (*Scheme 12.18*); the allylically-related species (10), (11) and (12) were formed in varying amounts, aliphatic aldehydes giving almost exclusively the vinylsilane (11) (*see also* Chapter 7).

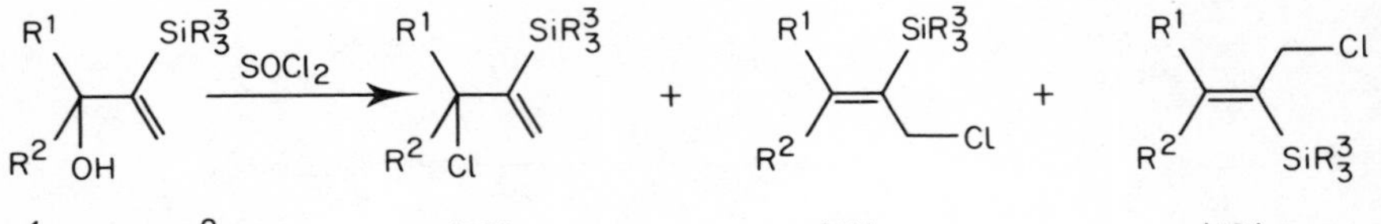

R^1 = Ar or R, R^2 = H
R^1, R^2 = Ar

(10) (11) (12)

Scheme 12.18

Treatment of these chlorides separately, or, more conveniently, together without purification, with fluoride ion resulted in smooth elimination to give the desired allenes (*Scheme 12.19*). In cases where the chlorides could not be prepared, the unrearranged trifluoroacetates could be used in a similar fashion.

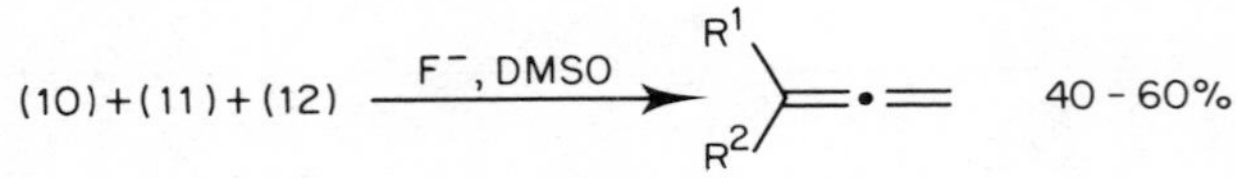

(10) + (11) + (12) —F^-, DMSO→ 40 - 60%

Scheme 12.19

The relative stability of such β-functionalized vinylsilanes can be put to good use. It proved possible to epoxidize the vinyl group of structures of type (13); subsequent treatment with fluoride ion then gave allene oxides[71], thus providing a reliable and easy entry into the allene oxide–oxyallyl zwitterion–cyclopropanone set of interconverting isomers (*Scheme 12.20*).

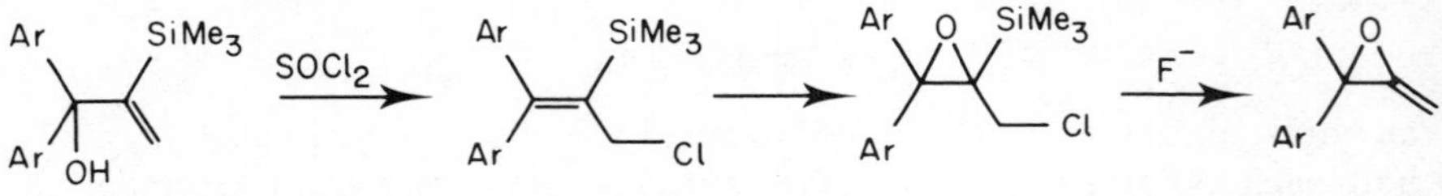

(13)

Scheme 12.20

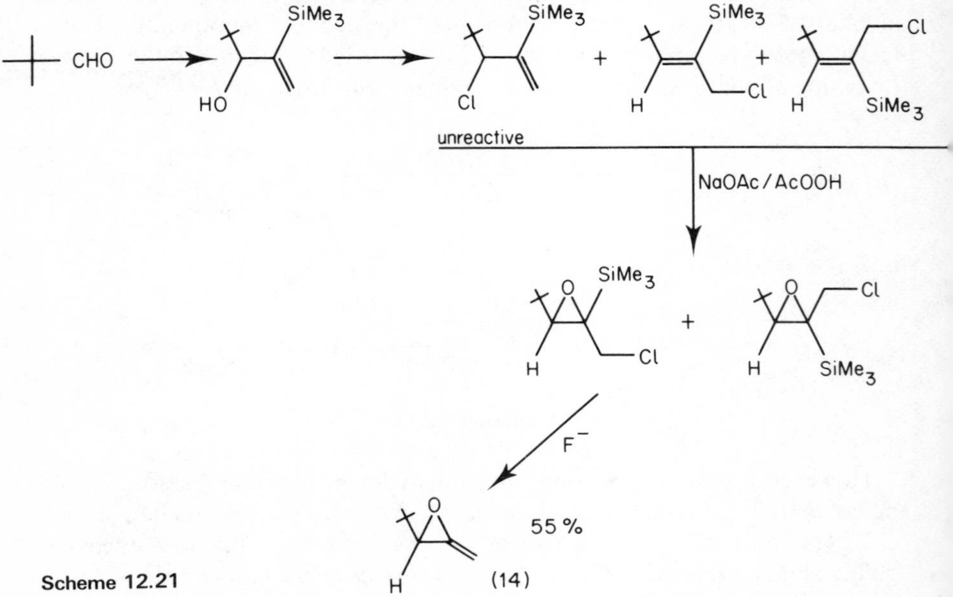

unreactive

NaOAc/AcOOH

F^-

55 %

(14)

Scheme 12.21

This methodology has culminated[72] in the preparation, isolation, and characterization for the first time of a monosubstituted allene oxide, t-butylallene oxide (14) (*Scheme 12.21*), and the generation and trapping of several others.

It should be noted at this point that simple $\alpha\beta$-epoxysilanes (Chapter 8) undergo desilylation on treatment with fluoride ion with retention[73] of stereochemistry (*Scheme 12.22*).

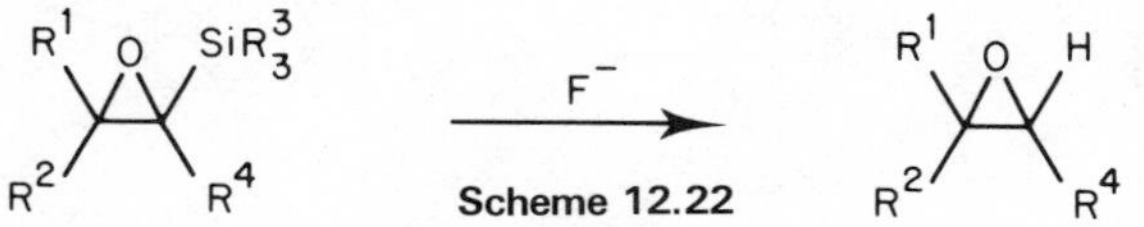

Scheme 12.22

An alternative manner by which one can attain the requisite 1,2-relationship of silicon and halogen is by addition of dihalogenocarbenes/carbenoids[74,75] to vinylsilanes (*Scheme 12.23*).

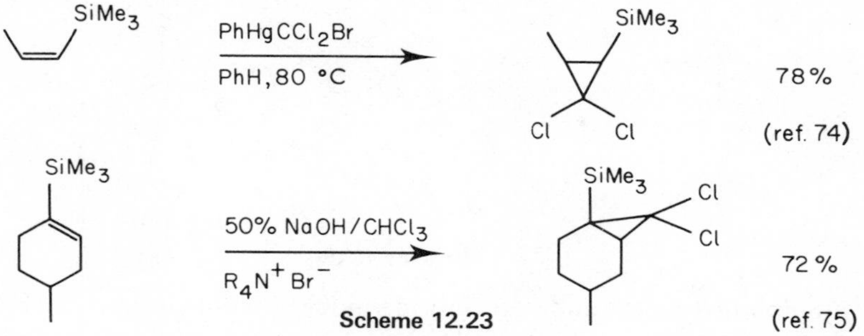

Scheme 12.23

Reaction of such species with ethoxide ion gave rise to products whose formation suggests[76] the intermediacy of cyclopropenes. By use of fluoride ion (*Scheme 12.24*), it proved possible to isolate in some cases, and to trap in all, a range of halogenocyclopropenes[77]; the advantage here is that the reagent, alkali fluoride, and the products, alkali halide and trimethylsilyl fluoride, are generally neutral and relatively inert to most other functionalities.

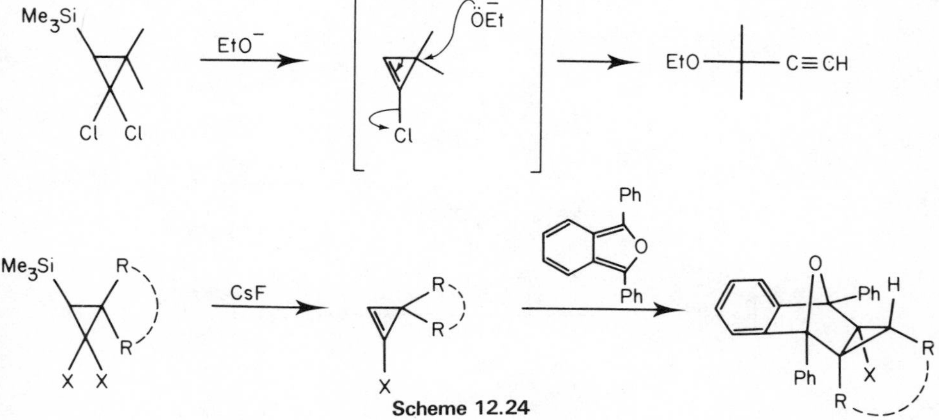

Scheme 12.24

Thus, fluoride ion-promoted 1,2-elimination of a 1,2-halogenosilane is a powerful method for the generation of strained alkenes. A further application of this philosophy is seen in the transient preparation[81] of the strained bridgehead[78-80] alkene (15), a derivative of bicyclo[2.2.2]oct-1-ene (16) (*Scheme 12.25*).

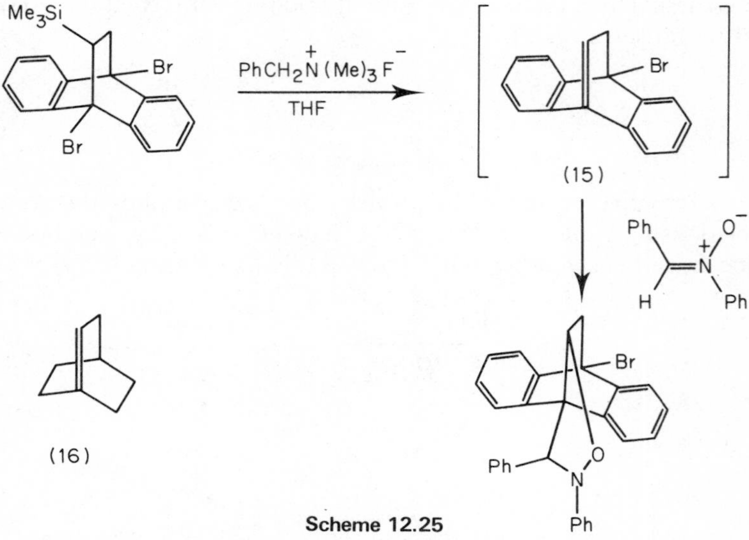

Me3Si Br Br $PhCH_2\overset{+}{N}(Me)_3\overset{-}{F}$ THF Br (15) Ph O^- + N H Ph (16) Br Ph O N Ph

Scheme 12.25

β-Sulphonylsilanes undergo fluoride-induced elimination to give (in the cases cited) terminal alkenes[82]. The requisite substrates are obtainable by α-alkylation of the sulphone (17), or by alkylation of elaborated sulphones with trimethylsilylmethyl iodide, as shown in *Scheme 12.26*.

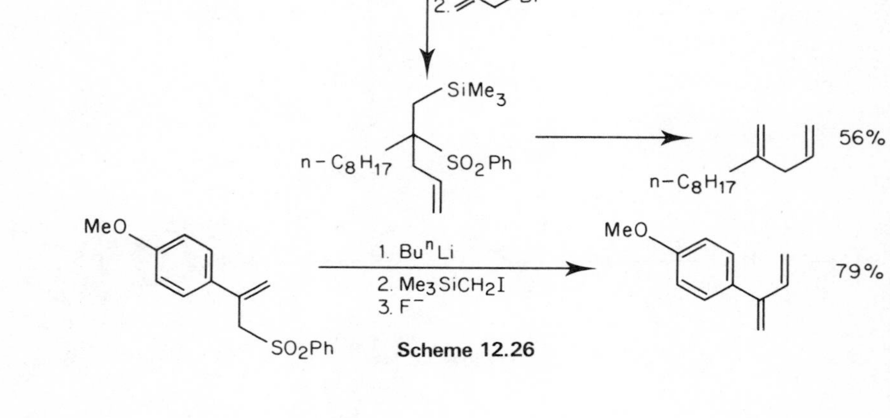

$SiMe_3$ SO_2Ph (17) 1. Bu^nLi 2. n-$C_8H_{17}Br$ $SiMe_3$ n-C_8H_{17} SO_2Ph $Bu^n_4N^+F^-$ THF n-C_8H_{17} 80%

1. Bu^nLi 2. Br

$SiMe_3$ n-C_8H_{17} SO_2Ph n-C_8H_{17} 56%

MeO SO_2Ph 1. Bu^nLi 2. Me_3SiCH_2I 3. F^- MeO 79%

Scheme 12.26

Whereas *β*-ketosilanes are extremely susceptible towards acid- or base-induced loss of the silyl group, *γ*-ketosilanes are not[83] except under the most drastic conditions, no equivalent mechanistic breakdown being possible. Indeed, ethyl trimethylsilylmethylacetoacetate (18) behaves[84] like any other monosubstituted acetoacetate (*Scheme 12.27*).

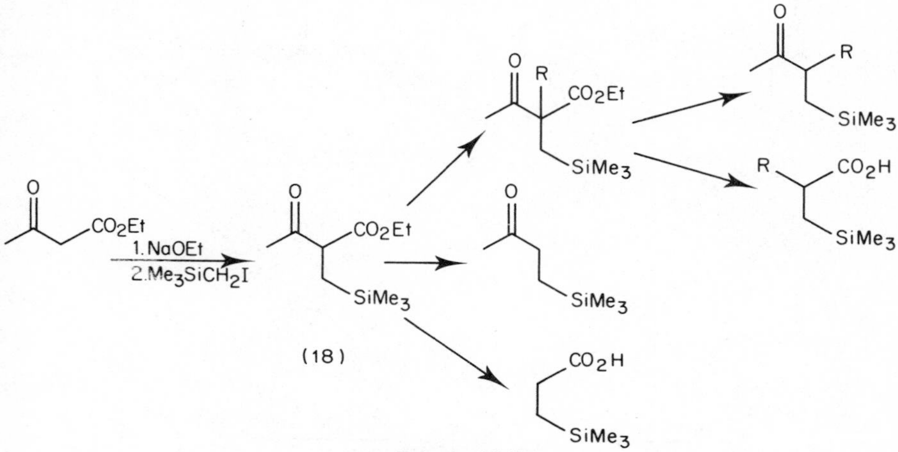

1. NaOEt
2. Me_3SiCH_2I

(18)

Scheme 12.27

The related malonate ester[85] (19) can be *α*-brominated (*Scheme 12.28*), to produce the requisite 1,2-juxtaposition of functionality for ready elimination. However, the presence of the two ethoxycarbonyl groups results in a tendency for ionization to occur in the 'wrong' sense, to give malonate ion and Br^+ [86], and the compound is relatively stable towards *β*-elimination. It does undergo thermal elimination, and gives a product on treatment with bromine suggestive of *β*-elimination having occurred.

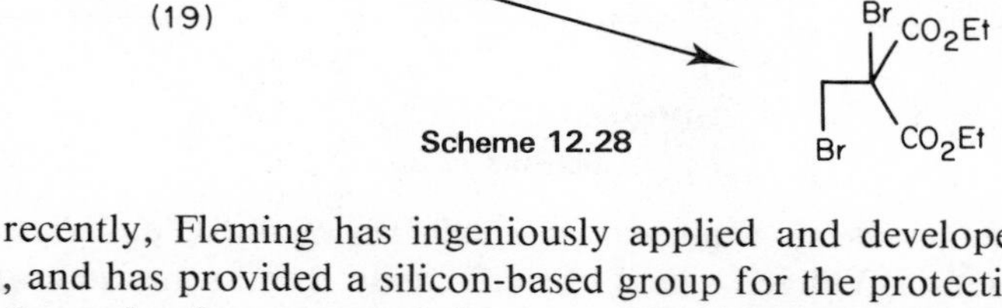

200 °C

Br_2

(19)

Scheme 12.28

More recently, Fleming has ingeniously applied and developed[87] these concepts, and has provided a silicon-based group for the protection of the unsaturation of *αβ*-unsaturated ketones. The basic philosophy is that outlined above, i.e., that it should be possible to manipulate the ketone function of a *γ*-ketosilane by use of the normal range of synthetic procedures without any risk to the silyl function. However, *α*-bromination will then set up the silyl and bromine functions for easy elimination, which will produce an $\alpha\beta$-unsaturated ketone. Possible lack of high regiospecificity in bromination is not important, as enolization or equilibration can provide equivalent pathways for elimination to occur (*Scheme 12.29*). The correctness of this reasoning is illustrated by the example shown.

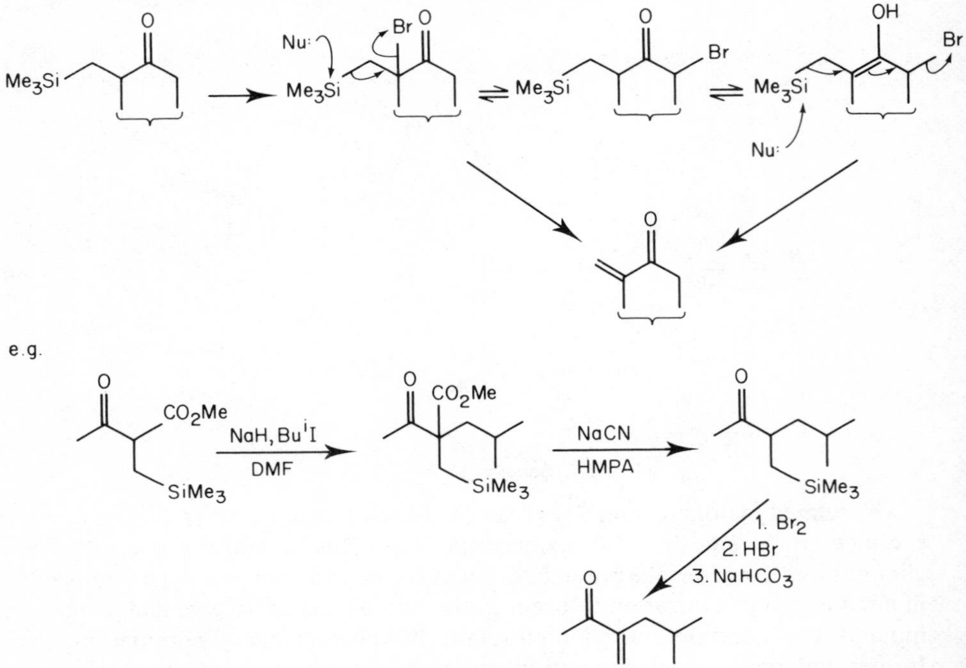

Scheme 12.29

Full application of this procedure to afford protection to an $\alpha\beta$-unsaturated ketone is illustrated in *Scheme 12.30*. The intermediate lithium enolates formed by conjugate addition of a silyl anion (Chapter 11) can be trapped with reactive halides such as allyl bromide or methyl iodide, as shown.

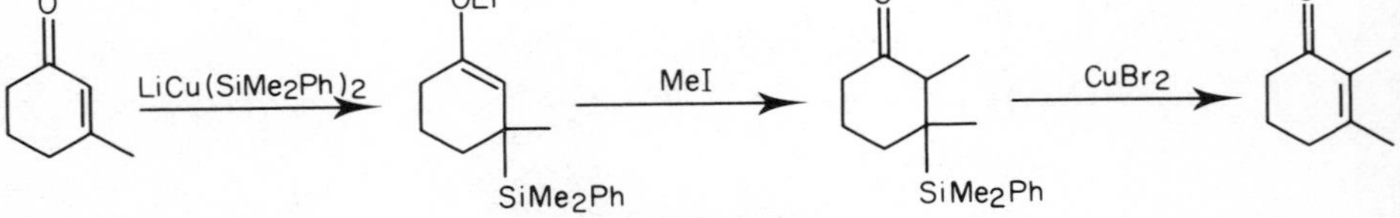

Scheme 12.30

γ-Ketosilanes such as those shown in *Schemes 29* and *30* undergo a silicon-assisted Baeyer–Villiger reaction, ultimately producing ω-unsaturated carboxylic acids[88] (*Scheme 12.31*). As in all Baeyer–Villiger reactions, the carbon which migrates is that one more capable of sustaining development of positive charge; stabilization is being provided here by the β-effect (Chapter 3).

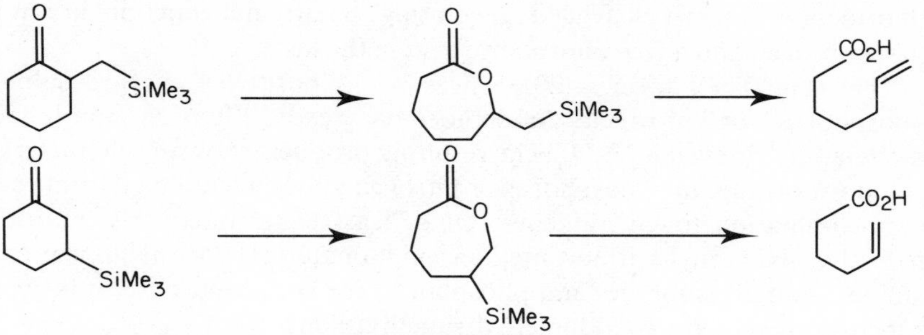

Scheme 12.31

Finally, some cases in which the unsaturated product is not necessarily the centre of attention will now be discussed. 2-Trimethylsilylethyl esters are stable to a wide variety of conditions used during coupling and isolation in peptide synthesis and other processes, but are readily cleaved[89, 90] by fluoride ion (*Scheme 12.32*).

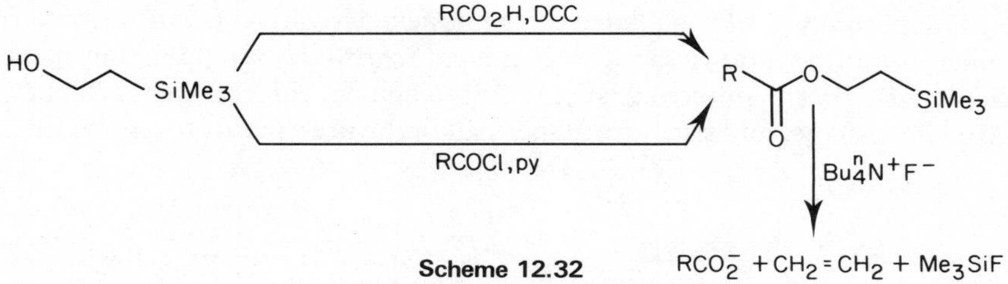

RCO_2H, DCC

RCOCl, py

$Bu^n_4N^+F^-$

$RCO_2^- + CH_2{=}CH_2 + Me_3SiF$

Scheme 12.32

Variants on this theme can be seen in methods for the protection of amines[91], alcohols[92], and phosphoric acid[93] derivatives (*Scheme 12.33*),

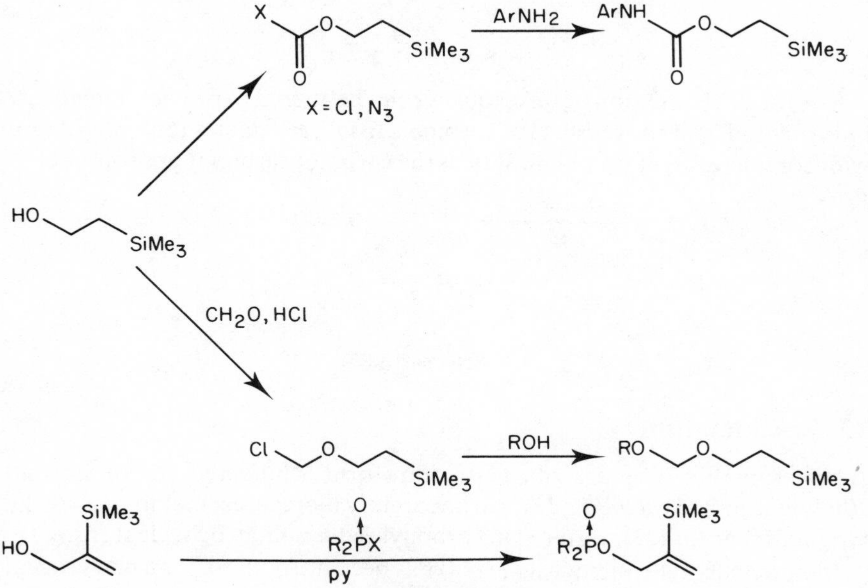

$ArNH_2$

X = Cl, N_3

CH_2O, HCl

ROH

$R_2P(O)X$, py

Scheme 12.33

fluoride ion-induced cleavage regenerating the original functionality in the first two cases and a free phosphoric acid in the last.

The generalized species (20), where X^+ is positively charged sulphur, phosphorus or nitrogen, undergoes a ready fluoride ion-induced desilylation[94] (*Scheme 12.34*). The resulting products show the characteristic reactions of sulphur, phosphorus or nitrogen ylides, including in some cases cycloelimination to give alkenes. Of at least equal interest is the use of trimethylsilylmethyl trifluoromethanesulphonate (21) for the alkylation of amines, imines, sulphides and phosphines; for such a purpose, it is greatly superior in reactivity to iodomethyltrimethylsilane.

$$X: + CF_3SO_3CH_2SiMe_3 \xrightarrow{MeCN} \overset{+}{X}-CH_2SiMe_3 \xrightarrow[MeCN]{CsF} \overset{+}{X}-\ddot{C}H_2^-$$

(21) (20)

X = SR_2, NR_3, PR_3

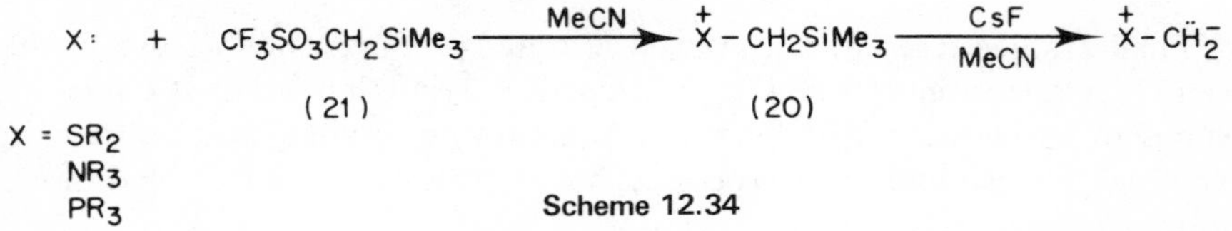

Scheme 12.34

Mild thermolysis of *N*-silylated carbamic acid derivatives (22) of secondary amines produces good yields of isocyanates (*Scheme 12.35*). Silylating agents which have been employed successfully include chlorotrimethylsilane[95], tetrachlorosilane, and trichlorosilane[96], all in the presence of triethylamine.

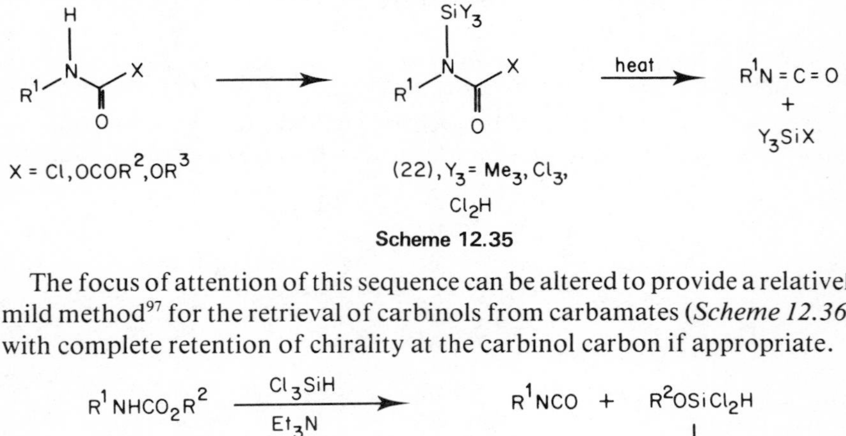

$$R^1N(H)C(O)X \longrightarrow R^1N(SiY_3)C(O)X \xrightarrow{heat} R^1N=C=O + Y_3SiX$$

X = Cl, $OCOR^2$, OR^3

(22), Y_3 = Me_3, Cl_3, Cl_2H

Scheme 12.35

The focus of attention of this sequence can be altered to provide a relatively mild method[97] for the retrieval of carbinols from carbamates (*Scheme 12.36*), with complete retention of chirality at the carbinol carbon if appropriate.

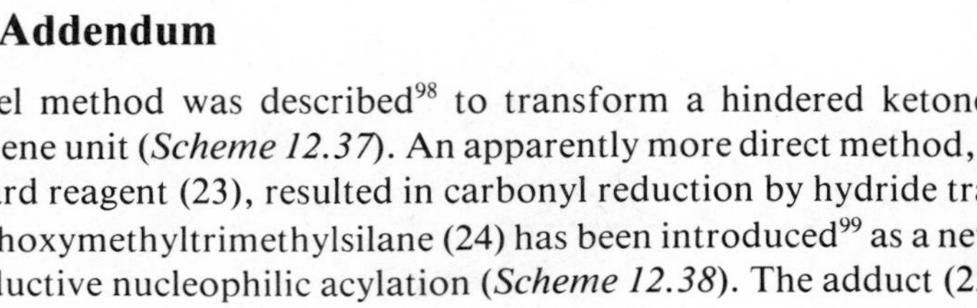

$$R^1NHCO_2R^2 \xrightarrow[Et_3N]{Cl_3SiH} R^1NCO + R^2OSiCl_2H \longrightarrow R^2OH$$

Scheme 12.36

12.3 Addendum

A novel method was described[98] to transform a hindered ketone into an ethylidene unit (*Scheme 12.37*). An apparently more direct method, using the Grignard reagent (23), resulted in carbonyl reduction by hydride transfer.

Methoxymethyltrimethylsilane (24) has been introduced[99] as a new reagent for reductive nucleophilic acylation (*Scheme 12.38*). The adduct (25) did not

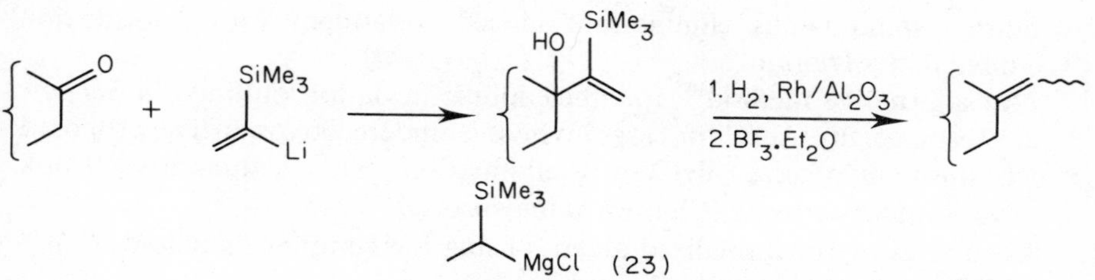

SiMe$_3$ HO SiMe$_3$ O + Li 1. H_2, Rh/Al_2O_3 2. $BF_3.Et_2O$

SiMe$_3$ MgCl (23)

Scheme 12.37

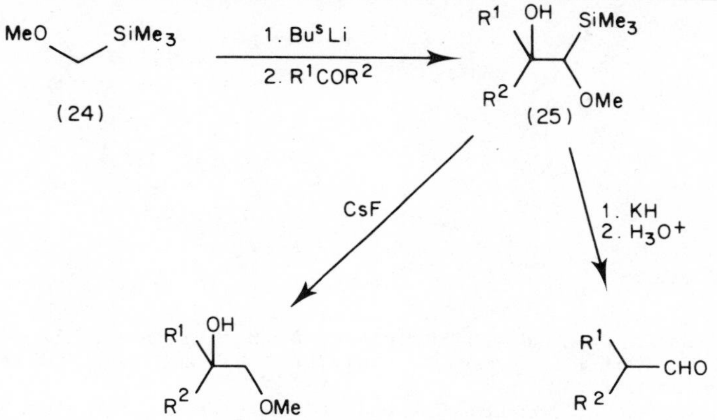

MeO SiMe$_3$ (24) 1. Bu^{s}Li 2. R^1COR^2 R^1 OH SiMe$_3$ R^2 OMe (25)

CsF 1. KH 2. H_3O^+

R^1 OH R^2 OMe R^1 CHO R^2

Scheme 12.38

SiMe$_3$ HO OSiMe$_3$

Me$_3$SiLi −35 °C LDA, 23 °C

Me$_3$SiO O 1. Me$_3$SiCN 2. Bu$^i_2$AlH CHO OSiMe$_3$ H_3O^+ CHO

Me$_3$SiLi −35 °C

Me$_3$SiO OH SiMe$_3$ LDA 23 °C Me$_3$SiO OSiMe$_3$

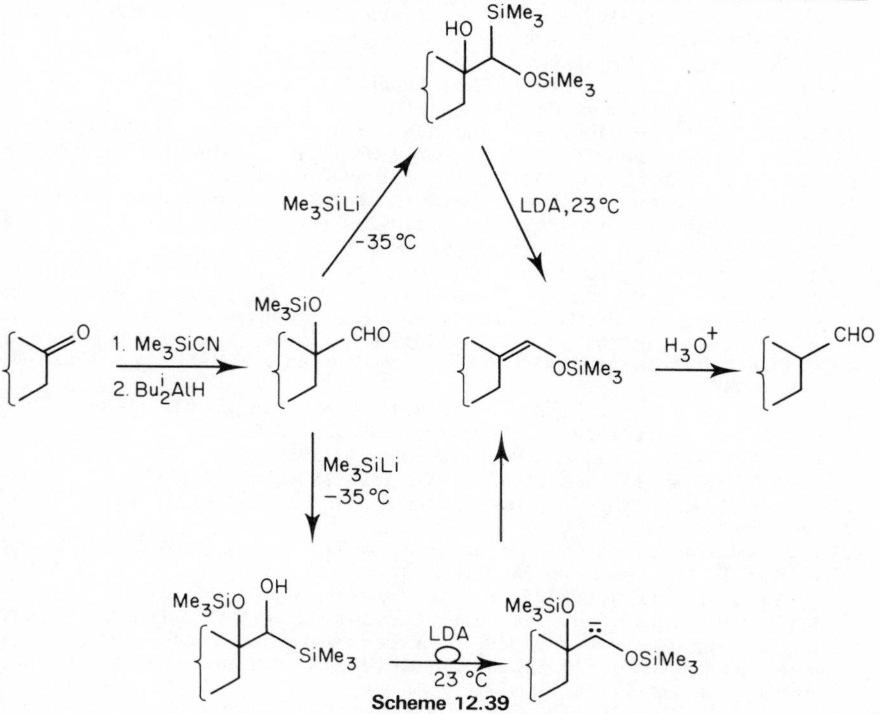

Scheme 12.39

undergo spontaneous elimination; indeed, treatment with fluoride ion resulted in desilylation.

An alternative method[100] for nucleophilic acylation outlined in *Scheme 12.39* is especially useful for cases involving hindered ketones. The proposed mechanism involves a silyl-Wittig elimination, but an alternative Brook rearrangement pathway (Chapter 5) may prevail.

Syntheses of functionalized alkenes using the complex metalloids (26)[101] and (27)[102] have been outlined.

SiMe$_3$ N Li

(26)

SiMe$_3$ O B O Li

(27)

References

1 SOMMER, L. H. and WHITMORE, F. C., *J. Am. chem. Soc.* **68,** 485 (1946)
2 WHITMORE, F. C., SOMMER, L. H., GOLD, J. and Van STRIEN, R. E., *J. Am. chem. Soc.* **69,** 1551 (1947)
3 SOMMER, L. H., BAILEY, D. L. and WHITMORE, F. C., *J. Am. chem. Soc.* **70,** 2869 (1948)
4 JARVIE, A. W. P., *Organometal. Chem. Rev. A* **6,** 153 (1970)
5 GILMAN, H. and TOMASI, R. A., *J. org. Chem.* **27,** 3647 (1962); *see also* SCHMIDBAUR, H., *Accts chem. Res.* **8,** 62 (1975)
6 PETERSON, D. J., *Organometal. Chem. Rev. A* **7,** 295 (1972)
7 HUDRLIK, P. F., *J. organometal. Chem. Library* **1,** 127 (1976)
8 CHAN, T. H., *Accts chem. Res.* **10,** 442 (1977)
9 HUDRLIK, P. F., PETERSON, D. and RONA, R. J., *J. org. Chem.* **40,** 2263 (1975)
10 HUDRLIK, P. F. and PETERSON, D., *Tetrahedron Lett.* 1785 (1972); 1133 (1974)
11 RUDEN, R. A. and GAFFNEY, B. L., *Synth. Communs* **5,** 15 (1975)
12 UTIMOTO, K., OBAYASHI, M. and NOZAKI, H., *J. org. Chem.* **41,** 2940 (1976)
13 SATO, T., ABE, T. and KUWAJIMA, I., *Tetrahedron Lett.* 259 (1978)
14 PETERSON, D. J., *J. org. Chem.* **33,** 780 (1968)
15 HUDRLIK, P. F. and PETERSON, D., *J. Am. chem. Soc.* **97,** 1464 (1975); YAMAMOTO, K., TOMO, Y. and SUZUKI, S., *Tetrahedron Lett.* 2861 (1980); for alternative routes to *β*-ketosilanes, *see* DEMUTH, M., *Helv. chim. Acta* **61,** 3136 (1978); KOWALSKI, C. J., O'DOWD, M. L., BURKE, M. C. and FIELDS, K. W., *J. Am. chem. Soc.* **102,** 5411 (1980)
16 BOECKMAN, R. K. and SILVER, S. M., *Tetrahedron Lett.* 3497 (1973); *J. org. Chem.* **40,** 1755 (1975)
17 CHAN, T. H., CHANG, E. and VINOKUR, E., *Tetrahedron Lett.* 1137 (1970)
18 CHAN, T. H. and CHANG, E., *J. org. Chem.* **39,** 3264 (1974)
19 CAREY, F. A. and TOLER, J. R., *J. org. Chem.* **41,** 1966 (1976)
20 CHAN, T. H. and MYCHAJLOWSKIJ, W., *Tetrahedron Lett.* 171 (1974)
21 LAU, P. W. K. and CHAN, T. H., *Tetrahedron Lett.* 2383 (1978)
22 Ref. 7, p. 142
23 For CNDO–MO comparison with the Wittig, *see* TRINDLE, C., HWANG, J.-T. and CAREY, F. A., *J. org. Chem.* **38,** 2664 (1973)
24 CRAM, D. J. and ELHAFEZ, F. A. A., *J. Am. chem. Soc.* **74,** 5828 (1952)
25 For the stereochemical features of the analogous Sn series, *see* DAVIS, D. D. and GRAY, C. E., *J. org. Chem.* **35,** 1303 (1970); for the related transformation of ketones into exomethylene units using Ph_3SnCH_2Li, *see* KAUFFMANN, T. and KRIEGESMAN, R., *Angew. Chem. int. Edn* **16,** 862 (1977)
26 HAUSER, C. R. and HANCE, C. R., *J. Am. chem. Soc.* **74,** 5091 (1952)

27 OBAYASHI, M., UTIMOTO, J. and NOZAKI, H., *Tetrahedron Lett.* 1807 (1977)
28 DERVAN, P. B. and SHIPPEY, M. A., *J. Am. chem. Soc.* **98,** 1265 (1976)
29 REETZ, M. T. and PLACHKY, M., *Synthesis* 199 (1976)
30 VEDEJS, E. and FUCHS, P. L., *J. Am. chem. Soc.* **95,** 822 (1973); BRIDGES, A. J. and WHITHAM, G. H., *J. chem. Soc. chem. Communs* 142 (1974); SONNET, P. E. and OLIVER, J. E., *J. org. Chem.* **41,** 3279 (1976); SONNET, P. E., *Tetrahedron* **36,** 557 (1980)
31 GILMAN, H., AOKI, D. and WITTENBERG, D., *J. Am. chem. Soc.* **81,** 1107 (1959); PEDDLE, G. J. D., *J. organometal. Chem.* **14,** 115 (1968); EISCH, J. J. and TRAINOR, J. T., *J. org. Chem.* **28,** 2870 (1963)
32 COLVIN, E. W. and HAMILL, B. J., *J. chem. Soc. chem. Communs* 151 (1973); *J. Chem. Soc. Perkin I* 869 (1977)
33 GILBERT, J. C. and WEERASOORIYA, U., *J. org. Chem.* **44,** 4997 (1979)
34 GILBERT, J. C., WEERASOORIYA, U. and GIAMALVA, D., *Tetrahedron Lett.* 4619 (1979)
35 STORK, G. and COLVIN, E., unpublished observations
36 CARTER, M. J. and FLEMING, I., *J. chem. Soc. chem. Communs* 679 (1976)
37 SAKURAI, H., NISHIWAKI, K. and KIRA, M., *Tetrahedron Lett.* 4193 (1973)
38 GRÖBEL, B.-Th. and SEEBACH, D., *Chem. Ber.* **110,** 852 (1977)
39 SEYFERTH, D., LEFFERTS, J. L. and LAMBERT, R. L., *J. organometal. Chem.* **142,** 39 (1977)
40 CAREY, F. A. and COURT, A. S., *J. org. Chem.* **37,** 939 (1972); CAREY, F. A. and HERNANDEZ, O., *J. org. Chem.* **38,** 2670 (1973)
41 GRIECO, P. A., WANG, C.-L. J. and BURKE, S., *J. chem. Soc. chem. Communs* 537 (1975); but *see* ref. 15
42 LANSBURY, P. T. and SERELIS, A. J., *Tetrahedron Lett.* 1909 (1978); LANSBURY, P. T., HANGAUER, D. G. and VACCA, J. P., *J. Am. chem. Soc.* **102,** 3964 (1980)
43 SHIMOJI, K., TAGUCHI, H., OSHIMA, K., YAMAMOTO, H. and NOZAKI, H., *J. Am. chem. Soc.* **96,** 1620 (1974); TAGUCHI, H., SHIMOJI, K., YAMAMOTO, H. and NOZAKI, H., *Bull. chem. Soc. Japan* **47,** 2529 (1974); GREENE, A. E., Le DRIAN, C. and CRABBÉ, P., *J. org. Chem.* **45,** 2713 (1980)
44 HARTZELL, S. L., SULLIVAN, D. F. and RATHKE, M. W., *Tetrahedron Lett.* 1403 (1974)
45 HARTZELL, S. L. and RATHKE, M. W., *Tetrahedron Lett.* 2757 (1976)
46 LUCAST, D. H. and WEMPLE, J., *Tetrahedron Lett.* 1103 (1977)
47 OJIMA, I., KUMAGAI, M. and NAGAI, Y., *Tetrahedron Lett.* 4005 (1974)
48 RATHKE, M. W. and WOODBURY, R. P., *J. org. Chem.* **42,** 1688 (1977); **43,** 1947 (1978); WOODBURY, R. P. and RATHKE, M. W., *Tetrahedron Lett.* 709 (1978)
49 HART, D. J., CAIN, P. A. and EVANS, D. A., *J. Am. chem. Soc.* **100,** 1549 (1978)
50 CHAN, T. H. and MORELAND, M., *Tetrahedron Lett.* 515 (1978)
51 HARTZELL, S. L. and RATHKE, M. W., *Tetrahedron Lett.* 2737 (1976)
52 KANO, S., EBATA, T., FUNAKI, K. and SHIBUYA, S., *Synthesis* 746 (1978)
53 COREY, E. J. and BOGER, D. L., *Tetrahedron Lett.* 5, 9, 13 (1978)
54 COREY, E. J. and ENDERS, D., *Tetrahedron Lett.* 3 (1976); COREY, E. J., ENDERS, D. and BOCK, M. G., *Tetrahedron Lett.* 7 (1976); COREY, E. J., CLARK, D. A., GOTO, G., MARFAT, A., MIOSKOWSKI, C., SAMUELSSON, B. and HAMMARSTRÖM, S., *J. Am. chem. Soc.* **102,** 1463 (1980)
55 KAUFFMANN, T., KOCH, U., STEINSEIFER, F. and VAHRENHORST, A., *Tetrahedron Lett.* 3341 (1977)
56 SACHDEV, K., *Tetrahedron Lett.* 4041 (1976)
57 CAREY, F. A. and COURT, A. S., *J. org. Chem.* **37,** 1926 (1972)
58 JONES, P. F. and LAPPERT, M. F., *J. chem. Soc. chem. Communs* 526 (1972)
59 SEEBACH, D., GRÖBEL, B.-Th., BECK, A. K., BRAUN, M. and GEISS, K.-H., *Angew. Chem. int. Edn* **11,** 443 (1972); SEEBACH, D., KOLB, M. and GRÖBEL, B.-Th., *Chem. Ber.* **106,** 2277 (1973); *Tetrahedron Lett.* 3171 (1974); GRÖBEL, B.-Th., BURSTINGHAUS, R. and SEEBACH, D., *Synthesis* 121 (1976)
60 ANDERSEN, N. H., YAMAMOTO, Y. and DENNISTON, A. D., *Tetrahedron Lett.* 4547 (1975)
61 VAN DER LEIJ, M., PORSKAMP, P. A. T. W., LAMMERINK, B. H. M. and ZWANENBURG, B., *Tetrahedron Lett.* 811, 3383 (1978)
62 FLEISCHMANN, C. and ZBIRAL, E., *Tetrahedron* **34,** 317 (1978)
63 COOKE, F., MAGNUS, P. and BUNDY, G. L., *J. chem. Soc. chem. Communs* 714 (1978)

64 LAU, P. W. K. and CHAN, T. H., *Tetrahedron Lett.* 2383 (1978)

65 CORRIU, R. J. P., LANNEAU, G. F., LECLERCQ, D. and SAMATE, D., *J. organometal. Chem.* **144,** 155 (1978) and references therein; AYALON-CHASS, D., EHLINGER, E. and MAGNUS, P., *J. chem. Soc. chem. Communs* 772 (1977)

66 CUNICO, R. F. and DEXHEIMER, E. M., *J. Am. chem. Soc.* **94,** 2868 (1972)

67 CUNICO, R. F. and DEXHEIMER, E. M., *J. organometal. Chem.* **59,** 153 (1973); *see also* CUNICO, R. F. and HAN, Y.-K., *J. organometal. Chem.* **105,** C29 (1976); CUNICO, R. F. and CHOU, B. B., *J. organometal. Chem.* **154,** C45 (1978)

68 GRÖBEL, B.-Th. and SEEBACH, D., *Chem. Ber.* 867 (1977)

69 CHAN, T. H. and MYCHAJLOWSKIJ, W., *Tetrahedron Lett.* 3479 (1974)

70 CHAN, T. H., MYCHAJLOWSKIJ, W., ONG, B. S. and HARPP, D. N., *J. org. Chem.* **43,** 1526 (1978); *J. organometal. Chem.* C1 (1976)

71 CHAN, T. H., LI, M. P., MYCHAJLOWSKIJ, W. and HARPP, D. N., *Tetrahedron Lett.* 3511 (1974); for a related cleavage, *see* DJURIC, S., SARKAR, T. and MAGNUS, P., *J. Am. chem. Soc.* **102,** 6885 (1980)

72 CHAN, T. H., ONG, B. S. and MYCHAJLOWSKIJ, W., *Tetrahedron Lett.* 3253 (1976); ONG, B. S. and CHAN, T. H., *Tetrahedron Lett.* 3257 (1976); CHAN, T. H. and ONG, B. S., *J. org. Chem.* **43,** 2994 (1978)

73 CHAN, T. H., LAU, P. W. K. and LI, M. P., *Tetrahedron Lett.* 2667 (1976)

74 SEYFERTH, D., BURLITCH, J. M., MINASZ, R. J., MUI, J. Y. P., SIMMONS, H. D., TREIBER, A. J. H. and DOWD, S. R., *J. Am. chem. Soc.* **87,** 4259 (1965)

75 MILLER, R. B., *Synth. Communs* **4,** 341 (1974)

76 SEYFERTH, D. and JULA, T. F., *J. organometal. Chem.* **14,** 109 (1968)

77 CHAN, T. H. and MASSUDA, D., *Tetrahedron Lett.* 3383 (1975); MASSUDA, D., Ph.D. Thesis, McGill University (1977)

78 BUCHANAN, G. L., *Chem. Soc. Rev.* **3,** 41 (1974)

79 KÖBRICH, G., *Angew. Chem. int. Edn* **12,** 464 (1973)

80 KEESE, R., *Angew. Chem. int. Edn* **14,** 528 (1975)

81 CHAN, T. H. and MASSUDA, D., *J. Am. chem. Soc.* **99,** 936 (1977)

82 KOCIENSKI, P. J., *Tetrahedron Lett.* 2649 (1979); *J. org. Chem.* **45,** 2037 (1980)

83 SOMMER, L. H., PIOCH, R. P., MARANS, N. S., GOLDBERG, G. M., ROCKETT, J. and KERLIN, J., *J. Am. chem. Soc.* **75,** 2932 (1953); PRICE, C. C. and SOWA, J. R., *J. org. Chem.* **32,** 4126 (1967)

84 SOMMER, L. H. and MARANS, N. S., *J. Am. chem. Soc.* **72,** 1935 (1950)

85 EBERSON, L., *Acta chem. scand.* **10,** 633 (1956)

86 *See,* for example, MELVIN, L. S. and TROST, B. M., *J. Am. chem. Soc.* **94,** 1790 (1972)

87 FLEMING, I. and GOLDHILL, J., *J. chem. Soc. chem. Communs* 176 (1978); AGER, D. J. and FLEMING, I., *J. chem. Soc. chem. Communs* 177 (1978); FLEMING, I. and PERCIVAL, A., *J. chem. Soc. chem. Communs* 178 (1978); FLEMING, I. and GOLDHILL, J., *J. chem. Soc. Perkin I* 1493 (1980)

88 HUDRLIK, P. F., HUDRLIK, A. M., NAGENDRAPPA, G., YIMENU, T., ZELLERS, E. T. and CHIN, E., *J. Am. chem. Soc.* **102,** 6894 (1980)

89 SIEBER, P., *Helv. chim. Acta* **60,** 2711 (1977)

90 GERLACH, H., *Helv. chim. Acta* **60,** 3039 (1977)

91 CARPINO, L. A. and TSAV, J. H., *J. chem. Soc. chem. Communs* 358 (1978); *see,* for example, MEYERS, A. I., COMINS, D. L., ROLAND, D. M., HENNING, R. and SHIMUZU, K., *J. Am. chem. Soc.* **101,** 7104 (1979)

92 LIPSHUTZ, B. H. and PEGRAM, J. J., *Tetrahedron Lett.* 3343 (1980)

93 CHAN, T. H. and Di STEFANO, M., *J. chem. Soc. chem. Communs* 761 (1978)

94 VEDEJS, E. and MARTINEZ, G. R., *J. Am. chem. Soc.* **101,** 6452 (1979); *see also* SATO, Y., YAGI, Y. and KOTO, M., *J. org. Chem.* **45,** 613 (1980); SATO, Y. and SAKAKIBARA, H., *J. organometal. Chem.* **166,** 303 (1979)

95 GREBER, G. and KRICHELDORF, H. R., *Angew. Chem. int. Edn* **7,** 941, 942 (1968)

96 PIRKLE, W. H. and HOEKSTRA, M. S., *J. org. Chem.* **39,** 3904 (1974)

97 PIRKLE, W. H. and HAUSKE, J. R., *J. org. Chem.* **42,** 2781 (1977); PIRKLE, W. H. and RINALDI, P. L., *J. org. Chem.* **43,** 3803 (1978)

98 JUNG, M. E. and HUDSPETH, J. P., *J. Am. chem. Soc.* **102,** 2463 (1980)

99 MAGNUS, P. and ROY, G., *J. chem. Soc. chem. Communs* 822 (1979)

100 COREY, E. J., TIUS, M. A. and DAS, J., *J. Am. chem. Soc.* **102,** 1742 (1980)

101 KONAKAHARA, T. and TAKAGI, Y., *Tetrahedron Lett.* 2073 (1980)

102 MATTESON, D. S. and MAJUMDAR, D., *J. chem. Soc. chem. Communs* 39 (1980)

Chapter 13

Alkynyl- and allenyl-silanes

13.1 Alkynylsilanes

Terminal alkynes are converted readily into alkynylsilanes, usually by reaction[1-3] of the alkyne anion or its equivalent with a suitable silyl chloride (*Scheme 13.1*). The reverse reaction, liberation of the terminal alkyne, is easily effected[4] by several reagent combinations, including hydroxide ion[5], methanolysis[6], silver(I) ion[7a] followed by cyanide ion[7b], fluoride ion[8], and the methyl-lithium–lithium bromide complex[9]. The degree of protection afforded can be modified by judicious selection of the triorganosilyl moiety (*Table 13.1*)[5].

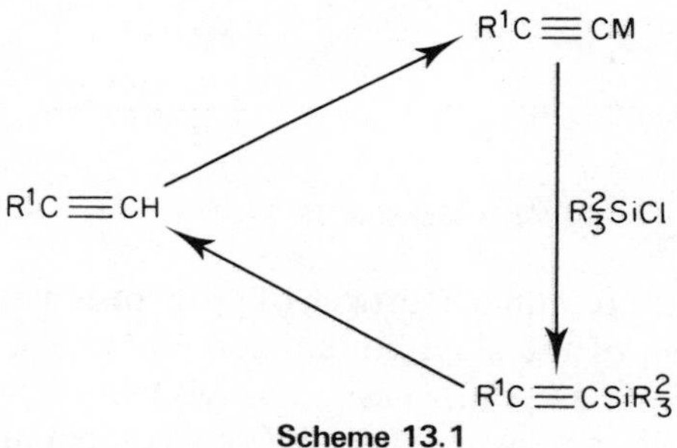

Scheme 13.1

Table 13.1 Relative rates of base-induced cleavage of triorganosilylphenylethynes

$$R_3SiC{\equiv}CPh + {}^-OH \xrightarrow[29.4\,°C]{MeOH} PhC{\equiv}CH + R_3SiO^-$$

R_3Si	Me_3Si	$EtMe_2Si$	Et_2MeSi	Et_3Si	Ph_3Si
Relative rate	277	49	7.4	1	11.8

There are two main purposes for the terminal silylation of alkynes: either to mask the potentially acidic ethynyl proton and thus, or otherwise, afford a degree of chemical protection to the triple bond, or to activate regioselectively the triple bond towards electrophilic attack (*see also* Chapters 7 and 10).

13.1.1 Terminal protection

The use of the trialkylsilyl group to afford protection to terminal alkynes is a most active area, important contributions having been made by Walton[6] and

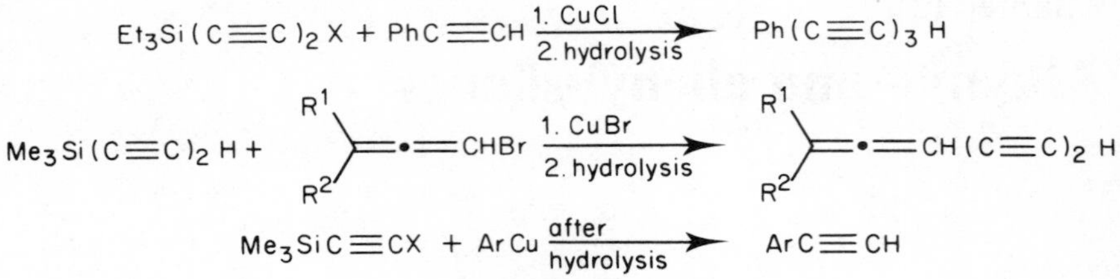

$$Et_3Si(C{\equiv}C)_2X + PhC{\equiv}CH \xrightarrow[\text{2. hydrolysis}]{\text{1. CuCl}} Ph(C{\equiv}C)_3H$$

$$Me_3Si(C{\equiv}C)_2H + R^1R^2C{=}C{=}CHBr \xrightarrow[\text{2. hydrolysis}]{\text{1. CuBr}} R^1R^2C{=}C{=}CH(C{\equiv}C)_2H$$

$$Me_3SiC{\equiv}CX + ArCu \xrightarrow[\text{hydrolysis}]{\text{after}} ArC{\equiv}CH$$

Scheme 13.2

others, resulting in routes to polyalkynes[10], allene diynes[11], and aryl alkynes[12] (*Scheme 13.2*) and a bis-dihydroannulene[13].

Selective reduction of non-terminal triple bonds in polyalkynes is possible if the terminal alkyne is protected first by silylation[7a], as illustrated by the semihydrogenation of alkyne (1) to give the terminal (*Z*)-enyne unit in an approach[14] to histrionicotoxin. The isomeric (*E*)-enyne system can be obtained[8] by reaction of an aldehyde with the ylide from the Wittig salt (2) (*Scheme 13.3*).

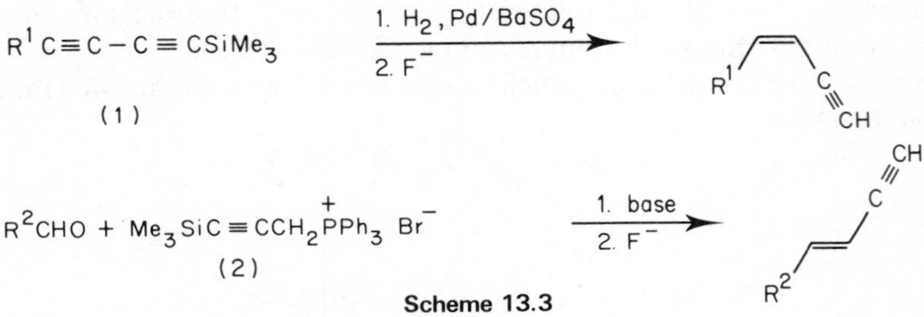

$$\underset{(1)}{R^1C{\equiv}C{-}C{\equiv}CSiMe_3} \xrightarrow[\text{2. } F^-]{\text{1. } H_2\text{, Pd/BaSO}_4} (Z)\text{-}R^1CH{=}CH{-}C{\equiv}CH$$

$$R^2CHO + \underset{(2)}{Me_3SiC{\equiv}CCH_2\overset{+}{P}Ph_3\ \overset{-}{Br}} \xrightarrow[\text{2. } F^-]{\text{1. base}} (E)\text{-}R^2CH{=}CH{-}C{\equiv}CH$$

Scheme 13.3

The relatively acidic terminal hydrogen of propyne is masked by silylation, allowing preparation of the alkyl-lithium compound (3), a species used in routes to homologous alkyl alkynes[7b], allenylsilanes[15], α-santalol[16], some triterpenoids[17], and in the classic synthesis[18] or Cecropia juvenile hormone.

The related organocopper species (4) adds (1,6) to penta-2,4-dienoate esters in a simple route[19] to functionalized 1,5-enynes and 1,4,5-en-allenes.

$$\underset{(3)}{Me_3SiC{\equiv}CCH_2Li} \qquad \underset{(4),\ R_3 = Me_3,\ Bu^tMe_2}{R_3SiC{\equiv}CCH_2Cu}$$

Similar terminal protection has been employed in synthetic routes[20] to α-acetylenic-α-aminoacids and acetylenic amines and diamines, all of which are of interest as potential or proven enzyme inhibitors. As a prelude to the next section, that of regioselective activation to electrophilic attack, the preparation of the key intermediate (6) should be noted in terms of its selectivity, considering the number of potential electrophilic sites in the precursor (5) (*Scheme 13.4*).

The ethynyl unit can be induced to add to *s-trans-αβ*-unsaturated ketones in the manner shown[21] in *Scheme 13.5*: previous methods were either multi-

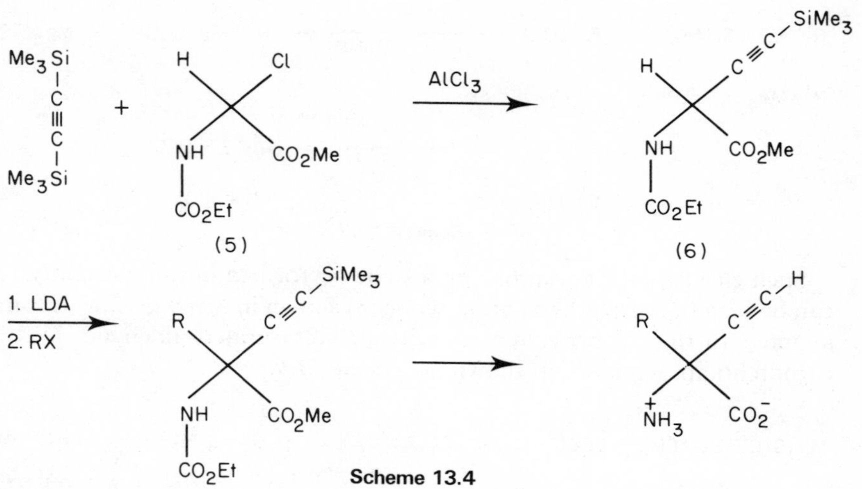

$Me_3SiC{\equiv}CSiMe_3$ + (5) $\xrightarrow{AlCl_3}$ (6)

(5) (6)

$\xrightarrow[2.\ RX]{1.\ LDA}$ → $RC(C{\equiv}CH)(\overset{+}{N}H_3)CO_2^-$

Scheme 13.4

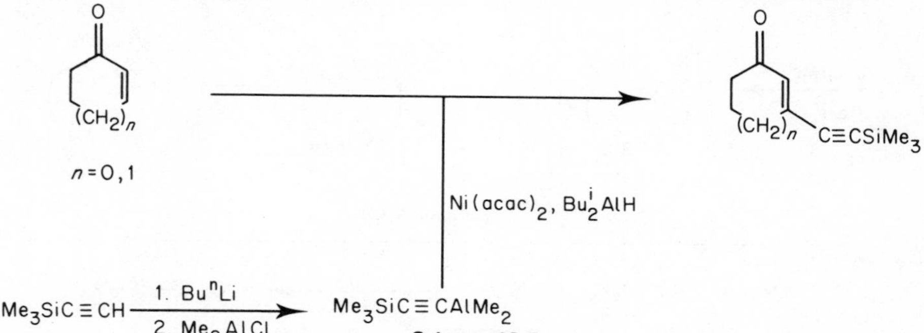

$n = 0, 1$

Ni(acac)$_2$, Bu$^i_2$AlH

$Me_3SiC{\equiv}CH \xrightarrow[2.\ Me_2AlCl]{1.\ Bu^nLi} Me_3SiC{\equiv}CAlMe_2$

Scheme 13.5

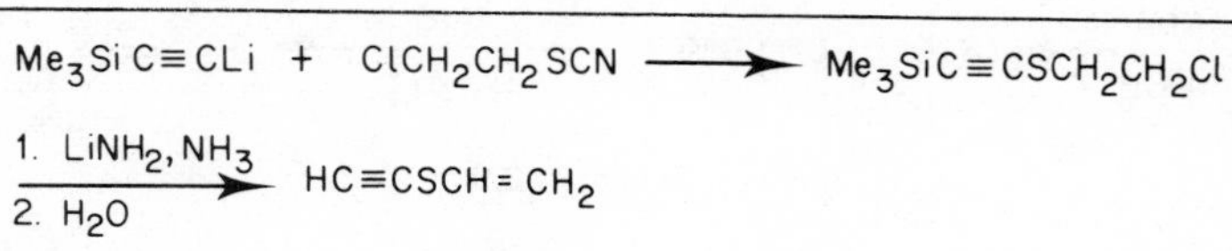

$Me_3SiC{\equiv}CLi + ClCH_2CH_2SCN \longrightarrow Me_3SiC{\equiv}CSCH_2CH_2Cl$

$\xrightarrow[2.\ H_2O]{1.\ LiNH_2,\ NH_3} HC{\equiv}CSCH{=}CH_2$

(7)

Scheme 13.6

step, or succeeded only with *s-cis* enones, i.e., when a cyclic transition state was possible.

The ethynyl vinyl thioether (7) is readily obtained[22] from ethynyl trimethylsilane (*Scheme 13.6*).

13.1.2 Regioselective electrophilic attack

Alkynylsilanes undergo electrophilic attack under generally mild conditions, and with terminal regioselectivity. For example, conversion into alkynyl ketones is readily and efficiently achieved[23] (*Scheme 13.7*).

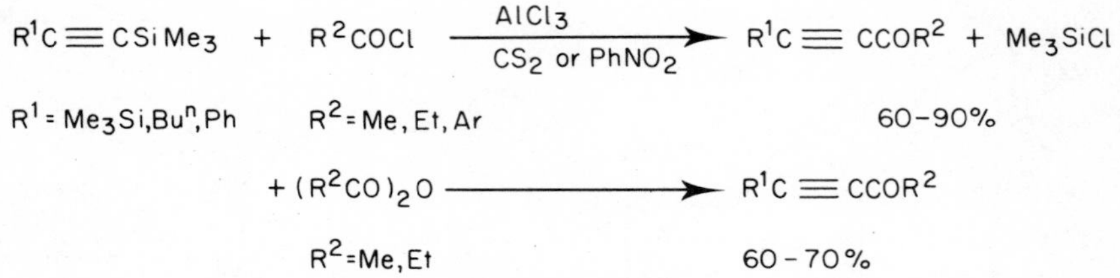

$$R^1C{\equiv}CSiMe_3 + R^2COCl \xrightarrow[CS_2 \text{ or } PhNO_2]{AlCl_3} R^1C{\equiv}CCOR^2 + Me_3SiCl$$

$R^1 = Me_3Si, Bu^n, Ph$ $\quad R^2 = Me, Et, Ar$ $\quad$ 60–90%

$$+ (R^2CO)_2O \longrightarrow R^1C{\equiv}CCOR^2$$

$R^2 = Me, Et$ $\quad$ 60–70%

Scheme 13.7

Such electrophilic acylation can also be performed intramolecularly[24], as can be seen in the synthesis of muscone (8) shown in *Scheme 13.8*. It can be adapted further to provide a route to $\alpha\beta$-unsaturated aldehydes by two-carbon homologation[25], as shown in *Scheme 13.9*.

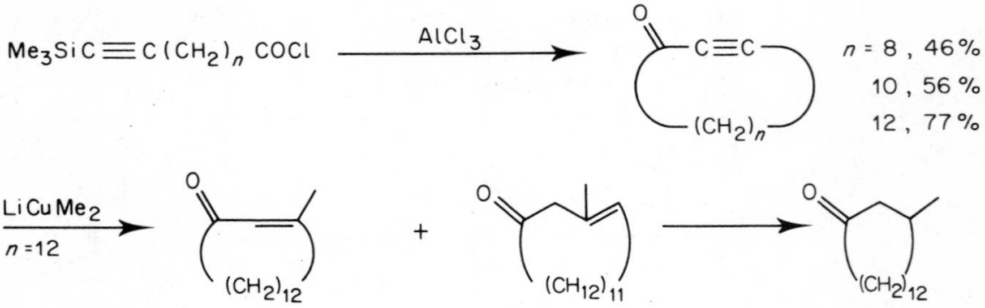

$$Me_3SiC{\equiv}C(CH_2)_nCOCl \xrightarrow{AlCl_3} \text{cyclic } C(=O){-}C{\equiv}C{-}(CH_2)_n$$

n = 8, 46%
10, 56%
12, 77%

$\xrightarrow[n=12]{LiCuMe_2}$ (CH2)12 enone + (CH12)11 enone ⟶ (CH2)12 ketone (8)

Scheme 13.8

$$Me_3SiC{\equiv}CSiMe_3 + RCOCl \xrightarrow[CH_2Cl_2]{AlCl_3} RCOC{\equiv}CSiMe_3$$

$$\xrightarrow[MeOH]{0.1M\ MeO^-} RCOCH_2CH(OMe)_2 \xrightarrow[2.\ H_3O^+]{1.\ NaBH_4} RCH{=}CHCHO$$

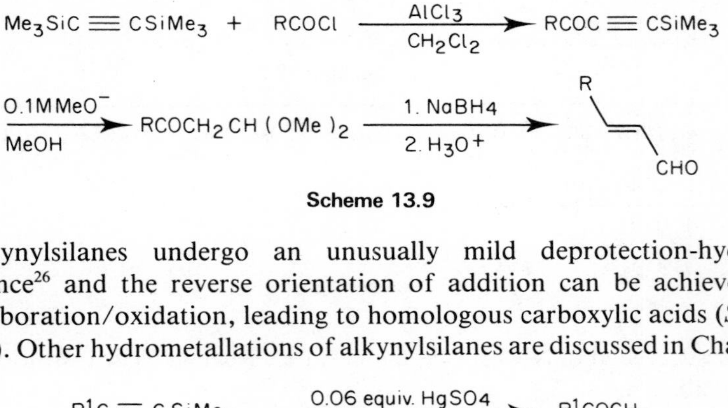

Scheme 13.9

Alkynylsilanes undergo an unusually mild deprotection-hydration sequence[26] and the reverse orientation of addition can be achieved[27] by hydroboration/oxidation, leading to homologous carboxylic acids (*Scheme 13.10*). Other hydrometallations of alkynylsilanes are discussed in Chapter 7.

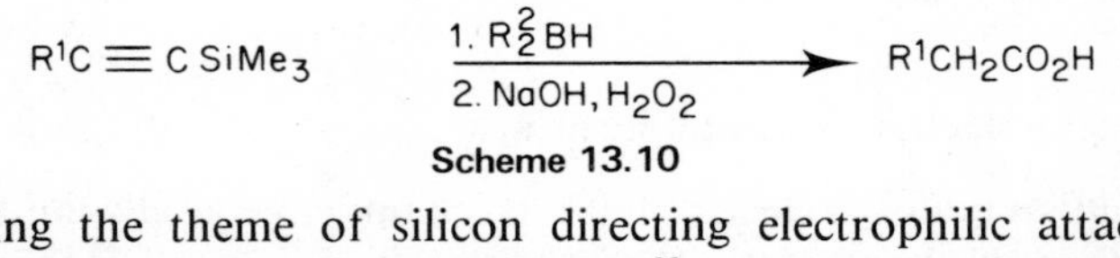

$$R^1C{\equiv}CSiMe_3 \xrightarrow[\text{trace } H^+,\ H_2O,\ THF]{0.06 \text{ equiv. } HgSO_4} R^1COCH_3$$

$$R^1C{\equiv}CSiMe_3 \xrightarrow[2.\ NaOH,\ H_2O_2]{1.\ R^2_2BH} R^1CH_2CO_2H$$

Scheme 13.10

Continuing the theme of silicon directing electrophilic attack to the α-position in alkynylsilanes, Heathcock[28] has reported a remarkable differentiation between trimethylsilyl and methyl terminal substituents on alkynes. When subjected to intramolecular carbonium ion attack, the alkynylsilane (9) cyclized to a bicyclo[3.2.2.]nonene (10), whereas the methylalkyne (11) gave a bicyclo[2.2.2.]octene (12) (*Scheme 13.11*). This can be ascribed to a variety of factors, including the preference of the methyl-substituted vinyl cation to be linear, the β-stabilizing effect of silicon, and the likely steric preference[29] for electrophilic attack on an alkyne bond to occur at that end which carries the larger substituent, because of the probable angle of approach to the *sp*-hybridized carbon.

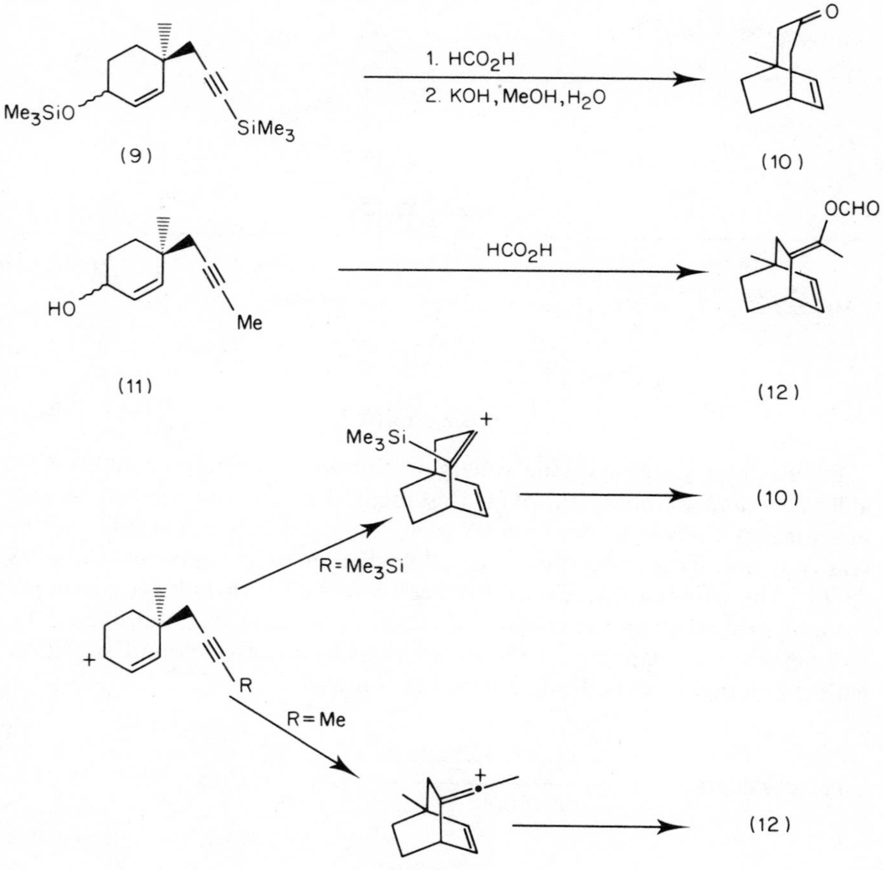

Me₃SiO (9) SiMe₃ — 1. HCO_2H 2. KOH, MeOH, H_2O → (10)

HO (11) Me — HCO_2H → OCHO (12)

Me_3Si + → (10); R = Me_3Si; + R; R = Me; + → (12)

Scheme 13.11

An identical cyclization differentiation was observed by Johnson[30], exploring the poly-olefin cyclization route to steroids and related species. Using the trimethylsilylalkyne unit as cation terminator, he observed exclusive production of a D-homosteroid, whereas the methylalkyne cyclized to the normal steroid skeleton (*Scheme 13.12*). Other cycloaddition reactions of alkynylsilanes are discussed in detail elsewhere (pp. 129–131).

β-Silylated ynamines have the ability to insert[31] electrophilic π-systems of various kinds into their C–Si single bonds, as exemplified in *Scheme 13.13*.

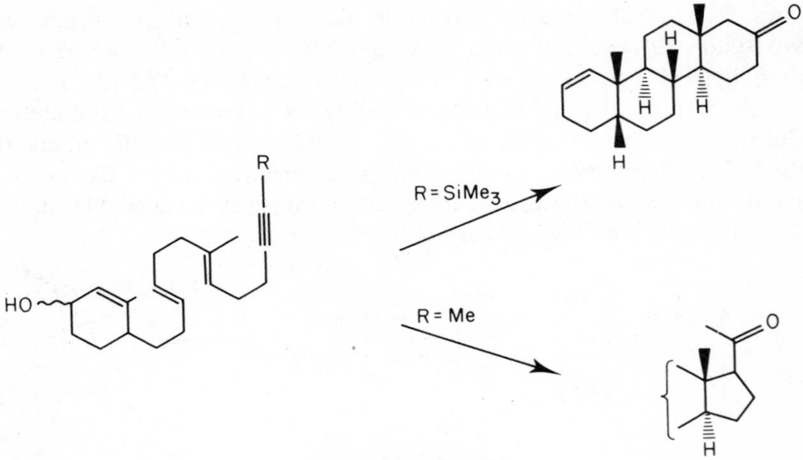

R = $SiMe_3$

HO

R = Me

H

H H

O

O

H

Scheme 13.12

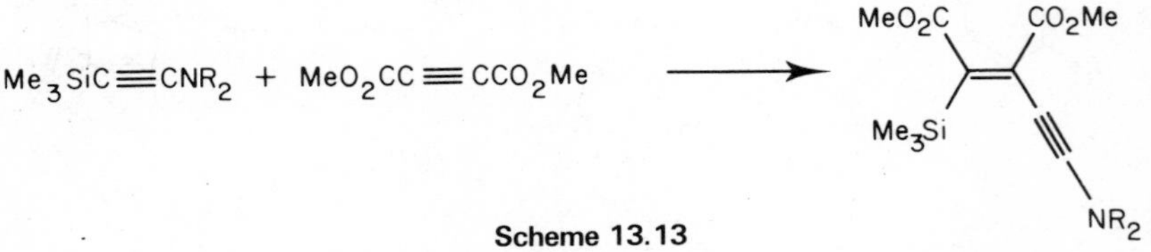

$Me_3SiC{\equiv}CNR_2$ + $MeO_2CC{\equiv}CCO_2Me$ ⟶

MeO_2C CO_2Me Me_3Si NR_2

Scheme 13.13

Finally, phenylethynylsilane undergoes fluoride ion-catalysed addition to aldehydes and ketones; although this method cannot be applied to $\alpha\beta$-unsaturated carbonyl compounds, it does provide[32] a remarkably mild, relatively non-basic generation of an alkynyl anion or its equivalent (*Scheme 13.14*). The scope of this process has been extended to include mono-anions from bis(trimethylsilyl)alkynes and dialkynes[9]; in such cases, it proved to be more efficient to generate the requisite anions by selective monodesilylation with the methyl-lithium–lithium bromide complex.

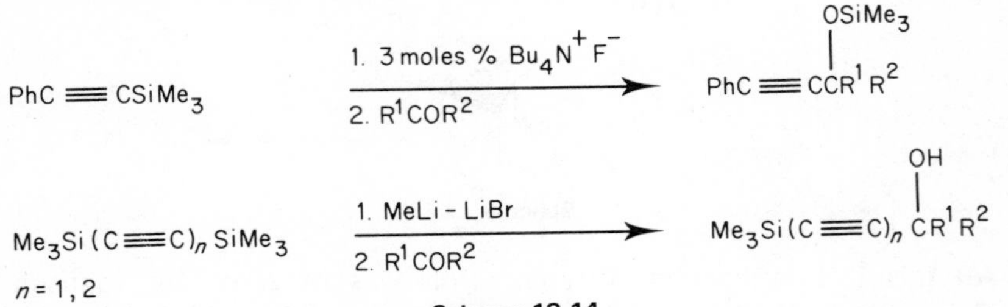

$PhC{\equiv}CSiMe_3$ → (1. 3 moles % $Bu_4N^+F^-$; 2. R^1COR^2) → $PhC{\equiv}CCR^1R^2(OSiMe_3)$

$Me_3Si(C{\equiv}C)_nSiMe_3$, $n = 1, 2$ → (1. MeLi - LiBr; 2. R^1COR^2) → $Me_3Si(C{\equiv}C)_nCR^1R^2(OH)$

Scheme 13.14

13.2 Allenylsilanes

Treatment of γ-acetoxyalkynylsilanes with dialkyl cuprates results[33] in direct replacement of the acetoxy group (*Scheme 13.15*). On the other hand, with

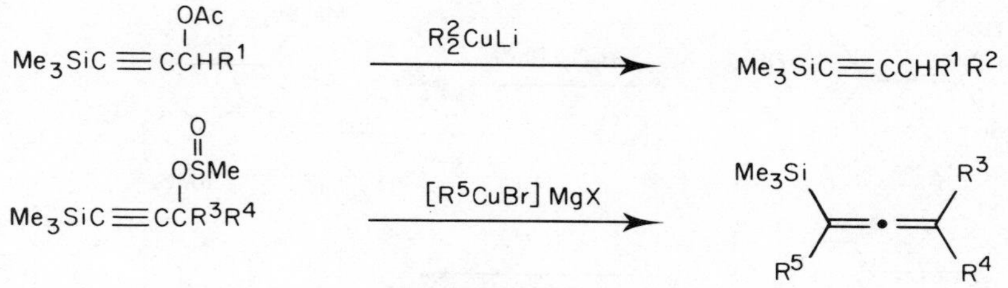

$$Me_3SiC{\equiv}CCH(OAc)R^1 \xrightarrow{R^2_2CuLi} Me_3SiC{\equiv}CCHR^1R^2$$

$$Me_3SiC{\equiv}CCR^3R^4(OS(O)Me) \xrightarrow{[R^5CuBr]\,MgX} (Me_3Si)(R^5)C{=}C{=}CR^3R^4$$

Scheme 13.15

methanesulphinyl derivatives and using a complex organometallic reagent, α-attack occurs[34] and allenylsilanes are produced in good yield.

Three alternative and somewhat more specialized routes[35,36] to these species are outlined in *Scheme 15.16.*

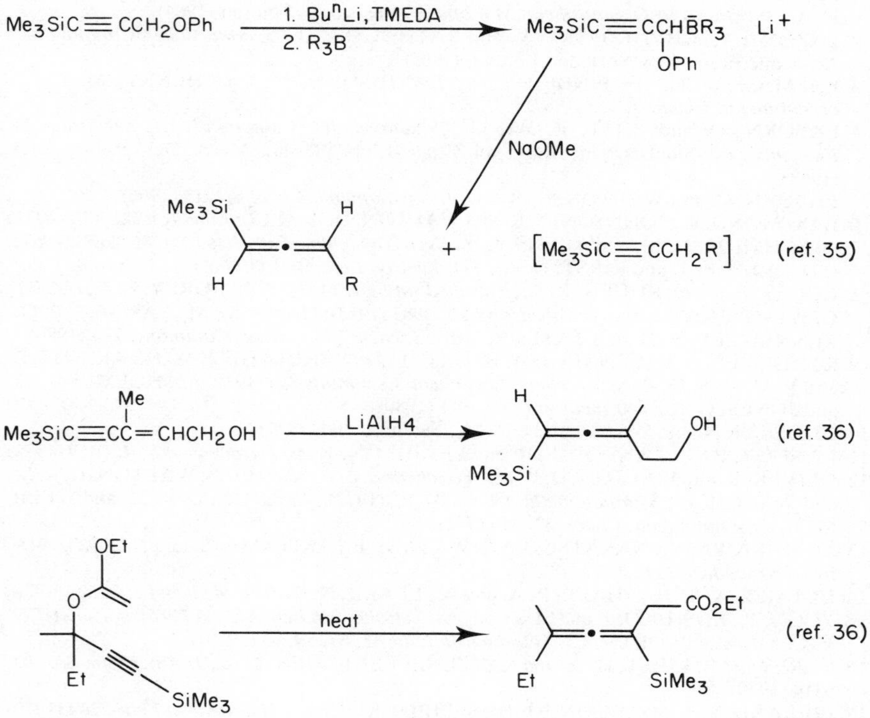

$$Me_3SiC{\equiv}CCH_2OPh \xrightarrow[2.\ R_3B]{1.\ Bu^nLi,\ TMEDA} Me_3SiC{\equiv}CCH(OPh)\bar{B}R_3\ Li^+ \xrightarrow{NaOMe} (Me_3Si)HC{=}C{=}CHR + [Me_3SiC{\equiv}CCH_2R] \quad \text{(ref. 35)}$$

$$Me_3SiC{\equiv}CC(Me){=}CHCH_2OH \xrightarrow{LiAlH_4} (H)(Me_3Si)C{=}C{=}C(Me)CH_2CH_2OH \quad \text{(ref. 36)}$$

$$EtC(Me)(C{\equiv}CSiMe_3)OC(OEt){=}CH_2 \xrightarrow{heat} (Me)(Et)C{=}C{=}C(SiMe_3)CH_2CO_2Et \quad \text{(ref. 36)}$$

Scheme 13.16

Allenylsilanes show little promise so far of major synthetic utility. They appear to be unreactive towards organocuprates, and Lewis acid-catalysed electrophilic reactions either do not proceed under mild conditions, or result in polymerization under more drastic ones[37]. Oxidation of (2-hydroxyethyl)allenylsilanes produces lactones, intermediate silyl-stabilized carbanions (Chapter 2) being invoked[36] to account for the observed regiospecificity (*Scheme 13.17*).

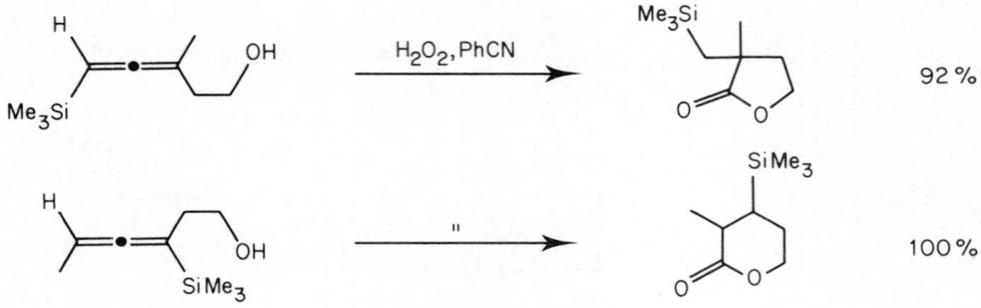

Scheme 13.17

References

1 PETROV, A. D., MIRONOV, B. F., PONOMARENKO, V. A. and CHERNYSHEV, E. A., *'Synthesis of Organosilicon Monomers'*, Heywood, London (1964)

2 BAZANT, V., CHVALOVSKY, V. and RATHOUSKY, J., *'Organosilicon Compounds'*, Academic Press, New York and London (1965)

3 For $Me_3SiC\equiv CH$, *see* FINDEISS, W., DAVIDSOHN, W. and HENRY, M. C., *J. organometal. Chem.* **9,** 435 (1967)

4 EABORN, C. and BOTT, R. W., in *'Organometallic Compounds of the Group IV Elements'*, Ed. MacDiarmid, A. G., vol. 1, part 1, pp. 398–402, Marcel Dekker, New York (1968)

5 EABORN, C. and WALTON, D. R. M., *J. organometal. Chem.* **4,** 217 (1966)

6 EASTMOND, R., JOHNSON, T. R. and WALTON, D. R. M., *Tetrahedron* **28,** 4601 (1972)

7 (*a*) SCHMIDT, H. M. and ARENS, J. F., *Recl Trav. chim. Pays-Bas Belg.* **86,** 1138 (1967); (*b*) COREY, E. J. and KIRST, H. A., *Tetrahedron Lett.* 5041 (1968)

8 COREY, E. J. and RUDEN, R. A., *Tetrahedron Lett.* 1495 (1973); COREY, E. J., FLEET, G. W. J. and KATO, M., *Tetrahedron Lett.* 3963 (1973); HANN, M. M., SAMMES, P. G., KENNEWELL, P. D. and TAYLOR, J. B., *J. chem. Soc. chem. Communs,* 234 (1980)

9 (*a*) HOLMES, A. B., JENNINGS-WHITE, C. L. D., SCHULTHESS, A. H., AKINDE, B. and WALTON, D. R. M., *J. chem. Soc. chem. Communs* 840 (1979); (*b*) HOLMES, A. B. and JONES, G. E., *Tetrahedron Lett.* 3111 (1980)

10 GHOSE, B. N. and WALTON, D. R. M., *Synthesis* 890 (1974)

11 LANDOR, P. D., LANDOR, S. R. and LEIGHTON, J. P., *Tetrahedron Lett.* 1019 (1973)

12 OLIVER, R. and WALTON, D. R. M., *Tetrahedron Lett.* 5209 (1972); WALTON, D. R. M. and WAUGH, F., *J. organometal. Chem.* **37,** 45 (1972); WALTON, D. R. M. and WEBB, M. J., *J. organometal. Chem.* **37,** 41 (1972)

13 YOSHIKAWA, Y., NAKATSUJI, S., IWATANI, F., AKIYAMA, S. and NAKAGAWA, M., *Tetrahedron Lett.* 1737 (1977)

14 HOLMES, A. B., RAPHAEL, R. A. and WELLARD, N. K., *Tetrahedron Lett.* 1539 (1976)

15 YOGO, T., KOSHINO, J. and SUZUKI, A., *Tetrahedron Lett.* 1781 (1979); *see also* SHEN, C. C. and AINSWORTH, C., *Tetrahedron Lett.* 83, 87, 89, 93 (1979)

16 COREY, E. J., KIRST, H. A. and KATZENELLENBOGEN, J. A., *J. Am. chem. Soc.* **92,** 6314 (1970)

17 IRELAND, R. E., DAWSON, M. I. and LIPINSKI, C. A., *Tetrahedron Lett.* 2247 (1970)

18 COREY, E. J., KATZENELLENBOGEN, J. A. and POSNER, G. A., *J. Am. chem. Soc.* **89,** 4245 (1967)

19 GANEM, B., *Tetrahedron Lett.* 4467 (1974)

20 METCALF, B. W., BEY, P., DANZIN, C., JUNG, M. J., CASARA, P. and VEVERT, J. P., *J. Am. chem. Soc.* **100,** 2551 (1978); CASARA, P. and METCALF, B. W., *Tetrahedron Lett.* 1581 (1978); METCALF, B. W. and CASARA, P., *J. chem. Soc. chem. Communs* 119 (1979)

21 HANSEN, R. T., CARR, D. B. and SCHWARTZ, J., *J. Am. chem. Soc.* **100,** 2244 (1978)

22 VERBOOM, W., MEIJER, J. and BRANDSMA, L., *Synthesis* 577 (1978)

23 BIRKOFER, L., RITTER, A. and UHLENBRAUCK, H., *Chem. Ber.* **96,** 3280 (1963)

24 UTIMOTO, K., TANAKA, M., KITAI, M. and NOZAKI, H., *Tetrahedron Lett.* 2301 (1978)

25 NEWMAN, H., *J. org. Chem.* **38,** 2254 (1973); for similar version with HC≡CH itself *see* WAKAYAMA, S., ITOH, S., YUI, S. and MAEKAWA, H., *Nippon kagaku zasshi* **78,** 1525 (1957)

26 McCRAE, D. A. and DOLBY, L., *J. org. Chem.* **42,** 1607 (1977)

27 ZWEIFEL, G. and BACKLUND, S. J., *J. Am. chem. Soc.* **99,** 3184 (1977)

28 KOZAR, L. G., CLARK, R. D. and HEATHCOCK, C. H., *J. org. Chem.* **42,** 1386 (1977)

29 BALDWIN, J. E., *J. chem. Soc. chem. Communs* 734 (1976)

30 JOHNSON, W. S., YARNELL, T. M., MYERS, R. F. and MORTON, D. R., *Tetrahedron Lett.* 2549 (1978); JOHNSON, W. S., BRINKMEYER, R. S., KAPOOR, V. M. and YARNELL, T. M., *J. Am. chem. Soc.* **99,** 8341 (1977); JOHNSON, W. S., YARNELL, T. M., MEYERS, R. F., MORTON, D. R. and BOOTS, S. G., *J. org. Chem.* **45,** 1254 (1980); *see also* DESPO, A. D., CHIU, S. K., FLOOD, T. and PETERSON, P. E., *J. Am. chem. Soc.* **102,** 5120 (1980); SCHMID, R., HUESMANN, P. L. and JOHNSON, W. S., *J. Am. chem. Soc.* **102,** 5122 (1980)

31 HIMBERT, G., *Angew. Chem. int. Edn* **15,** 51 (1976); *J. chem. Res. S* **3,** 104 (1978)

32 NAKAMURA, E. and KUWAJIMA, I., *Angew. Chem. int. Edn* **15,** 498 (1976)

33 BRINKMEYER, R. S. and MACDONALD, T. L., *J. chem. Soc. chem. Communs* 876 (1978)

34 WESTMIJZE, H. and VERMEER, P., *Synthesis* 390 (1979)

35 YOGO, T., KOSHINO, J. and SUZUKI, A., *Tetrahedron Lett.* 1781 (1979); *see also* SHEN, C. C. and AINSWORTH, C., *Tetrahedron Lett.* 83, 87, 89, 93 (1979)

36 BERTRAND, M., DULCERE, J.-P.and GIL, G., *Tetrahedron Lett.* 1945, 4271 (1980)

37 MONTURY, M., PSAUME, B. and GORÉ, J., *Tetrahedron Lett.* 163 (1980); but *see* DANHEISER, R. L. and CARINI, D. J., *J. org. Chem.* **45,** 3925 (1980); JELLAL, A. and SANTELLI, M., *Tetrahedron Lett.* 4487 (1980)

Chapter 14

Silylketenes

Silyl substitution appears to diminish greatly the tendency of ketenes to undergo [2 + 2]-cycloaddition. Trimethylsilylketene (1) itself is readily prepared[1] by mild pyrolysis of ethoxytrimethylsilylethyne (*Scheme 14.1*), and is a colourless, stable oil, with little or no tendency to dimerize. It does undergo a thermal [2 + 2]-cycloaddition reaction with tetraethoxyethene[2a], and an acid-catalysed cycloaddition process with saturated aldehydes[2b].

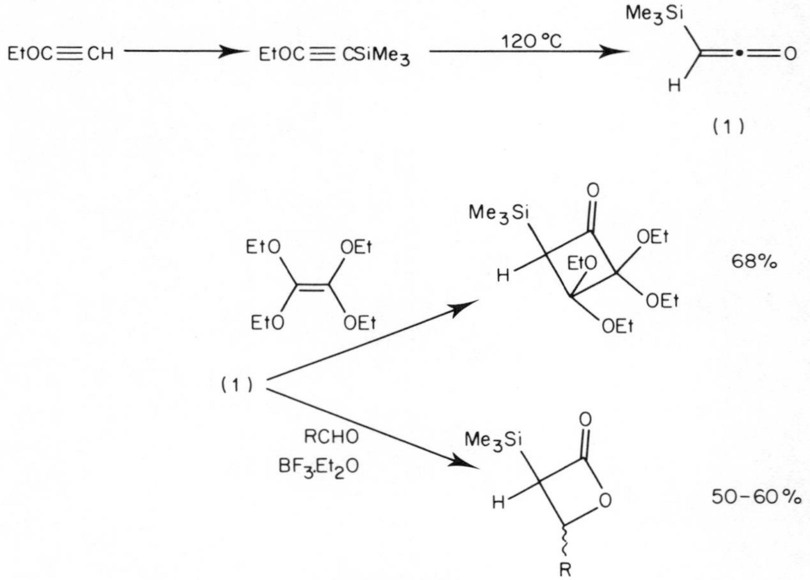

Scheme 14.1

It also acts as a potent acylating agent[3] for hindered amines and tertiary alcohols, and has been used for the preparation of trimethylsilyl substituted allenes and alkynes, as shown in *Scheme 14.2*.

Trimethylsilylbromoketene (2), on the other hand, is considerably more reactive[4], undergoing cycloaddition readily with imines and related species (*Scheme 14.3*).

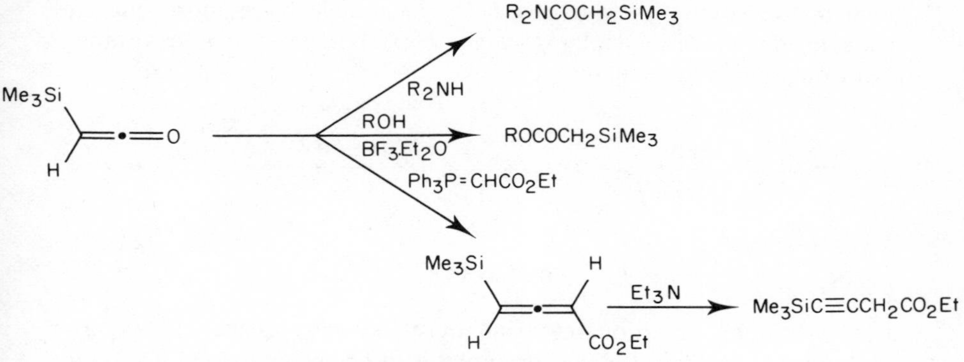

$R_2NCOCH_2SiMe_3$

$Me_3Si(H)C{=}C{=}O$ — R_2NH; ROH, $BF_3.Et_2O$; $Ph_3P{=}CHCO_2Et$

$ROCOCH_2SiMe_3$

$Me_3Si(H)C{=}C{=}C(H)CO_2Et$ —Et_3N→ $Me_3SiC{\equiv}CCH_2CO_2Et$

Scheme 14.2

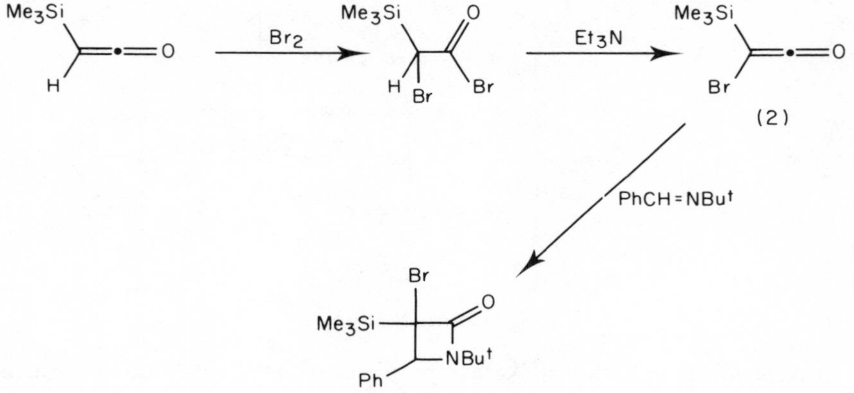

$Me_3Si(H)C{=}C{=}O$ —Br_2→ $Me_3SiCH(Br)COBr$ —Et_3N→ $Me_3Si(Br)C{=}C{=}O$ (2)

PhCH=NBu^t

Br, Me_3Si, O, NBu^t, Ph

Scheme 14.3

Trimethylsilylmethylketene (3) possesses normal ketene reactivity[5], with spontaneous dimerization taking place, and therefore must be generated *in situ* for appropriate reaction (*Scheme 14.4*).

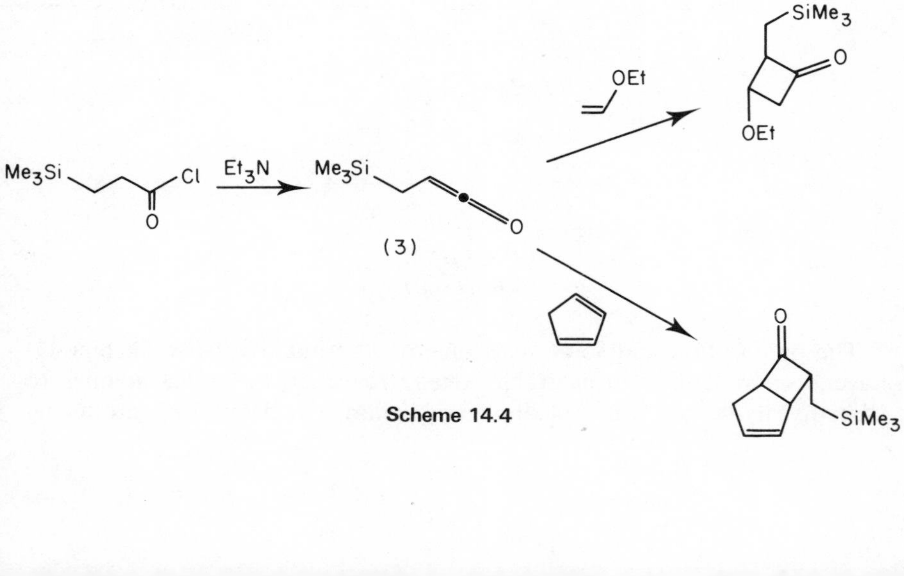

$Me_3SiCH_2CH_2COCl$ —Et_3N→ $Me_3SiCH_2CH{=}C{=}O$ (3)

OEt; SiMe$_3$, O, OEt

O; SiMe$_3$

Scheme 14.4

Bis(trimethylsilyl)ketene (4) has been generated[6] by a most unusual fragmentation of an ester enolate (*Scheme 14.5*). It appears to be very stable, existing once again as the monomer.

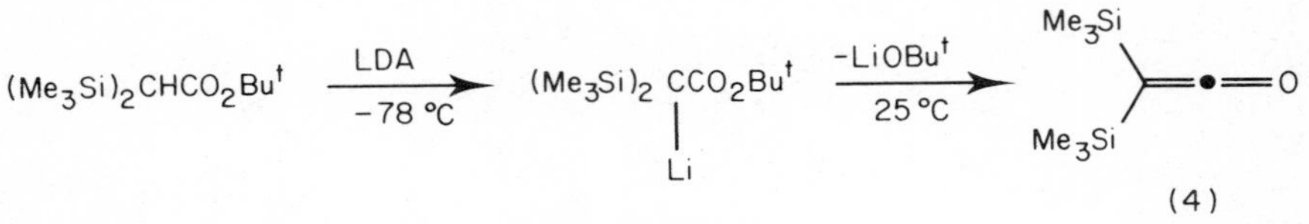

$$(Me_3Si)_2CHCO_2Bu^t \xrightarrow[-78\ °C]{LDA} (Me_3Si)_2C(Li)CO_2Bu^t \xrightarrow[25\ °C]{-LiOBu^t} (Me_3Si)_2C{=}C{=}O\ (4)$$

Scheme 14.5

n-Butyl-lithium deprotonates trimethylsilylketene; quenching experiments[7] provide evidence for the formation of a ketene enolate (5) (*Scheme 14.6*).

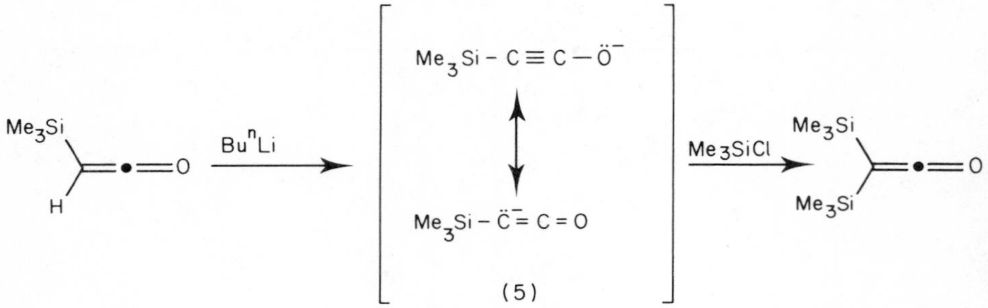

$$(Me_3Si)(H)C{=}C{=}O \xrightarrow{Bu^nLi} \left[Me_3Si{-}C{\equiv}C{-}\ddot{O}^- \longleftrightarrow Me_3Si{-}\ddot{C}^-{=}C{=}O \right] (5) \xrightarrow{Me_3SiCl} (Me_3Si)_2C{=}C{=}O$$

Scheme 14.6

Ozonolysis of trimethylsilylketene produces trimethylsilyl formate; since unreactive ketenes such as di-t-butylketene yield α-lactones under such conditions, it has been proposed[8] that a similar α-lactone (6) might be an intermediate in this reaction (*Scheme 14.7*).

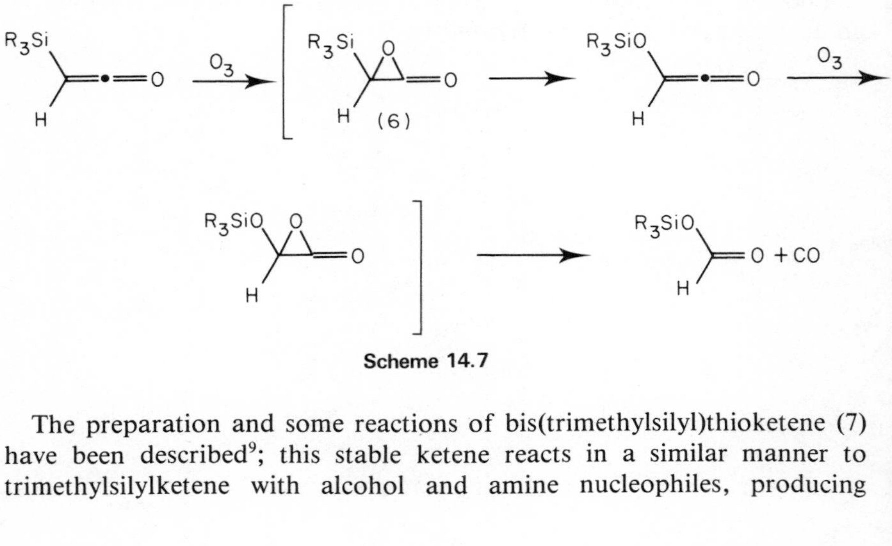

$$(R_3Si)(H)C{=}C{=}O \xrightarrow{O_3} \left[\alpha\text{-lactone } (6) \longrightarrow (R_3SiO)(H)C{=}C{=}O \xrightarrow{O_3} \alpha\text{-lactone} \right] \longrightarrow (R_3SiO)(H)C{=}O + CO$$

Scheme 14.7

The preparation and some reactions of bis(trimethylsilyl)thioketene (7) have been described[9]; this stable ketene reacts in a similar manner to trimethylsilylketene with alcohol and amine nucleophiles, producing

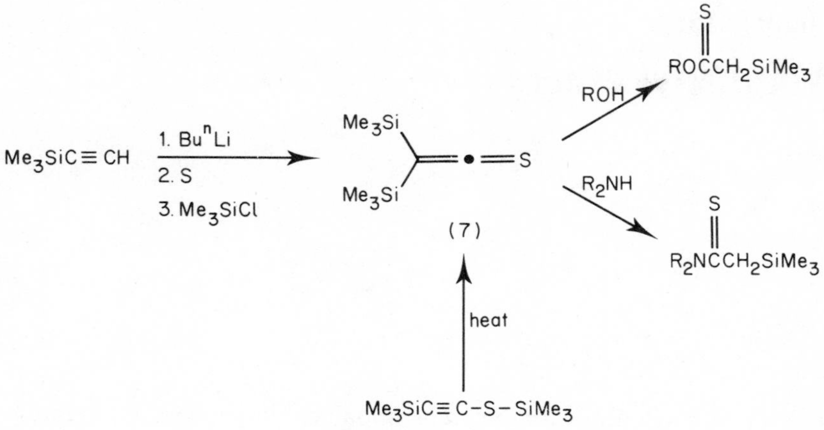

Scheme 14.8

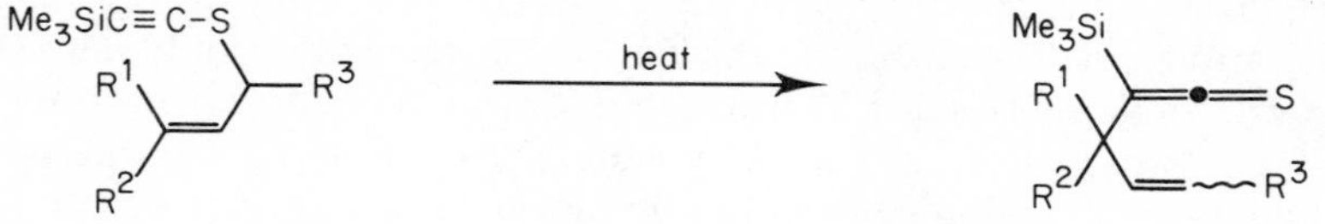

Scheme 14.9

O-alkylthioesters and thioamides (*Scheme 14.8*). Interestingly, the isomeric thioalkyne (8) rearranges thermally to this thioketene.

The thioalkynes (9) also isomerize[10] to thioketenes, although here the mechanism is clearly a thio-Claisen rearrangement (*Scheme 14.9*).

References

1 SHCHUKOUSKAYA, L. L., PAL'CHIK, R. I. and LAZAREV, A. N., *Dokl. Akad. Nauk SSSR* **164,** 357 (1965); *Chem. Abstr.* **63,** 18138 (1965)

2 (*a*) BRADY, W. T. and SAIDI, K., *J. org. Chem.* **45,** 727 (1980); (b) **44,** 733 (1979); (*c*) *see also* ZAITSEVA, G. S., BAUKOV, Yu. I., MAL'TSEV, V. V. and LUTSENKO, I. F., *J. gen. Chem. U.S.S.R.* **44,** 1389 (1974); ZAITSEVA, G. S., VINOKUROVA, N. G. and BAUKOV, Yu. I., *J. gen. Chem. U.S.S.R.* **45,** 1372 (1975)

3 RUDEN, R. A., *J. org. Chem.* **39,** 3607 (1974)

4 BRADY, W. T. and OWENS, R. A., *Tetrahedron Lett.* 1553 (1976)

5 BRADY, W. T. and CHENG, T. C., *J. org. Chem.* **42,** 732 (1977)

6 SULLIVAN, D. F., WOODBURY, R. P. and RATHKE, M. W., *J. org. Chem.* **42,** 2038 (1977)

7 WOODBURY, R. P., LONG, N. R. and RATHKE, M. W., *J. org. Chem.* **43,** 376 (1978)

8 BRADY, W. T. and SAIDI, K., *Tetrahedron Lett.* 721 (1978)

9 HARRIS, S. J. and WALTON, D. R. M., *J. chem. Soc. chem. Communs* 1008 (1976)

10 SCHAUMANN, E. and GRABLEY, F.-F., *Tetrahedron Lett.* 4307 (1977)

Chapter 15

Alkyl silyl ethers

This chapter will deal with the preparation, stability range, and cleavage of alkyl silyl ethers. It will not discuss these features in anything more than general terms, as the number of instances of use of silyl ethers is vast, and comprehensive coverage would be quite impossible. It will, however, highlight certain interesting applications. Specifically excluded are references to silylation as a derivatization procedure to confer GLC volatility or readily characterizable mass spectral fragmentation, both fields having been well reviewed[1,2] elsewhere. In addition, the preparation and properties of α-silyloxy-azides and -nitriles will be discussed subsequently (Chapter 18).

15.1 Solvolysis

Before exploring the preparative chemistry of silyl ethers, it is appropriate to consider the comparative rates of their acid- and base-catalysed solvolyses. A kinetic study[3] of the methanolysis [equations (1) and (2)] of a range of silyl ethers derived from phenol gave the second order rate constants shown in *Table 15.1.* For phenoxysilanes, it is apparent that base-catalysed solvolysis is faster than acid-catalysed solvolysis. A similar kinetic study[4] of the methanolysis of a range of silyl ethers derived from menthol showed the reverse trend (*Table 15.2*): alkoxysilanes tend to be more stable in base than in acid. In general, a *decrease* in the pK_a of the conjugate acid of the leaving group will bring about an *increase* in the rate of reaction with basic reagents, and an *increase* in the steric bulk of the substituents on silicon will result in a *decrease* in the rate of cleavage under either set of conditions.

$$R_3SiOR^1 + MeOH \xrightarrow{MeO^-} R_3SiOMe + R^1OH \quad (1)$$

$$R_3SiOR^1 + MeOH \xrightarrow{MeOH_2^+} R_3SiOMe + R^1OH \quad (2)$$

Table 15.1 Second order rate constants for base- (k_1) and acid- (k_2) catalysed methanolysis of R_3SiOPh

R_3Si	k_1/l mol^{-1} s^{-1}	k_2/l mol^{-1} s^{-1}
Me_3Si	330	10.4
Et_3Si	2.1	0.22
Pr^n_3Si	0.66	0.12
Bu^n_3Si	0.41	0.81
$(n-C_5H_{11})_3Si$	0.30	0.06
Bu^tMe_2Si	1.72×10^{-2}	5.9×10^{-4}

Table 15.2 Second order rate constants for base- (k_1) and acid- (k_2) catalysed methanolysis of R_3SiO-menthyl

R_3Si	k_1/l mol^{-1} s^{-1}	k_2/l mol^{-1} s^{-1}
Me_3Si	1.31×10^{-2}	1.4×10^{-3}
Et_3Si	1.01×10^{-5}	2.2×10^{2}
Pr^iMe_2Si	2.17×10^{-5}	16.25
Pr^i_3Si		1.8×10^{-3}
$PhMe_2Si$	3.6×10^{-2}	1.21×10^{3}
Ph_3Si	1.2×10^{-2}	3.66

15.2 Trimethylsilyl ethers, $ROSiMe_3$

15.2.1 Preparation

An excellent early paper[5] describes the most generally useful routes to alkyl (and aryl) silyl ethers. For simple primary and secondary alcohols, direct reaction with trimethylsilyl chloride (TMCS) in the presence of a stoichiometric amount of pyridine, with toluene or benzene as solvent, is the most convenient procedure; it has the minor disadvantage of requiring removal by filtration of the pyridine hydrochloride prior to evaporation of solvent and distillation of the product (*Scheme 15.1*). For tertiary and other hindered alcohols, the combination of equimolar amounts of hexamethyldisilazane (HMDS) and trimethylsilyl chloride in a non-polar solvent such as hexane is most effective, a minimum amount of precipitate being formed. For most purposes, one of these two methods should suffice.

$$ROH + Me_3SiCl + py \longrightarrow ROSiMe_3 + py.HCl$$

R = p or s-alkyl

$$ROH + Me_3SiCl + (Me_3Si)_2NH \longrightarrow ROSiMe_3 + NH_4Cl$$

Scheme 15.1

Table 15.3 Some silylating reagents

Reagent	*Name and symbol*	*Notes*	*References*
$X_3C-C(OSiMe_3)=NSiMe_3$, X = H	*N,O*-Bistrimethylsilylacetamide, BSA	1	7
$X_3C-C(OSiMe_3)=NSiMe_3$, X = F	*N,O*-Bistrimethylsilyltrifluoro-acetamide, BSTFA	1	7
$MeC(=O)NHSiMe_3$	*N*-Trimethylsilylacetamide	2	8
$Me_3SiNHC(=O)OSiMe_3$	*N,O*-Bistrimethylsilyl-carbamate	3	9
Me_3SiNEt_2	*N*-Trimethylsilyldiethylamide, TMSDEA	4	10
Me_3Si-N (imidazole)	*N*-Trimethylsilylimidazole, TMSI	5	11
$Me_3SiNHSO_3SiMe_3$	*N,O*-Bistrimethylsilylsulphamic acid	6	12
$(Me_3SiNH)_2CO$	*N,N'*-Bistrimethylsilylurea, BSU	7	13

[1]Universal application, very powerful, own solvent.
[2]Pyridine as cosolvent.
[3]Universal application, gaseous by-products NH_3 and CO_2.
[4]Selectively silylates equatorial alcohols and 11-hydroxy group of $PGF_{2\alpha}$.
[5]Universal application, very powerful, basic amino groups do not react.
[6]Sulphamic acid by-product is water-soluble, organic-insoluble.
[7]Universal, powerful, cheap.

Recently, a range of new silylating agents has been introduced[6], conferring higher reactivity, or more selectivity, or easier isolation (*Table 15.3*).

The general order of reactivity of the common silylating agents towards alcoholic hydroxyl groups is

$$\text{TMSI} > \text{silylamides} > \text{TMSDEA} > \text{HMDS/TMCS} > \text{TMCS/pyridine}$$

Considerable regio- and stereo-selectivity are readily attainable. The rate of silylation of secondary alcohols by HMDS in pyridine at 25 °C varies[14] over a factor of 10^3 from *endo*-fenchol to *exo*-norborneol. TMSDEA silylates equatorial hydroxyl groups, axial alcohols being unreactive under the conditions used; it selectively silylates[15] the prostaglandin F series at the 11-, and, if secondary, the 15-position, allowing clean conversion into the E series (*Scheme 15.2*). The demands made by the prostaglandins, both in synthesis and in interconversion, have done much to stimulate studies of techniques of protection which involve silyl ethers.

Methods have been devised to permit selective protection of hydroxyl groups in *C*-nucleosides[16] and other *C*-glycosides[17,18], although for such purposes a range of trialkylsilyl groups can be used.

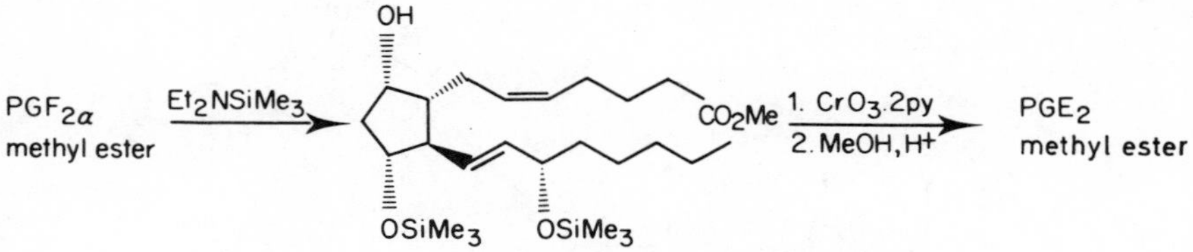

PGF$_{2\alpha}$ methyl ester $\xrightarrow{Et_2NSiMe_3}$ [bis-OSiMe$_3$ intermediate, CO_2Me] $\xrightarrow[2.\ MeOH,\ H^+]{1.\ CrO_3.2py}$ PGE$_2$ methyl ester

Scheme 15.2

15.2.2 Cleavage

Cleavage of trimethylsilyl ethers to the parent alcohols occurs readily on treatment with nucleophiles such as methanol, often with acidic or basic catalysis. Indeed, it is the solvolytic lability of such ethers which limits their utility, and has led to the introduction of the more hindered, more stable silyl ethers which will be discussed shortly. However, there are a number of cases where such lability can be tolerated.

15.2.3 Applications

The vast majority of applications of trimethylsilyl ethers is linked to GLC and mass spectrometry, although protection of hydroxy groups as their trimethylsilyl ethers has found use in the syntheses[19] of several natural products. Heathcock[20] has used the protected α-hydroxyketone (1) as a synthon in the preparation of 3-hydroxy-2-methyl- and 3-hydroxy-2,4-dimethyl-carboxylic acids. These acids can be produced with a high degree of acyclic stereoselection, and have obvious potential as macrolide and related ionophore synthons (*Scheme 15.3*). Here the silyloxy moiety not only performs a protecting function but also, by acting as a large, sterically congested group, encourages formation of the (*Z*)-enolate anion; the observed stereoselectivity[21] then results from the energetically favoured chair-like transition state believed to be involved in this aldol condensation.

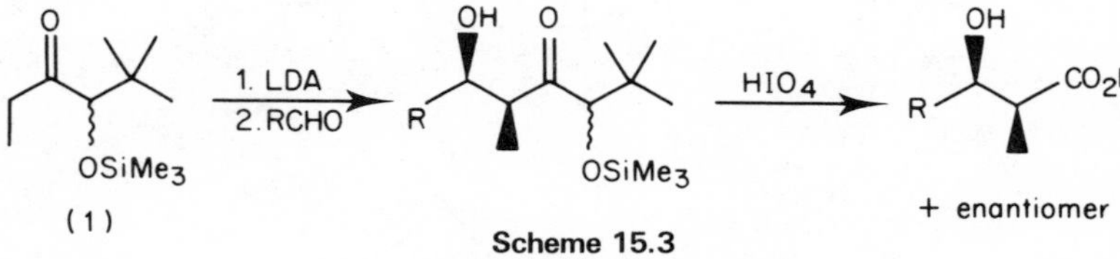

(1) [OSiMe$_3$] $\xrightarrow[2.\ RCHO]{1.\ LDA}$ [OH, O, R, OSiMe$_3$] $\xrightarrow{HIO_4}$ R–CH(OH)–CH(CH$_3$)–CO_2H + enantiomer

Scheme 15.3

Oxy-Cope and silyloxy-Cope rearrangements of the diene (2) gave quite different products in a dramatic demonstration[22] of the way in which hydroxyl masking can modify the course of reaction (*Scheme 15.4*).

Similarly, an intramolecular Diels–Alder route[23] to eudesmane sesquiterpenes failed with the free alcohol (3), but succeeded with the silyl ether (4) (*Scheme 15.5*).

The labile sulphenic acid partner in the reversible thermal rearrangement of penicillin sulphoxides was trapped[24] as the corresponding silyl ester (5);

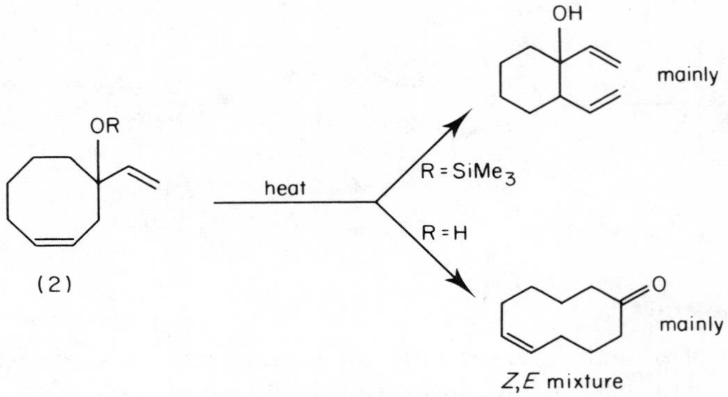

OH

mainly

OR

$R = SiMe_3$

heat

R = H

(2)

O

mainly

Z,E mixture

Scheme 15.4

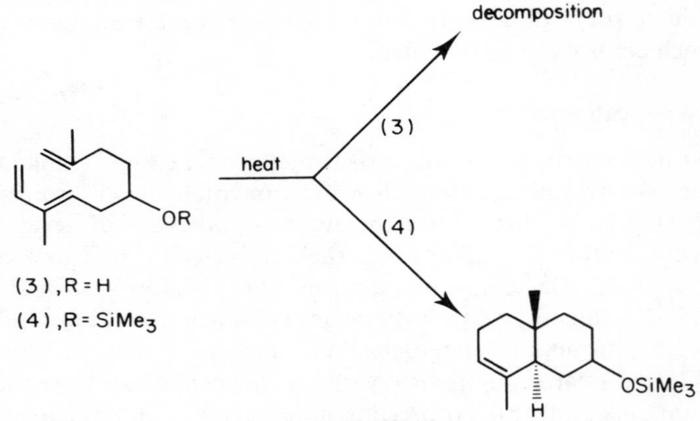

decomposition

(3)

heat

(4)

OR

(3), R = H

(4), $R = SiMe_3$

$OSiMe_3$

H

Scheme 15.5

H H O

Ft S

N

O

CO_2R

Me_3SiCl

heat

H H $OSiMe_3$

Ft S

N

O

CO_2R

(5)

$MeSO_3H$

H H

Ft S

N

O

CO_2R

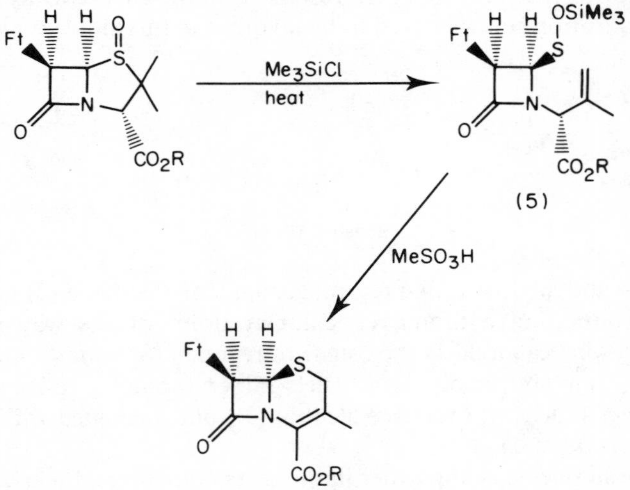

Scheme 15.6

this functions as a masked RS^+ species, as was shown by its acid-catalysed cyclization to give the cephem nucleus (*Scheme 15.6*).

Oxiranes react with trimethylsilyl trifluoromethanesulphonate (6) and a strong non-nucleophilic base to produce the corresponding allylic alcohol trimethylsilyl ethers in excellent yield[25]. This reaction (*Scheme 15.7*), which proceeds well with all except acyclic 2,3-disubstituted and monosubstituted oxiranes, also shows considerable selectivity. Ring opening occurs preferentially at the more substituted carbon; the second step, proton abstraction, takes place away from the silyloxy group, and shows *E*2 characteristics. Considerable regio- and chemo-selectivity are attainable, as illustrated [ketones and some esters react (p. 205) with this reagent in the presence of Et_3N to produce the corresponding enol ethers].

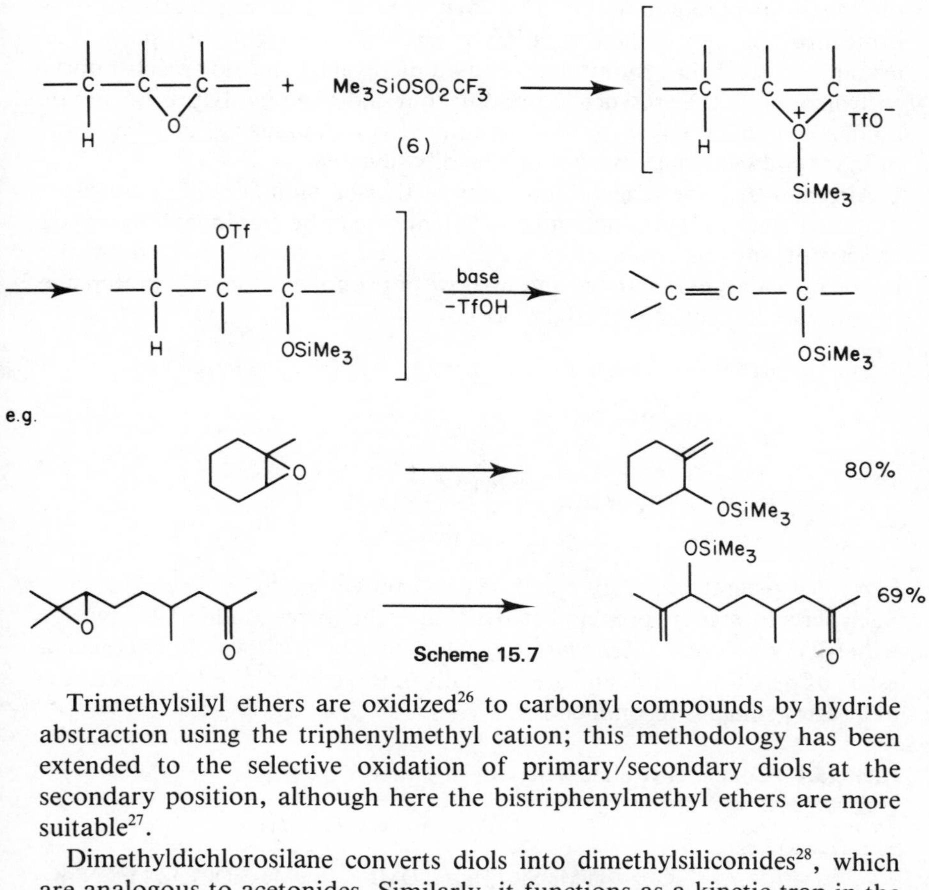

$Me_3SiOSO_2CF_3$ (6); TfO$^-$; SiMe$_3$; OTf; OSiMe$_3$; base, −TfOH

e.g. 80%; 69%

Scheme 15.7

Trimethylsilyl ethers are oxidized[26] to carbonyl compounds by hydride abstraction using the triphenylmethyl cation; this methodology has been extended to the selective oxidation of primary/secondary diols at the secondary position, although here the bistriphenylmethyl ethers are more suitable[27].

Dimethyldichlorosilane converts diols into dimethylsiliconides[28], which are analogous to acetonides. Similarly, it functions as a kinetic trap in the

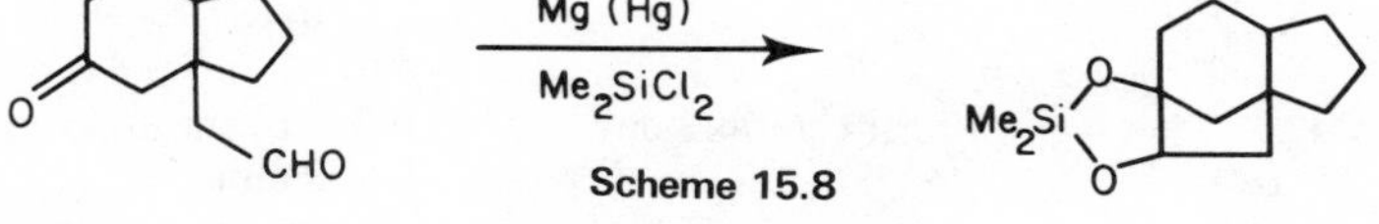

CHO; Mg (Hg), Me_2SiCl_2; Me_2Si

Scheme 15.8

gibberellin[29]-orientated pinacol cyclization[30] shown in *Scheme 15.8*, a complex mixture of products being formed in its absence.

15.3 t-Butyldimethylsilyl ethers, $ROSiMe_2Bu^t$, and related species

15.3.1 Preparation and cleavage

As stated earlier, the acid- and base-catalysed solvolytic lability of alkyl trimethylsilyl ethers significantly limits their synthetic utility. In response to this, a variety of more hindered silyl groups has been introduced; these are not only significantly more stable towards solvolysis (particularly base-induced), but also their steric bulk can afford protection to a particular area or face of the parent molecule. However, the steric bulk which affords such protection also resists the formation of the silyl ether in the first place. For example, Corey[31] has found that t-butyldimethylsilyl chloride reacts poorly with alcohols in the presence of pyridine, but under the catalytic influence of imidazole (which will form *N*-t-butyldimethylsilylimidazole) in DMF both primary and secondary alcohols are readily silylated.

Alternatively, the reagent combination of t-butyldimethylsilyl chloride, 4-(dimethylamino)pyridine, and triethylamine can be employed[32] in a wide variety of solvents; this gives yields as good as those achieved by the imidazole-catalysed method, and it shows a pronounced kinetic preference for primary as against secondary alcohols.

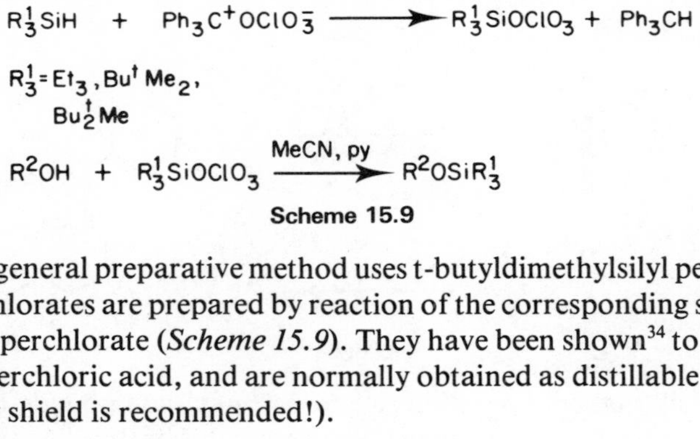

$$R^1_3SiH + Ph_3C^+OClO_3^- \longrightarrow R^1_3SiOClO_3 + Ph_3CH$$

$R^1_3 = Et_3, Bu^tMe_2, Bu^t_2Me$

$$R^2OH + R^1_3SiOClO_3 \xrightarrow{\text{MeCN, py}} R^2OSiR^1_3$$

Scheme 15.9

A third general preparative method uses t-butyldimethylsilyl perchlorate[33]. Such perchlorates are prepared by reaction of the corresponding silyl hydride with trityl perchlorate (*Scheme 15.9*). They have been shown[34] to be covalent esters of perchloric acid, and are normally obtained as distillable liquids (use of a safety shield is recommended!).

Table 15.4 Comparative effectiveness of silylation

Alcohol	*Silylating conditions*	
	Bu^tMe_2SiCl, ImH, DMF	$Bu^tMe_2SiOClO_3$, MeCN, py
Bu^tOH	30 per cent in 3 days	100 per cent in 5 min $t_{1/2} \sim 30$ s
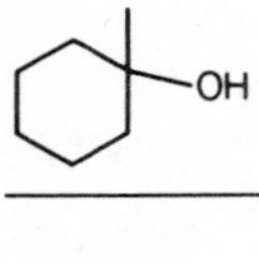 OH	10 per cent in 3 days	100 per cent in 20 min $t_{1/2} \sim 3$ min

They react with hindered alcohols, including tertiary ones, in the presence of pyridine (*Scheme 15.9*), to give the desired silyl ethers in much superior yields to those obtainable by alternative methods, as illustrated in *Table 15.4*.

The di-t-butylmethylsilyl ethers prepared by this method proved to be much more stable solvolytically than the corresponding t-butyldimethylsilyl analogues (*Table 15.5*).

Table 15.5 Comparison of silyl ether stability

Alkyl silyl ether	*Cleavage conditions*	
	1 per cent HCl in 95 per cent EtOH	5 per cent NaOH in EtOH
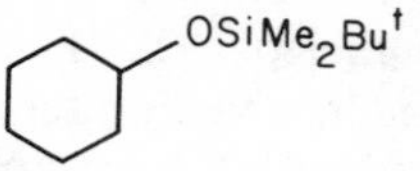 Cyclohexyl–$OSiMe_2Bu^t$	100 per cent cleavage in 15 min at 20 °C	15 per cent cleavage in 9 h at 80 °C
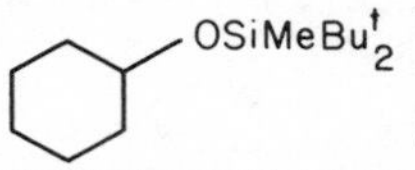 Cyclohexyl–$OSiMeBu^t_2$	No cleavage in 3 days at 20 °C 50 per cent cleavage in 24 h at 90 °C	No cleavage in 3 days at 80 °C

Indeed, it proved most difficult to cleave such highly hindered ethers, until it was discovered that boron trifluoride in methylene chloride quickly and cleanly effected cleavage at 0 °C (*Scheme 15.10*). Such cleavage conditions are not particularly mild, but the group is affording protection against both basic *and* acidic conditions.

$$ROSiMeBu^t_2 \xrightarrow[CH_2Cl_2]{BF_3} ROBF_2 \xrightarrow{NaHCO_3} ROH$$

Scheme 15.10

t-Butyldimethylsilyl ethers are up to 10^4 times less readily hydrolysed[3,4] than the corresponding trimethylsilyl ethers, and can survive several quite severe sequential synthetic operations. They are smoothly cleaved by fluoride ion[19a,31], normally provided by tetrabutylammonium fluoride in THF. Recently, the combination of tetrabutylammonium chloride and potassium fluoride dihydrate in acetonitrile has been suggested[35] for the same purpose. Mild acid treatment can also be effective, as can the use of boron trifluoride etherate in chloroform or dichloromethane[36]. Potassium superoxide in DMSO–DME in the presence of 18-crown-6 has been found to cleave such silyl ethers[37], but this method is unlikely to prove of general utility.

Iron(III) chloride in acetic anhydride converts t-butyldimethylsilyl ethers directly[38] into the corresponding acetates, with retention of configuration if applicable.

15.3.2 Applications

t-Butyldimethylsilyl ethers show considerable stability towards aqueous and alcoholic base, to mild reducing and oxidizing[39] agents, and to various organometallic species. The following short but representative list illustrates the diverse types of natural products and derived species whose recent syntheses have involved t-butyldimethylsilyl protection of hydroxyl groups: cytochalasin B[40], (±)-temsin[41], (±)-eriolanin[42], (±)-*N*-methylmaysenine[43], erythronolide (A)[44] and (B)[45], (±)-periplanone[46], (±)-pyrenophorin[47], (±)-serratenediol[48], (±)-Prelog–Djerassi lactone[49].

t-Butyldimethylsilyl protecting groups have been used extensively in nucleoside chemistry[16], selective protection of the hydroxy groups of the sugar portion being possible. This silyl protecting group can be directly replaced[50] by an acyl group under fluoride ion catalysis, in a process which proceeds well even with highly hindered acid anhydrides (*Scheme 15.11*). A cautionary note must be sounded at this point. Great care must be exercised in the use of this group, and presumably related silyl groups also, for partial protection of 1,2-diols such as those found in ribonucleosides. The structurally isomeric derivatives shown in *Scheme 15.11* undergo facile interconversion in methanol solution and in the presence of base; an intramolecular 1,4-silyl shift has been implicated[51].

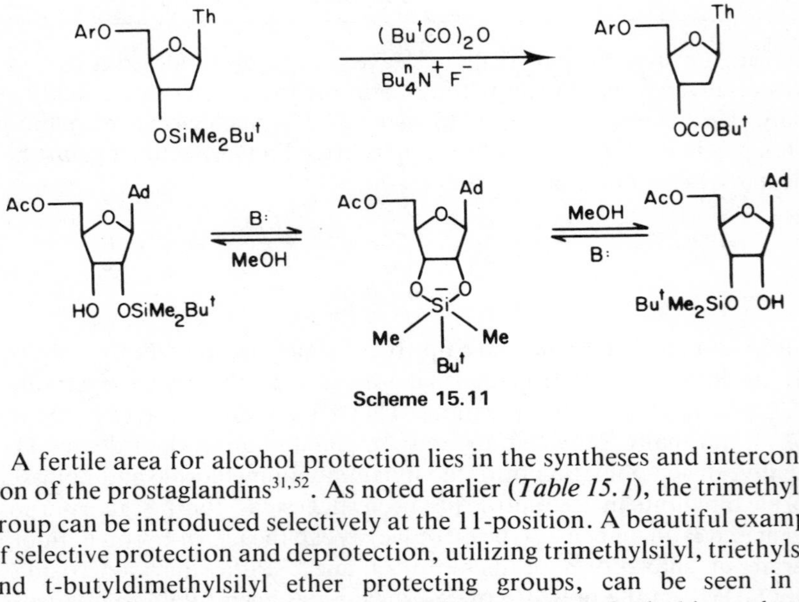

ArO, Th, $OSiMe_2Bu^t$ —$(Bu^tCO)_2O$, $Bu^n_4N^+F^-$→ ArO, Th, $OCOBu^t$

AcO, Ad, HO, $OSiMe_2Bu^t$ ⇌ (B:, MeOH) AcO, Ad, O, Si, O, Me, Me, Bu^t ⇌ (MeOH, B:) AcO, Ad, Bu^tMe_2SiO, OH

Scheme 15.11

A fertile area for alcohol protection lies in the syntheses and interconversion of the prostaglandins[31,52]. As noted earlier (*Table 15.1*), the trimethylsilyl group can be introduced selectively at the 11-position. A beautiful example[53] of selective protection and deprotection, utilizing trimethylsilyl, triethylsilyl, and t-butyldimethylsilyl ether protecting groups, can be seen in the conversion of $PGF_{1\alpha}$ into PGD_1, one of the by-products in the biosynthesis of the E series of prostaglandins (*Scheme 15.12*). The key mono-protected $PGF_{1\alpha}$ (7) was prepared by convergent synthesis.

Corey[54] has utilized the isopropyldimethylsilyl group to afford steric protection to the (13,14)-double bond of PGE_2 and thus allow selective hydrogenation to PGE_1 (*Scheme 15.13*). The isopropyldimethylsilyl group is up to 10^2—10^3 times less readily hydrolysed than the corresponding trimethylsilyl group, but is still smoothly cleaved by aqueous acetic acid.

A facile, base-induced 1,5-migration of a t-butyldimethylsilyl group has been observed[37] in a particular series of prostaglandin precursors

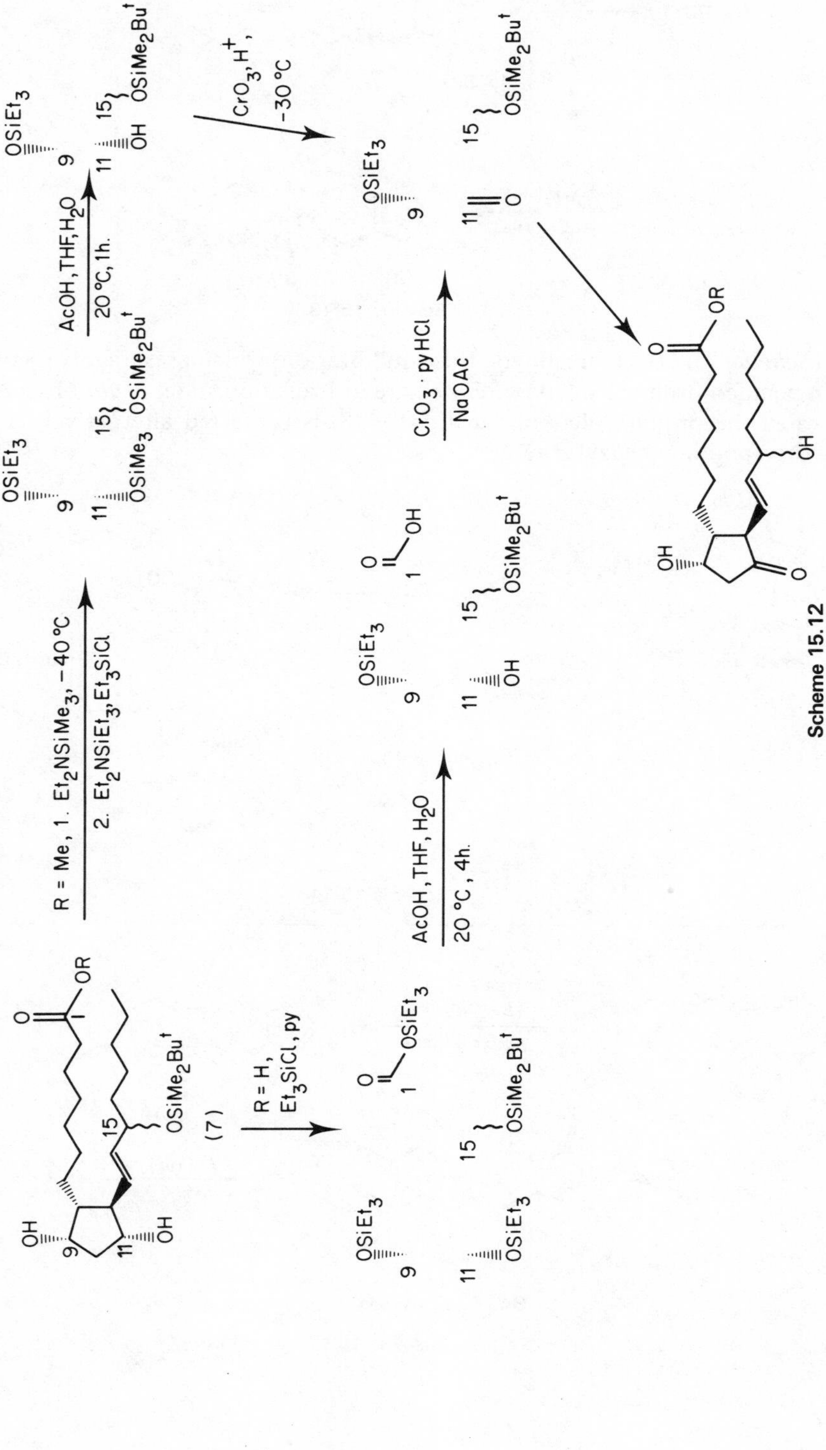

(7)

R = Me, 1. Et_2NSiMe_3, −40 °C
2. Et_2NSiEt_3, Et_3SiCl

AcOH, THF, H_2O
20 °C, 1h.

CrO_3, H^+,
−30 °C

R = H,
Et_3SiCl, py

AcOH, THF, H_2O
20 °C, 4h.

$CrO_3 \cdot$ pyHCl
NaOAc

Scheme 15.12

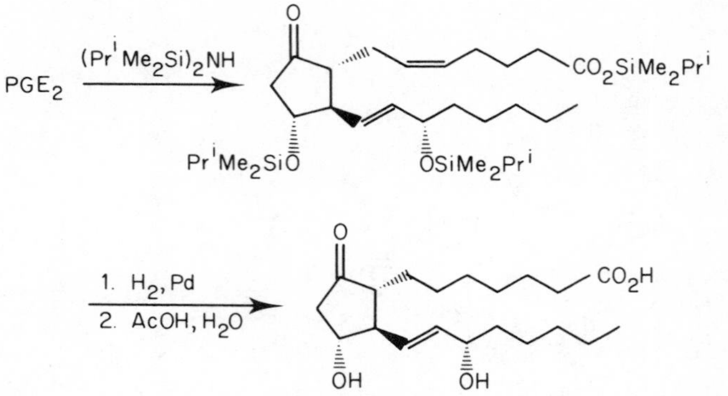

PGE_2 —($(Pr^iMe_2Si)_2NH$)→ [O; $CO_2SiMe_2Pr^i$; Pr^iMe_2SiO; $OSiMe_2Pr^i$]

—(1. H_2, Pd; 2. AcOH, H_2O)→ [O; CO_2H; OH; OH]

Scheme 15.13

(*Scheme 15.14*). Interestingly, this equilibration, which presumably involves a pentaco-ordinate silicon intermediate or transition state, is *not* observed[31] when the primary alcohol function of (8) is protected alternatively as its corresponding benzyl ether.

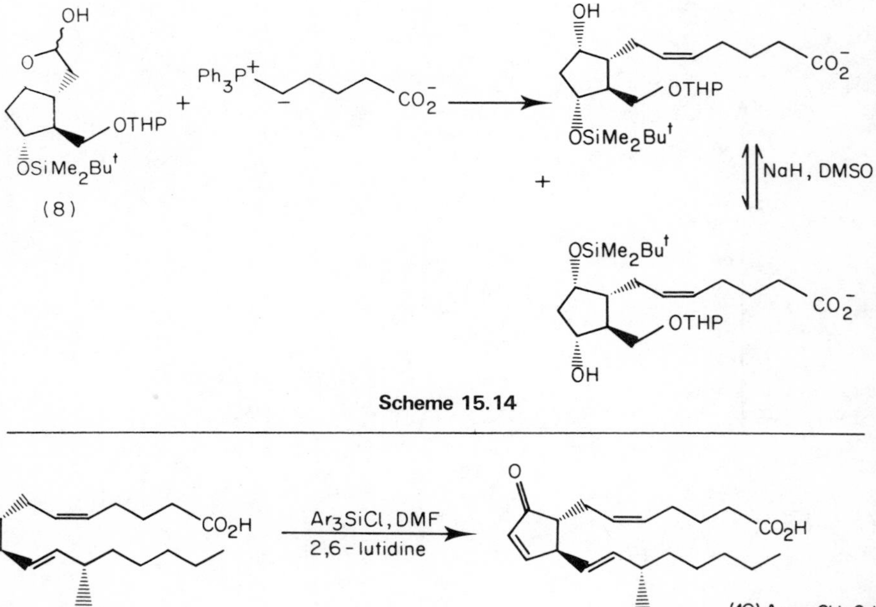

(8) [OH; O; OTHP; $OSiMe_2Bu^t$] + Ph_3P^+ – CO_2^- → [OH; CO_2^-; OTHP; $OSiMe_2Bu^t$] + [$OSiMe_2Bu^t$; CO_2^-; OTHP; OH] (NaH, DMSO)

Scheme 15.14

(9) [O; CO_2H; OH] —(Ar_3SiCl, DMF; 2,6-lutidine)→ [O; CO_2H; $OSiAr_3$] (10), Ar = p-$CH_3C_6H_4CH$

—(1. H_2O_2, HO^-; 2 AcOH, THF)→ [O; O; CO_2H; OH] —(Al(Hg))→ PGE_2

94 : 6 (β-oxiran)

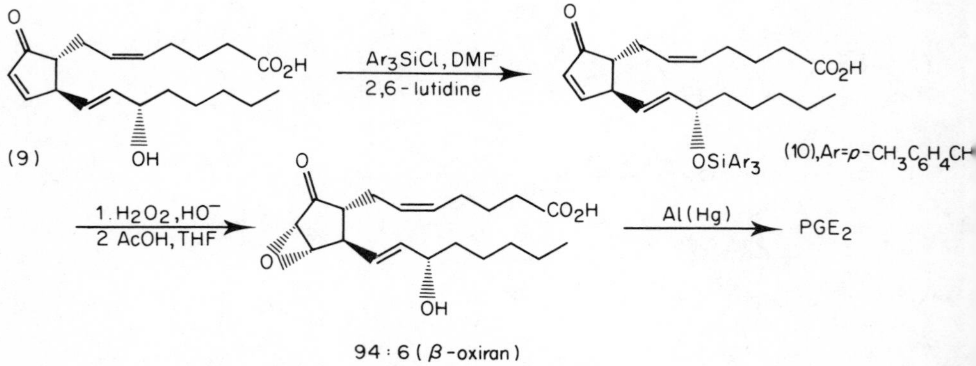

Scheme 15.15

Epoxidation of the prostaglandin (9) with alkaline hydrogen peroxide gives a mixture of α- and β-10,11-oxiranes. Attachment of a bulky 'remote controller' group to the hydroxy group at C-15 permits[55] stereoselective epoxidation, the highest degree being attained with the tri-(*p*-xylyl)silyl derivative (10), which screens the β face of the molecule because of the configuration of C-12. After regeneration of the hydroxy group, the α-oxirane was reductively cleaved with aluminium amalgam to afford PGE_2 (*Scheme 15.15*).

Hanessian[17,56] has recommended use of the t-butyldiphenylsilyl group for alcohol protection. Primary alcohols are protected preferentially, and the protection afforded is more effective than that afforded by t-butyldimethylsilyl in terms of acid stability. Cleavage can be achieved by one of the three methods shown (*Scheme 15.16*).

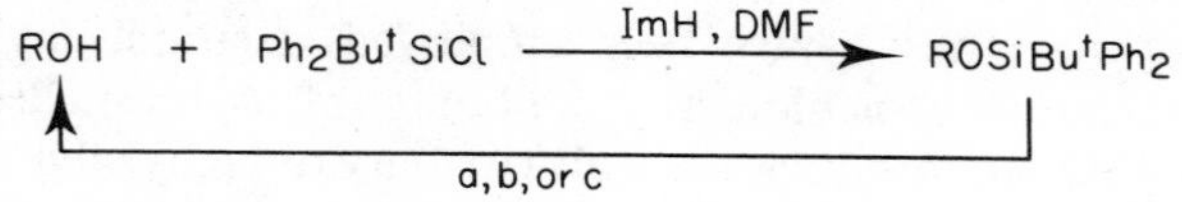

$$ROH + Ph_2Bu^tSiCl \xrightarrow{ImH,\ DMF} ROSiBu^tPh_2$$

$$ROSiBu^tPh_2 \xrightarrow{a,\ b,\ or\ c} ROH$$

a $Bu_4N^+F^-$, THF
b 3% HCl in MeOH
c 2 N NaOH in EtOH – H_2O

Scheme 15.16

Fleming[57] has introduced the trityldimethylsilyl protecting group. Although the derived ethers are more stable towards acid than the corresponding t-butyldimethylsilyl ethers, they are much more susceptible to nucleophilic C–Si cleavage, suitable reagents being fluoride ion in dioxan, or methanolic potassium carbonate (*Scheme 15.17*).

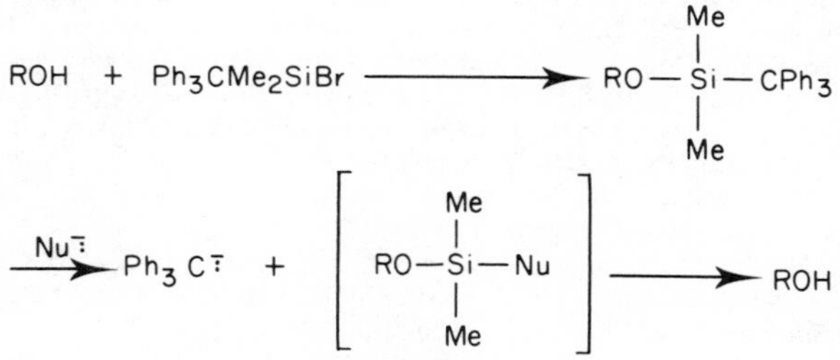

$$ROH + Ph_3CMe_2SiBr \longrightarrow RO{-}SiMe_2{-}CPh_3$$

$$\xrightarrow{Nu^-} Ph_3C^- + [RO{-}SiMe_2{-}Nu] \longrightarrow ROH$$

Scheme 15.17

15.4 Addendum

Ketones react[58] with alkyl silyl ethers in the presence of catalytic amounts of trimethylsilyl trifluoromethanesulphonate to give the derived ketals; remarkably, cyclohex-2-enone reacts without double bond migration (*Scheme 15.18*). t-Butyldimethylsilyl trifluoromethanesulphonate has been recommended[59] as a highly electrophilic silylating agent, as has a perfluorinated resinsulphonic acid trimethylsilyl ester[60].

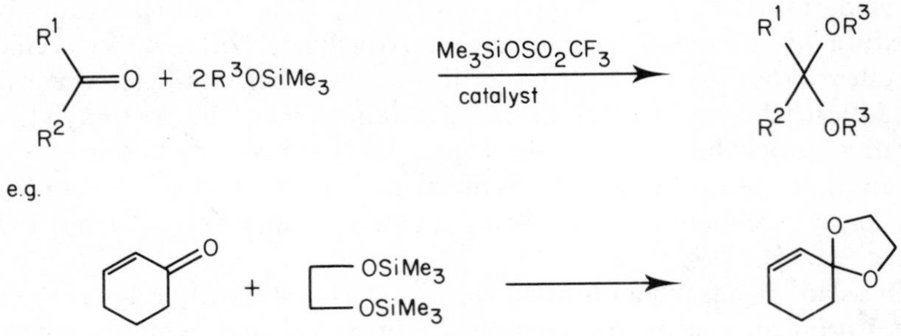

$$R^1R^2C{=}O + 2R^3OSiMe_3 \xrightarrow[\text{catalyst}]{Me_3SiOSO_2CF_3} R^1R^2C(OR^3)_2$$

e.g.

$$\text{cyclohexenone} + (CH_2OSiMe_3)_2 \longrightarrow \text{cyclohexenone ethylene ketal}$$

Scheme 15.18

A rather expensive pair of reagents, allyltrimethyl and allyl-t-butyldimethyl silane, transfer[61] their silyl groups to alcohols and carboxylic acids under acid catalysis. Dimethoxydiphenylsilane (11)[62] and dichlorotetraisopropyldisiloxane (12)[63,64] have been employed separately as bifunctional protecting groups. Stoichiometric amounts of *N*-bromosuccinimide in aqueous DMSO smoothly cleave[65] alkyl t-butyldimethylsilyl ethers; under such conditions, tetrahydropyranyl ethers are unaffected, as are double bonds.

$Ph_2Si(OMe)_2$

(11)

$Pr^i_2Si(Cl)—O—Si(Cl)Pr^i_2$

(12)

References

1 PIERCE, A. E., *'Silylation of Organic Compounds'*, Pierce Chemical Company, Rockford, Illinois (1968); BRITTAIN, G. D. and SULLIVAN, J. E., in *'Recent Advances in Gas Chromatography'*, Ed. Domsky, I. I. and Perry, J. A., Marcel Dekker, New York (1971); KLEBE, J. F., *Adv. org. Chem.* **8,** 97 (1972); *Accts chem. Res.* **3,** 299 (1970); BIRKOFER, L. and RITTER, A., in *'Newer Methods in Preparative Organic Chemistry'*, Ed. Foerst, W., vol. 5, p. 211, Academic Press, New York (1968)

2 *'Silylating Agents'*, Fluka, A. G., Switzerland (1977)

3 ACKERMAN, E., *Acta chem. Scand.* **11,** 373 (1957)

4 SOMMER, L. H., *'Stereochemistry, Mechanism, and Silicon'*, p. 132, McGraw-Hill, New York (1965) *see also* VORONKOV, M. G., MILESHKEVICH, V. P. and YUZHELEVSKII, Yu. A., *'The Siloxane Bond'*, Consultants Bureau, New York (1978)

5 LANGER, S. H., CONNELL, S. and WENDLER, I., *J. org. Chem.* **23,** 50 (1958)

6 COOPER, B. E., *Chemy Ind.* 794 (1978)

7 KLEBE, J. F., FINKBEINER, H. and WHITE, D. M., *J. Am. chem. Soc.* **88,** 3390 (1966); GALBRAITH, M. N., HORN, D. H. S., MIDDLETON, E. and HACKNEY, R. J., *Chem. Communs* 466 (1968); BIRKOFER, L., RITTER, A. and BENTZ, F., *Chem. Ber.* **97,** 2196 (1964)

8 BIRKOFER, L., RITTER, A. and DICKOPP, H., *Chem. Ber.* **96,** 1473 (1963)

9 BIRKOFER, L. and SOMMER, P., *J. organometal. Chem.* **99,** C1 (1975)

10 WEISZ, I., FELFÖLDI, K. and KOVACS, K., *Acta chim. hung.* **58,** 189 (1968); *Chem. Abstr.* **70,** 47 668 (1969)

11 BIRKOFER, L. and RITTER, A., *Angew. Chem. int. Edn* **4,** 417 (1965)

12 COOPER, B. E. and WESTALL, S., *J. organometal. Chem.* **118,** 135 (1976)

13 WANNAGAT, U., BURGER, H., KRUGER, C. and PUMP, J., *Z. anorg. allg. Chem.* **321,** 208 (1963)

14 SCHNEIDER, H. J. and HORNUNG, R., *Justus Liebigs Annln Chem.* 1864 (1974)

15 YANKEE, E. W., LIN, C. H. and FRIED, J., *J. chem. Soc. chem. Communs* 1120 (1972); YANKEE, E. W. and BUNDY, G. L., *J. Am. chem. Soc.* **94,** 3651 (1972); YANKEE, E. W., AXEN, U. and BUNDY, G. L., *J. Am. chem. Soc.* **96,** 5865 (1974)

16 OGILVIE, K. K., SADANA, K. L., THOMPSON, E. A., QUILLIAM, M. A. and WESTMORE, J. B., *Tetrahedron Lett.* 2861 (1974); OGILVIE, K. K., THOMPSON, E. A., QUILLIAM, M. A. and WESTMORE, J. B., *Tetrahedron Lett.* 2865 (1974); OGILVIE, K. K., THERIAULT, N. and SADANA, K. L., *J. Am. chem. Soc.* **99,** 7741 (1977); OGILVIE, K. K., BEAUCAGE, S. L., SCHIFMAN, A. L., THERIAULT, N. Y. and SADANA, K. L., *Can. J. Chem.* **56,** 2768 (1978); SADANA, K. L. and LOEWEN, P. C., *Tetrahedron Lett.* 5095 (1978)

17 HANESSIAN, S., *Accts chem. Res.* **12,** 159 (1979) and references therein

18 BIRKOFER, L., RITTER, A. and BENTZ, F., *Chem. Ber.* **97,** 2196 (1964)

19 (*a*) COREY, E. J. and SNIDER, B. B., *J. Am. chem. Soc.* **94,** 2549 (1972); (*b*) WIES, R. and PFAENDER, P., *Justus Liebigs Annln Chem.* 1269 (1973); (*c*) NEGISHI, E., LEW, G. and YOSHIDA, T., *J. chem. Soc. chem. Communs* 874 (1973)

20 BUSE, C. T. and HEATHCOCK, C. H., *J. Am. chem. Soc.* **99,** 8109 (1977)

21 HEATHCOCK, C. H., PIRRUNG, M. C., BUSE, C. T., HAGEN, J. P., YOUNG, S. D. and SOHN, J. E., *J. Am. chem. Soc.* **101,** 7076 (1979)

22 THIES, R. W., MILLS, M. T., CHIN, A. W., SCHICK, L. E. and WALTON, E. S., *J. Am. chem. Soc.* **95,** 5281 (1973); THIES, R. W. and BOLESTA, R. E., *J. org. Chem.* **41,** 1233 (1976)

23 WILSON, S. R. and MAO, D. T., *J. Am. chem. Soc.* **100,** 6289 (1978); *see also* OPPOLZER, W. and SNOWDEN, R. L., *Tetrahedron Lett.* 3505 (1978)

24 CHOU, T. S., *Tetrahedron Lett.* 725 (1974); CHOU, T. S., BURGTORF, J. R., ELLIS, A. L., LAMMERT, S. R. and KUKOLJA, S., *J. Am. chem. Soc.* **96,** 1609 (1974); *see also* DAVIS, F. A., RIZVI, S. Q. A., ARDECKY, R., GOSCINIAC, D. J., FRIEDMAN, A. J. and YOCKLOVICH, S. G., *J. org. Chem.* **45,** 1650 (1980)

25 MURATA, S., SUZUKI, M. and NOYORI, R., *J. Am. chem. Soc.* **101,** 2738 (1979); *see also* YAMAMOTO, H. and NOZAKI, H., *Angew. Chem. int. Edn* **17,** 169 (1978)

26 JUNG, M. E., *J. org. Chem.* **41,** 1479 (1976)

27 JUNG, M. E. and SPELTZ, L. M., *J. Am. chem. Soc.* **98,** 7882 (1976)

28 KELLY, R. W., *Tetrahedron Lett.* 967 (1969); *J. Chromatog.* **43,** 229 (1969)

29 COREY, E. J., DANHEISER, R. L., CHANDRASEKARAN, S., KECK, G. E., GOPALAN, B., LARSEN, S. D., SIRET, P. and GRAS, J.-L., *J. Am. chem. Soc.* **100,** 8034 (1978)

30 COREY, E. J. and CARNEY, R. L., *J. Am. chem. Soc.* **93,** 7318 (1971); *see also* COREY, E. J., DANHEISER, R. L. and CHANDRASEKARAN, S., *J. org. Chem.* **41,** 260 (1976)

31 COREY, E. J. and VENKATESWARLU, A., *J. Am. chem. Soc.* **94,** 6190 (1972)

32 CHAUDHARY, S. K. and HERNANDEZ, O., *Tetrahedron Lett.* 99 (1979)

33 BARTON, T. J. and TULLY, C. R., *J. org. Chem.* **43,** 3649 (1978)

34 WANNAGAT, U. and LIEHR, W., *Angew. Chem.* **69,** 783 (1957)

35 CARPINO, L. A. and SAU, A. C., *J. chem. Soc. chem. Communs* 514 (1979)

36 KELLY, D. R., ROBERTS, S. M. and NEWTON, R. F., *Synthetic Communs* **9,** 295 (1979); *see also* METCALF, B. W., BURKHART, J. P. and JUND, K., *Tetrahedron Lett.* 35 (1980)

37 TORISAWA, Y., SHIBASAKI, M. and IKEGAMI, S., *Tetrahedron Lett.* 1865 (1979)

38 GANEM, B. and SMALL, V. R., *J. org. Chem.* **39,** 3728 (1974)

39 TROST, B. and VERHOEVEN, T. R., *Tetrahedron Lett.* 2275 (1978)

40 STORK, G., NAKAHARA, Y., NAKAHARA, Y. and GREENLEE, W. J., *J. Am. chem. Soc.* **100,** 7775 (1978)

41 NISHIZAWA, M., GRIECO, P. A., BURKE, S. D. and METZ, W., *J. chem. Soc. chem. Communs* 78 (1978)

42 GRIECO, P. A., OGURI, T., GILMAN, S. and DETITTA, G. T., *J. Am. chem. Soc.* **100,** 1616 (1978)

43 COREY, E. J., WEIGEL, L. O., FLOYD, D. and BOCK, M. G., *J. Am. chem. Soc.* **100,** 2916 (1978)

44 COREY, E. J., HOPKINS, P. B., KIM, S., YOO, S., NAMBIAR, K. P. and FALCK, J. R., *J. Am. chem. Soc.* **101,** 7131 (1979)

45 COREY, E. J., TRYBULSKI, E. J., MELVIN, Jr., L. S., NICOLAOU, K. C., SECRIST, J. A., LETT, R., SHELDRAKE, P. W., FALCK, J. R., BRUNELLE, D. J., HASLINGER, M. F., KIM, S. and YOO, S., *J. Am. chem. Soc.* **100,** 4618 (1978) and subsequent paper

46 STILL, W. C., *J. Am. chem. Soc.* **101,** 2493 (1979)
47 GERLACH, H., OERTLE, K. and THALMANN, A., *Helv. chim. Acta* **60,** 2860 (1977)
48 PRESTWICH, G. D. and LABOWITZ, J. N., *J. Am. chem. Soc.* **96,** 7103 (1974)
49 WHITE, J. D. and FUKUYAMA, Y., *J. Am. chem. Soc.* **101,** 226 (1979)
50 BEAUCAGE, S. L. and OGILVIE, K. K., *Tetrahedron Lett.* 1691 (1977)
51 JONES, S. S. and REESE, C. B., *J. chem. Soc. Perkin I* 2762 (1979)
52 COREY, E. J. and SACHDEV, H. S., *J. Am. chem. Soc.* **95,** 8483 (1973); GILL, M. and RICKARDS, R. W., *J. chem. Soc. chem. Communs* 121 (1979); DIMSDALE, M. J., NEWTON, R. F., RAINEY, D. K., WEBB, C. F., LEE, T. V. and ROBERTS, S. M., *J. chem. Soc. chem. Communs* 716 (1977); CROSSLAND, N. M., ROBERTS, S. M. and NEWTON, R. F., *J. chem. Soc. chem. Communs* 886 (1977)
53 HART, T. W., METCALFE, D. A. and SCHEINMANN, F., *J. chem. Soc. chem. Communs* 156 (1979); *see also* STORK, G. and KRAUS, G., *J. Am. chem. Soc.* **98,** 6747 (1976)
54 COREY, E. J. and VARMA, R. K., *J. Am. chem. Soc.* **93,** 7319 (1971)
55 COREY, E. J. and ENSLEY, H. E., *J. org. Chem.* **38,** 3187 (1973)
56 HANESSIAN, S. and LAVALLEE, P., *Can. J. Chem.* **53,** 2975 (1975); for examples of use, *see* EVANS, D. A., SACKS, C. E., KLESCHICK, W. A. and TABER, T. R., *J. Am. chem. Soc.* **101,** 6789 (1979); STORK, G., TAKAHASHI, T., KAWAMOTO, I. and SUZUKI, T., *J. Am. chem. Soc.* **101,** 8272 (1979); JUST, G., LUTHE, C. and OH, H., *Tetrahedron Lett.* 1001 (1980); TAM, T. F. and FRASER-REID, B., *J. chem. Soc. chem. Communs* 556 (1980); WILLIAMS, R. M. and RASTETTER, W. H., *J. org. Chem.* **45,** 2625 (1980)
57 AGER, D. J. and FLEMING, I., *J. chem. Res. S* 6 (1977)
58 TSUNODA, T., SUZUKI, M. and NOYORI, R., *Tetrahedron Lett.* 1357 (1980)
59 RIEDIKER, M. and GRAF, W., *Helv. chim. Acta* **62,** 205 (1979)
60 MURATA, S. and NOYORI, R., *Tetrahedron Lett.* 767 (1980)
61 MORITA, T., OKAMOTO, Y. and SAKURAI, H., *Tetrahedron Lett.* 835 (1980)
62 JENNER, M. R. and KHAN, R., *J. chem. Soc. chem. Communs* 50 (1980)
63 MARKIEWICZ, W. T., *J. chem. Res. S* 24 (1979); MARKIEWICZ, W. T., SAMEK, Z. and SMRT, J., *Tetrahedron Lett.* 4523 (1980)
64 VERDEGAAL, C. H. M., JANSSE, P. L., De ROOIJ, J. F. M. and van BOOM, J. H., *Tetrahedron Lett.* 1571 (1980)
65 BATTEN, R. J., DOXON, A. J., TAYLOR, R. J. K. and NEWTON, R. F., *Synthesis* 234 (1980)

Chapter 16

Acyloxysilanes (silyl carboxylates)

The preparation and properties of acyloxysilanes (1) have been reviewed[1] up to 1960. Nucleophiles can react with acyloxysilanes in two distinct ways, either by attack at silicon or by attack at the carbonyl group (*Scheme 16.1*).

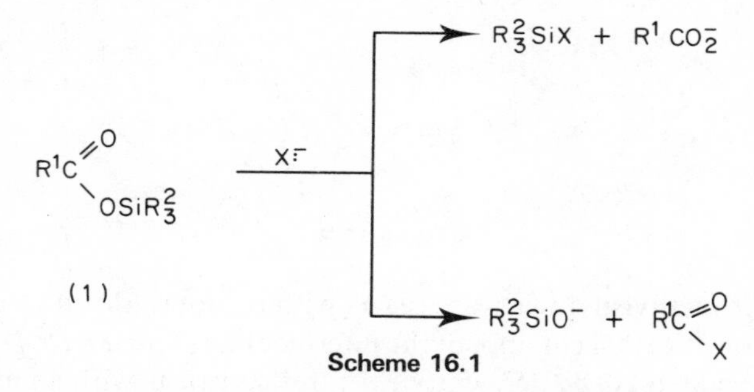

Scheme 16.1

In general terms, oxygen and nitrogen nucleophiles attack at silicon, whereas carbon nucleophiles attack at carbon[2], although if the silicon substituents are sufficiently bulky, oxygen nucleophiles can be induced[3] to attack at the carbonyl carbon.

However, t-butyl trimethylsilyl carbonate (2) is apparently attacked by amines and alcohols at the carbonyl group[4] to give urethanes and carbonates,

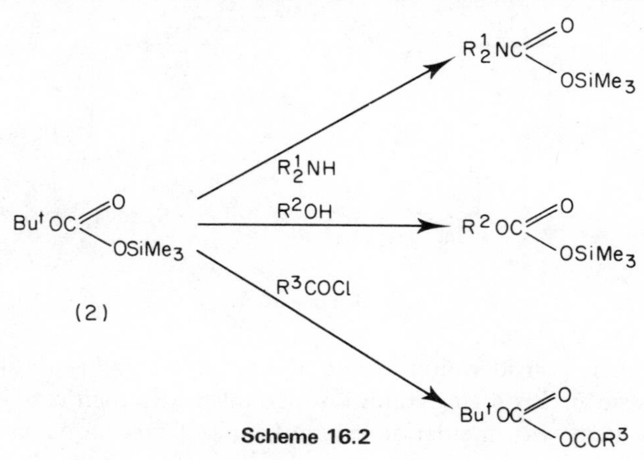

Scheme 16.2

respectively. Carboxylic acid chlorides, on the other hand, give mixed anhydrides resulting from effective chloride attack at silicon (*Scheme 16.2*).

α-Amino acid esters react analogously under mild conditions; more vigorous conditions result in complete silylation (*Scheme 16.3*). The potential of this type of nitrogen protection in peptide synthesis has been explored[5]. Another successful route[6] to such functionalized amino acids is by carboxylation of the *N*-silyl analogues.

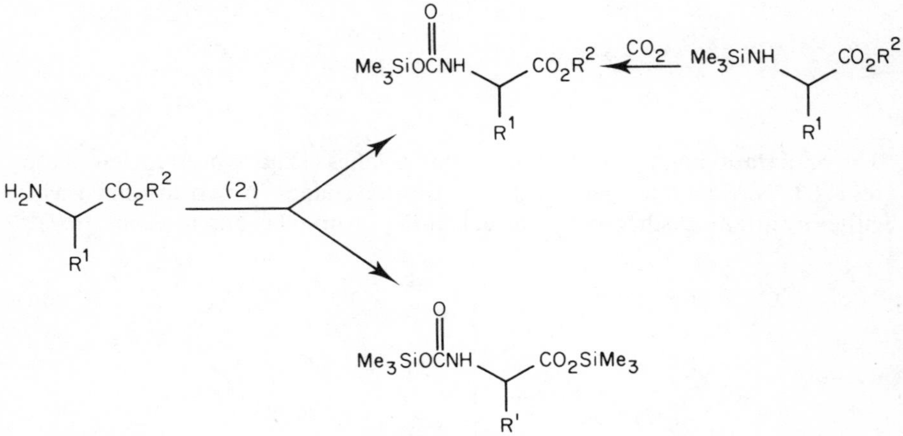

Scheme 16.3

Similarly, trimethylsilyl pyruvate reacts with oxaloyl chloride via halogen attack at silicon. Normal collapse of the intermediate (*Scheme 16.4*) produces pyruvoyl chloride in 50–80 per cent yield[7]; this contrasts with an alternative method[8] using oxaloyl chloride and anhydrous sodium carbonate, which gave the acid chloride in only 25 per cent yield, and that after a necessary and careful distillation.

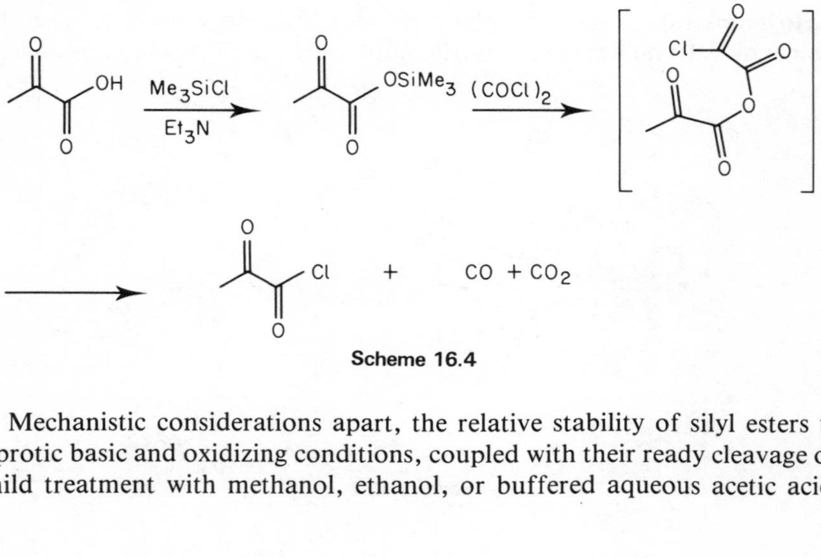

Scheme 16.4

Mechanistic considerations apart, the relative stability of silyl esters to aprotic basic and oxidizing conditions, coupled with their ready cleavage on mild treatment with methanol, ethanol, or buffered aqueous acetic acid, makes them attractive protecting groups for carboxyl functions (for another silicon-based carboxyl protecting group, see p. 159).

Corey recommends[9] isopropyldimethylsilyl esters to protect the prostaglandin carboxyl group during *N*-chlorosuccinimide/dimethyl sulphide mediated oxidation of a secondary alcohol to a ketone; removal is easily achieved with buffered acetic acid. The use of trimethylsilyl esters to protect the carboxyl group attached to C-3 in penicillins during side-chain cleavage[10] represented an important step in achieving a practical route to 6-aminopenicillanic acid (*Scheme 16.5*). Similar methodology has brought significant improvements to the preparation of 7-aminocephalosporanic acid.

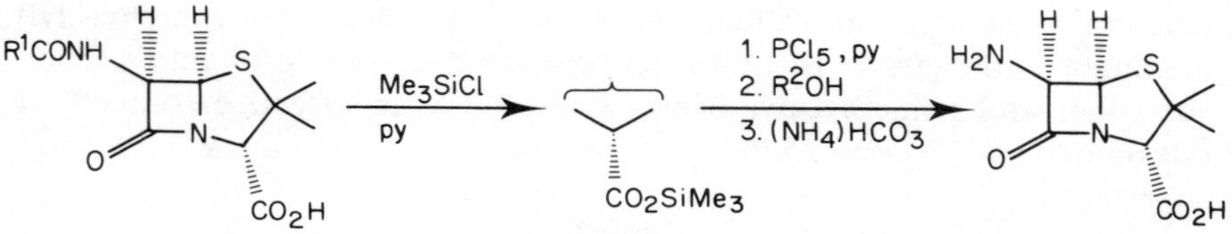

Scheme 16.5

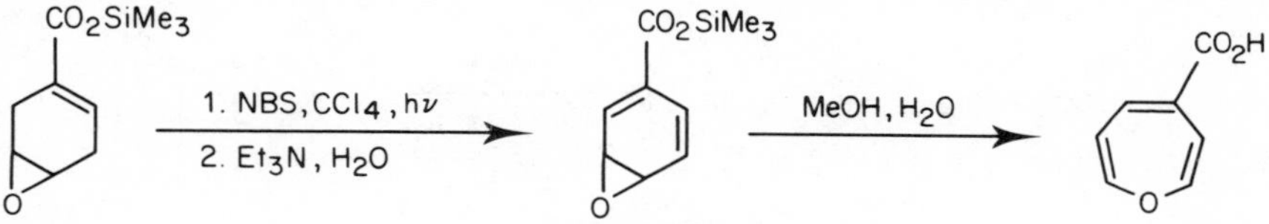

Scheme 16.6

The protecting group in the acid (3) survived[11] two sets of reagents before cleavage with methanol gave the oxepincarboxylic acid (4) (*Scheme 16.6*).

Pyrolysis[12] of the cyclobutene diester (5), followed by hydrolysis, yielded the hitherto elusive butadiene-2,3-dicarboxylic acid (6) in 90 per cent yield (*Scheme 16.7*); the previous route to this compound succeeded to the extent of 1 per cent.

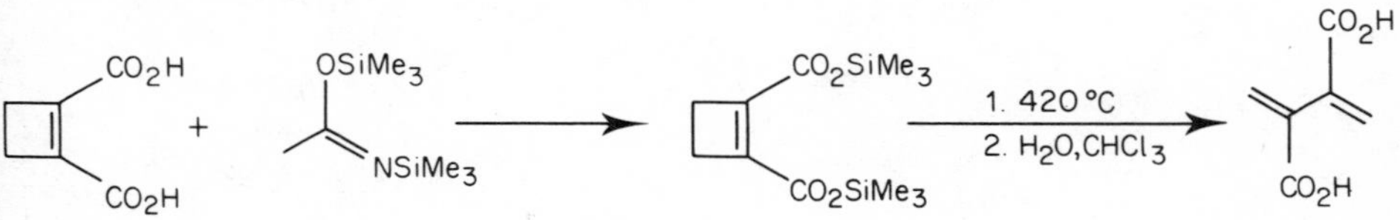

Scheme 16.7

Bis(trimethylsilyl) malonate[13] and alkyl trimethylsilyl malonates[14] can be converted, by treatment with n-butyl-lithium, into their respective monoanions, which may be used in synthesis in the expected way. Their main advantage over the more conventional malonate esters lies in the great ease of hydrolysis of the products. A similar advantage was claimed[15] for the Wadsworth–Emmons reagent (7), but yields are higher and manipulation is simplified by use[16] of the dianion (8) (*Scheme 16.8*).

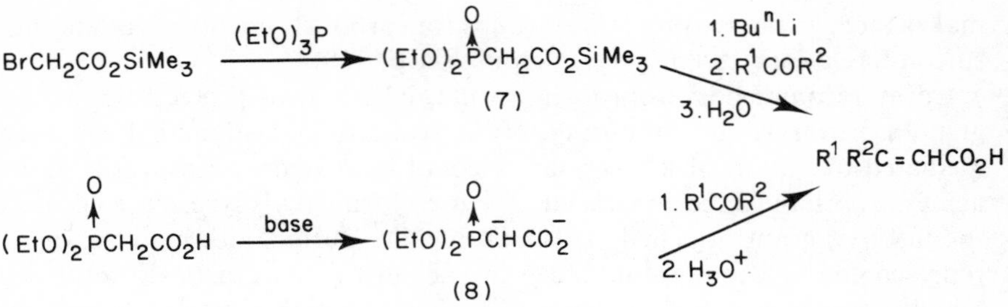

$BrCH_2CO_2SiMe_3 \xrightarrow{(EtO)_3P} (EtO)_2\overset{O}{\uparrow}PCH_2CO_2SiMe_3$ (7) $\xrightarrow{1.\ Bu^nLi;\ 2.\ R^1COR^2;\ 3.\ H_2O} R^1R^2C=CHCO_2H$

$(EtO)_2\overset{O}{\uparrow}PCH_2CO_2H \xrightarrow{base} (EtO)_2\overset{O}{\uparrow}P\bar{C}HCO_2^-$ (8) $\xrightarrow{1.\ R^1COR^2;\ 2.\ H_3O^+} R^1R^2C=CHCO_2H$

Scheme 16.8

Trimethylsilyl α-bromoesters are recommended[17] in the Reformatsky reaction when isolation of the β-hydroxyacid is desired. Bis(trimethylsilyl) malonate undergoes a thermal, acid-catalysed conversion[18] into carbon suboxide, and trimethylsilyl tribromoacetate is an effective source[19] of dibromoketene (*Scheme 16.9*).

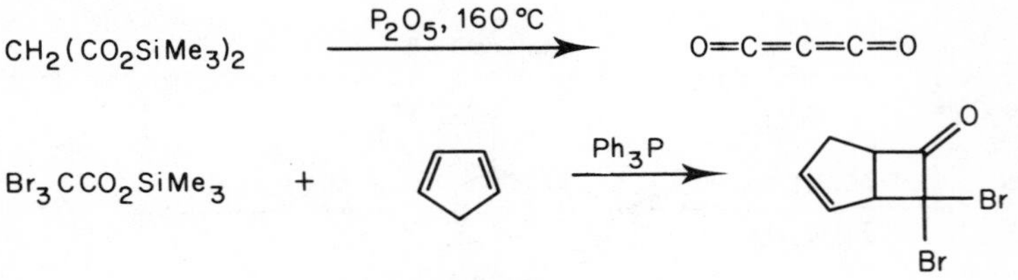

$CH_2(CO_2SiMe_3)_2 \xrightarrow{P_2O_5,\ 160\ °C} O{=}C{=}C{=}C{=}O$

$Br_3CCO_2SiMe_3$ + cyclopentadiene $\xrightarrow{Ph_3P}$ dibromobicyclic cyclobutanone (Br, Br, O)

Scheme 16.9

Finally, γ-, δ-, and ε-lactams can be prepared[20] in excellent yields by prior carboxyl activation as silyl esters. Subsequent *exo*-trigonal ring closure by attack by the terminal amino group produces the desired lactams under mild, non-racemizing conditions (*Scheme 16.10*).

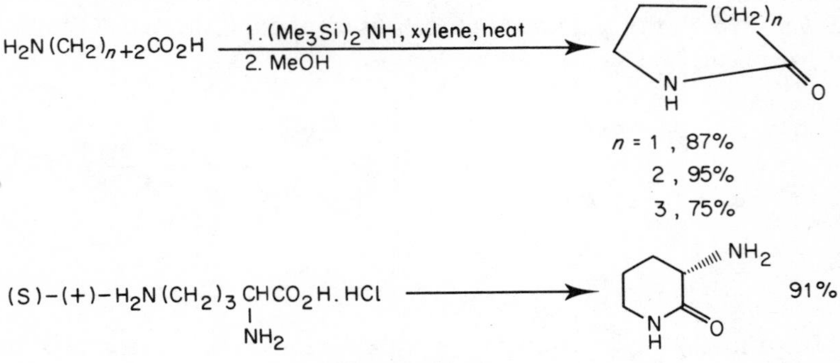

$H_2N(CH_2)_{n+2}CO_2H \xrightarrow{1.\ (Me_3Si)_2NH,\ xylene,\ heat;\ 2.\ MeOH}$ lactam ($(CH_2)_n$, N–H, C=O)

n = 1 , 87%
2 , 95%
3 , 75%

(S)-(+)-$H_2N(CH_2)_3CH(NH_2)CO_2H.HCl \longrightarrow$ 3-aminopiperidin-2-one (NH_2, N–H, O) 91%

Scheme 16.10

References

1 YUR'EV, Yu. K. and BELYAKOVA, Z. V., *Russ. chem. Revs* **29**, 383 (1960)
2 *See,* for example, GILMAN, H. and SMART, G. N. R., *J. org. Chem.* **19**, 441 (1954); SOMMER, L. H., PARKER, G. A. and FRYE, C. L., *J. Am. chem. Soc.* **86**, 3280 (1964);

GORNOWICZ, G. A. and WEST, R., *J. Am. chem. Soc.* **90,** 4478 (1968); ANDERSON, H. H. and FISCHER, H., *J. org. Chem.* **19,** 1296 (1954); RÜHLMANN, K., *J. prakt. Chem.* **16,** 172 (1962)
3 HUDRLIK, P. F. and FEASLEY, R., *Tetrahedron Lett.* 1781 (1972)
4 YAMAMOTO, Y. and TARBELL, D. S., *J. org. Chem.* **36,** 2954 (1971)
5 YAMAMOTO, Y., TARBELL, D. S., FEHLNER, J. R. and POPE, B. M., *J. org. Chem.* **38,** 2521 (1973)
6 KRICHELDORF, H. R., *Synthesis* 259 (1970); *Justus Liebigs Annln Chem.* **748,** 101 (1971) and references therein
7 HÄUSLER, J. and SCHMIDT, U., *Chem. Ber.* **107,** 145 (1974)
8 TANNER, D. D. and DAS, N. C., *J. org. Chem.* **35,** 3972 (1970)
9 COREY, E. J. and KIM, C. U., *J. org. Chem.* **38,** 1223 (1973)
10 HUBER, F. M., CHAUVETTE, R. C. and JACKSON, B. G., in *'Cephalosporins and Penicillins',* Ed. Flynn, E. H., chap. 2, Academic Press, New York (1972); WEISSENBURGER, H. W. O. and van der HOEVEN, M. G., *Recl Trav. chim. Pays-Bas Belg.* **89,** 1081 (1970); FECHTIG, B., PETER, H., BICKEL, H. and VISCHER, E., *Helv. chim. Acta* **51,** 1108 (1968)
11 RICHARDSON, J. D., BRUICE, T. C., WARASZKIEWICZ, S. M. and BERCHTOLD, G. A., *J. org. Chem.* **39,** 2088 (1974)
12 DOWD, P. and KANG, K., *Synthetic Communs* **4,** 151 (1974)
13 NAM, N. H., BEAUCOURT, J.-P., HOELLINGER, H. and PICHAT, L., *Bull. Soc. chim. Fr.* 1367 (1974); SCHMIDT, U. and SCHWOCHAU, M., *Mh. Chem.* **98,** 1492 (1967)
14 PICHAT, L. and BEAUCOURT, J.-P., *Synthesis* 537 (1973); TROST, B. M. and KUNZ, R. A., *J. org. Chem.* **39,** 2648 (1974)
15 LOMBARDO, L. and TAYLOR, R. J. K., *Synthesis* 131 (1978)
16 COUTROT, P., SNOUSSI, M. and SAVIGNAC, P., *Synthesis* 133 (1978)
17 HOREAU, A., *Tetrahedron Lett.* 3227 (1971)
18 BIRKOFER, L. and SOMMER, P., *Chem. Ber.* **109,** 1701 (1976)
19 OKADA, T. and OKAWARA, R., *Tetrahedron Lett.* 2801 (1971)
20 PELLEGATA, R., PINZA, M. and PIFFERI, G., *Synthesis* 614 (1978)

Chapter 17

Silyl enol ethers and silyl ketene acetals

Silyl enol ethers (1) have been known for some considerable time[1], and, from around 1968 until recently, have been those compounds of silicon to show major synthetic use. Their utility lies initially in providing regiostable, isolable enol derivatives which can, on demand and after purification and spectral identification, give rise to regio-pure enolate ions[2].

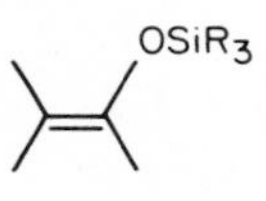

$OSiR_3$ (1)

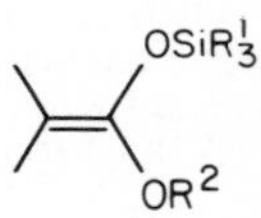

$OSiR^1_3$, OR^2 (2)

The synthetic development of silyl enol ethers can be divided into three distinct phases[3]. These separate phases involve

1. The use of silylation as a trap for equilibrium or kinetically generated enolate anions, with subsequent isolation, regeneration and reaction with electrophiles under *basic* conditions.
2. Direct reaction of the silyl enol ether with suitable electrophiles which are either good Lewis *acids* in their own right, or can be made so by addition of a Lewis acid catalyst.
3. The use of silyl enol ethers as synthons which give reaction products which are different from those obtainable by either of the first two phases.

A more or less equivalent development can be seen in the chemistry of silyl ketene acetals (*O*-silyl ester enols) (2); these species will also be discussed here.

17.1 Preparation of silyl enol ethers

One of the early, and still one of the most frequently used, routes to silyl enol ethers is the trapping of enolate anions[4] generated under conditions of either kinetic or equilibrium control. The products of trapping correspond accurately to those of the free enolates in a particular mixture, and considerable regioselectivity[5] can be attained (*Scheme 17.1*).

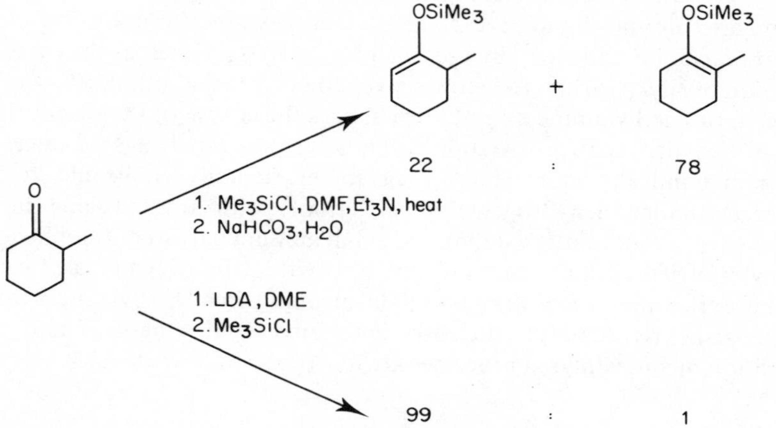
OSiMe3
OSiMe3
+
22
:
78
O
1. Me3SiCl, DMF, Et3N, heat
2. NaHCO3, H2O
1. LDA, DME
2. Me3SiCl
99
:
1

Scheme 17.1

Such mixtures of trimethylsilyl enol ethers are rapidly hydrolysed by water, but can be distilled, and thus particular components can be enriched; the t-butyldimethylsilyl analogues are, as expected, much more stable hydrolytically and to nucleophilic attack in general. The ability to obtain only one of the two regioisomeric silyl enol ethers derivable from an unsymmetrical ketone is of critical importance to further synthetic utility, and accordingly, much effort has been expended in this direction. Regiospecific generation can be achieved in a number of ways, including the trapping of the enolate ion formed from a cyclic $\alpha\beta$-unsaturated ketone by conjugate reduction[6], conjugate alkylation[7], or conjugate hydrocyanation[8] (*Scheme 17.2*). The conjugate addition of tin anions is discussed later

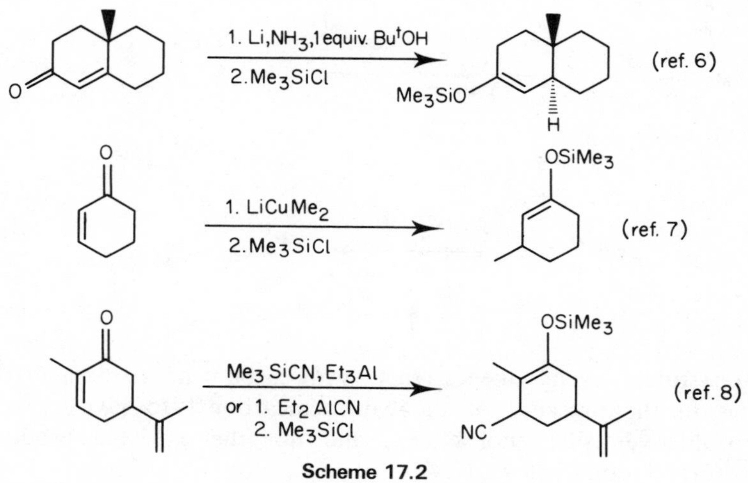
1. Li, NH3, 1 equiv. ButOH
2. Me3SiCl
O
Me3SiO
H
(ref. 6)
O
OSiMe3
1. LiCuMe2
2. Me3SiCl
(ref. 7)
O
OSiMe3
Me3 SiCN, Et3Al
or 1. Et2 AlCN
2. Me3SiCl
NC
(ref. 8)

Scheme 17.2

(p. 237), and conjugate additions of the formal anions of iodine, selenium and sulphur are revealed in Chapter 18.

$\alpha\beta$-Unsaturated carbonyl compounds undergo rhodium-catalysed hydrosilylation (Chapter 21) to generate silyl enol ethers regiospecifically[9]; only conjugated double bonds are affected, and asymmetric induction at the β-carbon can be achieved[10] by use of chiral catalysts. Whereas the catalysed addition of silyl hydrides to ketones gives ethers[11], the addition of 2–4 moles per cent of triethylamine or pyridine changes the course of the reaction into one of dehydrogenative silylation[12]; the latter process always produces the thermodynamically more stable regioisomer. Readily enolizable ketones undergo similar dehydrogenative silylation[13] with triethylsilane in the presence of a rhodium(I) catalyst and small amounts of either thiophenol or phenylthiotriethylsilane; alternatively, phenylthiotriethylsilane itself can be used in refluxing xylene. Primary allylic alcohol trimethylsilyl ethers can be isomerized[14] reasonably efficiently into silyl enol ethers. These four transition metal-catalysed processes are illustrated in *Scheme 17.3*.

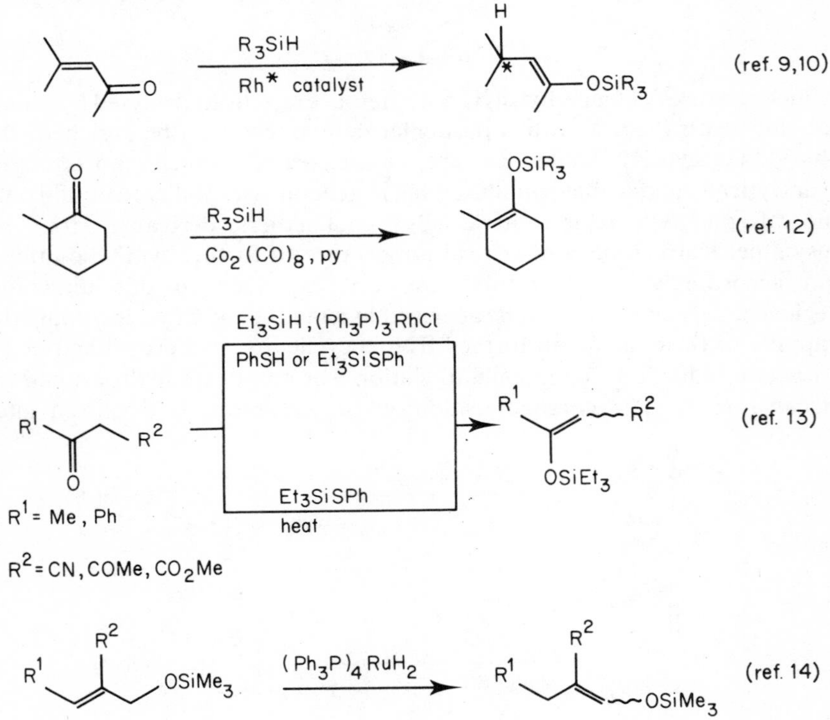

R_3SiH, Rh* catalyst → H, *, $OSiR_3$ (ref. 9,10)

O; R_3SiH, $Co_2(CO)_8$, py → $OSiR_3$ (ref. 12)

Et_3SiH, $(Ph_3P)_3RhCl$, PhSH or Et_3SiSPh; Et_3SiSPh, heat

R^1, R^2, O → R^1, R^2, $OSiEt_3$ (ref. 13)

R^1 = Me, Ph

R^2 = CN, COMe, CO_2Me

R^1, R^2, $OSiMe_3$; $(Ph_3P)_4RuH_2$ → R^1, R^2, $OSiMe_3$ (ref. 14)

Scheme 17.3

Interestingly, cyclic alkenes react[15] with carbon monoxide and trialkylsilanes in the presence of dicobalt octacarbonyl to give cycloalkane carboxaldehyde silyl enol ethers, and not the expected products of hydroformylation (*Scheme 17.4*).

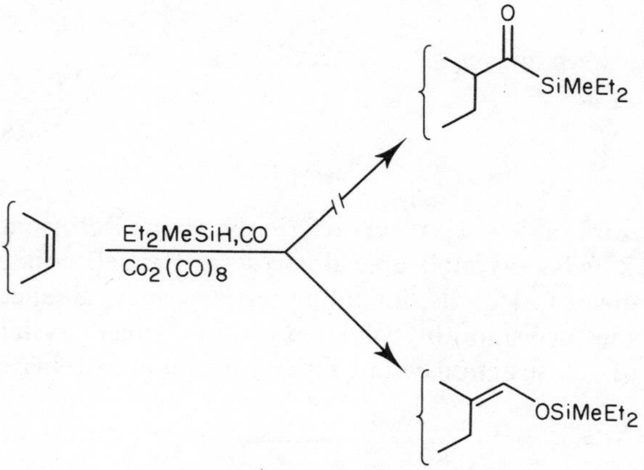

Et_2MeSiH, CO / $Co_2(CO)_8$ → SiMeEt$_2$ (acylsilane, not formed) ; OSiMeEt$_2$

Scheme 17.4

A reagent system[16] of general applicability which shows considerable regio-, stereo-, and chemo-selectivity employs stoichiometric amounts of ethyl trimethylsilylacetate (3), in the presence of 5–10 moles per cent of tetra-butylammonium fluoride. This silylating reagent is readily prepared[17] from ethyl bromoacetate (*Scheme 17.5*). It reacts with carbonyl compounds under mild conditions, innocuous ethyl acetate being co-produced.

$$BrCH_2CO_2Et + Me_3SiCl \xrightarrow{Zn} Me_3SiCH_2CO_2Et \quad (3)$$

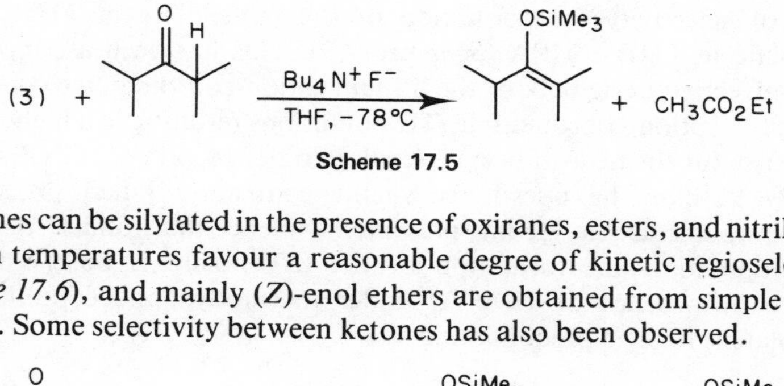

(3) + [ketone] $\xrightarrow[THF,\ -78\,°C]{Bu_4N^+F^-}$ [OSiMe$_3$ enol ether] + CH_3CO_2Et

Scheme 17.5

Ketones can be silylated in the presence of oxiranes, esters, and nitriles; low reaction temperatures favour a reasonable degree of kinetic regioselectivity (*Scheme 17.6*), and mainly (*Z*)-enol ethers are obtained from simple acyclic ketones. Some selectivity between ketones has also been observed.

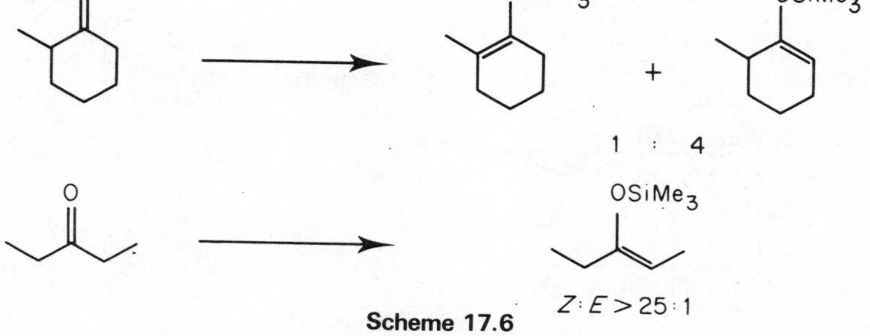

OSiMe$_3$ + OSiMe$_3$

1 : 4

OSiMe$_3$

$Z:E > 25:1$

Scheme 17.6

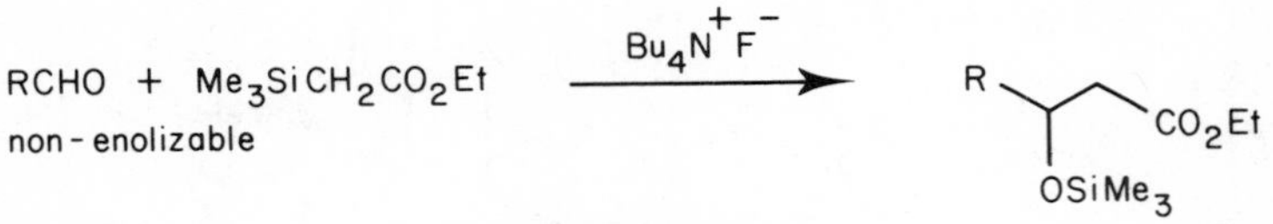

$$\underset{\text{non-enolizable}}{RCHO} + Me_3SiCH_2CO_2Et \xrightarrow{Bu_4\overset{+}{N}\,\overset{-}{F}} RCH(OSiMe_3)CH_2CO_2Et$$

Scheme 17.7

Non-enolizable aldehydes react differently, giving Reformatsky products[18] (*Scheme 17.7*), whereas enolizable aldehydes suffer self-condensation.

A wide range of alcohols, including tertiary ones, alkanethiols, phenols and arylalkynes are smoothly silylated by this reagent system, the general mechanism of whose action can be represented as shown in *Scheme 17.8*.

$$Me_3SiCH_2CO_2Et + Bu_4N^+ F^- \longrightarrow Bu_4N^+ \bar{C}H_2CO_2Et + Me_3SiF$$

$$Bu_4N^+ \bar{C}H_2CO_2Et + NuH \longrightarrow Bu_4\overset{+}{N}\, \overset{-}{Nu} + CH_3CO_2Et$$

$$Bu_4\overset{+}{N}\, \bar{Nu} + Me_3Si\,CH_2CO_2Et\ (\text{or } Me_3SiF) \longrightarrow Nu{-}SiMe_3 + Bu_4\overset{+}{N}\ \bar{C}H_2CO_2Et\ (\text{or } F^-$$

Scheme 17.8

To digress momentarily, it should be mentioned at this point that initial enolate geometry exerts considerable control on the stereochemical outcome of the aldol condensation[19] and of [3,3]-sigmatropic rearrangements of allyl ester silyl ketene acetals (*see* p. 260). Ireland[20a] has observed a dramatic solvent effect on enolate geometry. In particular, 3-pentanone shows a high degree of selectivity for formation of the (*E*)-enolate in THF and the (*Z*)-enolate in THF–HMPA (*Scheme 17.9*). This has been ascribed to the greater effective steric bulk of the lithium cation-coordinated oxygen in the former, less cation-coordinating THF conditions resulting in a higher activation energy for the transition state leading to production of the (*Z*)-enolate. Recently, Rathke[20b] has questioned whether competitive kinetic processes are involved in both cases. He has presented evidence that, under stict kinetic control, predominant formation of the (*E*)-enolate is observed in the presence or absence of HMPA; under thermodynamic conditions, the (*Z*)-enolate is formed predominantly.

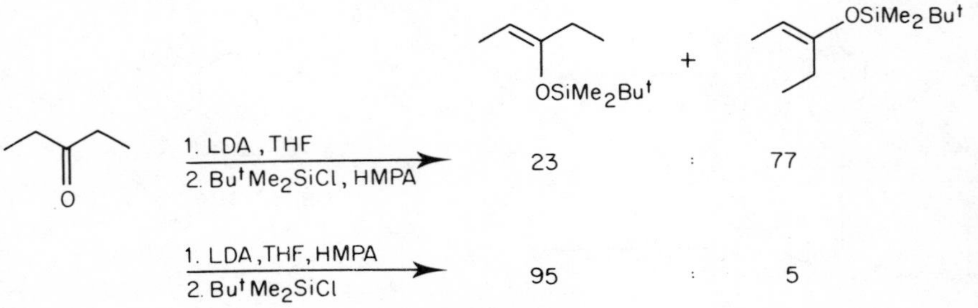

$$EtC(O)Et \longrightarrow CH_3CH{=}C(OSiMe_2Bu^t)Et + CH_3CH{=}C(OSiMe_2Bu^t)Et$$

Conditions	(*Z*)-isomer ($OSiMe_2Bu^t$ cis to Me)		(*E*)-isomer
1. LDA, THF; 2. Bu^tMe_2SiCl, HMPA	23	:	77
1. LDA, THF, HMPA; 2. Bu^tMe_2SiCl	95	:	5

Scheme 17.9

α-Halogenoalkyl ketones have been employed in three regiospecific routes. In a reaction of considerable generality[21], unsymmetrically substituted ketones are regioselectively α-chlorinated at the more substituted position with sulphuryl chloride, and the reductively derived zinc enolates are trapped *in situ* (*Scheme 17.10*), to give predominantly the more substituted (thermodynamic) silyl enol ether. The hitherto difficultly-obtainable trimethylsilyl enol ether (4) of camphor has been prepared[22] similarly.

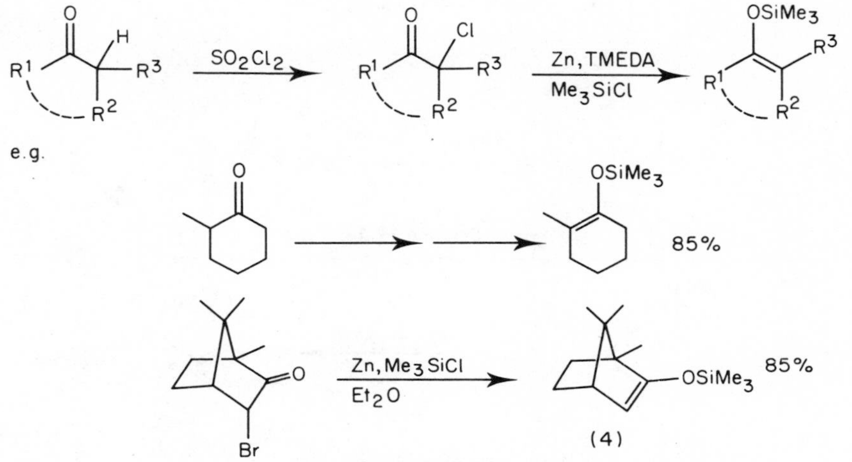

Scheme 17.10

Appropriately substituted oxiranes undergo regiospecific opening with HBr to give bromohydrins; these, after oxidation to the α-bromoketones, can give rise to regio-pure enol phosphites[23] or, as illustrated in *Scheme 17.11*, enol silyl ethers[24].

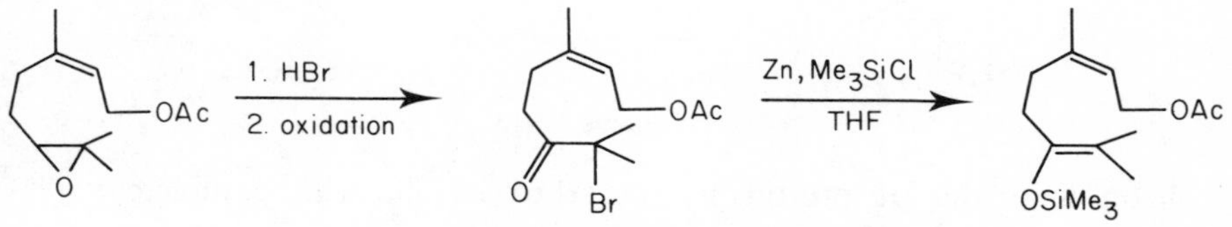

Scheme 17.11

The less-substituted (kinetic) enol ether can be obtained by treatment[25a] of an αα′-dibromoketone with excess of lithium dialkyl cuprate (*Scheme 17.12*); the full scope of this reaction is, at present, undefined. 2-(Trimethylsilyloxy)allyl halides undergo regiospecific attack[25b] when treated with dialkyl or diaryl cuprates, either direct or allylic alkylation taking place depending upon the substrate.

A conceptually different approach[26] relies on silatropic rearrangement[27] of trimethylsilyl β-ketoesters and related species (*Scheme 17.13*). The requisite silyl β-ketoesters are readily prepared[28] by acylation of the appropriate carboxylic acid dianion followed by silylation. Pyrolysis then effects the

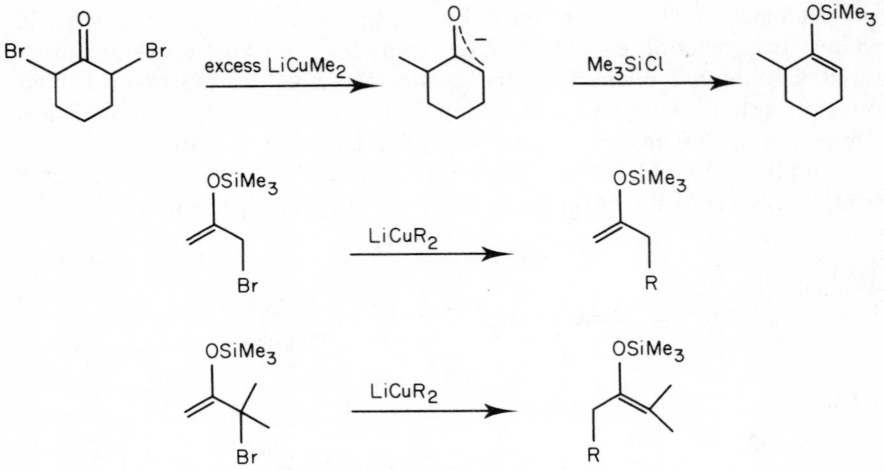

excess $LiCuMe_2$

Me_3SiCl

$OSiMe_3$

$LiCuR_2$

Scheme 17.12

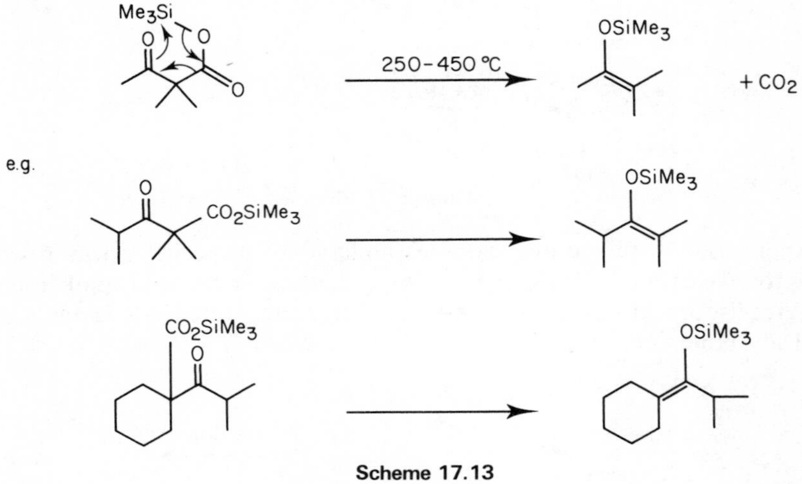

250–450 °C

+ CO_2

e.g.

CO_2SiMe_3

Scheme 17.13

silicon analogue of prototropic decarboxylation, with complete regio-specificity (*see also* Chapter 5).

Trimethylsilyl enol ethers of allylic β-ketoesters undergo the silyl analogue of the Carroll reaction, providing a relatively simple, regiospecific synthesis[26] of allyl-substituted trimethylsilyl enol ethers (*Scheme 17.14*).

The $\beta\gamma$-unsaturated cyclohexenone (5) on kinetic deprotonation gives

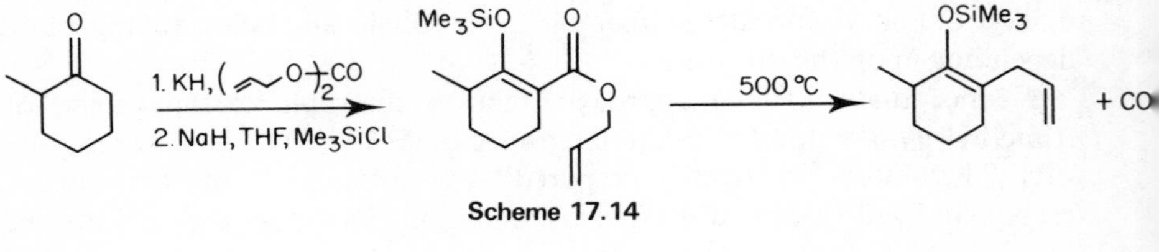

1. KH, $(\text{allyl-O})_2CO$
2. NaH, THF, Me_3SiCl

500 °C

+ CO

Scheme 17.14

mainly the conjugated enolate (6). A general route to species such as the isomeric (7) is revealed in the lithium–ammonia reduction[29] of isopropyldimethyl- and t-butyldimethyl-silyl aryl ethers (the corresponding trimethylsilyl aryl ethers are too labile under the reaction conditions employed). Such isopropyldimethylsilyl enol ethers are readily converted into the corresponding lithium enolates and undergo, either as such, or directly, the normal range of electrophilic addition reactions. Additionally, they provide convenient substrates for the synthesis of $\beta\gamma$-unsaturated cyclohexenones by fluoride ion-mediated hydrolysis, only the silyl ether being cleaved; the t-butyldimethylsilyl analogues were hydrolysed too slowly, and competitive reactions took place. Alternatively, taking advantage of the acid-lability of alkyl enol ethers, and the relative stability of t-butyldimethylsilyl ethers under such conditions, the complementary $\beta\gamma$-unsaturated cyclohexenones could be obtained (*Scheme 17.15*).

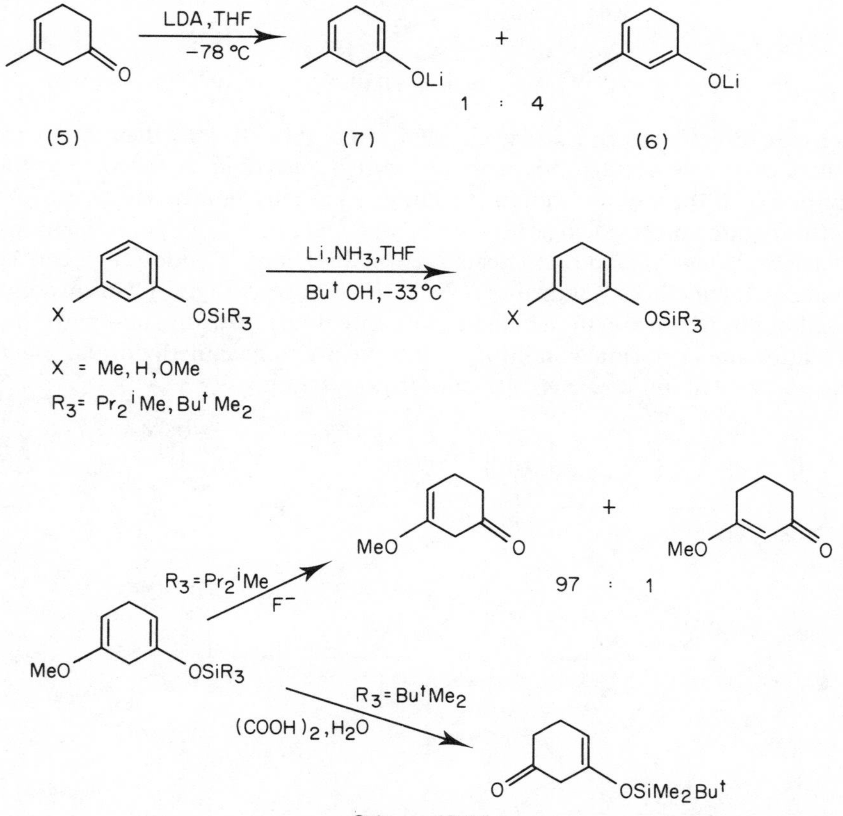

LDA, THF, −78 °C; (5) → (7) OLi + (6) OLi, 1 : 4

Li, NH_3, THF, Bu^tOH, −33 °C; X, $OSiR_3$

X = Me, H, OMe

R_3 = Pr^i_2Me, Bu^tMe_2

R_3 = Pr^i_2Me, F^- → MeO, O + MeO, O, 97 : 1

MeO, $OSiR_3$

R_3 = Bu^tMe_2, $(COOH)_2$, H_2O → O, $OSiMe_2Bu^t$

Scheme 17.15

Other silylating systems are known, but most show little if any regioselectivity. Trimethylsilyl trifluoromethanesulphonate is a most potent silylating agent, smoothly converting[30] (*Scheme 17.16*) a wide range of

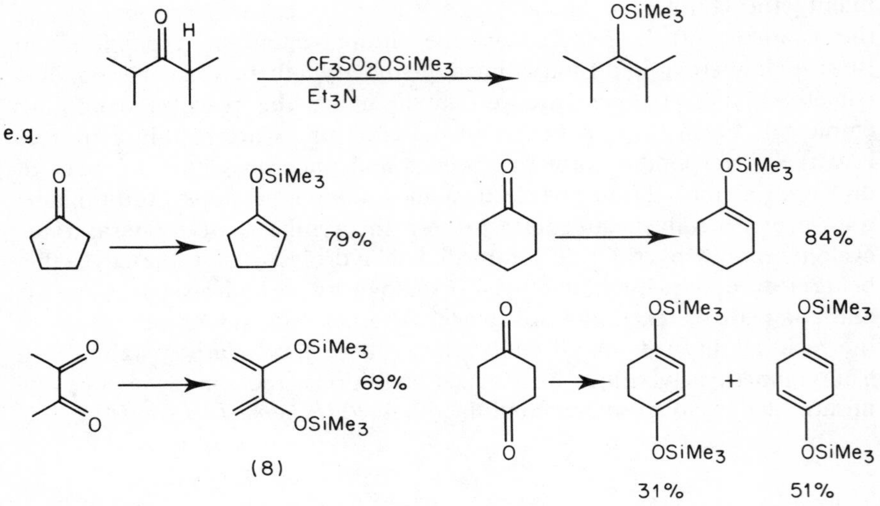

OSiMe3
CF3SO2OSiMe3
Et3N
e.g.
OSiMe3
79%
OSiMe3
84%
OSiMe3
69%
OSiMe3
OSiMe3
OSiMe3
+
OSiMe3
OSiMe3
31%
51%

(8) (9)

Scheme 17.16

ketones, diketones, and arylacetic acid esters directly into their silyl enol ethers or ketene acetals. No pronounced regioselectivity is shown, but the simplicity of the method should encourage its use in symmetrical cases.

Conjugated dienes such as (8) and (9) (*Scheme 17.16*) have potential utility in Diels–Alder cycloaddition reactions (p. 254), and a wide range can be made by the methods[31] shown in *Scheme 17.17*; surprisingly, 1,3-dicarbonyl compounds which are often high in enol content do not always undergo clean silylation under normal conditions[32], but the use of hexamethyldisilazane in the presence of imidazole effects smooth conversion.

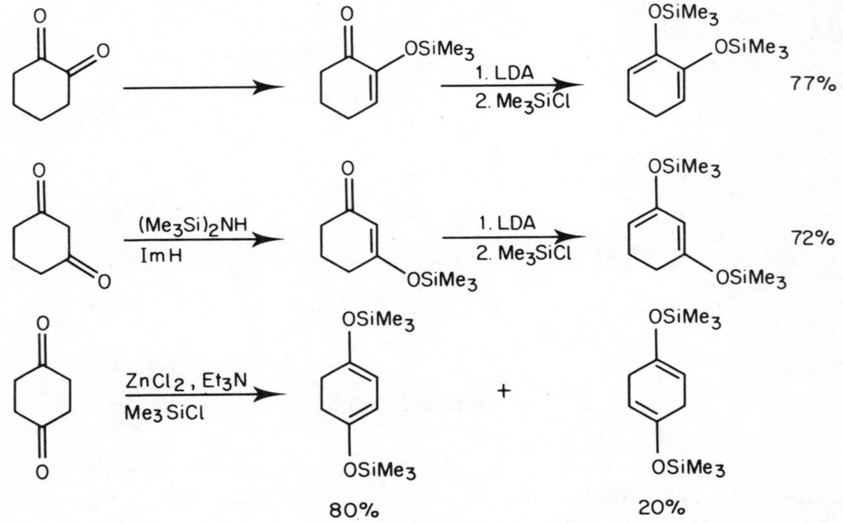

OSiMe3
1. LDA
2. Me3SiCl
OSiMe3
OSiMe3
77%
(Me3Si)2NH
ImH
OSiMe3
1. LDA
2. Me3SiCl
OSiMe3
OSiMe3
72%
ZnCl2, Et3N
Me3SiCl
OSiMe3
OSiMe3
+
OSiMe3
OSiMe3
80%
20%

Scheme 17.17

As discussed in Chapter 5, α-silyl alkoxides can undergo Brook rearrangement[33] to the isomeric silyloxy carbanions, the rearrangement being favoured by carbanion stabilizing substituents such as aryl and vinyl groups. Reich[34] reasoned that if a leaving group was present in the β-position, then elimination could lead to silyl enol ethers. Such a situation could arise by attack by a suitable organolithium species on an acylsilane, as shown in *Scheme 17.8*, which also illustrates the success of the method. Notably, 1-phenyl-2-butanone does not give pure enol ether (11) under kinetic conditions, a 30:70 ratio of (11) and its conjugated isomer being produced. Application of this present technique, however, allows it to be prepared regiospecifically. Further extension allows the separate preparation of the regioisomers (10) and (12). Acylsilanes, being rather unreactive compounds for both steric and electronic reasons, do not react cleanly with the more highly substituted lithium reagents required for the production of more complex silyl enol ethers.

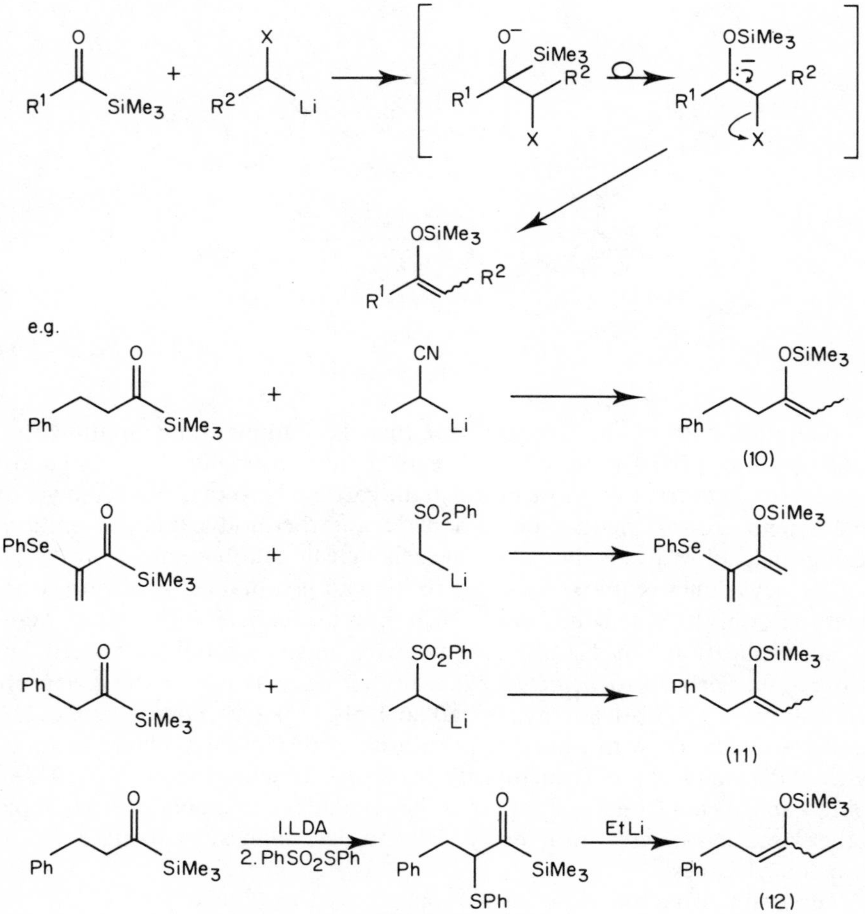

Scheme 17.18

Another regiospecific (and stereospecific) route[35a] to silyl enol ethers from acylsilanes which utilizes Brook rearrangement as its key step is shown in *Scheme 17.19.* Acylsilanes react with vinylmagnesium bromide to give alcohols (13), which, when converted into their alkoxides, rearrange to metallated silyl enol ethers (14). The latter species can be alkylated with primary alkyl iodides to give pure (*Z*)-silyl enol ethers in good yield. Alkynyl Grignard reagents have been employed analogously[35b, 35c] to produce silyloxyallenes.

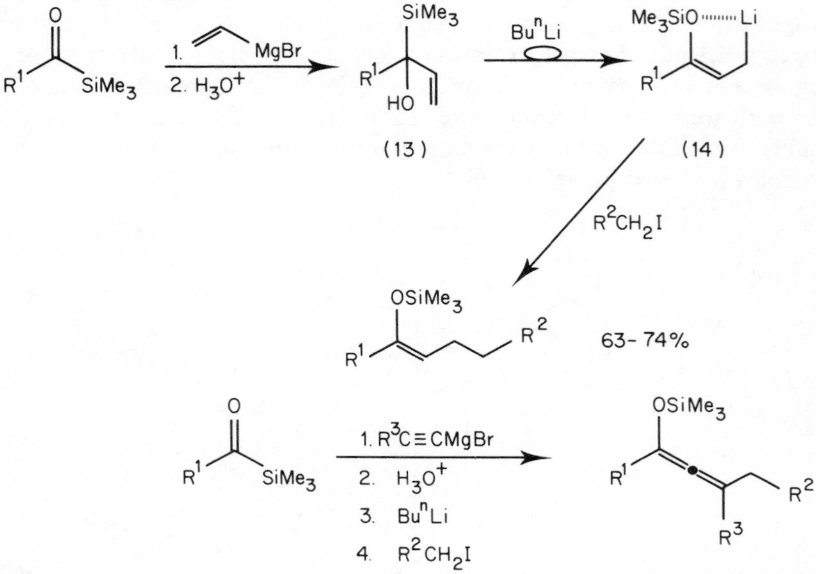

Scheme 17.19

Although both of these routes have their limitations, they do allow the regiospecific preparation of silyl enol ethers formally from pseudo-symmetric ketones, i.e., those in which the carbonyl group is flanked by two methylene groups; neither direct kinetic nor thermodynamic generation using such substrates would result in such a clean-cut outcome.

The acylsilanes required for these routes can be obtained by a variety of methods, only a limited number of which show useful flexibility. One of these is an application[36] of 1,3-dithiane anion chemistry, initially explored by Corey and Seebach (*Scheme 17.20*). Another route is due to the group of Kuwajima[37]. 1,1-Bis(trimethylsilyl)alkan-1-ols (15) (*see also* Chapter 20) undergo oxidation with t-butyl hypochlorite to produce acylsilanes in good yield (*Scheme 17.20*). α-Trimethylsilyl trimethylsilyl enol ethers (16) (p. 273), obtainable by controlled oxidation of the alkoxides corresponding to (15) or otherwise, can be cleaved hydrolytically to acyl silanes, or oxidatively to α-haloacylsilanes.

Three somewhat more specialized routes[38] are shown in *Scheme 17.21.*

The methodology involved in such approaches to silyl enol ethers is not

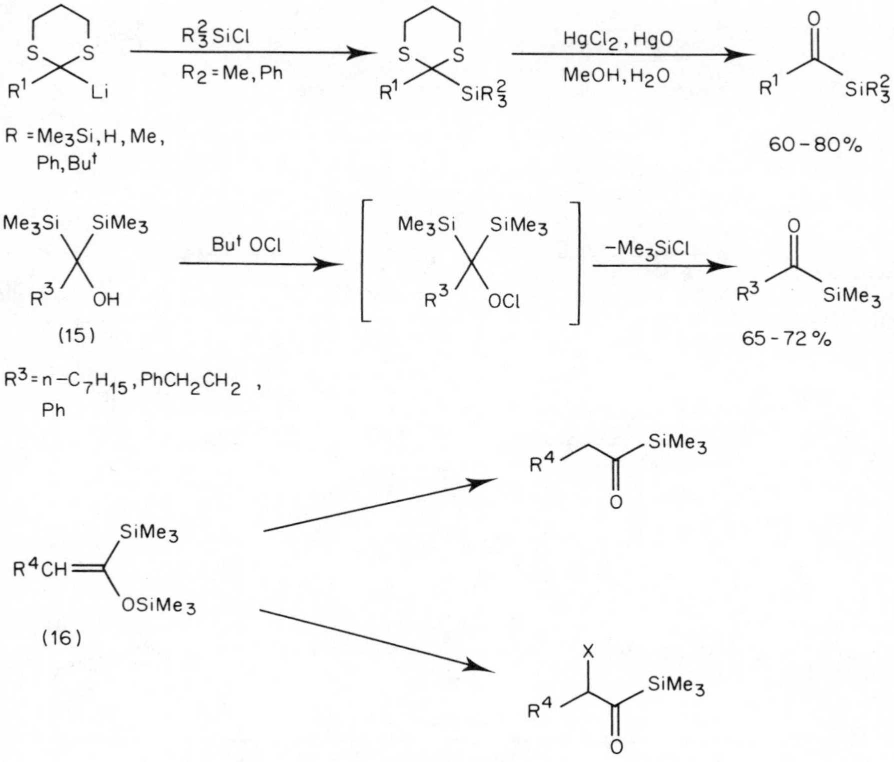

Scheme 17.20

without considerable precedent. Allyloxycarbanions such as (17), derived from allyl silyl ethers, are in rapid equilibrium[39] with the corresponding silyl alkoxide (18). These equilibrating species react with carbon electrophiles to give products of either α- or γ- carbon–carbon bond formation, whereas they are protonated and silylated[40] on oxygen. If the equilibrating mixture of anions is allowed to warm above −10 °C, then irreversible rearrangement to the lithium enolate (19) takes place (*Scheme 17.22*).

To expand on some of these processes, reaction of deprotonated allyl silyl ethers with alkylating agents[41,42] results in preferential γ-*C*-alkylation, therefore providing β-acyl carbanion equivalents (*Scheme 17.23*).

Reaction with carbonyl electrophiles[43], on the other hand, results in preferential α-attack, providing, *inter alia*, a route to 3,4-dihydroxy-alk-1-enes. α-Regioselectivity is enhanced by the presence of HMPA, and $\alpha\beta$-unsaturated carbonyl compounds undergo 1,2-addition, as illustrated in *Scheme 17.24*.

These principles can be extended to diallyl systems, although with reversal of regioselectivity. The pentadienyl silyl ether (20) on deprotonation gives[44] the symmetrical anion (21). This anion reacts with carbonyl compounds to give products of exclusive γ-attack, acting as the synthetic equivalent of the β-acyl carbanion (22), as illustrated in a recent synthesis[45] of

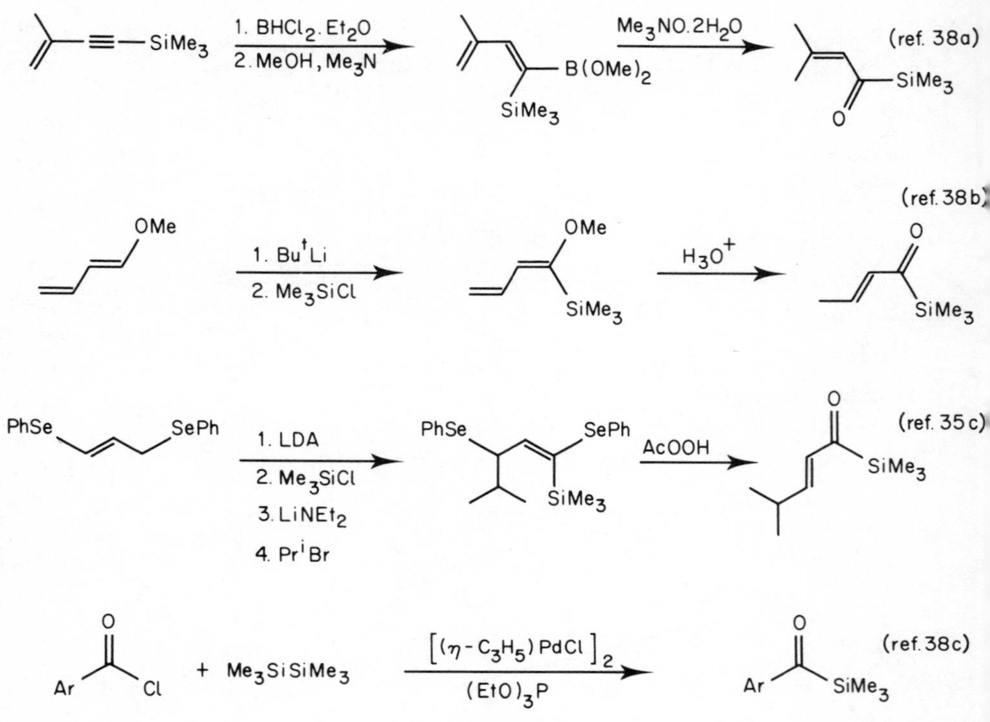

SiMe3 — 1. $BHCl_2.Et_2O$ 2. MeOH, Me_3N → $B(OMe)_2$, $SiMe_3$ — $Me_3NO.2H_2O$ → O, $SiMe_3$ (ref. 38a)

(ref. 38b)

OMe — 1. Bu^tLi 2. Me_3SiCl → OMe, $SiMe_3$ — H_3O^+ → O, $SiMe_3$

PhSe, SePh — 1. LDA 2. Me_3SiCl 3. $LiNEt_2$ 4. Pr^iBr → PhSe, SePh, $SiMe_3$ — AcOOH → O, $SiMe_3$ (ref. 35c)

O, Ar, Cl + $Me_3SiSiMe_3$ — $[(\eta\text{-}C_3H_5)PdCl]_2$, $(EtO)_3P$ → O, Ar, $SiMe_3$ (ref. 38c)

Scheme 17.21

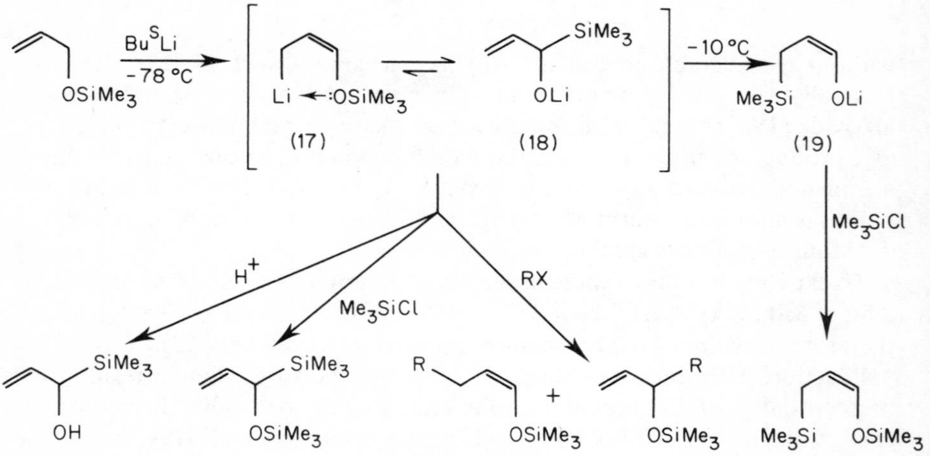

$OSiMe_3$ — Bu^sLi, −78 °C → [Li ←:$OSiMe_3$ (17) ⇌ $SiMe_3$, OLi (18)] — −10 °C → Me_3Si, OLi (19)

(19) — Me_3SiCl → Me_3Si, $OSiMe_3$

[(17) ⇌ (18)] — H^+ → $SiMe_3$, OH

[(17) ⇌ (18)] — Me_3SiCl → $SiMe_3$, $OSiMe_3$

[(17) ⇌ (18)] — RX → R, $OSiMe_3$ + R, $OSiMe_3$

Scheme 17.22

[(17) ⇌ (18)] + RX → R, $OSiMe_3$ — H_3O^+ → R, H, O

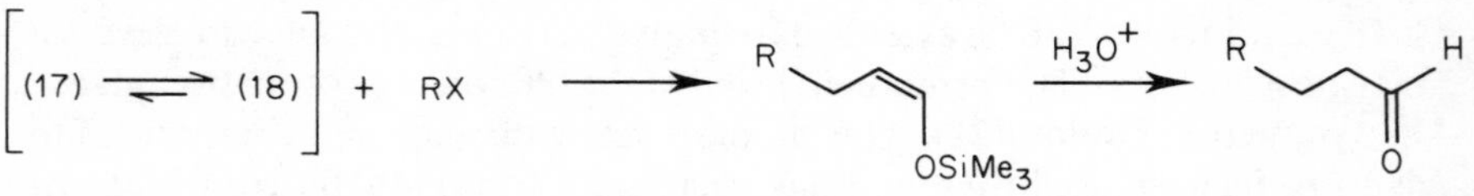

Scheme 17.23

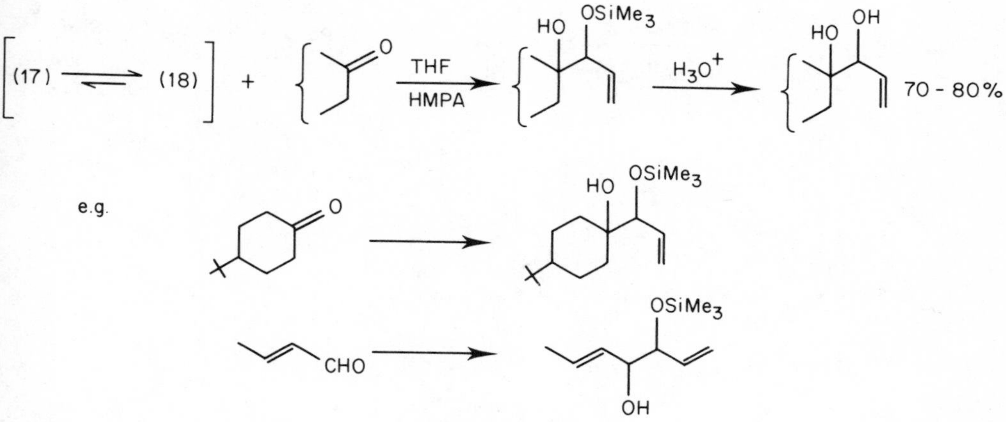

[(17) ⇌ (18)] + {ketone} $\xrightarrow[\text{HMPA}]{\text{THF}}$ HO, $OSiMe_3$ $\xrightarrow{H_3O^+}$ HO, OH 70 - 80%

e.g. → HO, $OSiMe_3$

CHO → $OSiMe_3$, OH

Scheme 17.24

($\pm$)-norpatchoulenol (23). The anion (21) reacts with alkyl halides to give mixtures resulting from α- and γ-attack, but alkyl tosylates and trifluoromethanesulphonates give high α-regioselectivity.

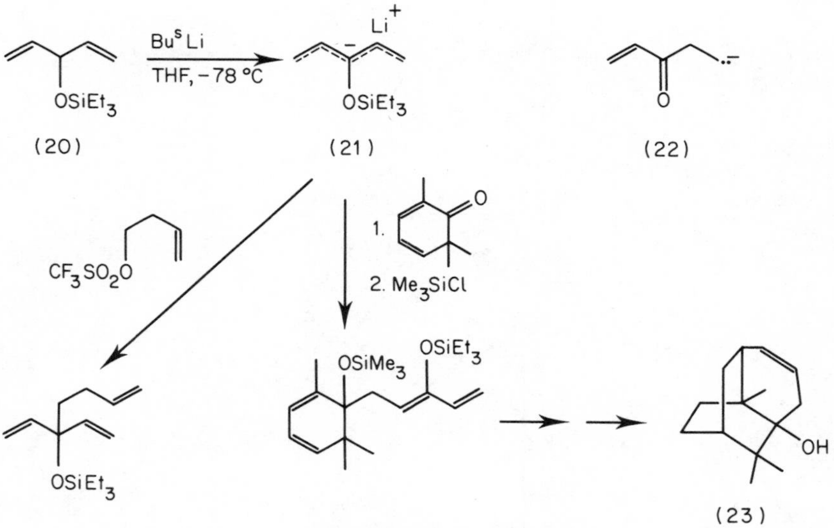

$OSiEt_3$ (20) $\xrightarrow[\text{THF, } -78\,°\text{C}]{\text{Bu}^s\text{Li}}$ Li^+ $OSiEt_3$ (21) — O (22)

CF_3SO_2O — $OSiEt_3$

1. (cyclohexadienone) 2. Me_3SiCl — $OSiMe_3$, $OSiEt_3$ → → OH (23)

Scheme 17.25

Finally, use can be made of a third product of *Scheme 17.22*. As was stated earlier, quenching of the equilibrating mixture of anions (17) and (18) with trialkylsilyl chlorides results in *O*-silylation. The product silyloxyallylsilanes (24) react[40] with acid chlorides in the presence of titanium(IV) chloride, displaying their allylsilane functionality (Chapter 9), and behaving as the synthetic equivalent of the homo-enolate anion (25) (*Scheme 17.26*). Competitive attack at oxygen, to give esters (26), can be minimized by increasing the steric bulk of the silyloxy silicon substituents.

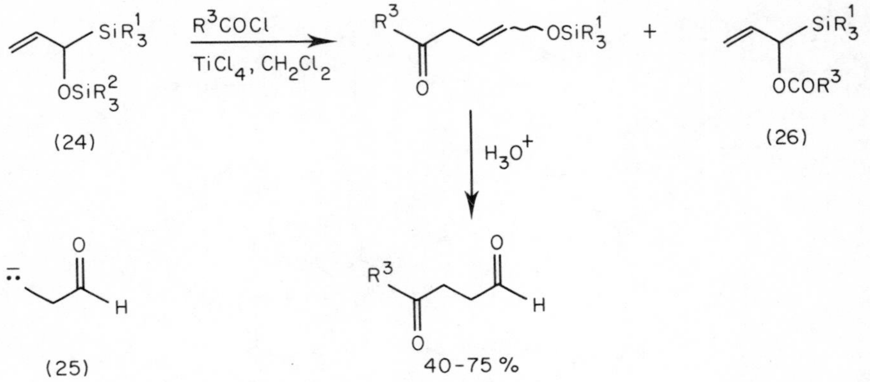

Scheme 17.26

α-Ketosilanes (acylsilanes)[46], β-ketosilanes[47], and αβ-epoxysilanes[48] all undergo more or less clean thermal rearrangement to silyl enol ethers (*Scheme 17.27*), although none of these routes is likely to compete directly with the more standard methods so far discussed (*see also* Chapter 5).

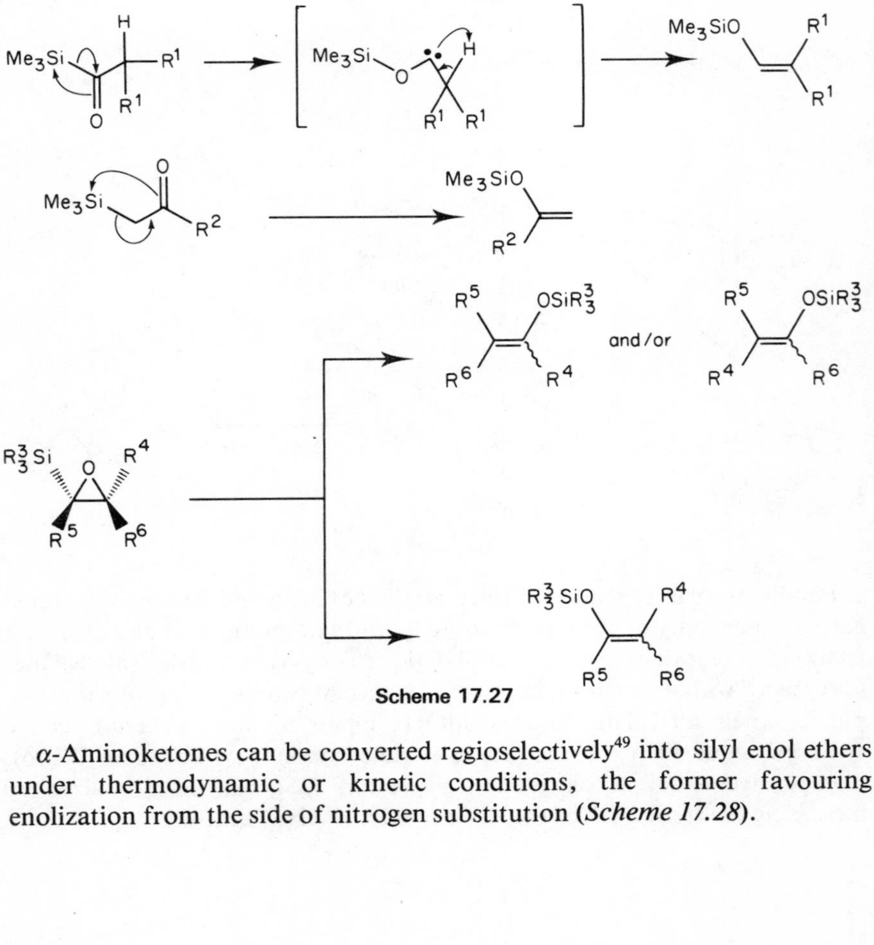

Scheme 17.27

α-Aminoketones can be converted regioselectively[49] into silyl enol ethers under thermodynamic or kinetic conditions, the former favouring enolization from the side of nitrogen substitution (*Scheme 17.28*).

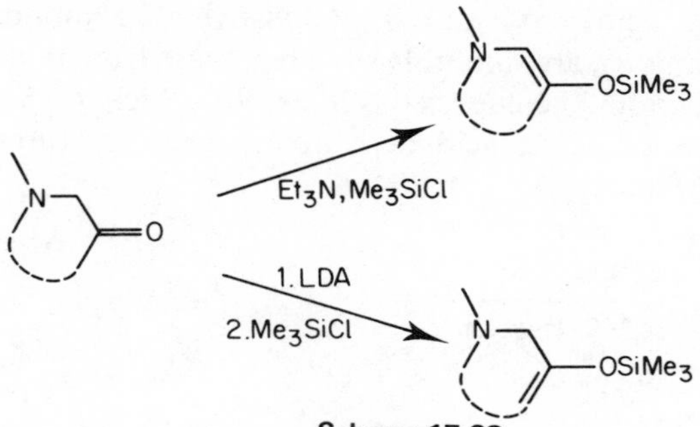

Et$_3$N, Me$_3$SiCl

1. LDA
2. Me$_3$SiCl

OSiMe$_3$

OSiMe$_3$

Scheme 17.28

Tetrahydrofuran is smoothly cleaved[50] by n-butyl-lithium to produce ethylene and the lithium enolate of acetaldehyde, in a process of α-hydrogen abstraction followed by a $[\pi^4s + \pi^2s]$ cycloreversion; extension to substituted tetrahydrofurans is limited to symmetrical 3,4-disubstituted species, as exocyclic β-cleavage occurs when 2-substituted species are employed. The lithium enolate generated in this way from non-carbonyl precursors can be trapped[51] by electrophiles such as trialkylsilyl chlorides (*Scheme 17.29*).

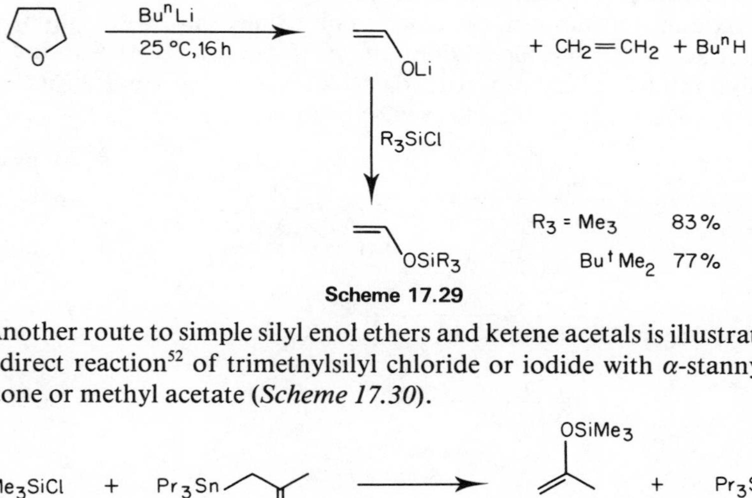

Bu^{n}Li
25 °C, 16 h

OLi + $CH_2{=}CH_2$ + Bu^{n}H

R$_3$SiCl

OSiR$_3$

R$_3$ = Me$_3$ 83%
Bu^{t}Me$_2$ 77%

Scheme 17.29

Another route to simple silyl enol ethers and ketene acetals is illustrated by the direct reaction[52] of trimethylsilyl chloride or iodide with α-stannylated acetone or methyl acetate (*Scheme 17.30*).

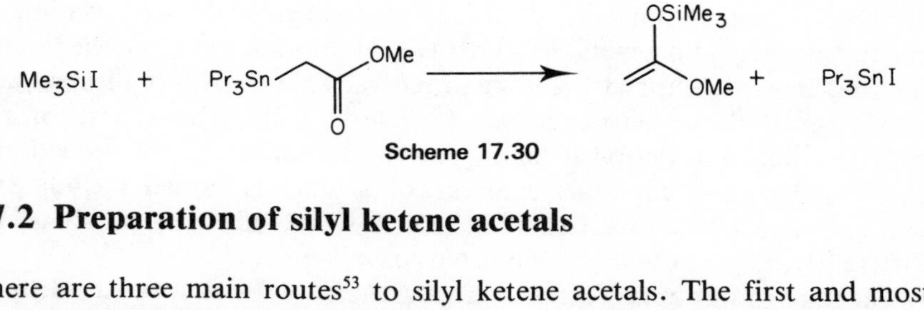

Me$_3$SiCl + Pr$_3$Sn–CH$_2$C(O)CH$_3$ → CH$_2$=C(OSiMe$_3$)CH$_3$ + Pr$_3$SnCl

Me$_3$SiI + Pr$_3$Sn–CH$_2$C(O)OMe → CH$_2$=C(OSiMe$_3$)OMe + Pr$_3$SnI

Scheme 17.30

17.2 Preparation of silyl ketene acetals

There are three main routes[53] to silyl ketene acetals. The first and most generally applicable of these mirrors one of the major routes to silyl enol ethers. Carboxylic acid alkyl and trimethylsilyl ester monoanions, and carboxylic acid dianions, are smoothly silylated[54] with trimethylsilyl chloride to give the corresponding ketene acetals in excellent yield (*Scheme 17.31*) in all cases bar those of acetic acid derivatives, when mixtures of *O*- and *C*-silylation are obtained.

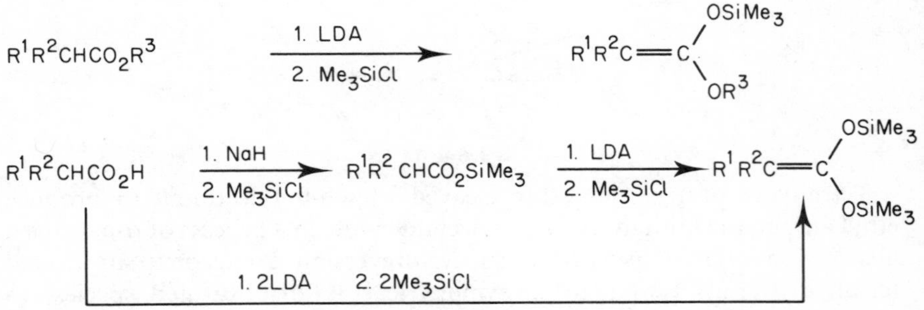

$$R^1R^2CHCO_2R^3 \xrightarrow[2.\ Me_3SiCl]{1.\ LDA} R^1R^2C{=}C(OSiMe_3)(OR^3)$$

$$R^1R^2CHCO_2H \xrightarrow[2.\ Me_3SiCl]{1.\ NaH} R^1R^2CHCO_2SiMe_3 \xrightarrow[2.\ Me_3SiCl]{1.\ LDA} R^1R^2C{=}C(OSiMe_3)(OSiMe_3)$$

$$R^1R^2CHCO_2H \xrightarrow{1.\ 2LDA\quad 2.\ 2Me_3SiCl} R^1R^2C{=}C(OSiMe_3)(OSiMe_3)$$

Scheme 17.31

Many functional variations can be tolerated, including monosubstituted malonates and γ-[55,56] and δ-[57] lactones (*Scheme 17.32*).

t-Butyldimethylchlorosilane reacts only sluggishly with lithio ester enolates at −78 °C, but the addition of *ca.* 10 per cent HMPA to the THF solution results in smooth reaction[54]. All substrates studied, including acetates, gave products of exclusive *O*-silylation.

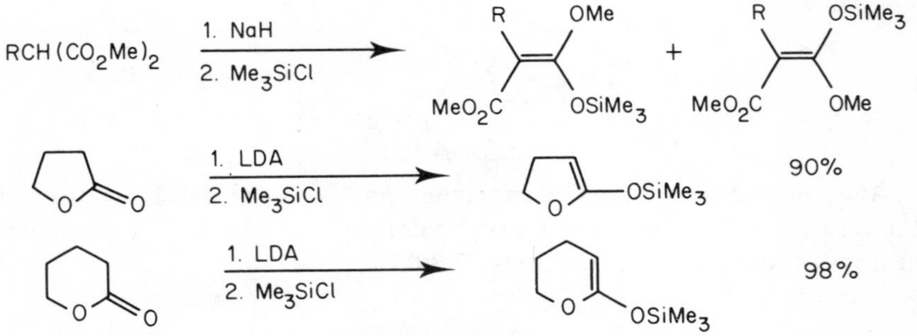

$$RCH(CO_2Me)_2 \xrightarrow[2.\ Me_3SiCl]{1.\ NaH} R(MeO_2C)C{=}C(OMe)(OSiMe_3) + R(MeO_2C)C{=}C(OSiMe_3)(OMe)$$

γ-Butyrolactone $\xrightarrow[2.\ Me_3SiCl]{1.\ LDA}$ 2-(trimethylsilyloxy)-4,5-dihydrofuran ($OSiMe_3$) 90%

δ-Valerolactone $\xrightarrow[2.\ Me_3SiCl]{1.\ LDA}$ 6-(trimethylsilyloxy)-3,4-dihydro-2H-pyran ($OSiMe_3$) 98%

Scheme 17.32

Disubstituted malonates cannot, of course, react in a similar manner. However, they do undergo a process of reductive decarboxylation on treatment with sodium metal. If trimethylsilyl chloride is present, the reactive intermediates are trapped[58], leading to generally good yields of disubstituted alkyl trimethylsilyl ketene acetals (*Scheme 17.33*). The course of this reductive decarboxylation is highly solvent-dependent. If the reduction is performed in liquid ammonia, a mixture of products is formed, including the cyclopropanediol derivative (27). Diethyl oxalate reacts somewhat differently, to produce the isomeric tetraoxyalkenes (28).

The third route duplicates the hydrosilylation of $\alpha\beta$-unsaturated aldehydes and ketones discussed earlier (p. 200). A variety of $\alpha\beta$-unsaturated

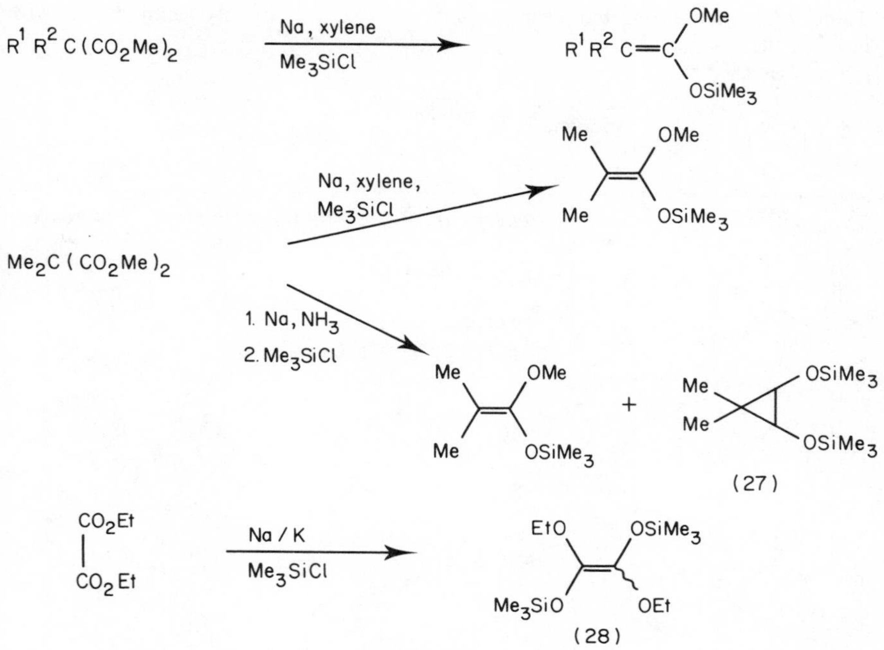

$R^1R^2C(CO_2Me)_2 \xrightarrow[Me_3SiCl]{Na,\ xylene} R^1R^2C{=}C(OMe)(OSiMe_3)$

$Me_2C(CO_2Me)_2 \xrightarrow[Me_3SiCl]{Na,\ xylene,} Me_2C{=}C(OMe)(OSiMe_3)$

$Me_2C(CO_2Me)_2 \xrightarrow{1.\ Na,\ NH_3;\ 2.\ Me_3SiCl} Me_2C{=}C(OMe)(OSiMe_3) + $ (27)

$(CO_2Et)_2 \xrightarrow[Me_3SiCl]{Na/K} (EtO)(Me_3SiO)C{=}C(OSiMe_3)(OEt)$ (28)

Scheme 17.33

carboxylic acid esters undergo rhodium(I)-catalysed 1,4-hydrosilylation[59], to give silyl ketene acetals in excellent yields (*Scheme 17.34*) in all cases except that of methyl acrylate itself, when competitive 1,2-addition takes place.

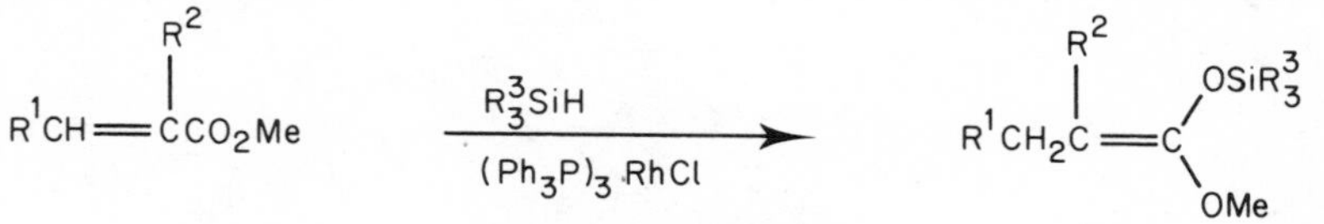

$R^1CH{=}C(R^2)CO_2Me \xrightarrow[(Ph_3P)_3RhCl]{R^3_3SiH} R^1CH_2C(R^2){=}C(OSiR^3_3)(OMe)$

Scheme 17.34

17.3 Reactions

It would be impossible to cover here all cases of application of silyl enol ethers in organic synthesis. The number of such cases is truly vast, and the area has been well reviewed[1,2,4] and discussed on various occasions. Instead, emphasis will be given to more recent applications, including new methods for the liberation of specific enolate anions and those reactions which involve direct attack (albeit often catalysed) by electrophiles on silyl enol ethers themselves. *Table 17.1* summarizes some of those reactions with electrophiles where Lewis acid catalysis is *not* required; trimethylsilyloxycyclohexene is merely representative. Some of the more useful of these transformations will be discussed individually, as will other more recent applications.

Table 17.1 Some selected examples of the reaction of silyl enol ethers with electrophiles

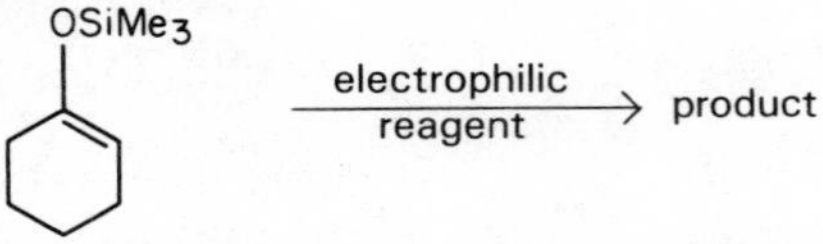

$OSiMe_3$ (cyclohexenyl) $\xrightarrow[\text{reagent}]{\text{electrophilic}}$ product

Product	*Reagent(s)*	*Notes*	*References*
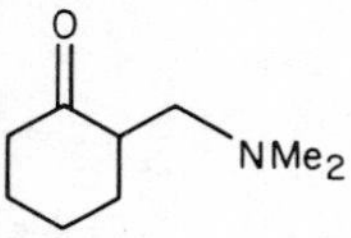 NMe_2	$Me_2\overset{+}{N}{=}CH_2\ I^-$	1	60, 61
	1. MeLi 2. $Me_2\overset{+}{N}{=}CH_2\ CF_3CO_2^-$	2	56
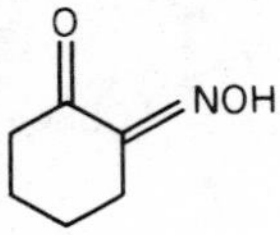 NOH	NOCl	3	55
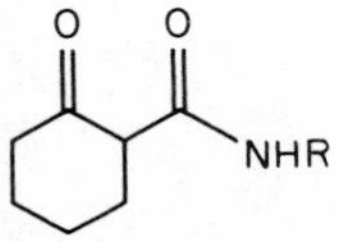 NHR	RNCO, Et_3N		62
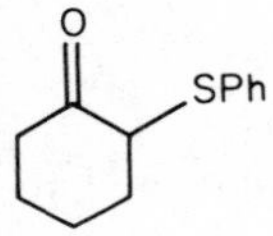 SPh	PhSCl		63
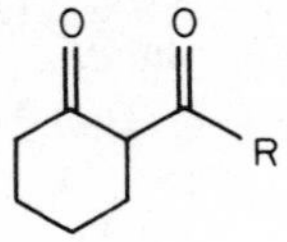 R	RCOCl	4, 5	63, 64, 65
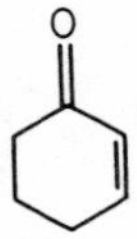	1. $^1O_2$ 2. Ph_3P		66
	chloranil		67
	DDQ		68, 69
	$Ph_3C^+BF_4^-$		68
	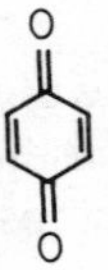 (benzoquinone), PdX_2		70
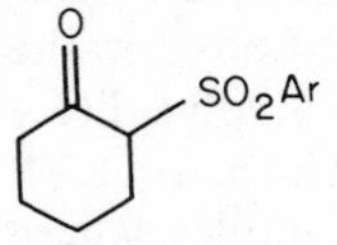 SO_2Ar	$ArSO_2Cl$		71
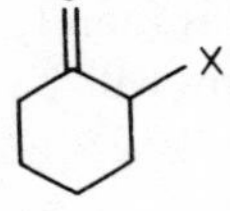 X	Br_2 or Cl_2		72

Table 17.1—*cont.*

Product	*Reagent(s)*	*Notes*	*References*
	1. I_2, AgOAc 2. $Et_3\overset{+}{N}H\ F^-$		73
OSiMe3, Br	1. Br_2 2. R_3N		74
O, OH	1. MCPBA 2. $Et_3\overset{+}{N}H\ F^-$	6	75, 76
O, OSiMe3	$h\nu$	7	77
O, OCOPh	1. $Pb(OCOPh)_4$ 2. $Et_3\overset{+}{N}H\ F^-$		78
O, OAc	1. MCPBA 2. Ac_2O/R_3N		79
O, OOSiMe3	$^1O_2$	8, 9	80, 81, 82

[1]For an intramolecular application, see ref. 61.
[2]Applicable also to ester- and lactone-derived silyl ketene acetals, after generation of the free enolate ion by treatment with methyl-lithium.
[3]Applicable also to ester- and lactone-derived silyl ketene acetals.
[4]Good yields with polychloroacid chlorides.
[5]Applicable also to α-unhindered ester-derived silyl ketene acetals, to produce β-ketoesters, and to malonate-derived silyl ketene acetals, to produce acyl malonates.
[6]Applicable also to acid-derived bistrimethylsilyl ketene acetals, to produce α-hydroxy carboxylic acids.
[7]Ring-D steroids.
[8]With simple silyl enol ethers, product of 'ene' reaction is major.
[9]Applicable also to acid-derived bistrimethylsilyl ketene acetals, to produce α-peroxy acids.

17.3.1 Generation of specific enolate anions

In earlier studies[83] of specific structural and/or geometric isomers of enolate anions, such anions were generated by reaction of the appropriate purified enol acetates with methyl-lithium (*Scheme 17.35*). The co-produced, strongly basic, lithium t-butoxide frequently complicated alkylation of the enolate ions by promoting further alkylation of the initially formed product. Trimethylsilyl enol ethers, on the other hand, were found to undergo smooth cleavage on treatment with methyl-lithium[4,84,85] to give the corresponding lithium enolates and innocuous tetramethylsilane, which does not exchange its hydrogen atoms under the reaction conditions (*Scheme 17.35*), and can act as a useful internal standard for NMR spectroscopic characterization of the enolate ions in question.

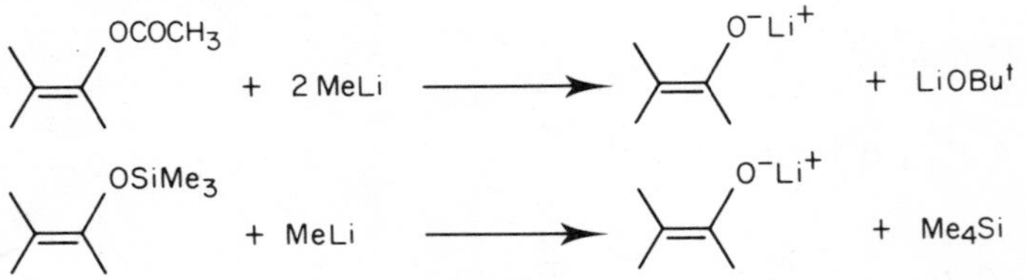

OCOCH$_3$ + 2 MeLi ⟶ O$^-$Li$^+$ + LiOBu^t

OSiMe$_3$ + MeLi ⟶ O$^-$Li$^+$ + Me$_4$Si

Scheme 17.35

Parenthetically, it should be noted that methylmagnesium bromide is not nearly as effective as methyl-lithium for this purpose. The silyl enol ether (29) is cleaved by methyl-lithium in six minutes at room temperature, whereas it takes 24 hours at reflux to achieve the same cleavage with methylmagnesium bromide[84], the solvent in both cases being DME (*Scheme 17.36*). Indeed, it has been shown[86] that methylmagnesium bromide will attack ketonic carbonyl groups in preference to silyl enol ethers.

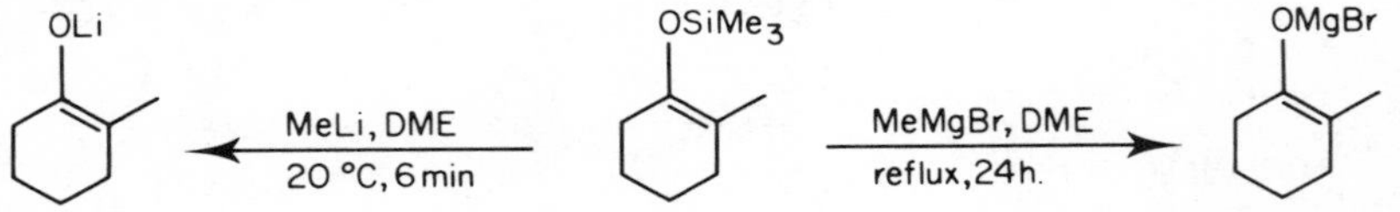

OLi ⟵ (MeLi, DME; 20 °C, 6 min) OSiMe$_3$ (29) ⟶ (MeMgBr, DME; reflux, 24h.) OMgBr

Scheme 17.36

Ketone trimethylsilyl enol ethers are cleaved by fluoride ion to give enolate anions or their equivalents. Initially, and for the purposes of alkylation[87], fluoride ion was present as benzyltrimethylammonium fluoride, and was used in slight stoichiometric excess. In all cases reported, the crude reaction mixture obtained after filtration contained only the regiospecifically monoalkylated ketone, unreacted alkyl halide, and the ketone resulting from simple hydrolysis of the starting silyl enol ether; no product of polyalkylation or of regioisomeric alkylation was detected (*Scheme 17.37*). The formation of 2-methyl-6-butylcyclohexanone (30) is particularly noteworthy. The isomeric 2-methyl-2-butylcyclohexanone is the sole product when the lithium enolate derived from the enol phosphite[88] corresponding to (31) is used, enolate equilibration being more rapid than alkylation under such conditions; a high degree of regiospecificity in butylation *can* be obtained by generating the appropriate lithium enolate by silyl enol ether cleavage with lithium amide in ammonia–THF[89].

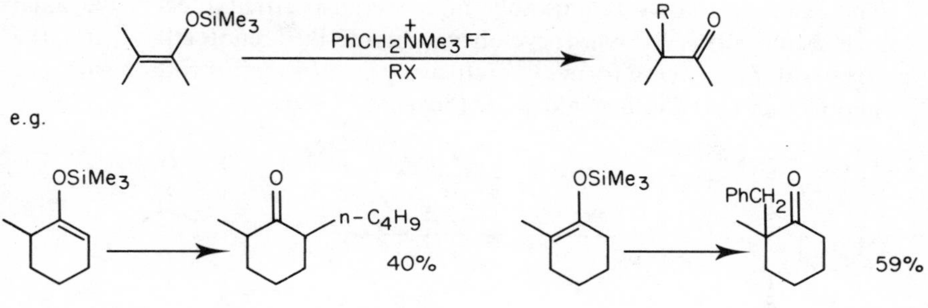

OSiMe3 — $PhCH_2\overset{+}{N}Me_3F^-$ / RX → (ketone, R)

e.g.

OSiMe3 → O, n-C_4H_9 40%   OSiMe3 → O, $PhCH_2$ 59%

(31)   (30)

Scheme 17.37

It may be of value to summarize at this point the techniques of regiospecific alkylation via regioisomerically pure silyl enol ethers. If it is desired to alkylate an unsymmetrical ketone at the more substituted α-position (thermodynamic enolate), then lithium enolates can be employed satisfactorily. On the other hand, quaternary ammonium enolates are more suitable if it is the less substituted α-position (kinetic enolate) which is to be alkylated.

Silyl enol ethers derived from a wide range of ketones react smoothly with aromatic and aliphatic aldehydes, but not ketones, in the presence of a *catalytic* amount[90] of fluoride ion, present this time as tetrabutylammonium fluoride. The reaction proceeds regiospecifically (*Scheme 17.38*); when $\alpha\beta$-unsaturated aldehydes are employed, products of 1,2-addition only are formed.

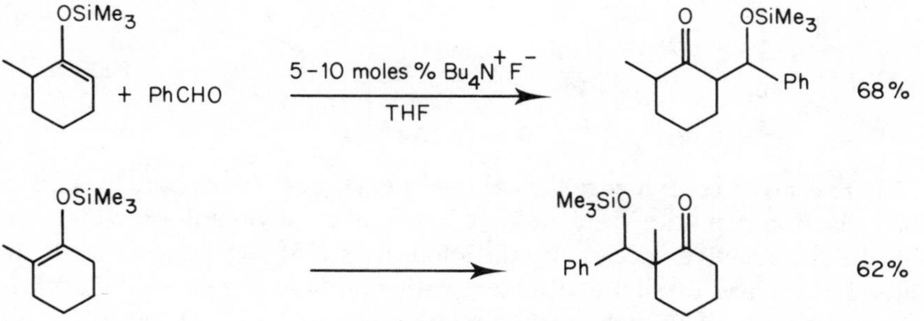

OSiMe3 + PhCHO — 5–10 moles % $Bu_4\overset{+}{N}F^-$ / THF → O, OSiMe3, Ph 68%

OSiMe3 → Me_3SiO, O, Ph 62%

Scheme 17.38

Combined with the use of ethyl trimethylsilylacetate to prepare the requisite silyl enol ethers, as discussed earlier (p. 201), this sequence allows crossed-aldol condensations to be performed without the isolation of any intermediates (*Scheme 17.39*).

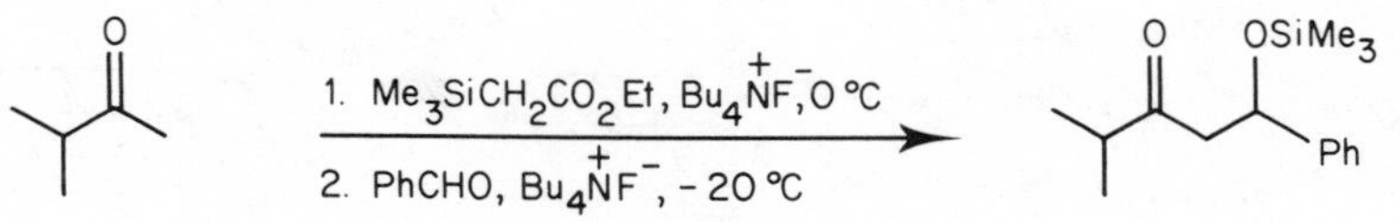

O — 1. $Me_3SiCH_2CO_2Et$, $Bu_4\overset{+}{N}\overset{-}{F}$, 0 °C; 2. PhCHO, $Bu_4\overset{+}{N}\overset{-}{F}$, −20 °C → O, OSiMe3, Ph

Scheme 17.39

The reaction shows remarkably high stereospecificity, exclusive axial attack being observed with cyclohexanone-derived enol ethers and the *erythro*-aldol (32) being formed[19] exclusively as the kinetic product from the (*Z*)-enol ether (33) and benzaldehyde (*Scheme 17.40*).

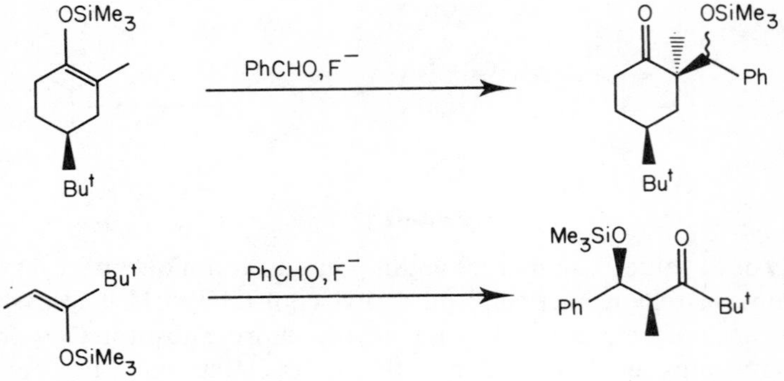

Scheme 17.40

The process appears to proceed through a catalytic cycle involving several reversible steps. The ratio of diastereoisomeric products varies with time, and in some cases the addition of fluorotrimethylsilane (but not chloro-) can enhance yields. Additionally, separately prepared ketone/ketone aldol products, as their silyl ethers, suffer cleavage when treated with catalytic amounts of fluoride ion (*Scheme 17.41*).

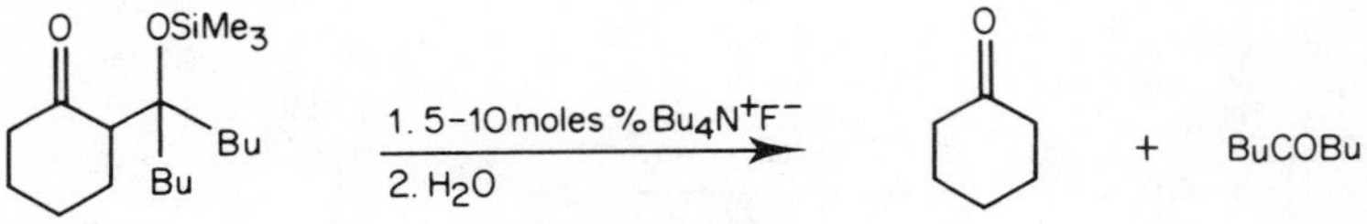

Scheme 17.41

There is no direct evidence for the intermediacy of free enolate ions under such reaction conditions; the anionic hypervalent silyl enol ether (34) may well be the reactive species. Free enolate ions (35) *can* be generated[91] by fluoride cleavage of silyl enol ethers with evaporative removal of fluorotrimethylsilane. Although such enolate ions undergo *O*-acylation and

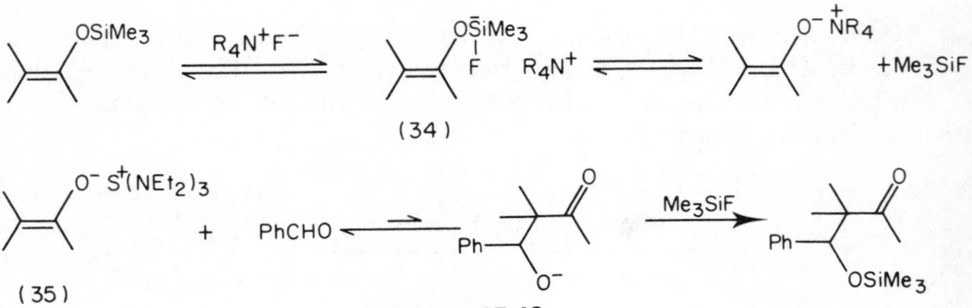

Scheme 17.42

C-alkylation, they do not react with benzaldehyde, owing to an unfavourable equilibrium position; if fluorotrimethylsilane is present in the reaction medium, the aldol product is trapped and the equilibrium so displaced (*Scheme 17.42*).

17.3.2 Lewis acid-catalysed alkylation

Primary and secondary alkyl groups can be introduced into the α-position of a carbonyl group by enolate alkylation with reasonably high regioselectivity[4, 83, 84, 85], but attempted introduction of tertiary alkyl groups using enolate anions as nucleophiles results in predominant 1,2-elimination processes occurring on the alkylating agent, e.g., a tertiary alkyl halide. If, on the other hand, the *electrophilicity* of the alkylating agent is enhanced via Lewis acid catalysis, then a wide range of tertiary alkyl groups can be smoothly and regiospecifically introduced[92a], even in those cases which result in the establishment of adjacent quaternary carbon atoms (*Scheme 17.43*).

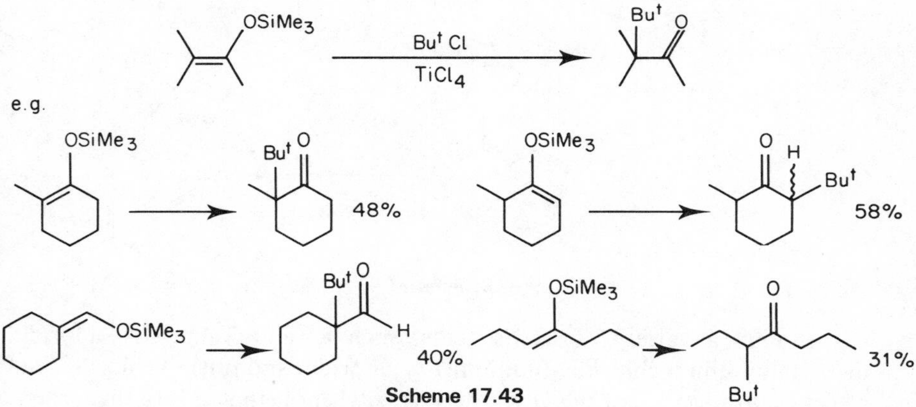

Scheme 17.43

A variety of α-adamantyl ketones and esters can be prepared[92b] in this manner (*Scheme 17.44*); such compounds are currently of great interest as potential antiviral agents. With ketone-derived enol ethers, titanium(IV) chloride is the most effective catalyst, whereas ester-derived ketene silyl acetals require the milder zinc(II) chloride.

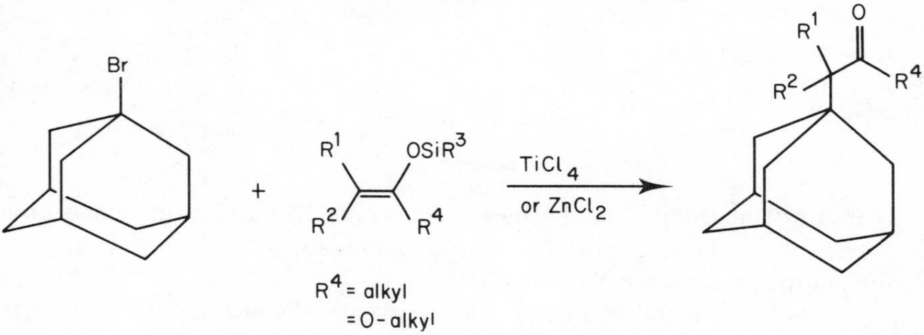

Scheme 17.44

This general method of activation of the electrophilic alkylating agent fails, as one might expect, with simple primary and secondary alkyl halides. It can, however, be extended[93] to a wide range of S_N1-type alkylating moieties, such as benzylic secondary, allylic, and α-methoxy halides, and α-chloroalkyl phenyl sulphides. Judicious selection of catalyst permits regiospecific introduction of the appropriate grouping. It also works well with ester- and lactone-derived silyl acetals, and has resulted in a short synthesis of the sesquiterpene (±)-*ar*-turmerone (36) (*Scheme 17.45*).

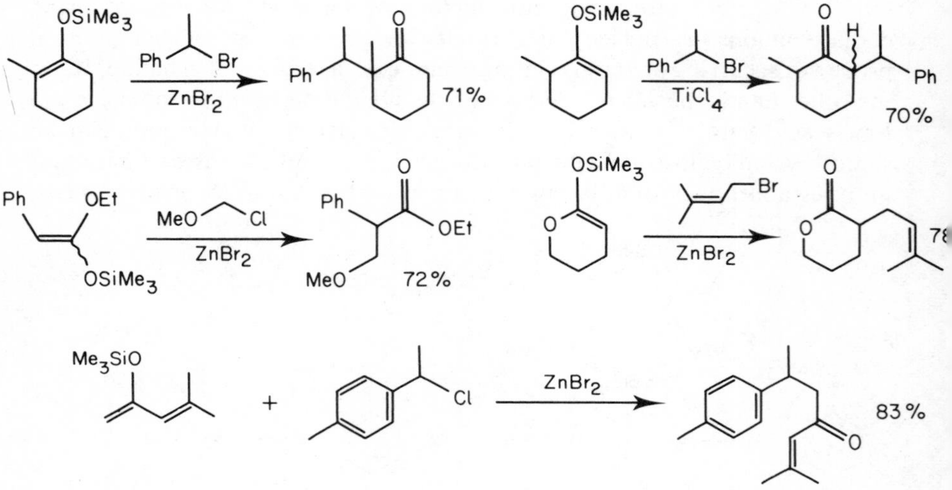

Scheme 17.45

Although a variety of Lewis acids, such as aluminium(III) chloride, diethylaluminium chloride, titanium(IV) chloride, and tin(IV) chloride, can effect cyclization[25,94] of the allylic acetate/silyl enol ether (37) to the terpene karahanaenone (38), the best yields are attained by using excess of methylaluminium bis(trifluoroacetate) (*Scheme 17.46*); other organoaluminium reagents give quite different products.

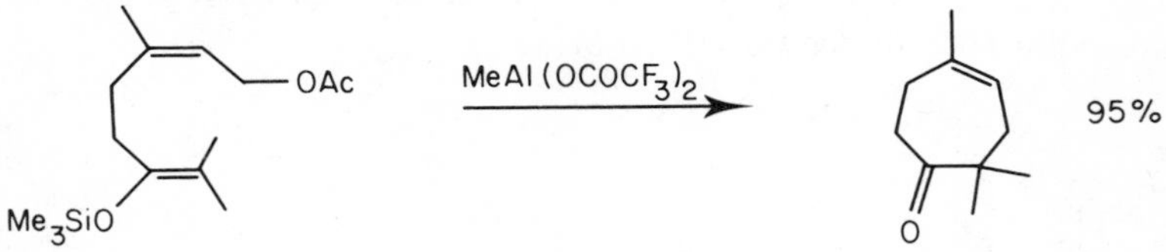

Scheme 17.46

Certain trimethylsilyl enol ethers react with polyhalogenomethanes and related species under copper(I) chloride catalysis[95], with the production of $\alpha\beta$-unsaturated ketones (*Scheme 17.47*).

Silyl enol ethers undergo Lewis acid-catalysed Michael addition with $\alpha\beta$-unsaturated nitro-compounds[96]. After addition of water, the corresponding

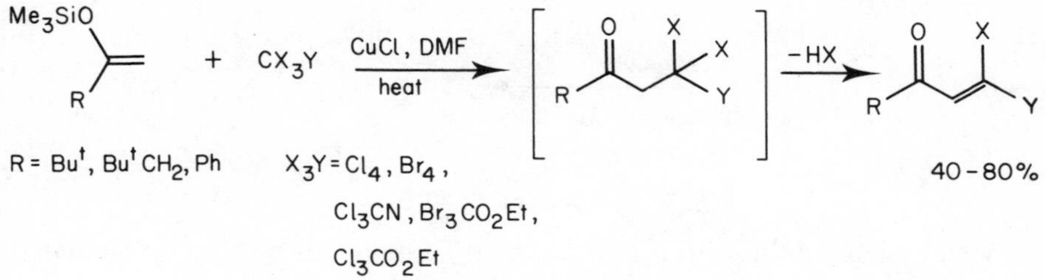

Me_3SiO / R + CX_3Y —CuCl, DMF, heat→ [R–CO–CH$_2$–CX$_2$Y] —(−HX)→ R–CO–CH=CXY

R = Bu^t, Bu^{t}CH$_2$, Ph X_3Y = Cl_4, Br_4, Cl_3CN, Br_3CO_2Et, Cl_3CO_2Et 40–80%

Scheme 17.47

1,4-diketones are obtained directly, possibly by way of a nitronate ester such as (39). The reaction appears to be of wide generality, and is regiospecific, leading, after aldol closure and dehydration, to a variety of substituted cyclopentenones (*Scheme 17.48*).

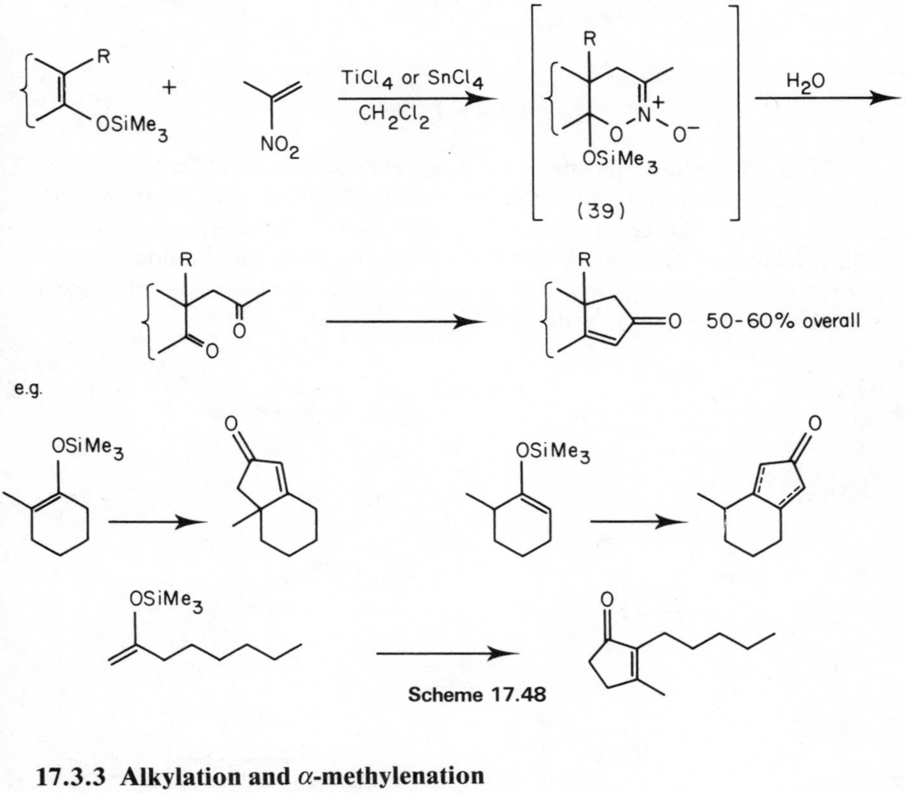

$TiCl_4$ or $SnCl_4$, CH_2Cl_2; OSiMe$_3$; NO_2; (39); H_2O; 50–60% overall

e.g.

OSiMe$_3$ OSiMe$_3$ OSiMe$_3$

Scheme 17.48

17.3.3 Alkylation and α-methylenation

Ketone-derived silyl enol ethers give products of Mannich condensation regiospecifically and in high yield when treated with dimethyl-(methylene)ammonium iodide[60]. Similar regiospecificity is observed when ketone-derived silyl enol ethers and acid-, ester- and lactone-derived silylketene acetals are first cleaved with methyl-lithium, then treated with dimethyl(methylene)ammonium trifluoroacetate[56]. The enolates generated

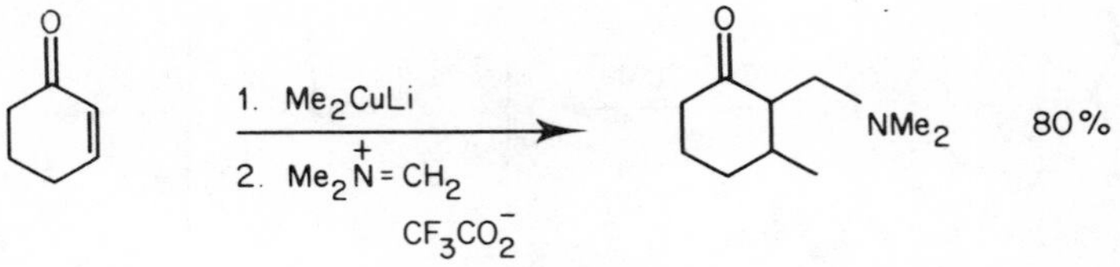

1. Me_2CuLi
2. $Me_2\overset{+}{N}=CH_2$ $CF_3CO_2^-$

NMe_2 80%

Scheme 17.49

by conjugate addition to $\alpha\beta$-unsaturated ketones can be trapped similarly (*Scheme 17.49*).

Such Mannich products are readily converted into α-methylene carbonyl compounds, as illustrated by the preparation of α-methylene-γ-butyrolactone[97] (*Scheme 17.50*).

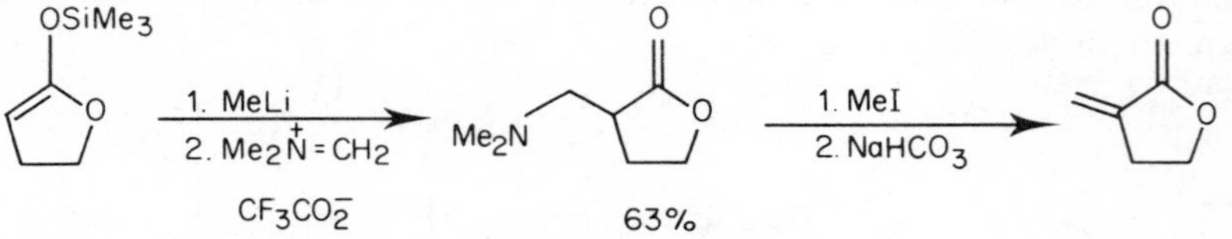

$OSiMe_3$ 1. MeLi 2. $Me_2\overset{+}{N}=CH_2$ $CF_3CO_2^-$ Me_2N 63% 1. MeI 2. $NaHCO_3$

Scheme 17.50

A different method of α-methylenation relies on regiospecific phenylthiomethylation[98], followed by oxidative removal[99] of sulphur. With ketone-derived silyl enol ethers, titanium(IV) chloride is the preferred Lewis acid, while powdered anhydrous zinc(II) bromide is employed as a milder catalyst in those cases involving ester- and lactone-derived silyl ketene acetals. Some examples are shown in *Scheme 17.51*.

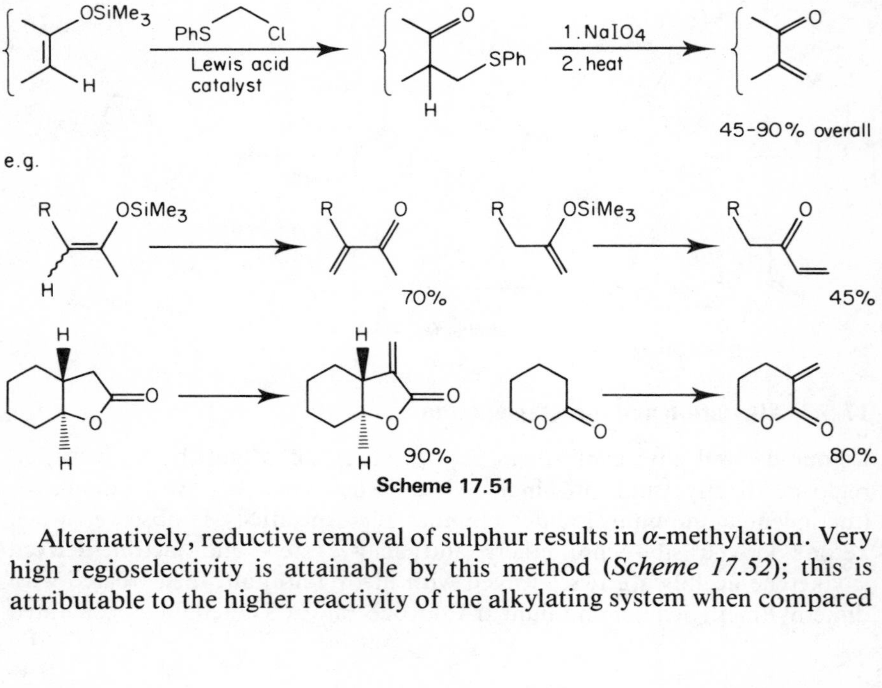

$OSiMe_3$ H PhS Cl Lewis acid catalyst SPh H 1. $NaIO_4$ 2. heat

45-90% overall

e.g.

R $OSiMe_3$ H → R O 70% R $OSiMe_3$ → R O 45%

H H → H H 90% → 80%

Scheme 17.51

Alternatively, reductive removal of sulphur results in α-methylation. Very high regioselectivity is attainable by this method (*Scheme 17.52*); this is attributable to the higher reactivity of the alkylating system when compared with methyl iodide, and also to the change to *acidic* conditions. Angular methylation to give the decalone (40) indicates a considerable improvement in regioselectivity; an alternative method of generation[100] of the required specific lithium enolate by conjugate reduction and then alkylation with methyl iodide resulted in a significant proportion (*ca.* 30 per cent) of the regioisomeric product.

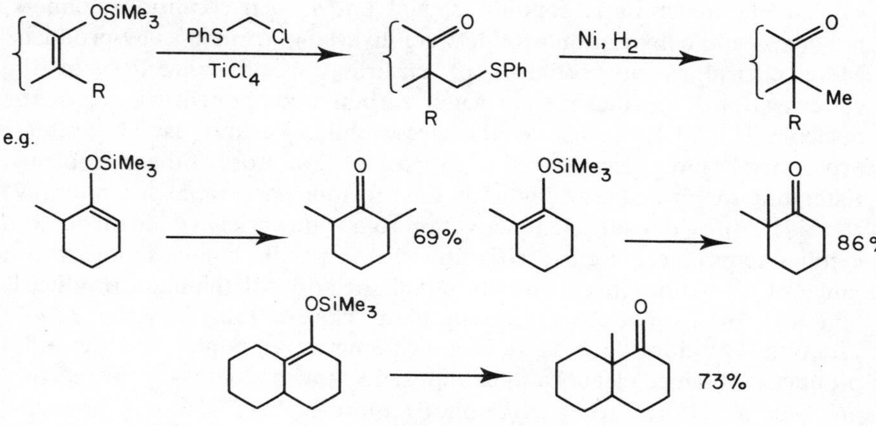

Scheme 17.52 (40)

A simple extension[101] of these principles permits efficient and regiospecific introduction of simple α-CH_2 alkyl groups, including neopentyl, into the α-positions of aldehydes, ketones, esters and δ-lactones, using readily available α-chloroalkyl phenyl sulphides and subsequent reductive desulphurization (*Scheme 17.53*). Alternatively, oxidative removal of sulphur leads to the (*E*)-alkene isomer of a formal directed aldol condensation.

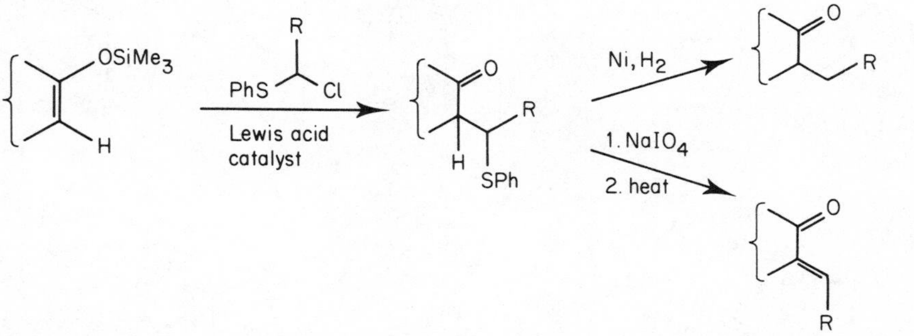

Scheme 17.53 R = Me, Pr^n, Pr^i, Bu^t

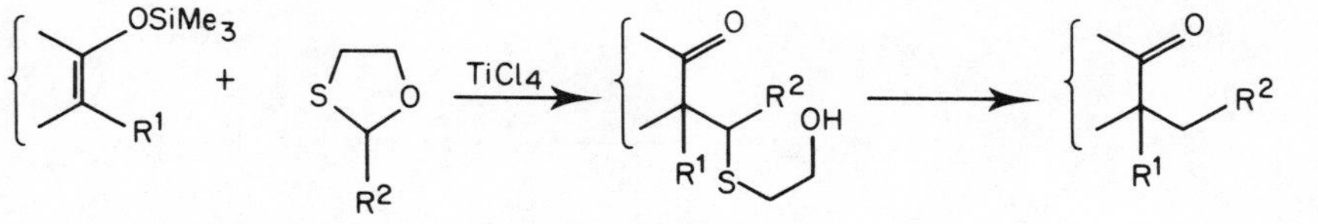

(41) **Scheme 17.54**

Alkyl groups of this type can also be introduced by use of hemithioacetals[3] (41) (*Scheme 17.54*).

17.3.4 Hydroxyalkylation and related reactions

The aldol condensation[102] between two carbonyl compounds is normally carried out under basic conditions, and under such conditions, dimers, polymers, and dehydration products are invariably formed as by-products. More critical is the problem of ensuring specific direction in the condensation, i.e., that one particular carbonyl component will act as the nucleophile and the other as the electrophile. Various useful synthetic procedures[103] have been devised to overcome all or most of these problems. Extending the principles outlined in the previous paragraphs, titanium(IV) chloride[104] provides sufficient activation to a wide variety of aldehydes and ketones to promote regiospecific and chemospecific condensation with a range of silyl enol ethers; undesirable dissociation of the aldol product is inhibited by formation of the titanium chelate (42) (*Scheme 17.42*). Titanium(IV) chloride is used in stoichiometric amounts, and the aldol product is obtained cleanly and in high yield[105]; with aldehydes, the reaction proceeds at −78 °C, whereas ketones require higher, 0 °C and upwards, temperatures. The regiospecificity can be illustrated by the examples shown in *Scheme 17.55*. A high degree of chemoselectivity is observed, the reactivity order being aldehydes > ketones ≫ esters; additionally, formaldehyde gives 1:1 adducts exclusively, no poly(hydroxymethyl) product being detected. Silyl enol ethers react with (−)-menthyl pyruvates under similar conditions; the products are formed[106] with a reasonably high degree of asymmetric induction.

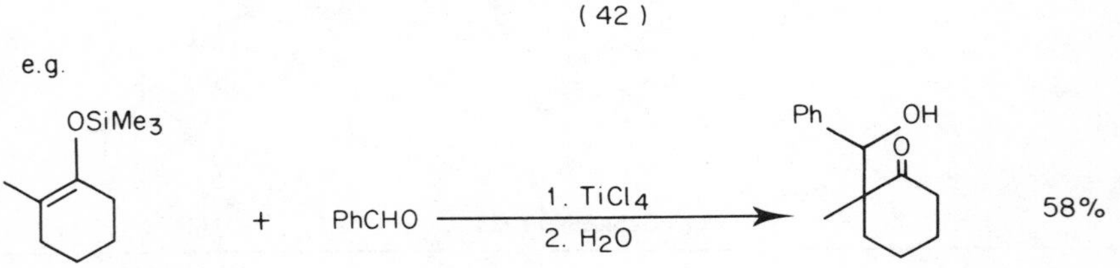

OSiMe$_3$ R^1 —(R^2COR^3; $TiCl_4$, CH_2Cl_2)→ [O···$TiCl_3$, O, R^1, R^2, R^3] —(H_2O)→ O, OH, R^1, R^2, R^3

(42)

e.g.

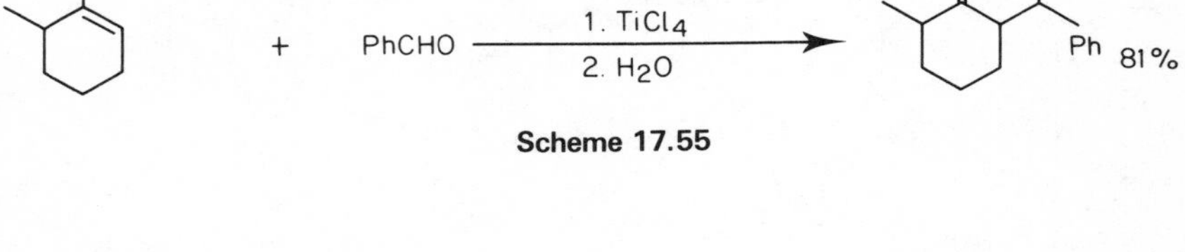

OSiMe$_3$ + PhCHO —(1. $TiCl_4$; 2. H_2O)→ Ph, OH, O — 58%

OSiMe$_3$ + PhCHO —(1. $TiCl_4$; 2. H_2O)→ O, OH, Ph — 81%

Scheme 17.55

The Reformatsky reaction[107] is similarly prone to the formation of artifacts, including those arising from dehydration. Silyl ketene acetals react thermally[108] with aromatic aldehydes to yield the corresponding Reformatsky products (*Scheme 17.56*), but this method is quite unsuccessful with aliphatic aldehydes and ketones.

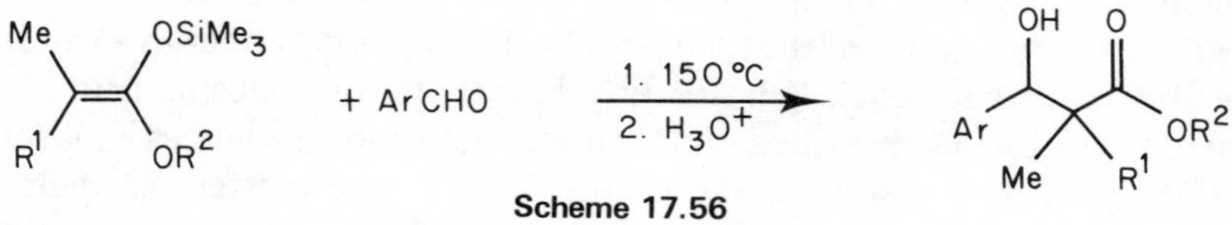

Me, $OSiMe_3$, R^1, OR^2 + ArCHO → 1. 150 °C 2. H_3O^+ → OH, O, Ar, OR^2, Me, R^1

Scheme 17.56

In the presence of titanium(IV) chloride, aldehydes and ketones react with ester- and lactone-derived silyl ketene acetals to give β-hydroxy esters and the related lactones in excellent yield[109], without formation of $\alpha\beta$-unsaturated species (*Scheme 17.57*).

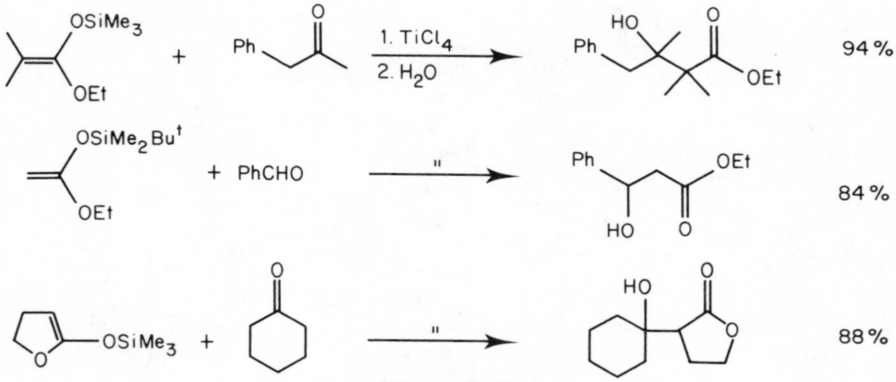

$OSiMe_3$, OEt + Ph, O → 1. $TiCl_4$ 2. H_2O → HO, Ph, O, OEt 94%

$OSiMe_2Bu^t$, OEt + PhCHO → " → Ph, OEt, HO, O 84%

$OSiMe_3$, O + O → " → HO, O, O 88%

Scheme 17.57

$OSiMe_3$, R^1 + MeO, OMe, R^2, R^3 → 1. $TiCl_4$ 2. H_2O → O, R^2, R^3, R^1, OMe 60-90%

e.g.

Ph, $OSiMe_3$, OMe + MeO, OMe → " → MeO, Ph, OMe, O 92%

$OSiMe_3$, O + Ph, OMe, OMe → " → OMe, Ph, O, O 75%

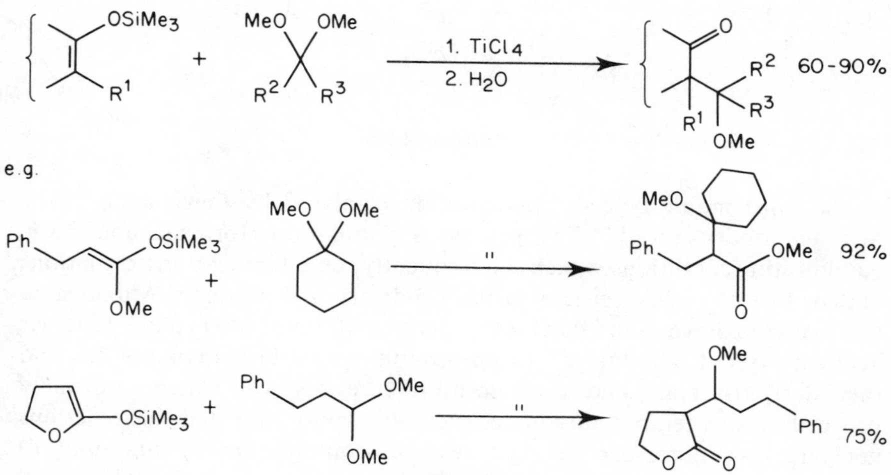

Scheme 17.58

Acetals and derivatives can be similarly activated by titanium(IV) chloride, when they react at low temperatures with ketone-derived silyl enol ethers[110] and ester- and lactone-derived silyl ketene acetals[111] to give the appropriate β-alkoxycarbonyl compound, as illustrated in *Scheme 17.58.*

In the presence of titanium(IV) chloride or boron trifluoride etherate, acetals and 1,2-bis(trimethylsilyloxy)cyclobut-1-ene undergo condensation to give cyclobutanones of the type (43). Such pinacols[112] undergo a variety of specific rearrangements. On treatment with trifluoroacetic acid[112] they rearrange to 2,2-disubstituted cyclopentane-1,3-diones, whereas exposure to tin(IV) chloride[113] results in the production of γ-keto-esters as their silyl enols. Prior reaction with a Grignard reagent followed by acid treatment provides a route[114] to 1,4-diketones and the aldol-related 3-hydroxycyclopentanones (*Scheme 17.59*). Such versatility has been illustrated further in a synthesis[115] of showdomycin.

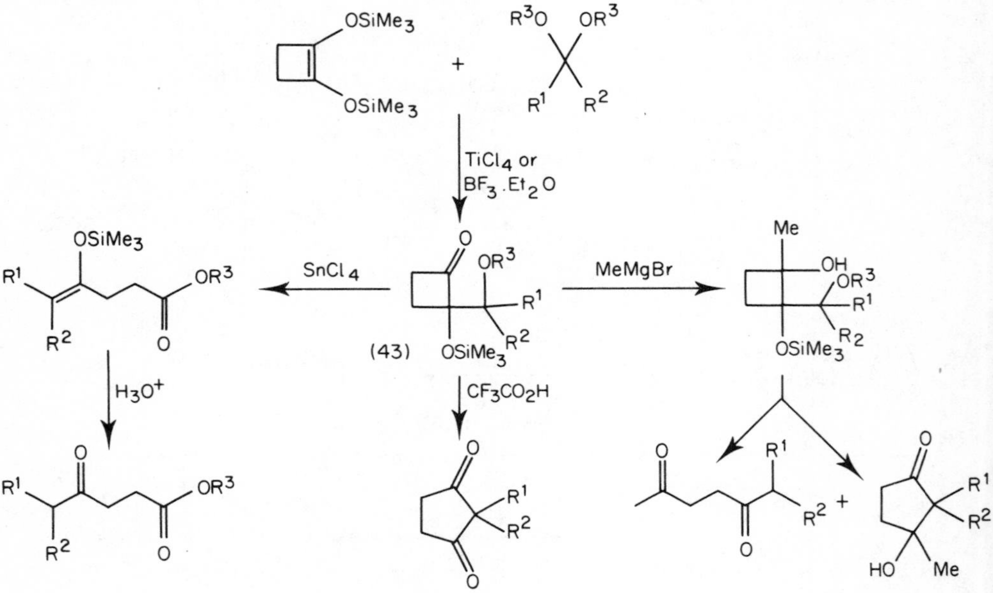

Scheme 17.59

The third major type of condensation catalysed by Lewis acids is the Michael condensation[116]. This process is normally performed under basic equilibrating conditions, which can frequently cause further transformations and/or loss of regiospecificity in the 1,5-dicarbonyl products. Mukaiyama and his group have found that ketone-derived silyl enol ethers undergo Lewis acid-catalysed 1,4-addition[117] to $\alpha\beta$-unsaturated carbonyl compounds and their derived acetals, and to $\alpha\beta$-unsaturated esters. The reaction normally proceeds at low temperature under such mild conditions that side reactions are largely suppressed; the less reactive combination of titanium(IV) chloride/titanium(IV) isopropoxide is appropriate when labile $\alpha\beta$-

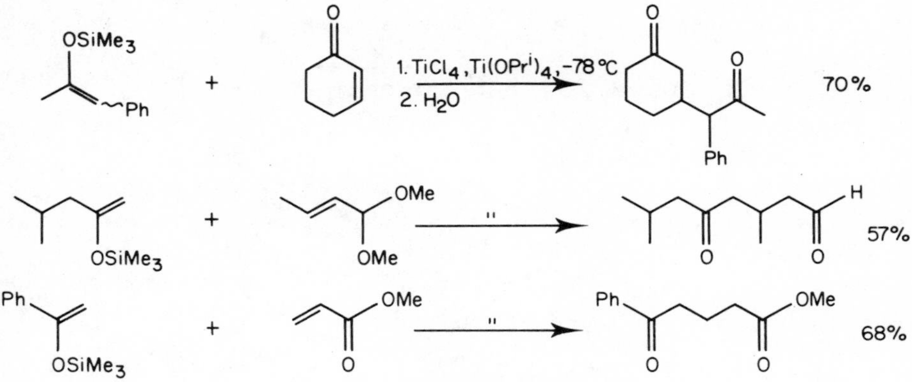

OSiMe3 + [cyclohexenone] 1. $TiCl_4$, $Ti(OPr^i)_4$, −78 °C; 2. H_2O → 70%

57%

68%

Scheme 17.60

unsaturated substrates are employed (*Scheme 17.60*). Ester-derived silyl ketene acetals react similarly.

17.3.5 Silyl dienol ethers and bis(silylenol) ethers

Lithium dienolates (44) undergo electrophilic attack under kinetic conditions to give, generally speaking, products of α-substitution. Under thermodynamic, equilibrating conditions[118], γ-substituted products are usually more favoured. Silyl dienol ethers (45), on the other hand, being neutral, have a lower electron density at the α-position, and accordingly show a marked preference for kinetic electrophilic attack at the γ-position.

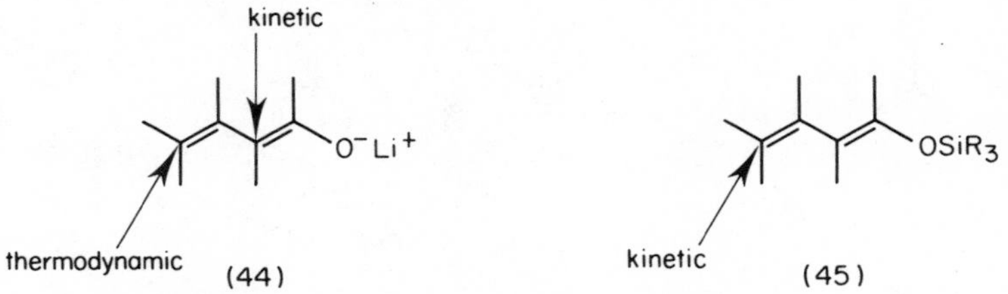

kinetic

$O^- Li^+$

thermodynamic (44)

$OSiR_3$

kinetic (45)

Mukaiyama and his group have shown that the dienol ethers derived from crotonaldehyde and its homologues react with acetals exclusively at the γ-position[119]. The so-produced δ-alkoxy-$\alpha\beta$-unsaturated aldehydes undergo a high-yield elimination reaction when treated with DBU or DBN to give polyenals. This methodology has found use in the syntheses of several natural products (*Scheme 17.61*).

Dienol ethers, and their $\alpha\beta$-unsaturated ester-derived analogues, also react with the S_N1-type electrophiles (46) and (47) in the presence of zinc(II) bromide[120] to give products of predominant, sometimes exclusive, γ-attack (*Scheme 17.62*); these chlorides were shown earlier (p. 224) to be useful primary alkylating agents (after reductive desulphurization) and alkylidene-introducing synthons (after sulphoxide elimination) for silyl enol ethers

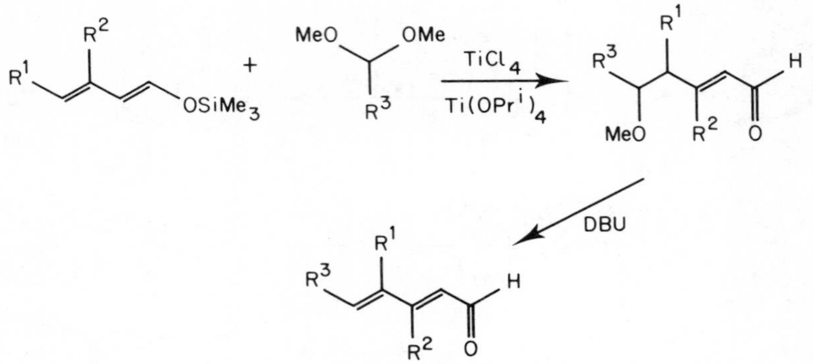

Scheme 17.61

derived from saturated carbonyl compounds. The $\alpha:\gamma$ ratios obtained are dependent upon the diene substitution pattern, the particular silyl group chosen, the steric bulk of the ester alkoxy group in the ketene acetals examined, and the nature of the electrophile. With highly γ-selective silyl dienol ethers, other carbon electrophiles such as acetyl chloride, methyl orthoformate, and S_N1-type species such as α-methoxy halides, and benzylic secondary and allylic bromides, all show a high tendency to attack the γ-position under zinc(II) bromide catalysis.

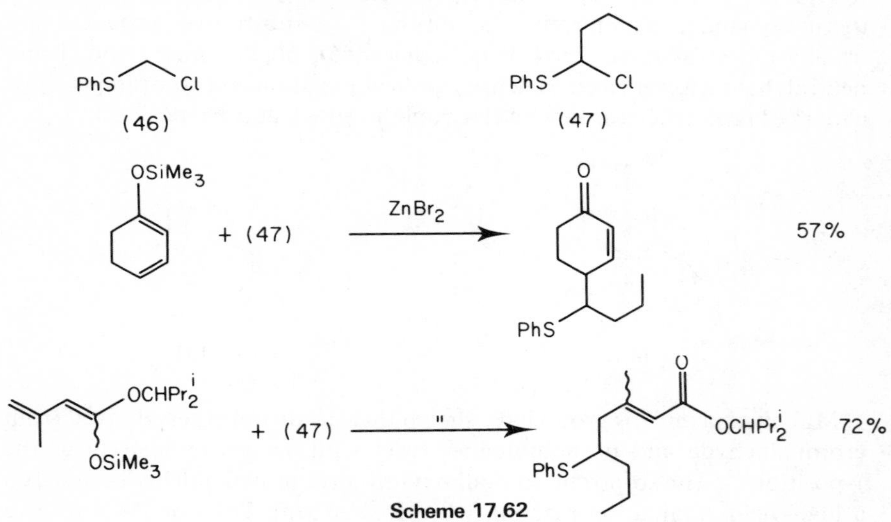

(46)

(47)

Scheme 17.62

Phenyl sulphenyl chloride reacts with the same wide range of silyl dienol ethers with high or exclusive γ-regioselectivity[121]; one example of the utility of such a sequence is illustrated in *Scheme 17.63*.

Succinic anhydride and methylsuccinic anhydride can be converted into bis(silylenol) ethers of the type (48); such ethers, although rather labile towards oxygen and self-addition, behave[3] as silyl dienol ethers in their reactions with electrophiles (*Scheme 17.64*). Regiospecific condensation is

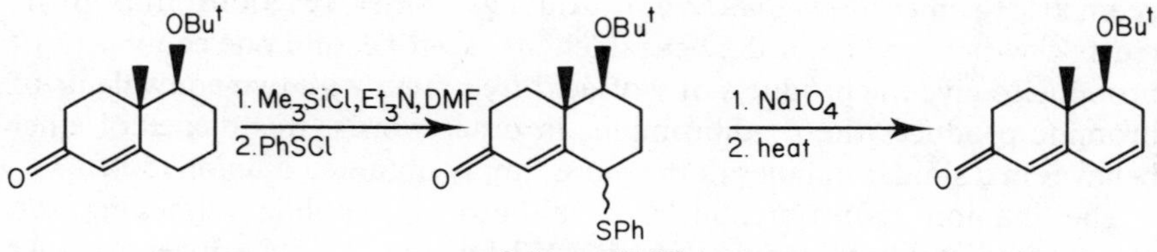

Scheme 17.63

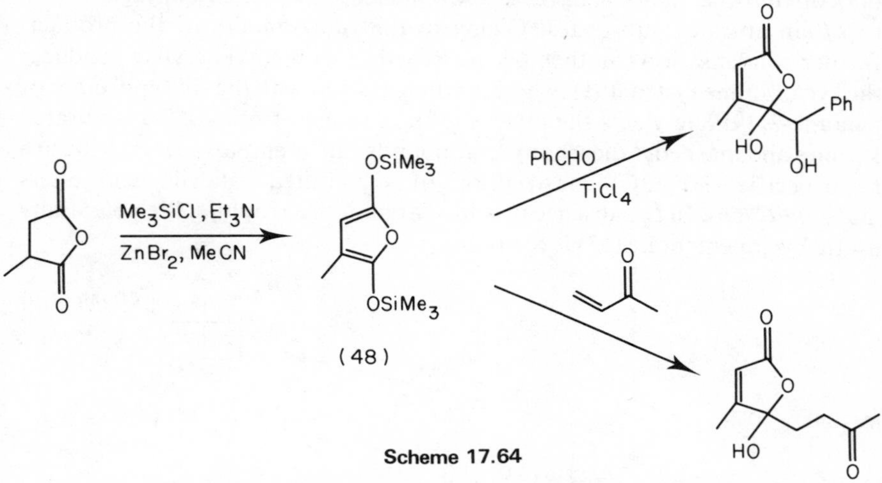

Scheme 17.64

observed, with aliphatic $\alpha\beta$-unsaturated ketones undergoing conjugate 1,4-addition.

Direct silylation of the dianion of methyl acetoacetate results in *C*-silylation. However, a two-step procedure[122] allows the preparation of the bis(silylenol) ether (49) of methyl acetoacetate as a single, probably (*E*),

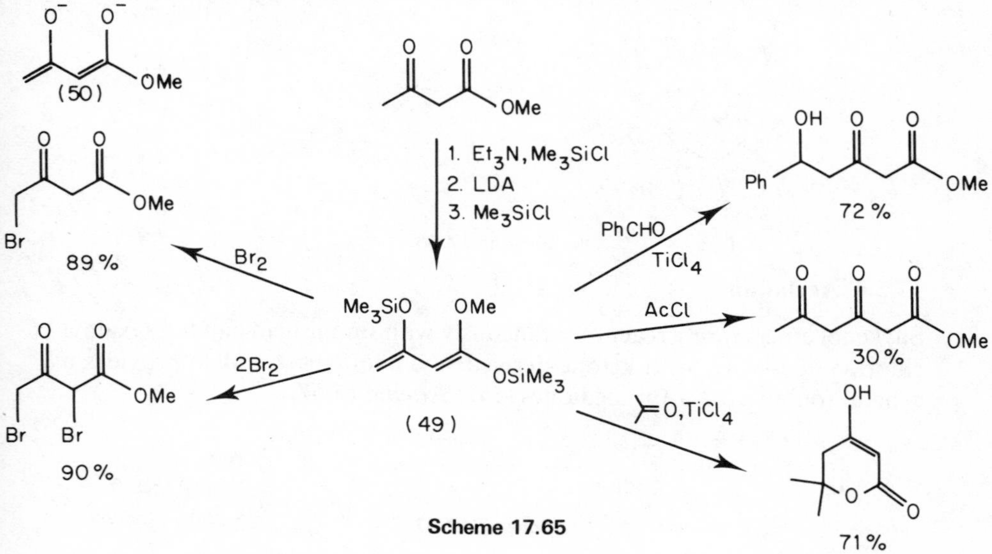

Scheme 17.65

geometric isomer (*Scheme 17.65*). Amongst other reactions, this diene reacts[123] with acetyl chloride, benzaldehyde, acetone, and one equivalent of bromine to give the products of γ-attack; treatment with two equivalents of bromine produces the $\alpha\gamma$-dibromide. In other words, this dienediol ether behaves in a similar manner to the more simply obtained dianion (50).

The dianion (50) functions[124] as a hard nucleophile, attacking $\alpha\beta$-unsaturated carbonyl compounds in a 1,2-fashion. The bis(silylenol) ether (49), on the other hand, behaves as a soft nucleophile, adding conjugatively[123] to $\alpha\beta$-unsaturated substrates. Owing to the functionality of the product, further condensation can then occur. Reaction with ethyl acrylate produces the cyclobutane system (51), whereas condensation with the silyl enol ether of pentane-2,4-dione yields the phenol (52). A range of other $\alpha\beta$-unsaturated ketones and masked β-dicarbonyl compounds can be employed, resulting in a regiospecific (3C + 3C) construction of substituted salicylic acid esters (*Scheme 17.66*). In the absence of added organic electrophile, the potentially useful cyclopentenone (53) is formed.

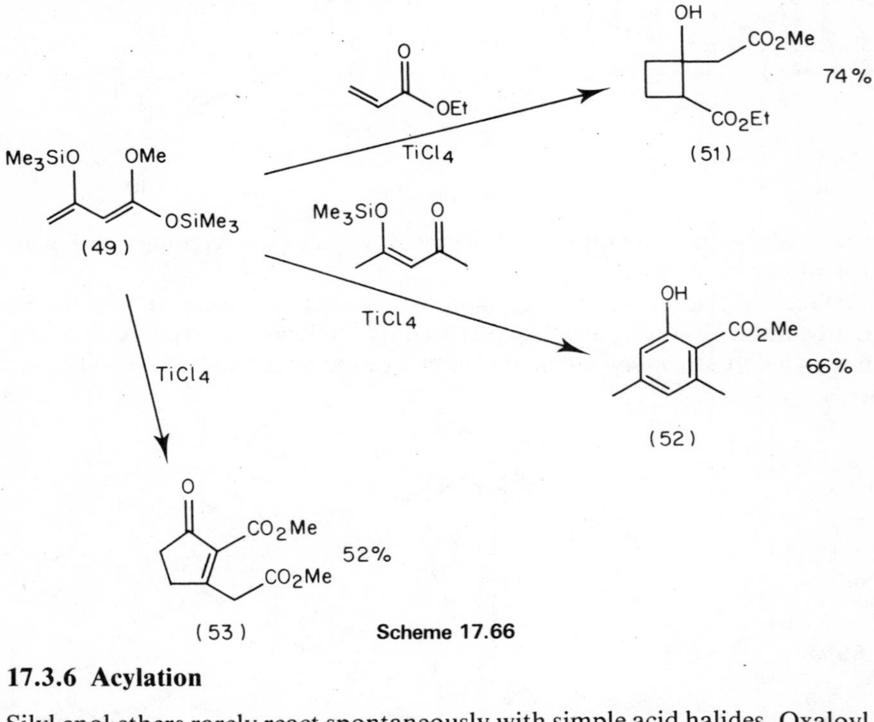

OH
CO_2Me
O
74%
OEt
CO_2Et
Me_3SiO OMe
$TiCl_4$
(51)
$OSiMe_3$
Me_3SiO O
(49)
OH
$TiCl_4$
CO_2Me
$TiCl_4$
66%
(52)
O
CO_2Me
52%
CO_2Me
(53)

Scheme 17.66

17.3.6 Acylation

Silyl enol ethers rarely react spontaneously with simple acid halides. Oxaloyl chloride does react with ketone-derived silyl enol ethers and so provides a general route[125] to 2,3-furanediones (54) (*Scheme 17.67*).

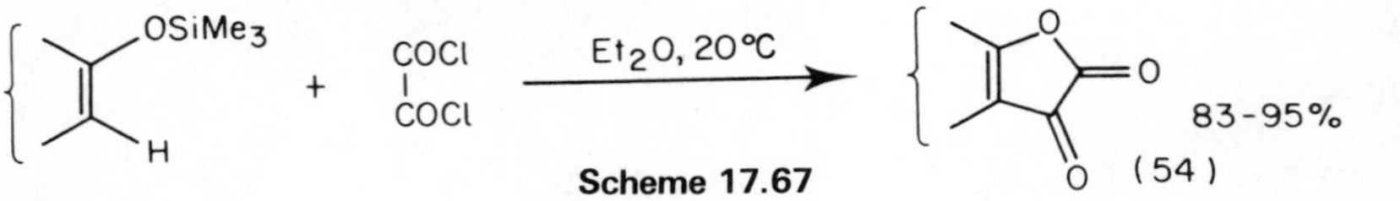

$OSiMe_3$
H
+
COCl
COCl
Et_2O, 20°C
O
O
83-95%
O (54)

Scheme 17.67

Polyhalogeno acid halides[63, 126] and acid anhydrides[127] react exothermically with silyl enol ethers to give 1,3-diketones; the use of anhydrides is of particular value where the halide is not too accessible, such as in the case of trifluoroacetyl halides (*Scheme 17.68*).

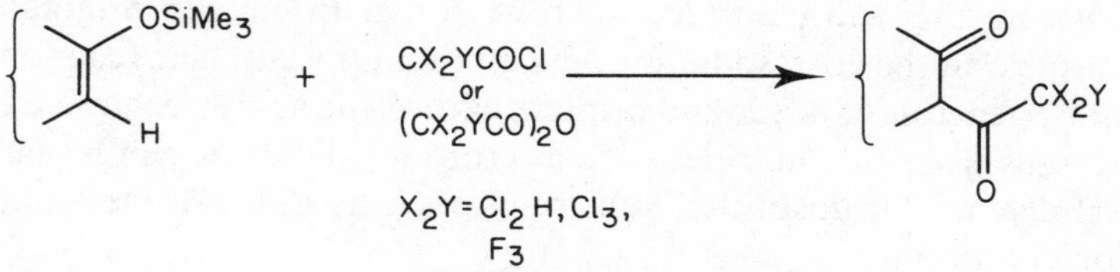

$X_2Y = Cl_2H, Cl_3, F_3$

Scheme 17.68

Simple acid halides do react with silyl enol ethers under Lewis acid catalysis, but not always to give products of *C*-acylation. With mercury(II) chloride as catalyst, enol carboxylates[126] are formed (*Scheme 17.69*). This reaction may involve initial cleavage of the silyl enol ether to an α-chloromercury ketone or its enol equivalent; no substrate has been utilized which might determine the degree of regioselectivity, if any, of this process. Acetic anhydride is reported[2] to perform the same transformation under basic conditions. As discussed earlier (p. 230), silyl dienol ethers are acylated preferentially at the γ-position under zinc(II) bromide catalysis.

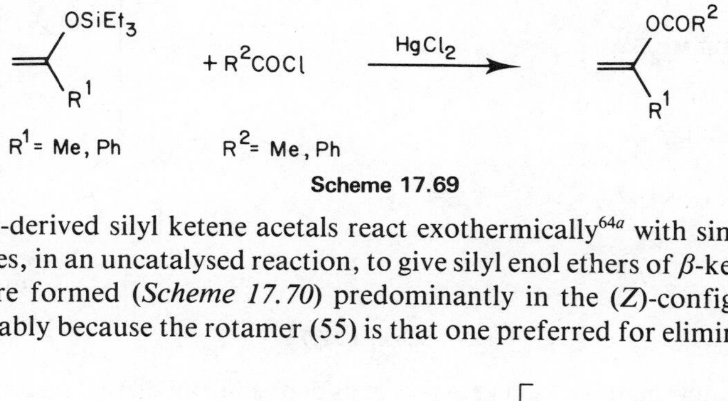

R^1 = Me, Ph R^2 = Me, Ph

Scheme 17.69

Ester-derived silyl ketene acetals react exothermically[64a] with simple acid chlorides, in an uncatalysed reaction, to give silyl enol ethers of β-ketoesters; these are formed (*Scheme 17.70*) predominantly in the (*Z*)-configuration, presumably because the rotamer (55) is that one preferred for elimination of HCl.

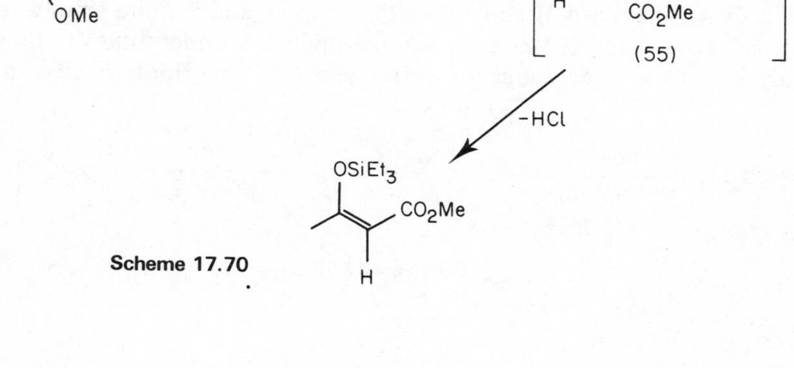

Scheme 17.70

This sequence has been developed to provide a reasonably good route[64b] to β-ketoesters of various types. Here, t-butyldimethylsilyl ketene acetals[54] are employed, and the reaction is performed in the presence of triethylamine (*Scheme 17.71*). The enol ether isomer (56) probably arises via the ketene derived from the acid chloride, whereas isomer (57) could originate from direct attack on the corresponding acyl ammonium salt. The reaction seems to be quite general for acetic and primary carboxylic acid ester-derived ketene acetals; sensitive acid chlorides such as crotonoyl chloride can be employed successfully. $\alpha\alpha$-Disubstituted substrates such as (58) are inert under the reaction conditions.

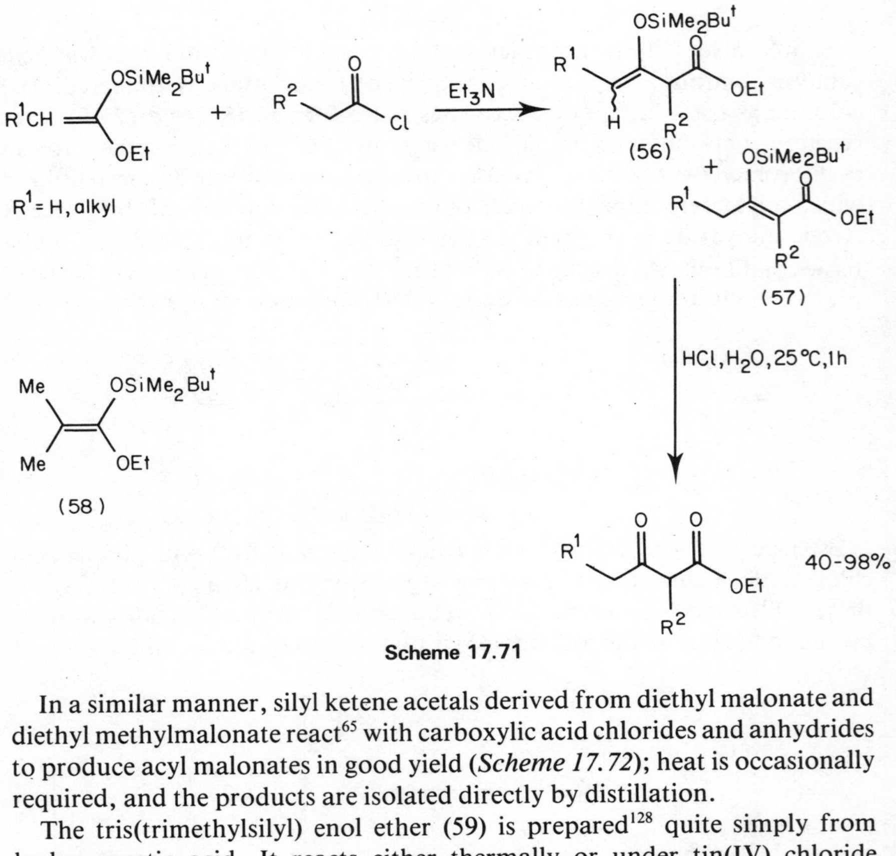

$R^1CH{=}C(OSiMe_2Bu^t)(OEt)$ + R^2CH_2COCl $\xrightarrow{Et_3N}$ (56) + (57) $\xrightarrow{HCl, H_2O, 25\,°C, 1\,h}$ $R^1CH_2COCH(R^2)CO_2Et$ 40–98%

R^1 = H, alkyl

(58) $Me_2C{=}C(OSiMe_2Bu^t)(OEt)$

Scheme 17.71

In a similar manner, silyl ketene acetals derived from diethyl malonate and diethyl methylmalonate react[65] with carboxylic acid chlorides and anhydrides to produce acyl malonates in good yield (*Scheme 17.72*); heat is occasionally required, and the products are isolated directly by distillation.

The tris(trimethylsilyl) enol ether (59) is prepared[128] quite simply from hydroxyacetic acid. It reacts either thermally or under tin(IV) chloride catalysis with a wide range of carboxylic acid chlorides to give, after

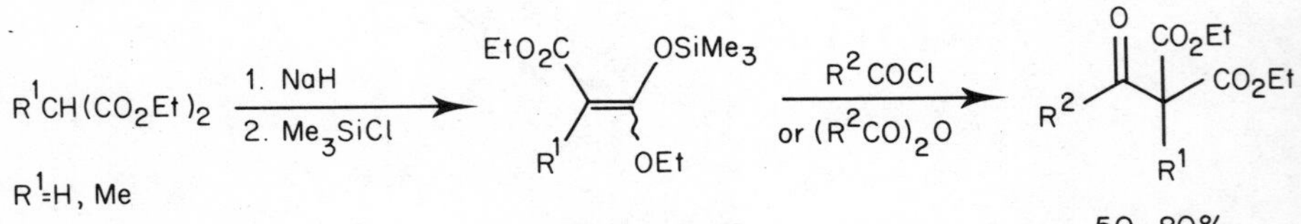

$R^1CH(CO_2Et)_2$ $\xrightarrow[2.\ Me_3SiCl]{1.\ NaH}$ $EtO_2C(R^1)C{=}C(OSiMe_3)(OEt)$ $\xrightarrow[\text{or } (R^2CO)_2O]{R^2COCl}$ $R^2COC(R^1)(CO_2Et)_2$ 50–80%

R^1 = H, Me

Scheme 17.72

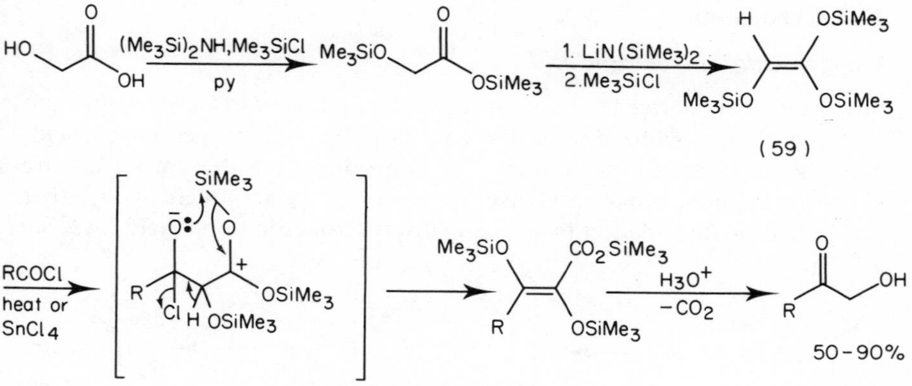

$$HOCH_2COOH \xrightarrow[py]{(Me_3Si)_2NH, Me_3SiCl} Me_3SiOCH_2COOSiMe_3 \xrightarrow[2. Me_3SiCl]{1. LiN(SiMe_3)_2} (Me_3SiO)HC{=}C(OSiMe_3)_2 \quad (59)$$

$$\xrightarrow[\text{heat or } SnCl_4]{RCOCl} [\text{intermediate}] \longrightarrow R(Me_3SiO)C{=}C(CO_2SiMe_3)(OSiMe_3) \xrightarrow[-CO_2]{H_3O^+} RCOCH_2OH \quad 50\text{–}90\%$$

Scheme 17.73

hydrolysis and decarboxylation, hydroxymethyl ketones in good yield (*Scheme 17.73*). This sequence fails with tertiary carboxylic acid chlorides of the trimethylacetic type.

17.3.7 Hydroboration — oxidation

Trimethylsilyl enol ethers react with diborane[129] with introduction of boron at the β-carbon atom. In acyclic systems, such β-silyloxyboranes undergo facile *syn*-elimination[130], and the resulting alkenes react further with diborane, with the ultimate, and laboriously realized, production of alkanols. With cyclic systems, *syn*-elimination does not occur with such facility, and the intermediates can be treated *in situ* with alkaline hydrogen peroxide to afford *trans*-1,2-diols[129, 131]; the monosilyl diol precursors are isolable under appropriate conditions (*Scheme 17.74*). Alternatively, and after excess of diborane has been quenched, treatment with acid produces alkenes[132]. Both of these reactions proceed regiospecifically.

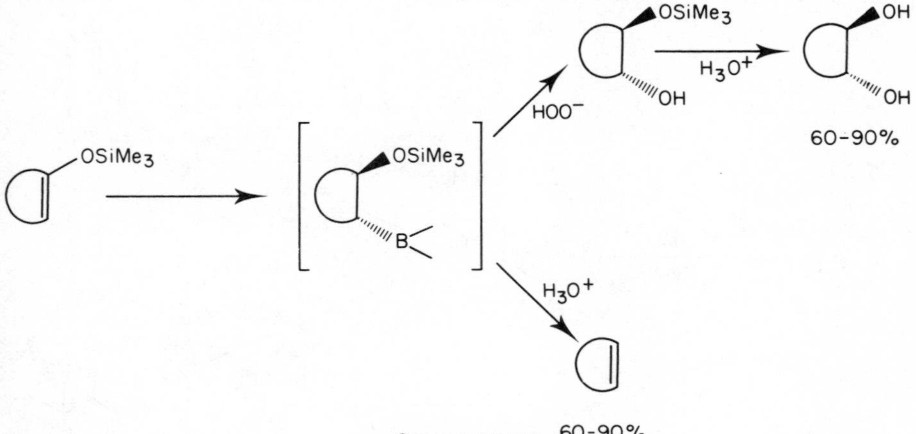

OSiMe3 → [OSiMe3 / B<] —HOO⁻→ OSiMe3 / OH —H_3O^+→ OH / OH 60-90%; —H_3O^+→ alkene 60-90%

Scheme 17.74

17.3.8 Oxidation

Using m-*chloroperbenzoic acid*

Silyl enol ethers derived from both aldehydes and ketones are smoothly and regiospecifically oxidized[75,79] at the α-position by *m*-chloroperbenzoic acid. With ketone-derived substrates, the immediate isolable products are α-silyloxyketones, which are easily deprotected by treatment with either dilute acid or fluoride ion to give α-hydroxyketones in good yield (*Scheme 17.75*).

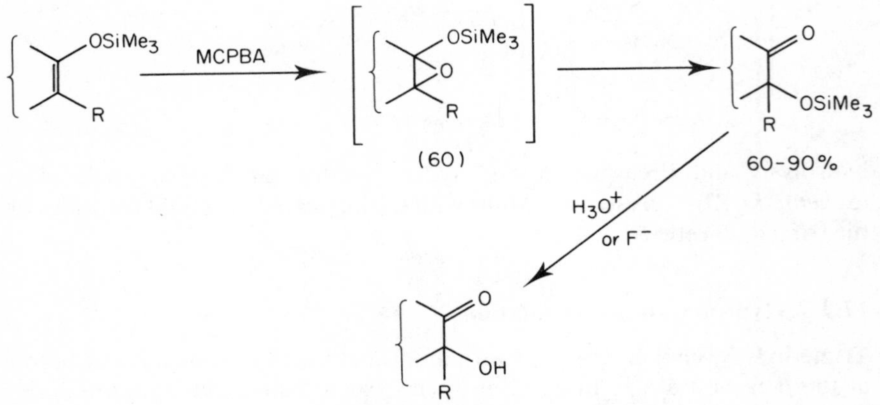

OSiMe3 MCPBA OSiMe3 O R (60) O OSiMe3 R 60-90% H_3O^+ or F^- O OH R

Scheme 17.75

The electron-richness of the silyl enol ether double bond when compared with that of a simple alkene can be seen, *inter alia*, in the route[133] to α-hydroxy- and α-acetoxy-αβ-unsaturated ketones exemplified in *Scheme 17.76.*

The immediate products obtained from aldehyde-derived substrates, on the other hand, are acetals (61), arising[79] from *m*-chlorobenzoic acid opening of the siloxy-oxirane (60). Fortunately, such acetals can be converted directly into α-acetoxyaldehydes by treatment with acetic anhydride. Once again,

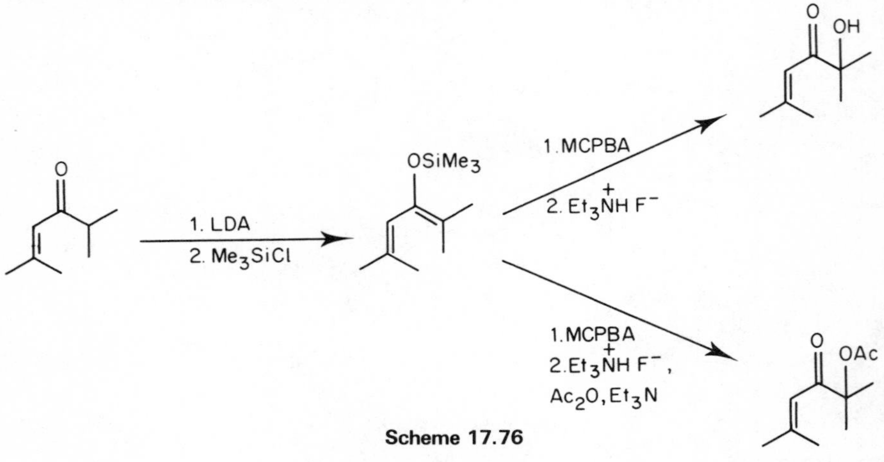

O 1. LDA 2. Me_3SiCl OSiMe3 1. MCPBA 2. $Et_3\overset{+}{N}H\ F^-$ O OH 1. MCPBA 2. $Et_3\overset{+}{N}H\ F^-$, Ac_2O, Et_3N O OAc

Scheme 17.76

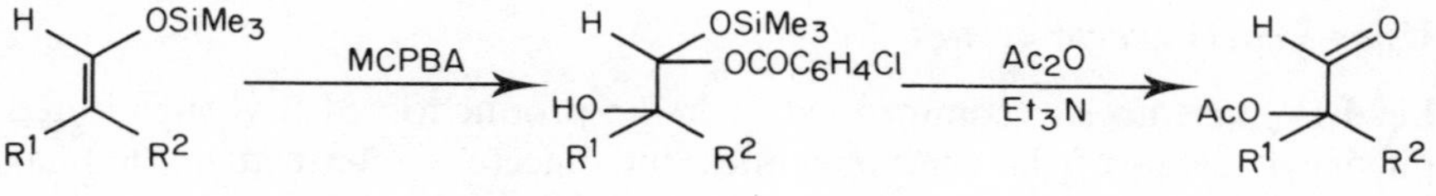

(61) 30–40%

e.g.

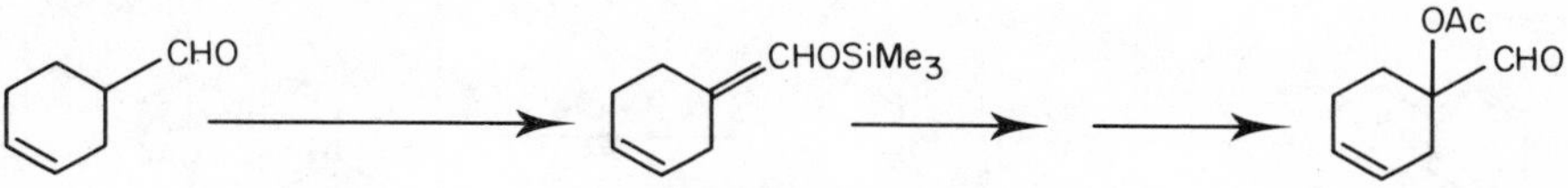

Scheme 17.77

isolated alkene double bonds survive these reaction conditions intact (*Scheme 17.77*).

This sequence can be extended[76] to ketene bis(trimethylsilyl) acetals, from which α-hydroxyacids are produced (*Scheme 78*).

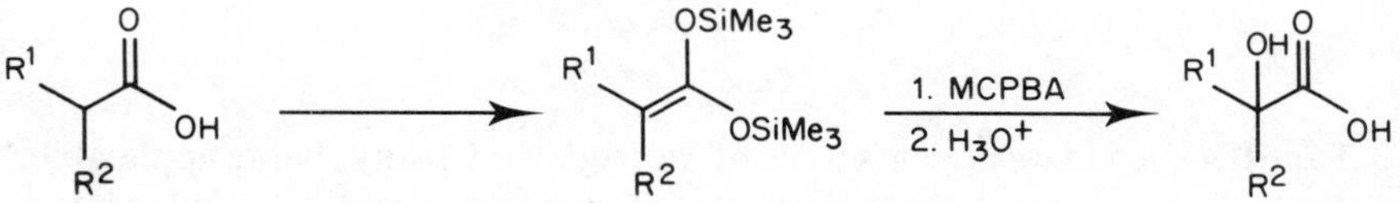

Scheme 17.78 50–83%

Still has described[134] an elegant method for the protection of αβ-unsaturated ketones against attack by, *inter alia*, lithium dimethylcuprate. β-Stannyl silyl enol ethers (62), obtained from αβ-unsaturated ketones as shown in *Scheme 17.79*, are relatively unreactive towards most nucleophiles, but are smoothly reconverted into enones by mild oxidation. The particular example shown is taken from a synthesis of the pheromone (±)-periplanone, and illustrates once again the relative electron richness of silyloxyalkenes.

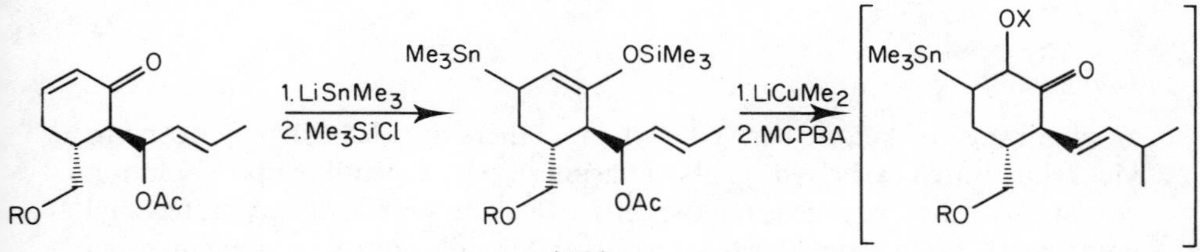

(62)

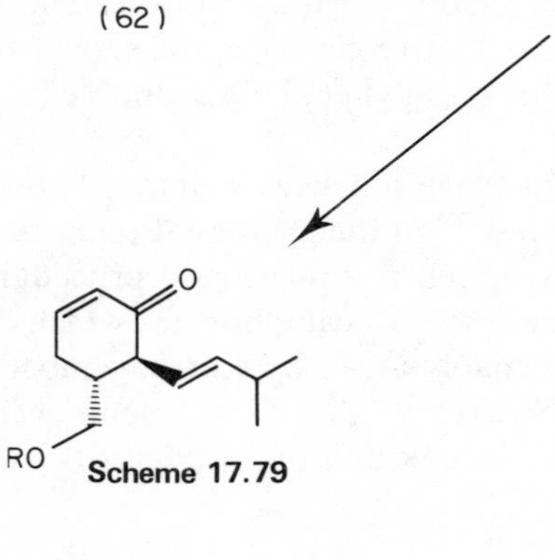

Scheme 17.79

Using lead(IV) carboxylates

Lead(IV) acetate is recommended[78] for the production of aryl-substituted phenacyl acetates from acetophenones; the diacetoxy intermediate (63) has been isolated (*Scheme 17.80*) in some cases.

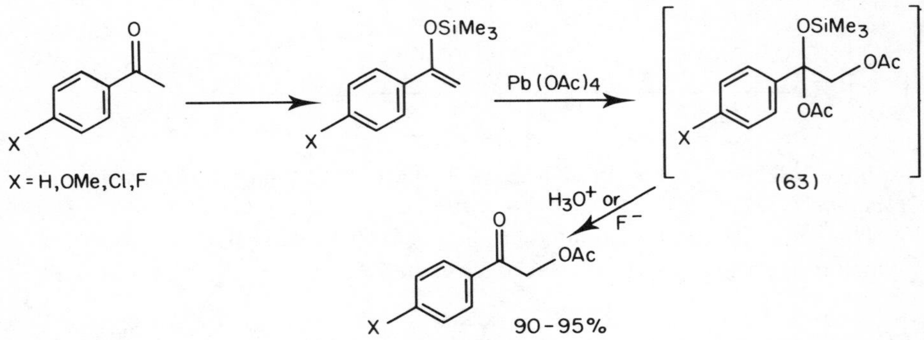

X = H, OMe, Cl, F

Pb(OAc)4

(63)

H_3O^+ or F^-

90–95%

Scheme 17.80

Lead(IV) benzoate appears to be of more general utility, being applicable[78] to both aldehyde- and ketone-derived silyl enol ethers; isolated double bonds elsewhere in the molecule are unaffected (*Scheme 17.81*).

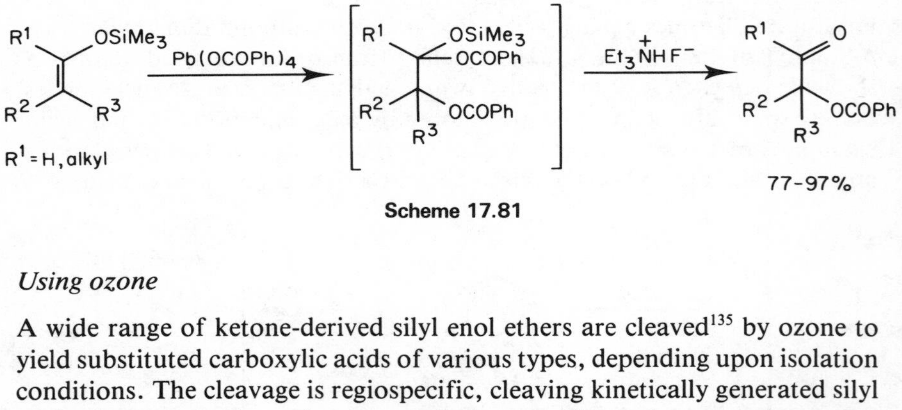

R^1 = H, alkyl

Pb(OCOPh)4

$Et_3\overset{+}{N}HF^-$

77–97%

Scheme 17.81

Using ozone

A wide range of ketone-derived silyl enol ethers are cleaved[135] by ozone to yield substituted carboxylic acids of various types, depending upon isolation conditions. The cleavage is regiospecific, cleaving kinetically generated silyl enol ethers *away* from the more substituted side, and so complementing Baeyer–Villiger oxidative cleavage. Selective ozonolysis, based on the high nucleophilicity of the silyl enol ether double bond, is also possible (*Scheme 17.82*).

Ozonolytic cleavage has been employed recently by Stork, in a stereo-controlled synthesis[136] of the Djerassi–Prelog lactone (64) (*Scheme 17.83*).

A notable exception to this general procedure is seen in the case of the trimethylsilyl enol ether of camphor, for which α-oxygenation occurs[135]. The product (66) probably arises by rearrangement of an intermediate oxirane such as (65) (*Scheme 17.84*). Silyl ketene acetals react in an anomalous manner also, giving mixtures of products of cleavage and of α-oxygenation.

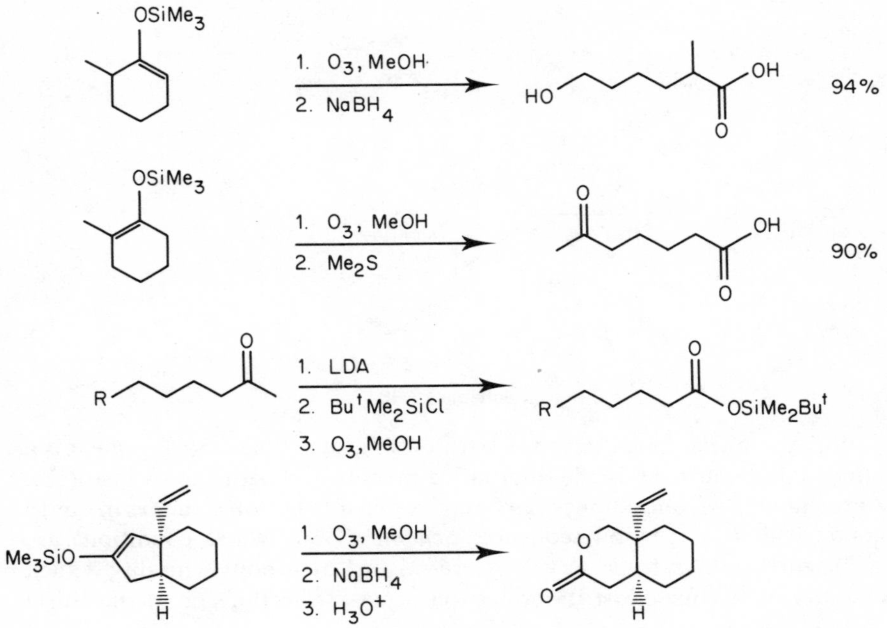

OSiMe3
1. O3, MeOH
2. NaBH4
HO
OH
O
94%
OSiMe3
1. O3, MeOH
2. Me2S
O
OH
O
90%
R
O
1. LDA
2. Bu^t Me2SiCl
3. O3, MeOH
R
O
OSiMe2Bu^t
Me3SiO
H
1. O3, MeOH
2. NaBH4
3. H3O+
O
O
H

Scheme 17.82

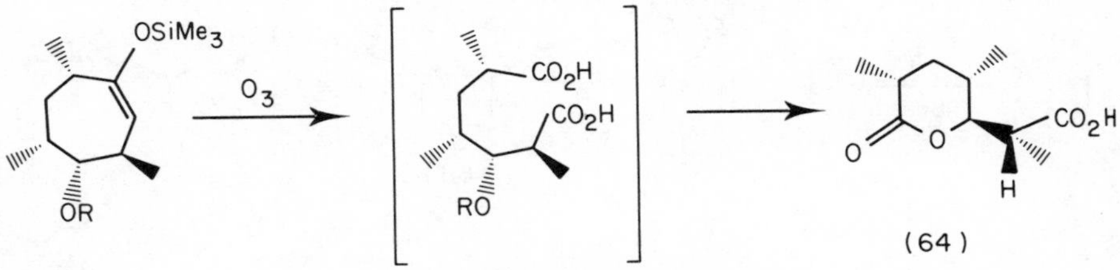

OSiMe3
OR
O3
CO2H
CO2H
RO
O
O
CO2H
H

(64)

Scheme 17.83

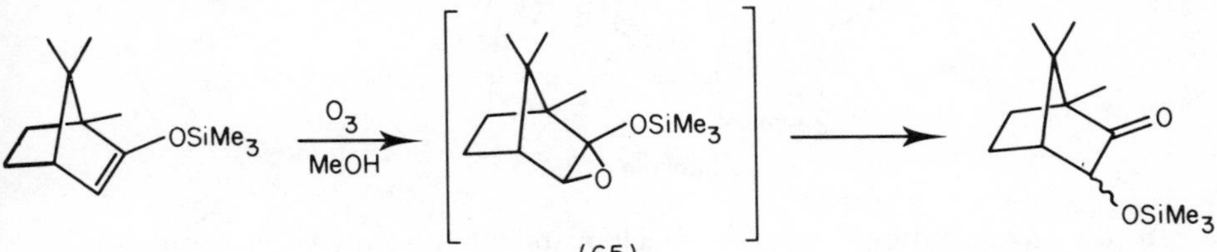

OSiMe3
O3
MeOH
OSiMe3
O
O
OSiMe3

(65)

(66) 100%

Scheme 17.84

Using singlet oxygen

Ketone-derived silyl enol ethers undergo sensitized photo-oxygenation to give products arising from competitive prototropic and silatropic 'ene' reactions[80]. The product formed by the prototropic pathway can be converted further into an $\alpha\beta$-unsaturated ketone, albeit in modest yield in acyclic cases (*Scheme 17.85*).

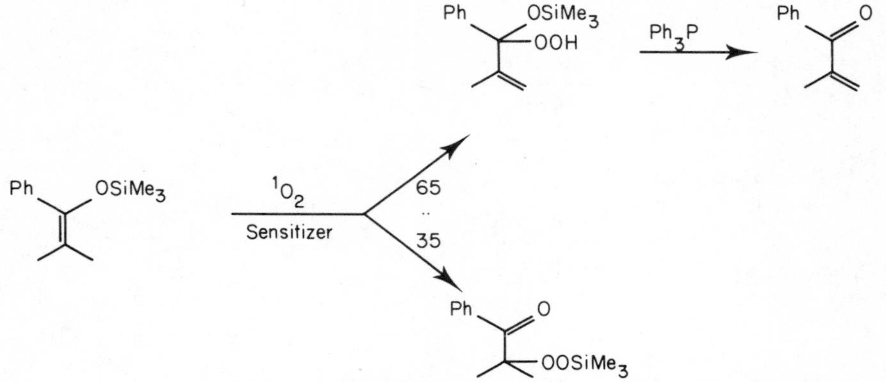

Scheme 17.85

With cyclohexanone-derived substrates (67), silatropic rearrangement is a minor competitor, as is the alternative prototropic pathway to give (68). Accordingly, cyclohexanones can be converted into cyclohexenones in fair to good yield in a reaction sequence (*Scheme 17.86*) whose conditions are sufficiently mild as to preserve any pre-existing α-carbonyl chirality[66]; such chirality is of course lost if enolization occurs from the side of the chiral carbon.

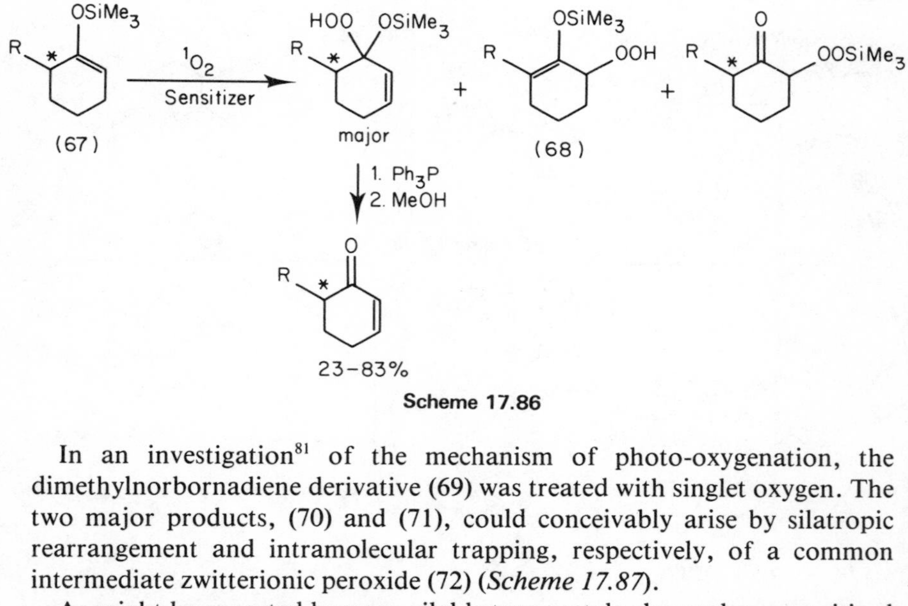

Scheme 17.86

In an investigation[81] of the mechanism of photo-oxygenation, the dimethylnorbornadiene derivative (69) was treated with singlet oxygen. The two major products, (70) and (71), could conceivably arise by silatropic rearrangement and intramolecular trapping, respectively, of a common intermediate zwitterionic peroxide (72) (*Scheme 17.87*).

As might be expected by now, silyl ketene acetals also undergo sensitized photo-oxygenation[82, 137]. By selecting substrates with no β-protons, and thus excluding the possibility of prototropic 'ene' reactions, certain carboxylic acids can be converted into α-peroxycarboxylic acids in excellent yield (*Scheme 17.88*).

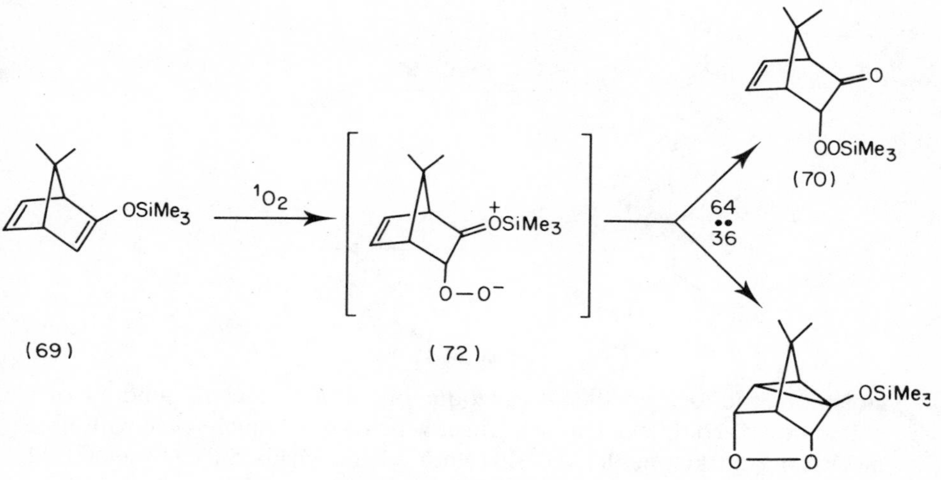

(69) (72) (70) (71)

Scheme 17.87

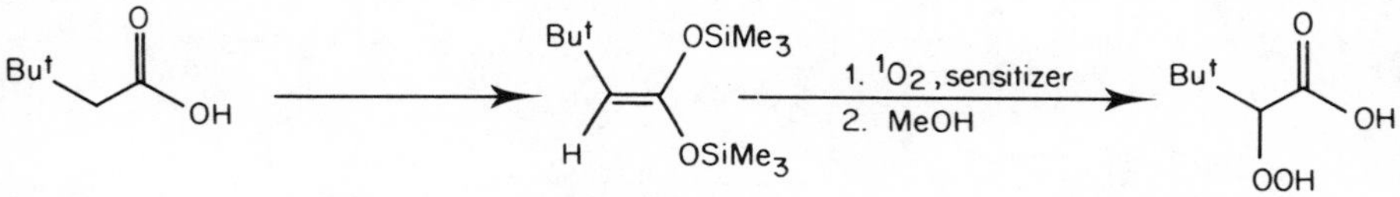

Scheme 17.88

Silyl ketene acetals of the general type (73), derived from azetidinecarboxylic acids, are cleaved[138] by singlet oxygen to β-lactams (*Scheme 17.89*); the homologous pyrrolidine analogues behave differently.

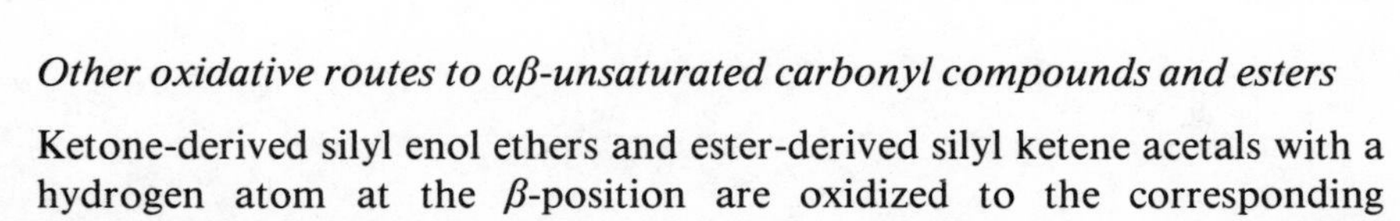

(73)

Scheme 17.89

Other oxidative routes to $\alpha\beta$-unsaturated carbonyl compounds and esters

Ketone-derived silyl enol ethers and ester-derived silyl ketene acetals with a hydrogen atom at the β-position are oxidized to the corresponding $\alpha\beta$-unsaturated derivatives by treatment[68] with the trityl cation (*Scheme 17.90*). The process is highly regioselective, the enol ether of 2,6-dimethylcyclohexanone producing less than 10 per cent of the exocyclic α-methylene ketone. Chloranil (tetrachloro-1,4-benzoquinone)[67] and DDQ (2,3-dichloro-5,6-dicyano-1,4-benzoquinone)[69] can be used for the same purpose.

Transfer-hydrogenation takes place[70] with the combination of a

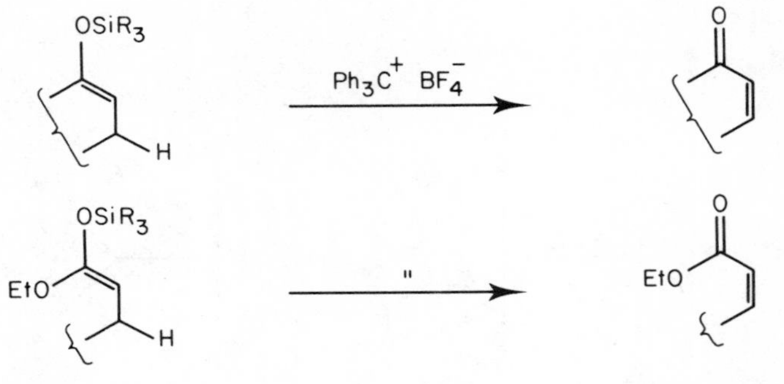

$OSiR_3$ — $Ph_3C^+ BF_4^-$ →

EtO, $OSiR_3$ — " →

$R_3 = Me_3, Bu^tMe_2$ **Scheme 17.90** 10 - 85%

stoichiometric amount of 1,4-benzoquinone and a catalytic amount of a palladium(II) complex. This sequence, which works equally well with both aldehyde- and ketone-derived silyl enol ethers, yields only (*E*) geometric isomers in acyclic cases, regardless of the original geometric isomeric ratio of enol ethers (*Scheme 17.91*).

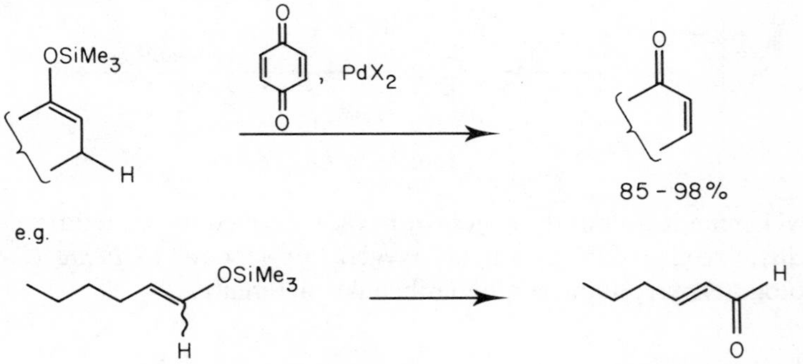

$OSiMe_3$ — 1,4-benzoquinone, PdX_2 → 85 - 98%

e.g. $OSiMe_3$ → 92%

Scheme 17.91

Oxidative coupling

Silyl enol ethers derived from methyl ketones undergo efficient regiospecific oxidative coupling[139] when treated with silver(I) oxide in DMSO (*Scheme 17.92*). With more heavily substituted examples, yields are poor,

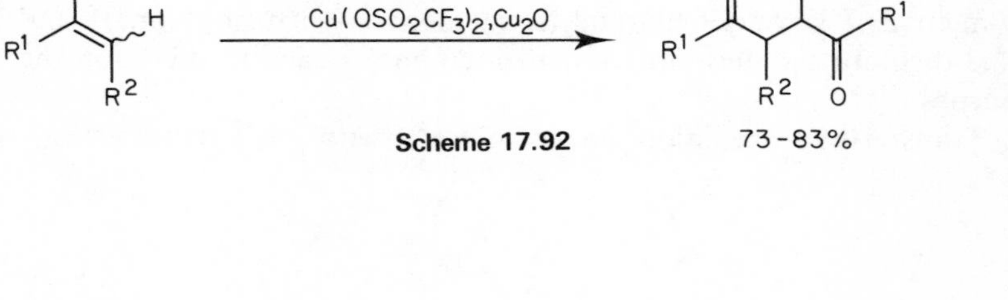

Ag_2O, DMSO, heat → 40-80%

$OSiMe_3$, R^1, R^2 — $Cu(OSO_2CF_3)_2, Cu_2O$ → 73-83%

Scheme 17.92

but the reaction becomes of more general utility with the employment[140, 141] of copper(II) trifluoromethanesulphonate. Under suitable conditions, good yields of cross-coupled products can be obtained from binary mixtures of silyl enol ethers.

The coupling process is also applicable to ester-derived ketene silyl acetals, yielding[142] substituted succinate esters in excellent yields, regardless of the degree of substitution (*Scheme 17.93*).

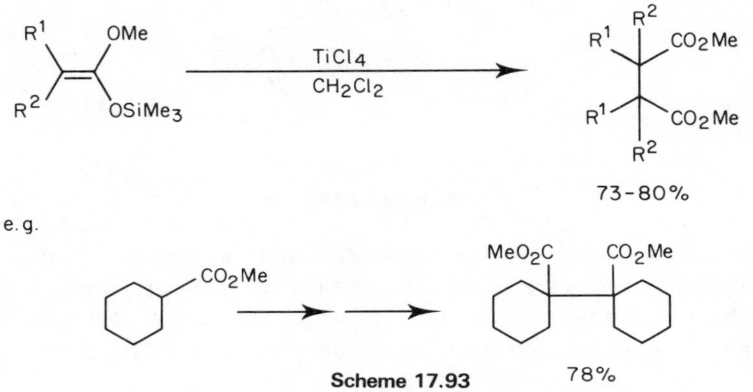

Scheme 17.93

17.3.9 Cycloaddition

The following reactions of silyl enol ethers are classified simply by the overall outcome of the cycloaddition processes. No inference is intended on the concertedness or non-concertedness of particular reactions unless specifically discussed.

[2 + 1]-Cycloaddition, cyclopropanation, silyloxycyclopropanes

Silyl enol ethers undergo ready reaction[143–145] with Simmons–Smith reagent combinations to give cyclopropane silyl ethers or the derived free alcohols (*Scheme 17.94*). Various improvements have been made to the original conditions, including the use of a zinc–silver couple[146] and the employment of pyridine in place of an aqueous isolation procedure; if the cyclopropanol is the ultimate target, the addition of an acid chloride can provide advantageous trapping[147] as an acyl ester.

As will be seen in later examples, regioselectivity is observed with polyolefin substrates, the relatively electron-rich enol ether double bond being attacked preferentially by both Simmons–Smith reagents and other carbenoid species.

The involvement of such cyclopropane silyl ethers in synthetic sequences has permitted attainment of, *inter alia,* α-monomethylation of saturated aldehydes and ketones (with regiospecificity being possible in the cases of the latter), regiospecific monomethylation of cyclic enones, rearrangement to cyclobutanones or cyclopentanones in certain cases, ring expansion to $\alpha\beta$-

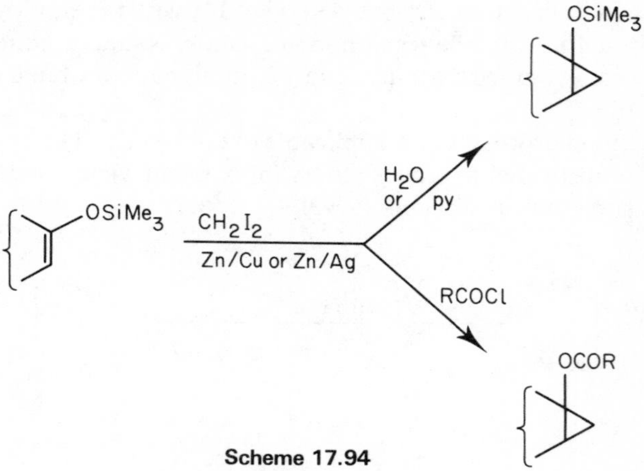

Scheme 17.94

unsaturated ketones or 1,3-diketones, and oxidative cleavage to ω-unsaturated carboxylic acids. This variety of transformations[148] arises from the possibility of '*a*' bond cleavage, '*b*' bond cleavage, or a combination of both occurring in the cyclopropane ring (74).

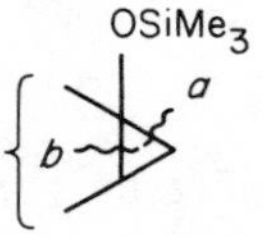

(74)

Processes involving 'a' bond cleavage. The first two of the above synthetic sequences derive from the observation[149] that treatment of cyclopropane silyl ethers with boiling ethanolic sodium hydroxide resulted in '*a*' bond cleavage, producing α-methyl carbonyl compounds in high yield (*Scheme 17.95*). This sequence seems to be applicable generally to aldehydes and cyclic and acyclic ketones; kinetic or thermodynamic regioselective generation of silyl enol ethers from unsymmetrical ketones, followed by purification, leads to regiospecific α-methylation, as illustrated in *Scheme 17.95*.

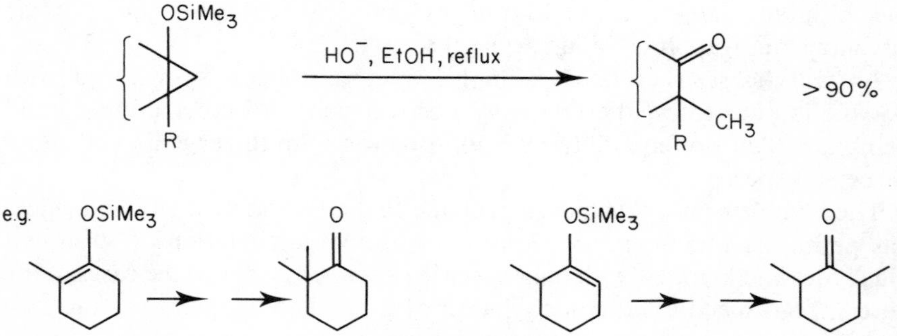

Scheme 17.95

Silyloxycyclopropanes react with bromine to give β-bromoketones in reportedly[150] quantitative yield (*Scheme 17.96*); although the free alcohols undergo the same cleavage process, they are labile in other directions under the normal reaction conditions, and yields of desired product are often poor.

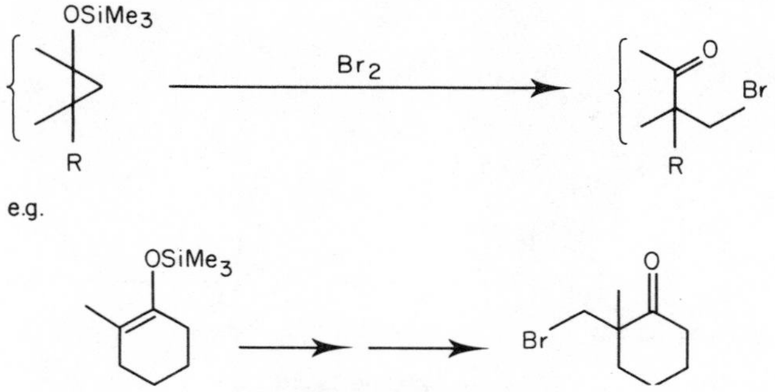

OSiMe$_3$ R $\xrightarrow{Br_2}$ O Br R

e.g.

OSiMe$_3$ ⟶ ⟶ Br O

Scheme 17.96

When the Simmons–Smith cyclopropanation of cyclic silyl enol ethers is carried out in concentrated solution, the initially formed silyloxycyclopropanes undergo zinc(II) iodide-induced isomerization[151] to protected 2-methylenecycloalkanols (75). With excess of diiodomethane, and the employment of diethylzinc as co-reagent[152], further cyclopropanation occurs, leading to 4-silyloxy-spiro[*n*.2]alkanes (76) in good yield; hydrolytic or oxidative cleavage then proceeds efficiently, giving the derived alcohols or ketones in good overall yields (*Scheme 17.97*).

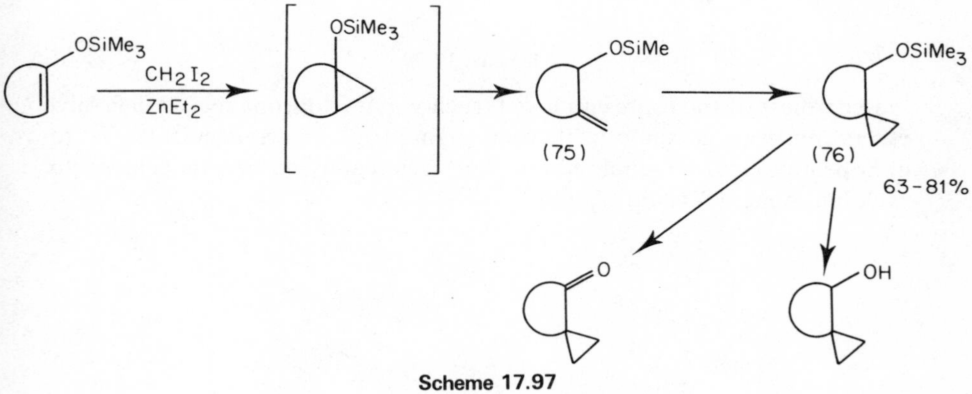

OSiMe$_3$ $\xrightarrow[ZnEt_2]{CH_2I_2}$ [OSiMe$_3$] ⟶ OSiMe (75) ⟶ OSiMe$_3$ (76) 63–81%

O OH

Scheme 17.97

Conia has additionally described[153] a process for the selective α- or α'-methylation of steroidal $\alpha\beta$-unsaturated ketones, as shown in *Scheme 17.98*; this sequence is not fully applicable to simple cyclohexenones, being useful only in those cases involving kinetically generated silyl enol ethers.

When cisoid or labile enones are precursors, a different course of ring opening[154] takes place. At this point, mention must be made of an alternative access to vinyl-substituted cyclopropanols devised by Trost[155] and his co-

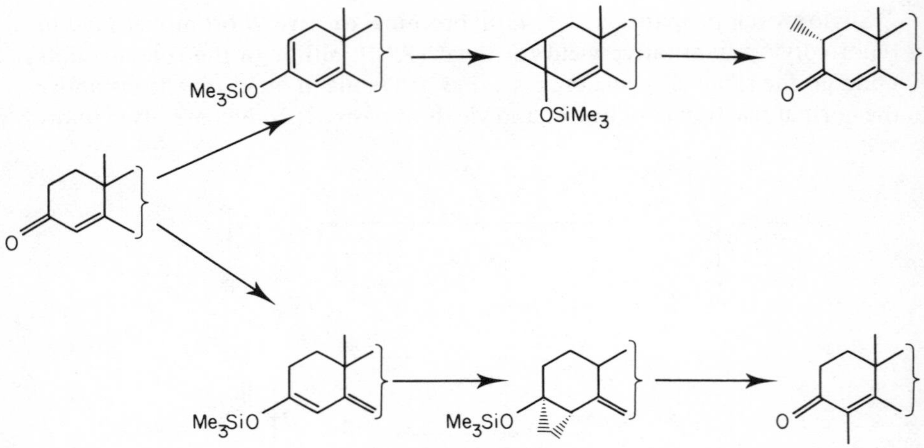

Me$_3$SiO OSiMe$_3$ O
O
Me$_3$SiO Me$_3$SiO O

Scheme 17.98

workers. Ketones react with diphenylsulphonium cyclopropylide or the lithium salt of cyclopropyl phenyl sulphide to give oxaspiropentanes; these species can then be converted into vinylcyclopropanol derivatives by treatment either with strong non-nucleophilic base, or with phenyl selenide anion followed by selenoxide elimination, as shown in *Scheme 17.99*.

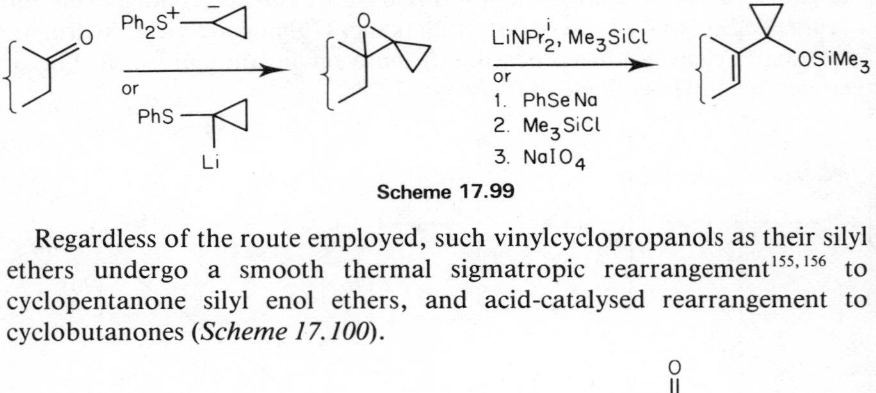

O Ph$_2$S$^+$– or PhS Li O LiNPr$^i_2$, Me$_3$SiCl or 1. PhSeNa 2. Me$_3$SiCl 3. NaIO$_4$ OSiMe$_3$

Scheme 17.99

Regardless of the route employed, such vinylcyclopropanols as their silyl ethers undergo a smooth thermal sigmatropic rearrangement[155, 156] to cyclopentanone silyl enol ethers, and acid-catalysed rearrangement to cyclobutanones (*Scheme 17.100*).

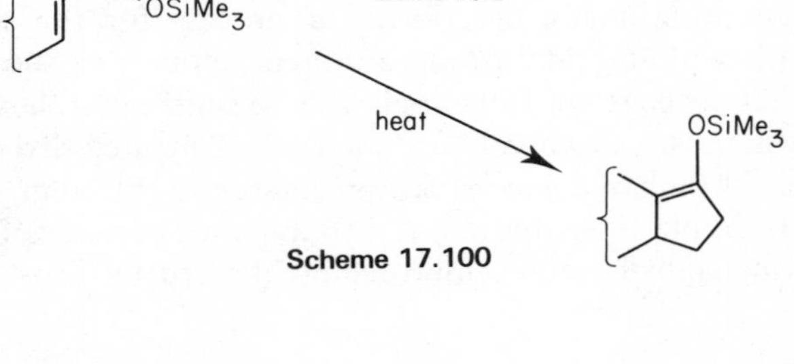

O OSiMe$_3$ H$^+$ or Lewis acid heat OSiMe$_3$

Scheme 17.100

1-Silyloxy-2,2-dichlorobicyclo[*n*.1.0]alkanes of the type (77), derived from the silyl enol ethers of cyclohexanones and higher homologues, are electrolysed in alcohols in the presence of iron(III) nitrate to give β-ketoesters[157] in good yield (*Scheme 17.101*). With cyclopentanone-derived substrates, '*b*' bond cleavage, and consequent ring expansion, takes place to give α-chloro-$\alpha\beta$-unsaturated cyclohexenones, but these can be obtained more simply otherwise (p. 248).

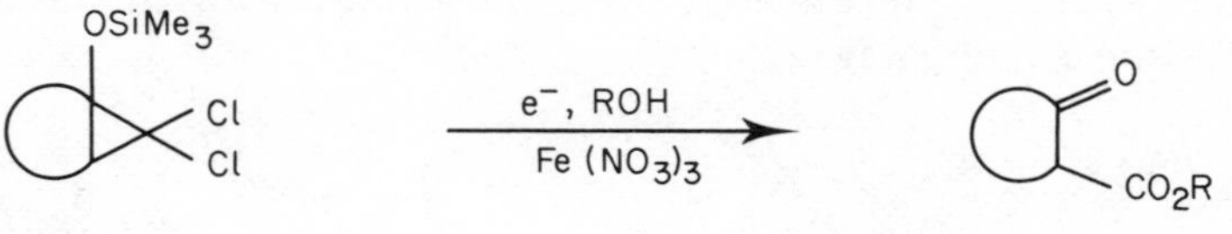

OSiMe3 Cl Cl (77) — e⁻, ROH / Fe (NO3)3 → CO2R 39-79%

Scheme 17.101

Carbonyl-substituted silyloxycyclopropanes, obtained by α-carbonyl-carbenoid addition to silyl enol ethers, undergo exclusive '*a*' bond cleavage to yield 1,4-dicarbonyl compounds or their derivatives. Cleavage can be effected thermally, via a silatropic retro-aldol reaction[26] (*Scheme 17.102*), or in the presence of acid[158], whereupon the free 1,4-dicarbonyl compound is isolated. The latter conditions have been utilized in an efficient route to cyclopentenone ring systems[159], and have been applied to syntheses of a variety of natural products.

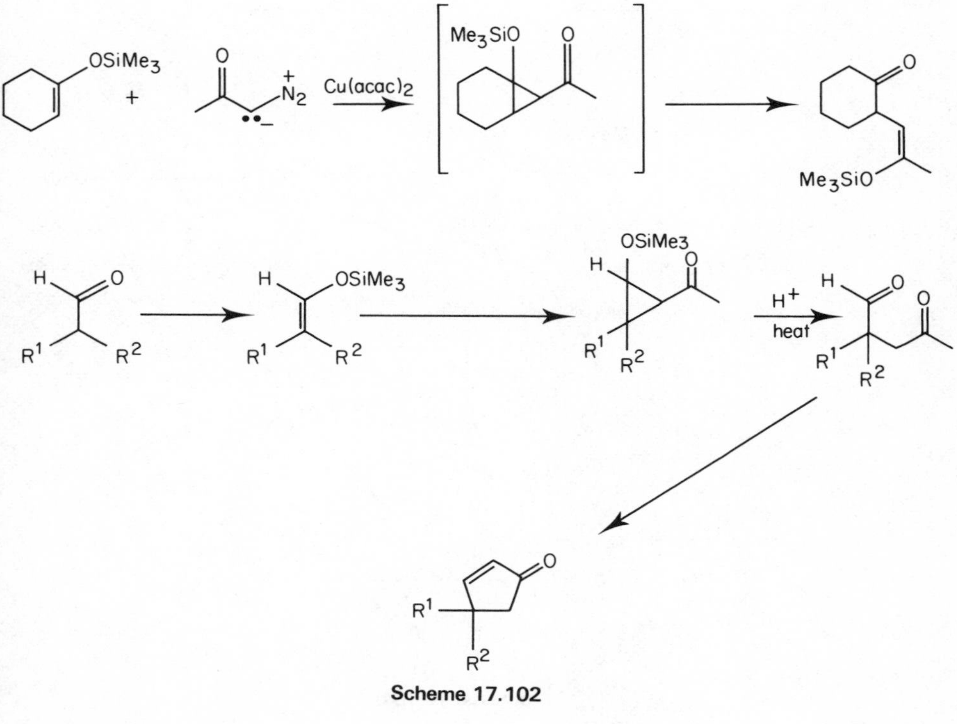

OSiMe3 + N2 — Cu(acac)2 → [Me3SiO] → Me3SiO

H R1 R2 → H OSiMe3 R1 R2 → H OSiMe3 R1 R2 — H+ / heat → H R1 R2 → R1 R2

Scheme 17.102

Processes involving 'b' *bond cleavage.* The 1,1-dihalogenocyclopropane silyl ethers obtained by dihalogenocarbene (or metal carbenoid) addition to silyl enol ethers of aldehydes and cyclic and acyclic ketones undergo smooth thermal or acid-catalysed rearrangement[160] to α-halogeno-αβ-unsaturated carbonyl compounds (*Scheme 17.103*). The reaction seems quite general, giving access to a hitherto poorly characterized group of compounds. Advantage can be taken of regioselective silyl enol ether generation, as illustrated; when (*E,Z*) mixtures of silyl enol ethers are involved, the product is a single isomer, probably (*Z*).

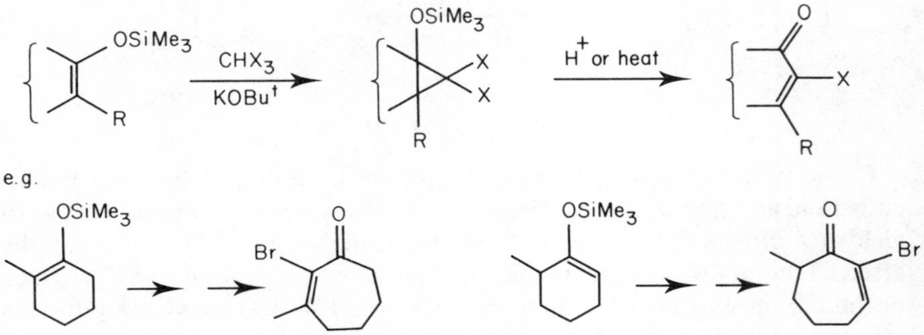

$OSiMe_3$ R CHX_3 $KOBu^t$ $OSiMe_3$ X X R H^+ or heat O X R

e.g.

$OSiMe_3$ O Br $OSiMe_3$ O Br

Scheme 17.103

$OSiMe_3$ $FeCl_3$ DMF O Cl NaOAc O 60–98%

$(CH_2)_n(CO_2Me)_2$ Me_3SiO $OSiMe_3$ $(CH_2)_n$ $(CH_2)_n$ Me_3SiO $OSiMe_3$ Me_3SiO $OSiMe_3$ $(CH_2)_n$ $(CH_2)_n$ Me_3SiO $OSiMe_3$ $FeCl_3$ DMF O O $(CH_2)_n$ $(CH_2)_n$ O O

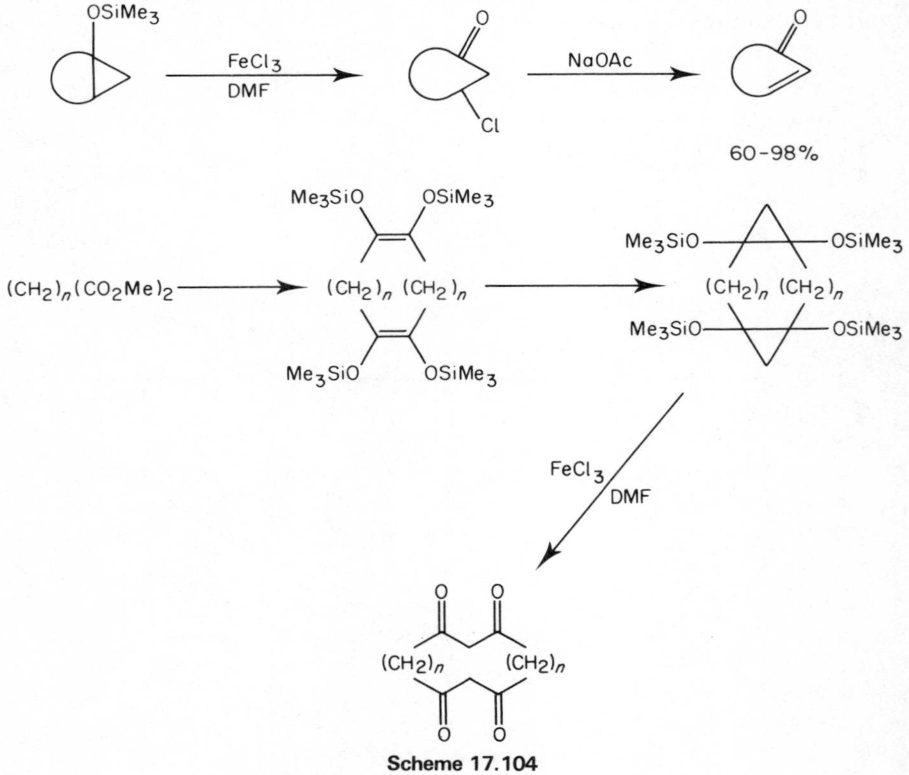

Scheme 17.104

1-Silyloxybicyclo[*n*.1.0]-alkanes derived from a range of cyclic ketones are cleaved by iron(III) chloride in DMF to β-chloroketones, which undergo a ready elimination to give ring-expanded cyclic enones[161] in good yield (*Scheme 17.104*). This reaction, which probably involves a 1-electron transfer process, produces 1,3-diketones when applied to bis(silyloxy)cyclopropanes [derived in turn from methylene addition to the products of the silyl acyloin reaction (*see* p. 267)]; such an overall sequence has been applied to a synthesis[162] of ($\pm$)-muscone, and to the preparation[163] of some macrocyclic tetraketones, as illustrated in *Scheme 17.104*.

Processes involving both 'a' *and* 'b' *bond cleavage.* 1-Silyloxybicyclo[*n*.1.0]-alkanes are oxidatively cleaved[164] by lead(IV) acetate to ω-unsaturated carboxylic acids in high yield; the cleavage itself probably takes place via the free alcohol, and, as usual, advantage can be taken of specific silyl enol ethers, as illustrated in *Scheme 17.105*. 1-Aryl-1-silyloxycyclopropanes (78) react somewhat differently as they are unable to suffer both '*a*' and '*b*' bond cleavage.

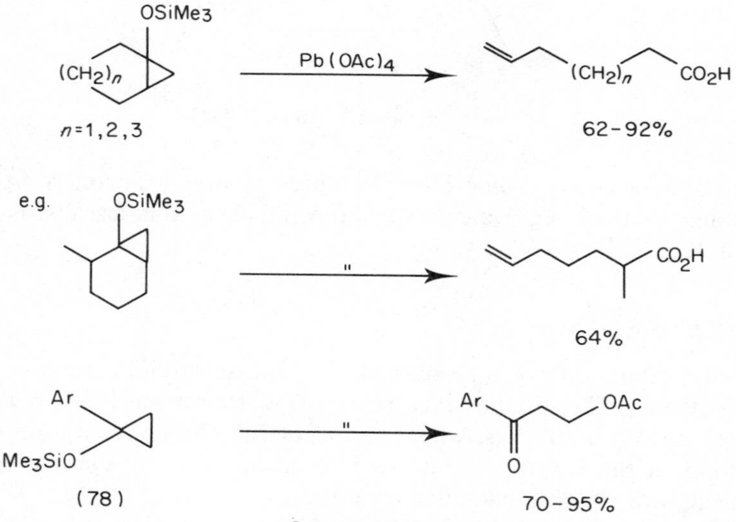

Scheme 17.105

Silyloxycyclopropanes as homoenolate anion precursors. 1-Ethoxy-1-trimethylsilyloxycyclopropane (79) is easily prepared[165], and acts[166] as a synthetic equivalent of the β-anion (80) of ethyl propionate. Such homoenolate anions are not normally of significant utility, since reaction with electrophiles normally takes place at oxygen in the ring-tautomer equivalent to (81). The cyclopropane (79) reacts with a wide range of saturated aliphatic and aromatic aldehydes in the presence of titanium(IV) chloride (other Lewis acids and fluoride ion are ineffective) to produce γ-hydroxyesters, usually isolated as the derived γ-lactones (*Scheme 17.106*). Ketones cannot be employed as such, but do react satisfactorily as the

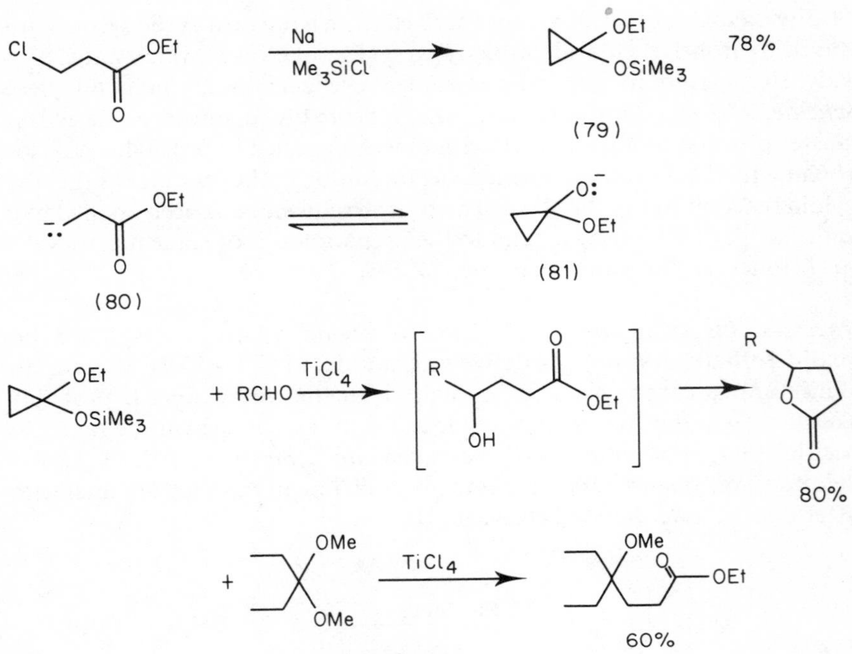

Cl–CH$_2$CH$_2$–C(=O)–OEt $\xrightarrow[Me_3SiCl]{Na}$ (79) 1-ethoxy-1-(trimethylsilyloxy)cyclopropane, 78%

(80) $\rightleftharpoons$ (81)

1-ethoxy-1-(trimethylsilyloxy)cyclopropane + RCHO $\xrightarrow{TiCl_4}$ [R–CH(OH)–CH$_2$CH$_2$–C(=O)–OEt] $\longrightarrow$ γ-lactone, 80%

+ 2,2-dimethoxypropane $\xrightarrow{TiCl_4}$ Me$_2$C(OMe)–CH$_2$CH$_2$–C(=O)–OEt, 60%

Scheme 17.106

corresponding ketals. Some other instances of organosilicon compounds providing synthetic equivalents of homoenolate anions are discussed on page 211.

[2 + 2]-Cycloaddition

Silyl enol ethers do not take part in [2 + 2]-cycloaddition reactions with simple alkenes. They do, however, react well with electron-deficient alkenes, alkynes and heteroalkenes, where partial or full charge separation can be stabilized in the transition state or intermediate, under thermal, photochemical, or Lewis acid-catalysed conditions.

An early investigation[167] showed that silyl enol ethers behaved in the photoexcited state as synthetic equivalents of enols[168]. The silyl enol ethers derived from α-tetralone and α-indanone both undergo photosensitized reaction with Michael acceptors. The regiospecifically produced silyloxycyclobutanes (82) undergo hydrolytic retro-aldolization, yielding the products (83) of conventional Michael addition (*Scheme 17.107*). The conditions employed in this overall photo-Michael reaction preclude the secondary condensation reactions frequently encountered in the more usual basic media.

In a mechanistic investigation[169] of zwitterion intermediates, it was noted that the silyl enol ether of acetaldehyde reacted spontaneously with highly electron-deficient alkenes such as tricyanoethene. Greater synthetic utility

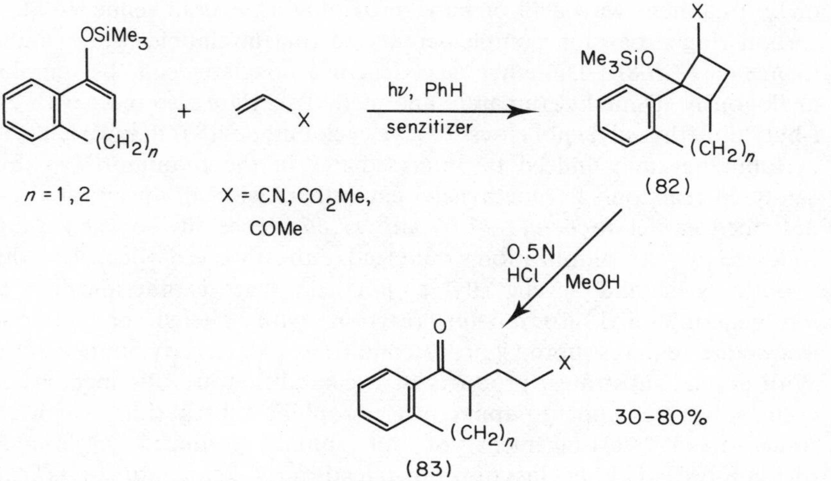

$OSiMe_3$ + X $\xrightarrow[\text{senzitizer}]{h\nu,\ PhH}$ (82)

n = 1, 2 X = CN, CO_2Me, COMe

(82) $\xrightarrow{0.5N\ HCl,\ MeOH}$ (83) 30–80%

Scheme 17.107

can be seen in the titanium(IV) chloride catalysed reaction[170] of a wide range of silyl enol ethers with ethyl propiolate and its analogues, with dimethyl fumarate, and with methyl crotonate. In those cases involving ethyl propiolate, yields of cyclobutene adducts are generally high when t-butyldimethylsilyl enol ethers are employed. When silyl enol ethers from cyclic ketones are involved, the cyclo-adducts can be opened either thermally

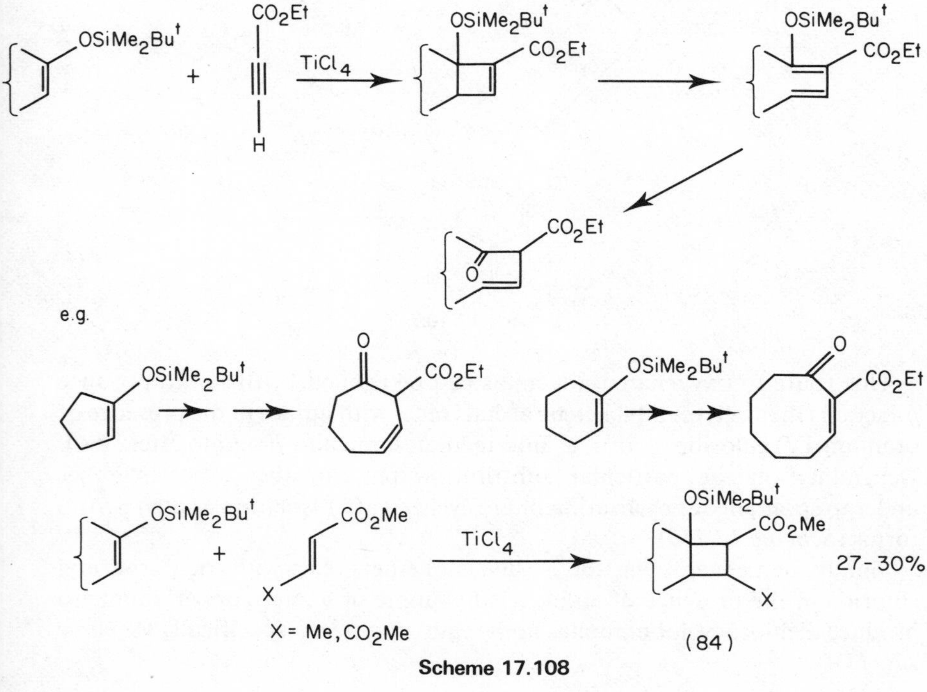

$OSiMe_2Bu^t$ + $HC{\equiv}CCO_2Et$ $\xrightarrow{TiCl_4}$ $OSiMe_2Bu^t$, CO_2Et ⟶ $OSiMe_2Bu^t$, CO_2Et ⟶ CO_2Et

e.g.

$OSiMe_2Bu^t$ ⟶ ⟶ CO_2Et $OSiMe_2Bu^t$ ⟶ ⟶ CO_2Et

$OSiMe_2Bu^t$ + CO_2Me, X $\xrightarrow{TiCl_4}$ $OSiMe_2Bu^t$, CO_2Me, X (84) 27–30%

X = Me, CO_2Me

Scheme 17.108

or by treatment with acid or base, providing an overall sequence of two-carbon ring expansion complementary to that attainable using enamines (*Scheme 17.108*). Dimethyl acetylenedicarboxylate can be employed analogously. Dimethyl fumarate and methyl acrylate also react with cyclic t-butyldimethyl silyl enol ethers to give cyclobutanes (84) in low yields; such cyclobutanes may indeed be intermediates in the titanium(IV) chloride catalysed reaction of trimethylsilyl enol ethers with $\alpha\beta$-unsaturated esters described earlier (*Schemes 17.60* and *17.66*), since silyl–oxygen cleavage followed by retro-aldolization would lead to the observed Michael products.

Both cyclic and acyclic silyl enol ethers react exothermically[62] with *p*-toluenesulphonyl isocyanate; reaction with phenyl or α-naphthyl isocyanate requires more vigorous conditions and tertiary amine catalysis. With acyclic substrates, products of cycloaddition are obtained, whereas cyclic substrates undergo apparent electrophilic substitution, which could result from rearrangement of an initially-produced cyclo-adduct. Methanolysis of either class of product leads[62] to β-ketoamides (85) (*Scheme 17.109*). The related reaction of allylsilanes with chlorosulphonyl isocyanate is discussed in Chapter 9.2

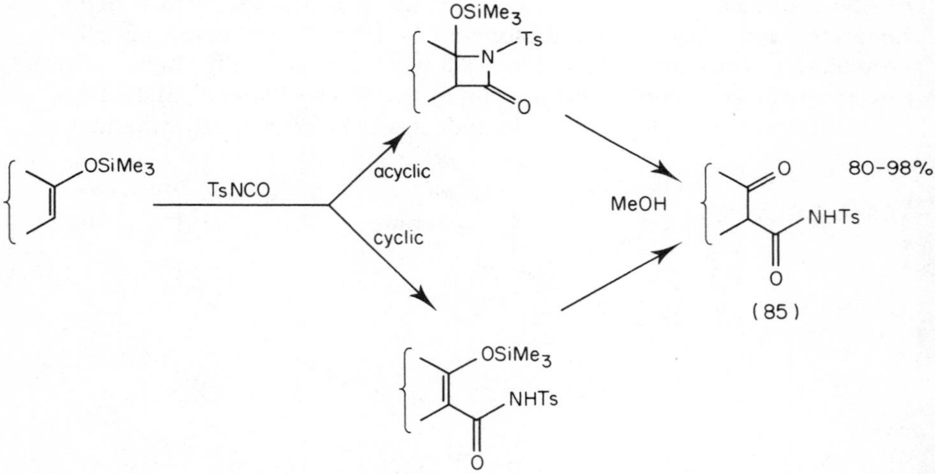

OSiMe3 TsNCO acyclic cyclic OSiMe3 N Ts O MeOH O 80–98% NHTs O (85) OSiMe3 NHTs O

Scheme 17.109

This route to oxygenated β-lactams can be refined further[172] to produce β-lactams themselves. Silyl ketene acetals react with imines in the presence of titanium(IV) chloride to give, as intermediates, metallo-β-amino esters (86). Depending on the particular substitution pattern, these species either undergo spontaneous cyclization or are cyclized after isolation as their protio forms (*Scheme 17.110*).

Finally, in a general reaction[173], silyl enol ethers react with trichloroacetyl chloride in the presence of either triethylamine or a zinc–copper couple to produce dichlorocyclobutanones both regio- and stereo-specifically (*Scheme 17.111*).

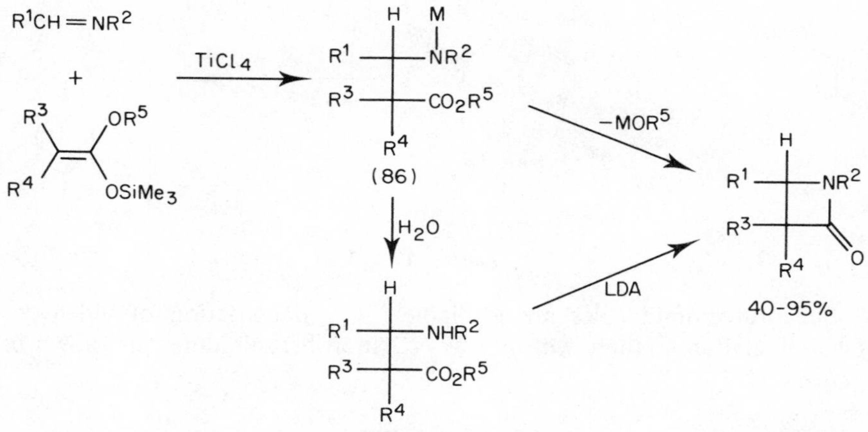

Scheme 17.110

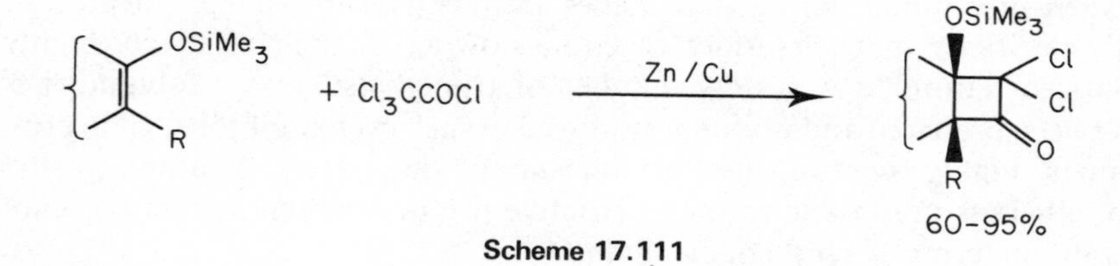

Scheme 17.111

[3 + 2]-Cycloaddition

Only two examples of such behaviour have been reported to date. Cyclic silyl enol ethers react with arenesulphonyl azides to give, after alcoholysis, ring-contracted carboxylic acid amide derivatives (87) in good to excellent yields[174]. A plausible mechanistic pathway involves the dihydrotriazole (88), which, by loss of nitrogen and a Wagner–Meerwein 1,2-shift, could lead *via* the imidate (89) to the observed products (*Scheme 17.112*).

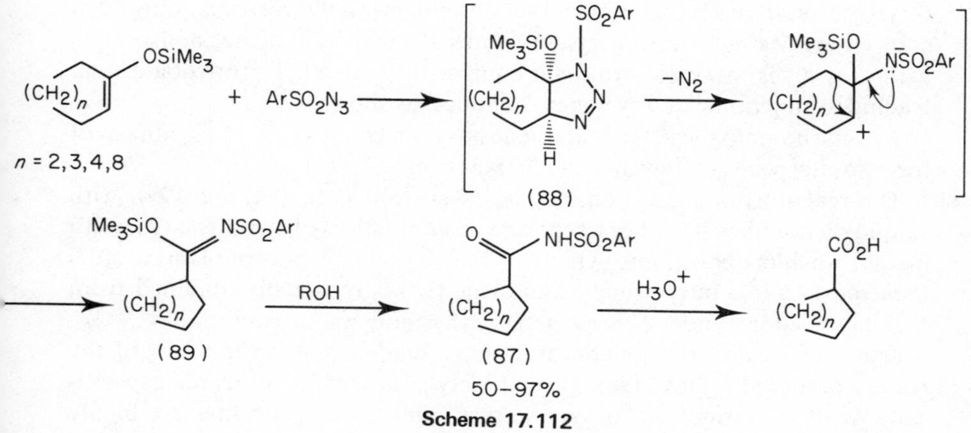

Scheme 17.112

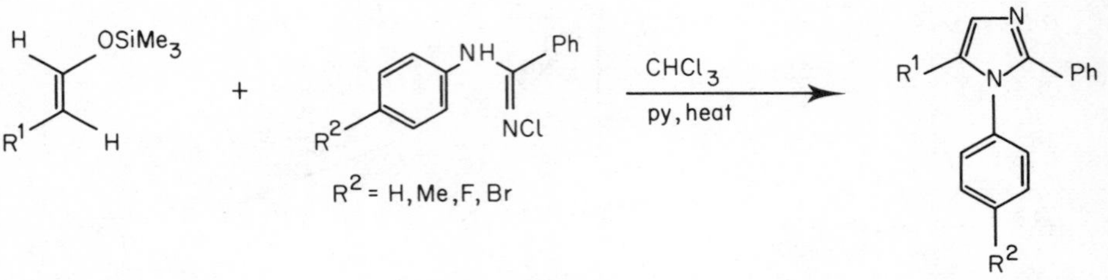

Scheme 17.113

Substituted imidazoles are available[175] by condensation of aldehyde-derived silyl enol ethers with *N*-aryl-*N′*-chlorobenzamidines, as shown in *Scheme 17.113*.

[4 + 2]-Cycloaddition, Diels–Alder reactions

Oxygen-substituted conjugated dienes frequently show a high degree of regiospecificity in Diels–Alder reactions, owing to their relatively highly polarized nature[176]. A goodly number of trimethylsilyloxy-1,3-butadienes have been prepared and usefully employed in such cycloadditions, giving rise to many highly functionalized cyclohexanone derivatives. Masking as silyl enol ethers is preferable to the alternative use of alkyl enol ethers or enol acetates in terms of synthetic flexibility.

Simple silyloxy-1,3-butadienes have been available[177], although poorly characterized, for some time. Earlier studies involved their cyclopropanation, as has been discussed (p. 245). For such purposes they were prepared[153] from $\alpha\beta$-unsaturated compounds by silyl enol ether formation under either kinetic or thermodynamic conditions. This section will now discuss briefly the current range of silyloxy-1,3-butadienes, arranging them in order of increasing complexity.

2-Trimethylsilyloxy-1,3-butadiene[178] (90) reacts in a regio- and stereospecific manner with a range of dienophiles[179], as depicted in *Scheme 17.114*; this regioselectivity is higher than that observed with the corresponding ethyl enol ether. Methyl 3-nitroacrylate condenses with this diene, and others, acting as a regiospecific synthetic equivalent[180] of ethyl propynoate, and leading to functionalized cyclohex-3-enones, as shown.

A cyclic homologue (91) of this diene has been employed[181] in a synthesis of the sesquiterpene (±)-seychellene (92) (*Scheme 17.115*).

The reactions of other homologues[182], such as (93), (94) and (95), with simple dienophiles have been reported; combination of the diene (96) with methyl dichlorophosphine yields[183] the bicyclic 3-phospholanone (97) (*Scheme 17.116*). Interestingly, the diene (93) is reportedly obtained from cyclohex-2-enone under *either* kinetic *or* thermodynamic conditions[181].

Danishefsky and his collaborators have made extensive use[180, 184] of the readily prepared[185] diene (98). Being highly polarized, it reacts regiospecifically with a variety of unsymmetrical dienophiles, leading to highly

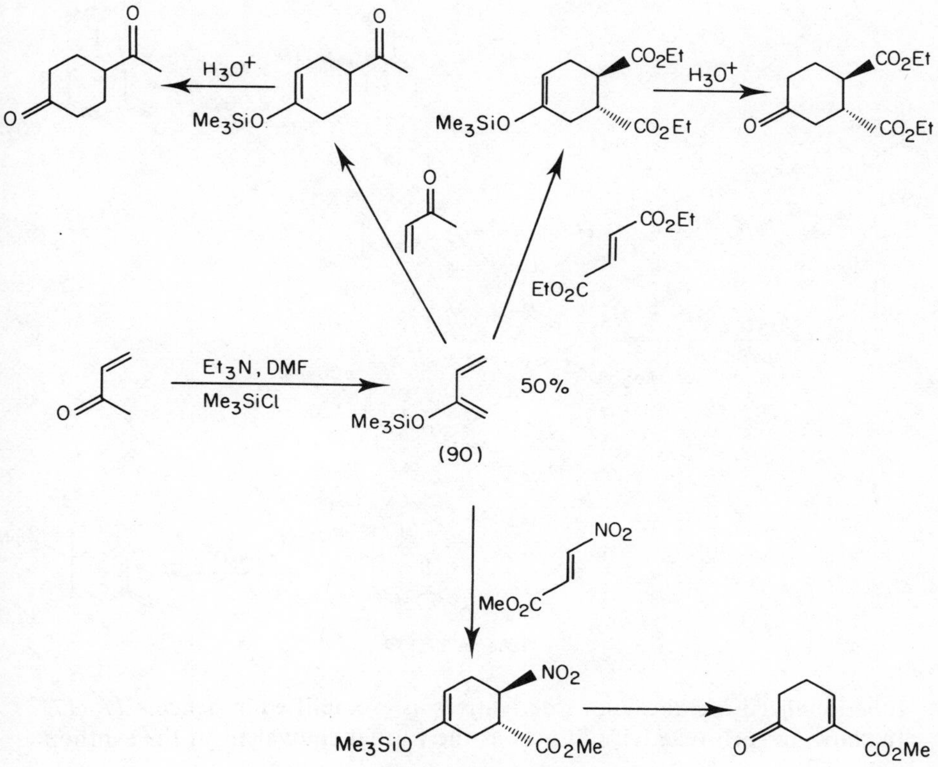

H_3O^+

Me_3SiO CO_2Et CO_2Et H_3O^+ CO_2Et CO_2Et

CO_2Et

EtO_2C

Et_3N, DMF

Me_3SiCl

Me_3SiO 50%

(90)

NO_2

MeO_2C

Me_3SiO NO_2 CO_2Me CO_2Me

Scheme 17.114

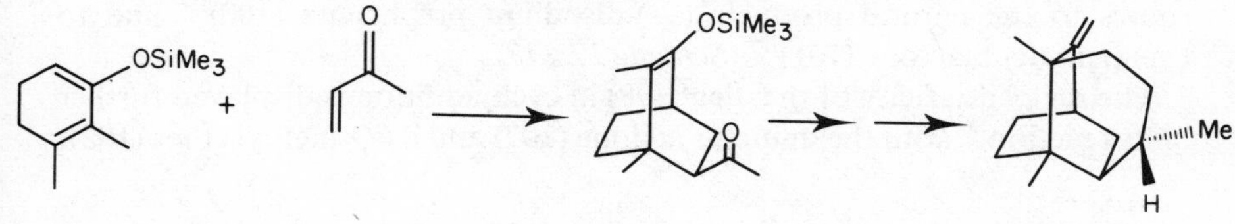

$OSiMe_3$ + $OSiMe_3$ Me H

(91) (92)

Scheme 17.115

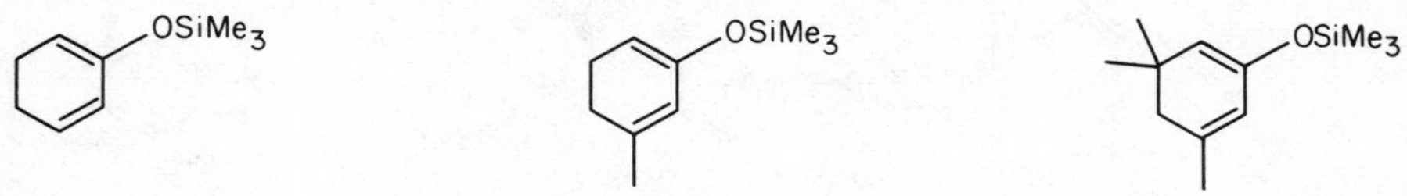

$OSiMe_3$ $OSiMe_3$ $OSiMe_3$

(93) (94) (95)

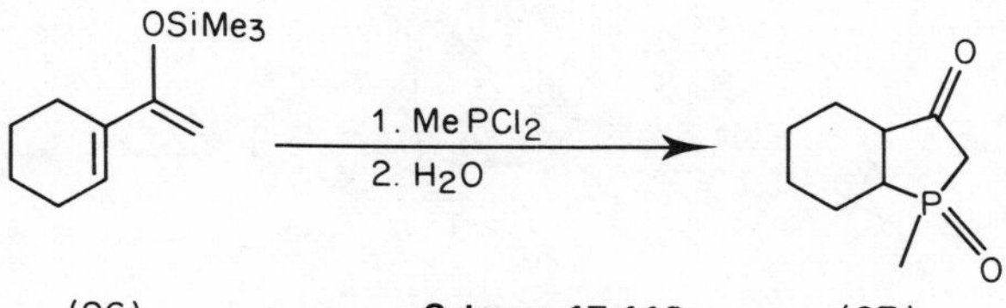

$OSiMe_3$

1. $MePCl_2$

2. H_2O

(96) (97)

Scheme 17.116

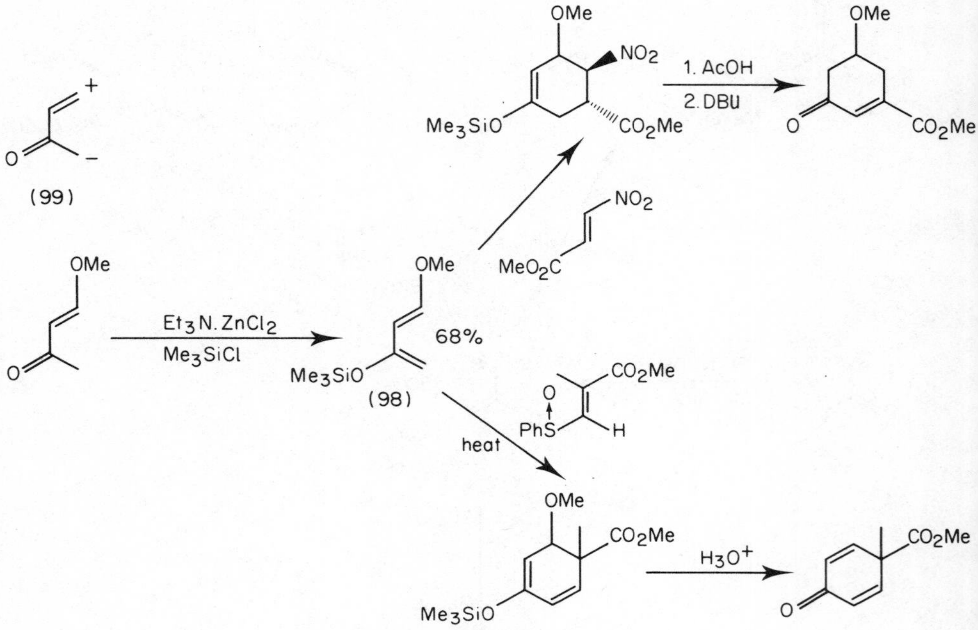
OMe
NO2
1. AcOH
2. DBU
Me3SiO
CO2Me
O
(99)
Et3N. ZnCl2
Me3SiCl
68%
MeO2C
(98)
PhS
H
heat
H3O+

Scheme 17.117

functionalized cyclohexanone derivatives as exemplified in *Scheme 17.117*. In showing such reactivity, it acts as the formal equivalent of the synthon (99).

Danishefsky has also used the diene (98) in elegant Diels–Alder-based routes to the natural products (±)-disodium prephenate (100)[186] and to (±)-pentalenolactone (101)[187] (*Scheme 17.118*).

The regiospecificity of the diene (98) in cycloaddition is displayed further in its reaction[188] with the quinone juglone (102) and its *O*-methyl ether (103),

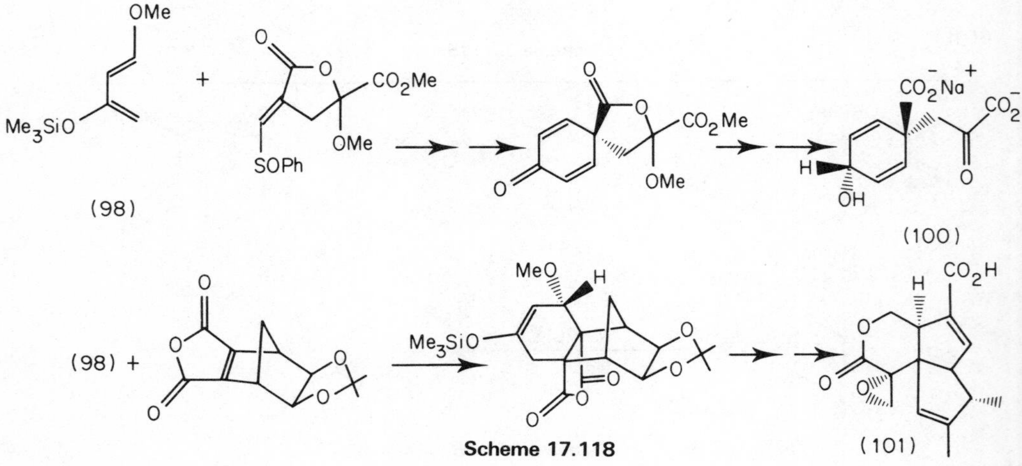
OMe
Me3SiO
(98)
CO2Me
SOPh
CO2 Na
CO2
OH
(100)
(98) +
MeO
H
Me3SiO
CO2H
(101)

Scheme 17.118

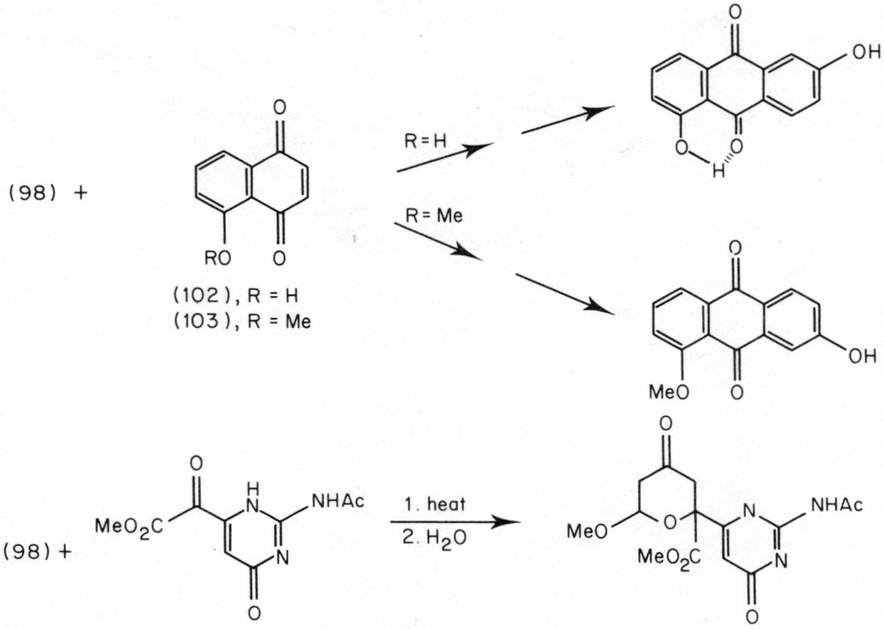

(98) + (102), R = H; (103), R = Me — R=H; R= Me

(98) + (104) — 1. heat; 2. H_2O

Scheme 17.119

unique products being formed in each case (*Scheme 17.119*). It also reacts with the pyrimidone (104), adding across the ketonic carbonyl group[189].

The diene (105) is readily accessible also. It reacts regiospecifically[190] with 2-halogeno-1,4-naphthaquinones; the juglone derivative (106) shows the same regiospecificity both as the free phenol and as its methyl ether (*Scheme 17.120*). The substituted anthraquinones prepared here and in *Scheme 17.119* are of obvious value as synthons in syntheses of the tetracyclines and related pharmacologically active substances. It also reacts with a range of other dienophiles, to produce[191] variously substituted aromatic systems (*Scheme*

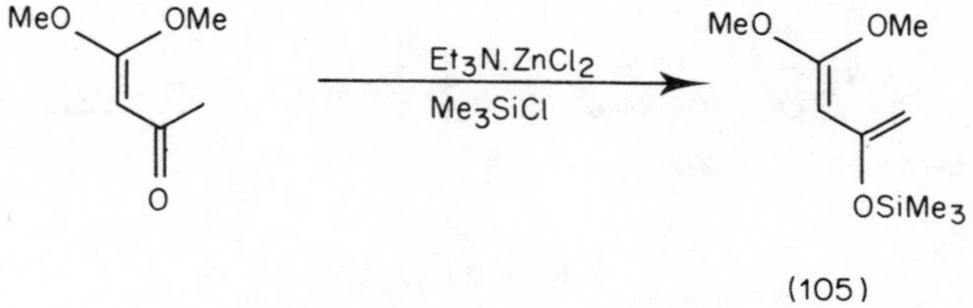

$Et_3N.ZnCl_2$, Me_3SiCl — (105)

(105) 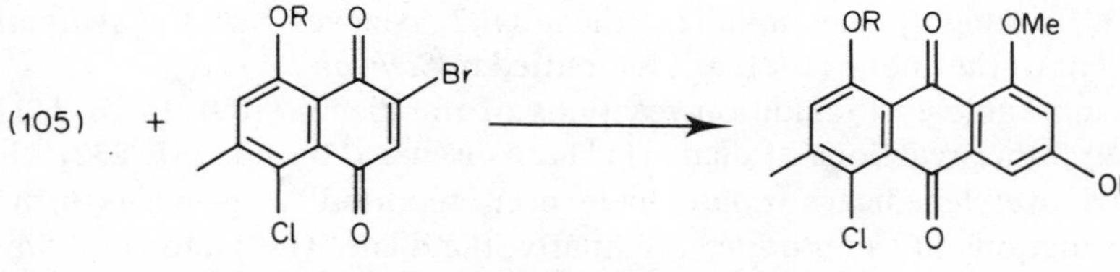 + (106), R = H or Me

Scheme 17.120

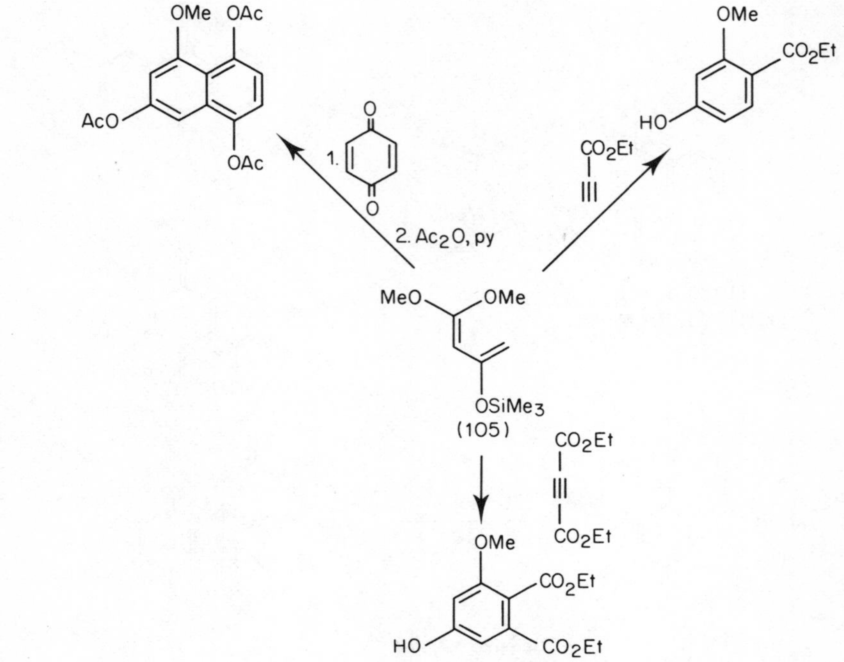

Scheme 17.121

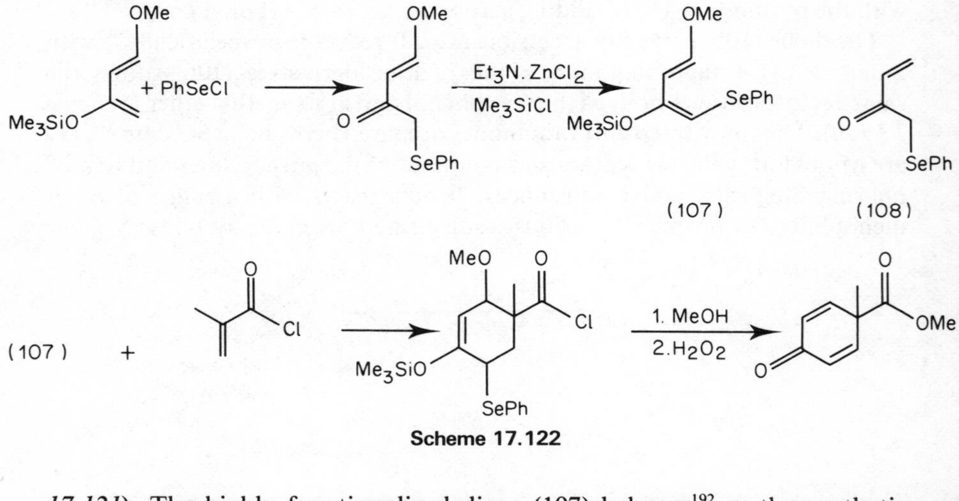

Scheme 17.122

17.121). The highly functionalized diene (107) behaves[192] as the synthetic equivalent of the enone (108), as exemplified in *Scheme 17.122*.

Unremarkable cyclo-addition reactions of the dienes (109), (110), (111) and (49) [other reactions of diene (111) are discussed on pp. 231, 232] with dimethyl acetylenedicarboxylate have been reported[193], as have further transformations of the products. Similarly, the dienes (112) and (113) have been generated[194] as shown (*Scheme 17.123*) and some of their properties

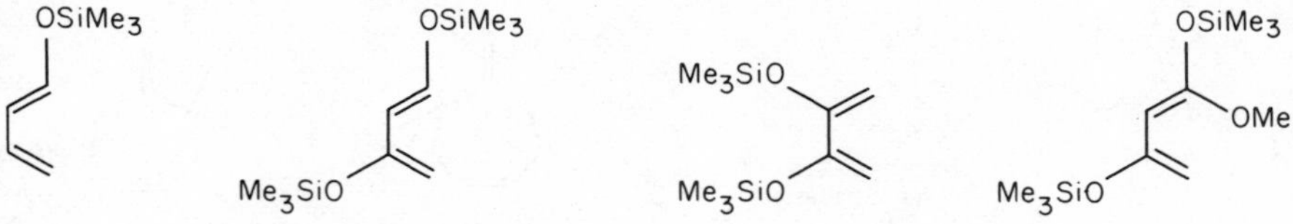

OSiMe$_3$ (109)   OSiMe$_3$ Me$_3$SiO (110)   Me$_3$SiO Me$_3$SiO (111)   OSiMe$_3$ OMe Me$_3$SiO (49)

studied. Finally, the dienes (114) and (110) have been separately employed[195] in stereoselective syntheses of (±)-pumiliotoxin C, a toxin with neuromuscular activity (*Scheme 17.124*).

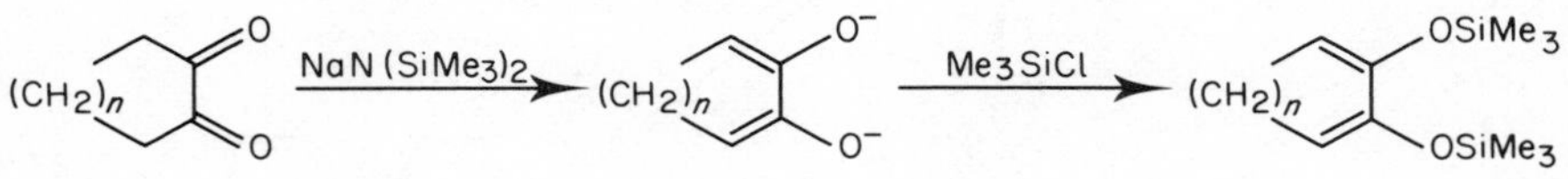

(CH$_2$)$_n$ O O → NaN(SiMe$_3$)$_2$ → (CH$_2$)$_n$ O$^-$ O$^-$ → Me$_3$SiCl → (CH$_2$)$_n$ OSiMe$_3$ OSiMe$_3$

(112), $n = 1$
(113), $n = 2$

Scheme 17.123

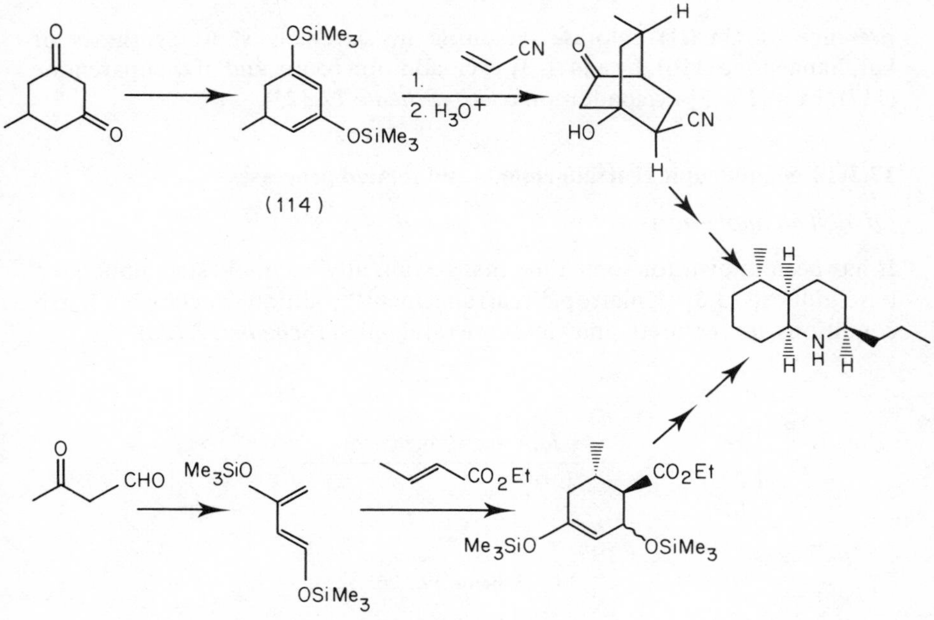

O O → OSiMe$_3$ OSiMe$_3$ (114) → 1. CN, 2. H$_3$O$^+$ → H O HO CN H → H H N H H H

O CHO → Me$_3$SiO OSiMe$_3$ (110) → CO$_2$Et → CO$_2$Et Me$_3$SiO OSiMe$_3$

Scheme 17.124

[4 + 3]- and [2 + 3]-Cycloaddition

In the presence of a zinc–copper couple[196] or of diiron nonacarbonyl[197], *αα'*-dibromoketones behave formally as oxyallyl zwitterions, providing the three-atom component for [4 + 3]- and [2 + 3]-cycloaddition processes. 3-Bromo-2-silyloxypropenes such as (115) function[198] similarly in the

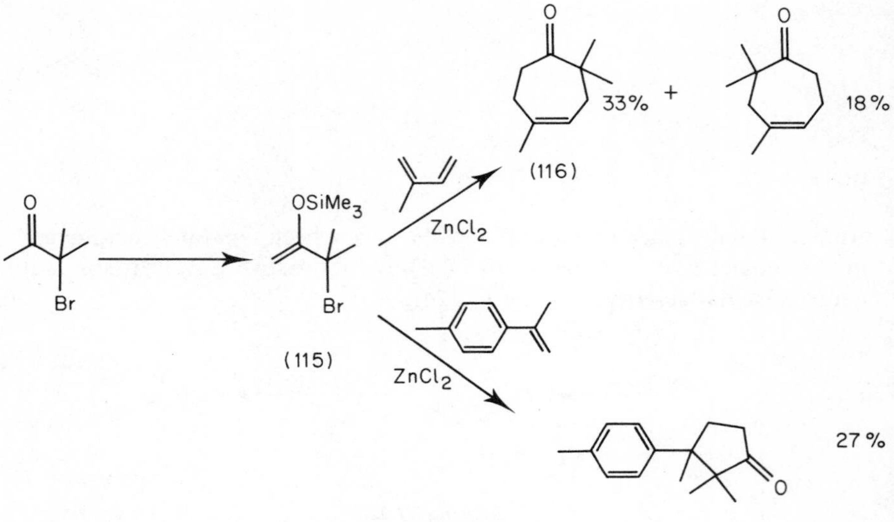

O OSiMe3 ZnCl2 (116) 33% + 18% Br Br (115) ZnCl2 27% O (117)

Scheme 17.125

presence of zinc(II) chloride, resulting in extremely short syntheses of karahanaenone (116), by a [4 + 3]-cycloaddition route, and of α-cuparenone (117), by a [2 + 3]-cycloaddition route (*Scheme 17.125*).

17.3.10 Sigmatropic rearrangements and related processes

Of allyl alcohol esters

It has been known for some time that certain allyl alcohol esters undergo a base-induced [3,3]-sigmatropic rearrangement[199], although relatively harsh conditions are required, and yields are rarely high (*Scheme 17.126*).

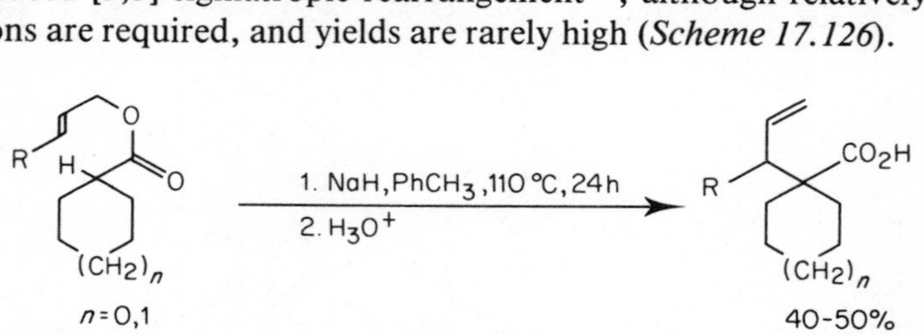

R H O O $(CH_2)_n$ n=0,1 R=H, Ph 1. NaH, $PhCH_3$, 110 °C, 24h 2. H_3O^+ R CO_2H $(CH_2)_n$ 40–50%

Scheme 17.126

Low-temperature conversion of such esters into their corresponding trimethylsilyl or t-butyldimethylsilyl ketene acetals (p. 213), followed by mild thermolysis, effects a great improvement[200] on this overall transformation, and also considerably widens its scope. Although trimethylsilyl ketene acetals rearrange at much lower temperatures than the corresponding t-butyldimethylsilyl analogues, use of the latter species is often preferable for two reasons. Not only is competitive α-*C*-silylation minimized if not excluded in their preparation, but also the resulting silyl ketene acetals can be isolated and purified if desired; rearrangement then proceeds at moderate temperatures, normally below 70 °C. This sequence is illustrated[20, 201] by two examples in *Scheme 17.127.* The second example shows its extension to provide a method for the construction of prostanoid and other functionalized cyclopentane ring systems.

Scheme 17.127

An alternative mild method[20] of generation of the requisite t-butyldimethylsilyl ketene acetal is by reaction of a suitable α-bromoester with zinc and t-butyldimethylsilyl chloride in THF–HMPA, as shown in *Scheme 17.128.*

Scheme 17.128

In the course of this Claisen rearrangement, two new carbon–carbon bonds are formed, one single and one double. The stereochemistries about both of these bonds are determined to a very large extent by the fact that, in the absence of any unusual steric constraints, [3,3]-sigmatropic rearrangements proceed through a chair-like transition state. For example, and considering the particular case in hand of allyl alcohol esters, the ester (118) could rearrange via either of the two possible transition states (119) or (120). Examination of non-bonded interactions indicates that the equatorial conformation of the substituent R in conformation (120) should result in a lower energy for this transition state, and hence encourage formation of the (*E*)-alkene isomer (121), as shown in *Scheme 17.129.*

The degree of stereoselectivity attained on alkene production is normally extremely high; for example, the ester (122) rearranges[202] to the acid (123) with greater than 98 per cent stereoselective production of the (*E*)-alkene isomer (*Scheme 17.130*).

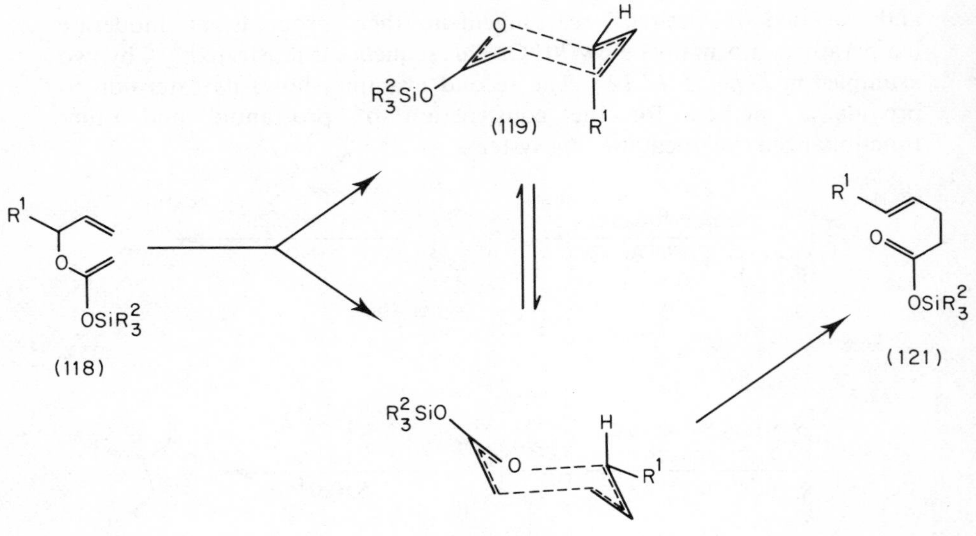

R^2_3SiO (119) R^1 H

R^1 O $OSiR^2_3$ (118)

R^1 O $OSiR^2_3$ (121)

R^2_3SiO H O R^1 (120)

Scheme 17.129

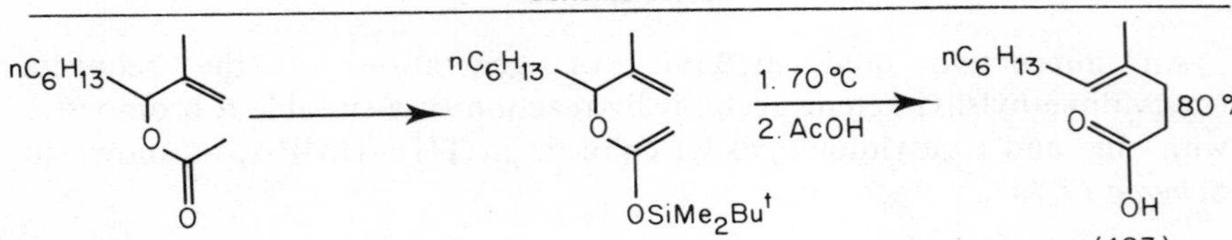

nC_6H_{13} O O (122) → nC_6H_{13} O $OSiMe_2Bu^t$ → 1. 70 °C 2. AcOH → nC_6H_{13} O OH (123) 80 % $E{:}Z > 98{:}2$

Scheme 17.130

A second consequence of a chair-like transition state is that the relative stereochemistry about the new carbon–carbon single bond should be predictable from the geometries of the double bonds in the starting 1,5-diene. This can be illustrated by considering the ester (124). This ester could give rise to either of the enolate isomers (125) or (126), which would then rearrange in a predictable manner to the diastereoisomeric acids (127) and (128), respectively, as shown in *Scheme 17.131*.

Such a postulate is indeed true, and the argument used is generally applicable. The principle is greatly enhanced by the observation[200] that solvent polarity can control the stereochemistry of the kinetically generated enolate double bond, and hence the geometry of the derived silyl ketene acetals (*see,* however, p. 202).

Ireland and his co-workers have discovered that when an ester is deprotonated in THF, and the resulting enolate ion silylated, the resulting ketene acetal is produced predominantly as the (*E*)-isomer (129), corresponding to (126). On the other hand, when the solvent system is THF–HMPA, enolization takes the alternative course, and the (*Z*)-ketene acetal (130) corresponding to (125) is produced, as exemplified in *Scheme 17.132*.

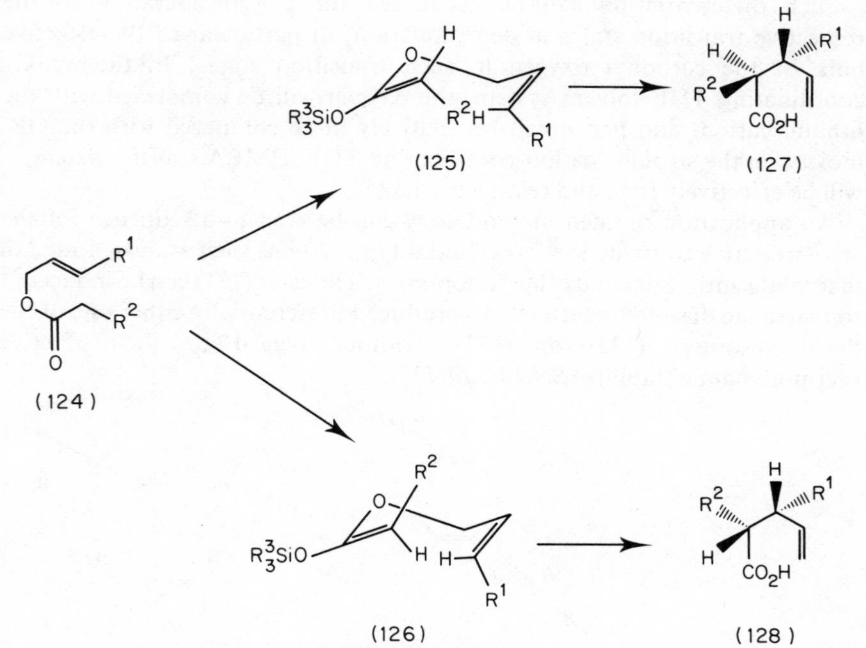

R^3_3SiO R^2H R^1 (125) → CO_2H (127)

(124)

R^3_3SiO R^2 R^1 (126) → CO_2H (128)

Scheme 17.131

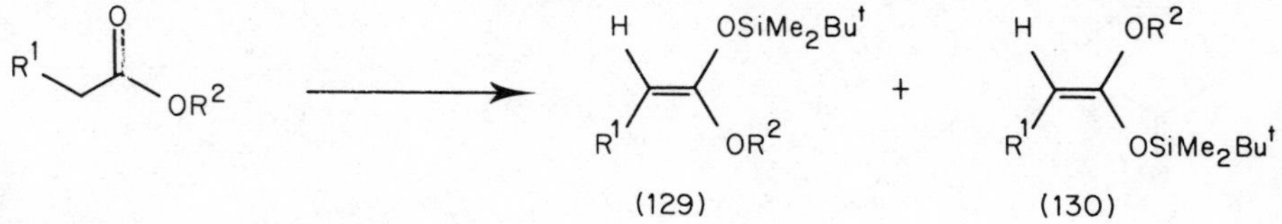

$R^1CH_2CO_2R^2$ → (129) $OSiMe_2Bu^t$, OR^2 + (130) OR^2, $OSiMe_2Bu^t$

Conditions:

		(129)	(130)
a.	1. LDA, THF 2. Bu^tMe_2SiCl, HMPA	major	minor
b.	1. LDA, THF, HMPA 2. Bu^tMe_2SiCl	minor	major

e.g.

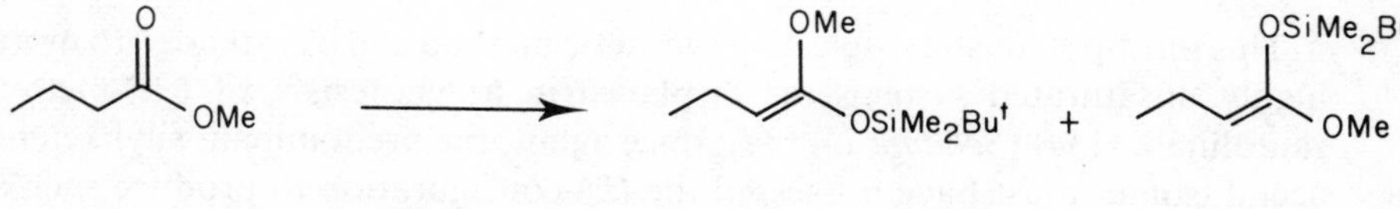

OMe → OMe, $OSiMe_2Bu^t$ + $OSiMe_2Bu^t$, OMe

Conditions: a. 91 : 9

b. 16 : 84

Scheme 17.132

Such differentiation can be accounted for by consideration of the respective transition states in deprotonation, in particular of the effective bulk of the carbonyl oxygen in such transition states. In the weakly coordinating THF solvent system, the oxygen will be complexed with the lithium cation, and hence will be relatively large compared with the OR^1 moiety; in the strongly cation-coordinating THF–HMPA solvent system, it will be effectively free, and relatively small.

An application of such methodology can be seen in a sequence for the stereoselective construction[19, 203] of aldol-type systems such as those found in macrolide antibiotics and other ionophores. The ester (131) can be induced[204] to rearrange diastereoselectively to produced preferentially either one of the diastereoisomers (132) or (133), without regard to their relative thermodynamic stability (*Scheme 17.133*).

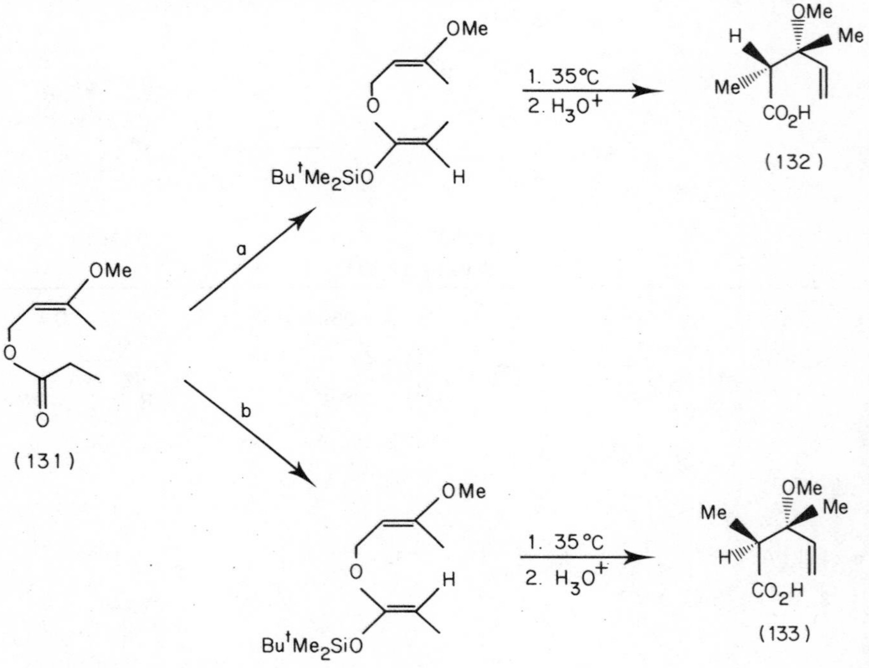

(131) —a→ [Bu^tMe_2SiO silyl ketene acetal] —1. 35 °C, 2. H_3O^+→ (132)

(131) —b→ [Bu^tMe_2SiO silyl ketene acetal] —1. 35 °C, 2. H_3O^+→ (133)

enolate generation conditions:		(132) : (133)
	a.	83 : 17
	b.	23 : 77

Scheme 17.133

This most potent stereoselective synthetic method can be extended to more highly unsaturated systems, as displayed in a synthesis[205] of (±)-methyl santolinate (134) (*Scheme 17.134*); once again, the predominant silyl ketene acetal isomer must have possessed the (*E*)-configuration to produce such a diastereoisomerically enriched mixture.

When α-chiral allyl alcohols are involved as substrates, the stereoelectronic control apparent in this rearrangement ensures transfer[206] of this relative

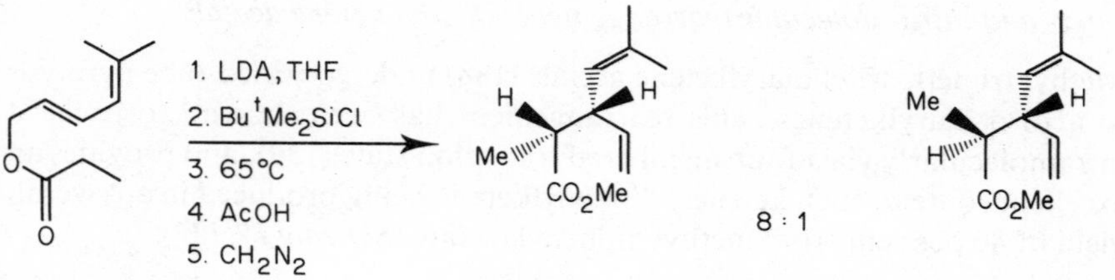

1. LDA, THF
2. Bu^tMe_2SiCl
3. 65°C
4. AcOH
5. CH_2N_2

8 : 1

(134)

Scheme 17.134

chirality to one of the termini of the newly-formed σ-bond (*Scheme 17.135*). A synthesis[207] of the sesquiterpene (±)-frullanolide (135) illustrates this point. In this particular synthesis, the comparatively inexpensive triethylsilyl chloride was used.

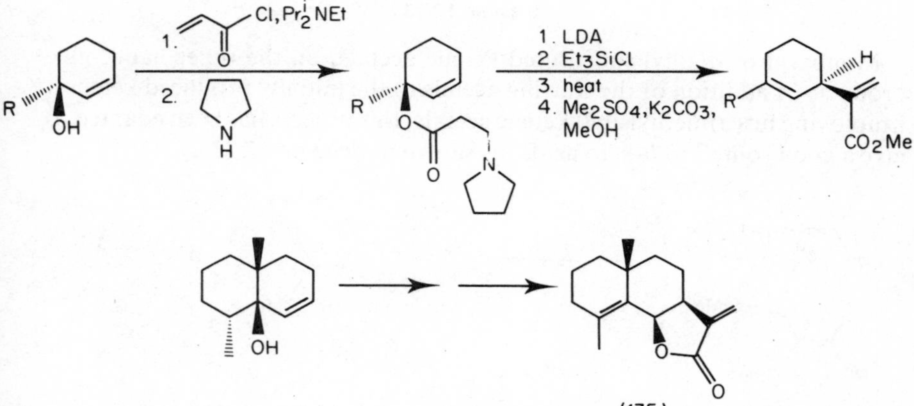

1. Cl, Pr^i_2NEt
2.

1. LDA
2. Et_3SiCl
3. heat
4. Me_2SO_4, K_2CO_3, MeOH

(135)

Scheme 17.135

Similar [3,3]-sigmatropic rearrangements have been applied in a stereochemically undemanding way to provide routes to certain allyl[208] and allenyl[209] ketones via ketone silyl enol ethers. A variety of preparative routes to the substrates (136) and (137) have been described.

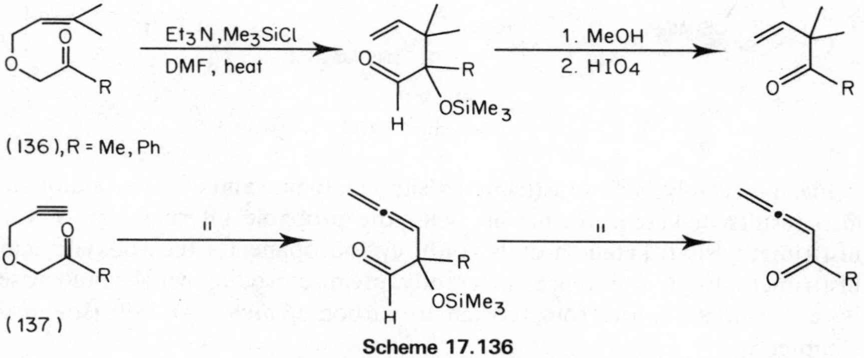

Et_3N, Me_3SiCl
DMF, heat

1. MeOH
2. HIO_4

(136), R = Me, Ph

(137)

Scheme 17.136

Inter- and intra-molecular rearrangements of silyl ketene acetals

Methyl trimethylsilyl diarylketene acetals (138) undergo sealed-tube pyrolysis to afford diarylketenes. This rearrangement has been shown[53] to proceed intramolecularly via a four-membered transition state (139), and provides an excellent route to such ketenes, diphenylketene being produced in an overall yield of 80 per cent from methyl diphenylacetate (*Scheme 17.137*).

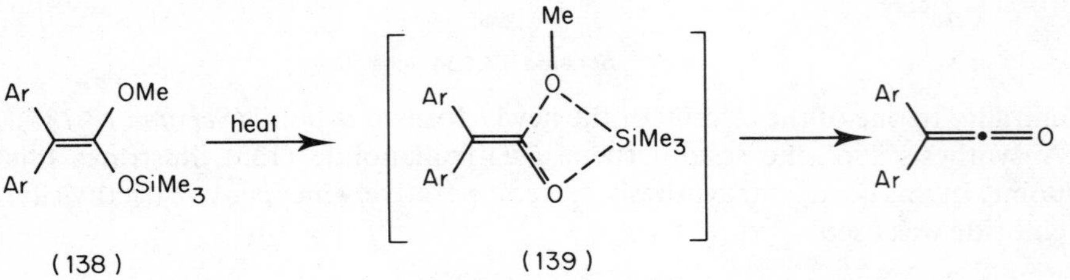

Scheme 17.137

Monoaryl or dialkyl substituted ketene acetals, on the other hand, give products of addition of the starting acetals to the initially produced ketenes. Employing bis(trimethylsilyl) ketene acetals, this process has been adapted to give a good route[53] to β-keto acids, as shown in *Scheme 17.138*.

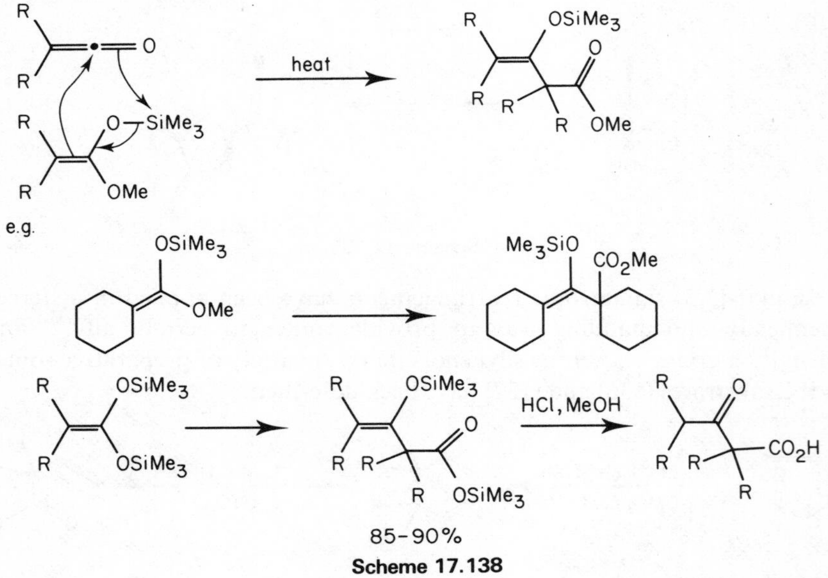

Scheme 17.138

Flash thermolysis[210] of bis(trimethylsilyl) malonate and dimethylmalonate also results in ketene formation, with the probable intermediacy of the bis(trimethylsilyl) ketene acetals (140); cyclopropane 1,1-dicarboxylic acid bis(trimethylsilyl) ester reacts differently, giving a product which could arise by a 1,3-silicon shift from oxygen to carbon (*Scheme 17.139*) (*see also* Chapter 5).

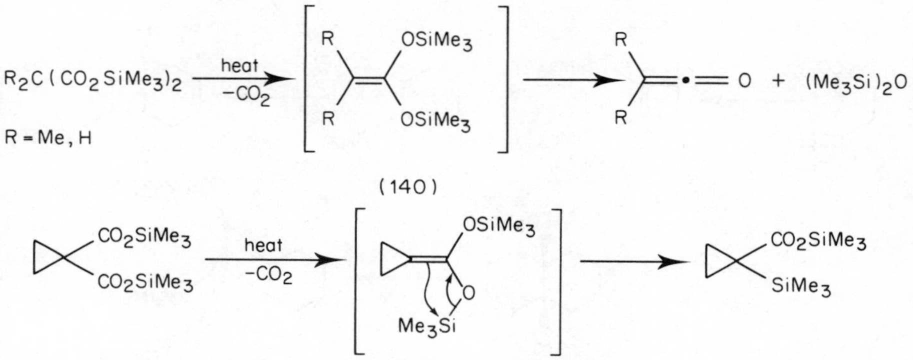

$$R_2C(CO_2SiMe_3)_2 \xrightarrow[-CO_2]{heat} \left[R_2C{=}C(OSiMe_3)_2 \right] \longrightarrow R_2C{=}C{=}O + (Me_3Si)_2O$$

R = Me, H

(140)

Scheme 17.139

17.3.11 Modified acyloin condensations and related reactions

Carboxylic acid esters react with metallic sodium in inert hydrocarbon solvents to give a variety of products, depending on the particular substrate and conditions employed. Processes such as the Claisen, Dieckmann and acyloin condensations, and side reactions such as β-elimination, can all take place under the same basic conditions[211]. If trimethylsilyl chloride is present, it reacts with any basic species formed, thus keeping the reaction mixture neutral and preventing most of the above reactions from taking place. Additionally, since trimethylsilyl chloride is a poor electron acceptor[213], acyloin condensation processes[165,212,214] are unaffected. Indeed, they proceed in greatly improved yields and the reaction products are trapped as 1,2-bis(trimethylsilyloxy)alkenes (141) (*Scheme 17.140*), which can be transformed further as desired. For example, the diesters (142) and (143) both give considerable amounts of the Dieckmann products under normal acyloin condensation conditions; under 'Rühlmann' conditions, the acyloins are the sole products, and are formed in high yields[215].

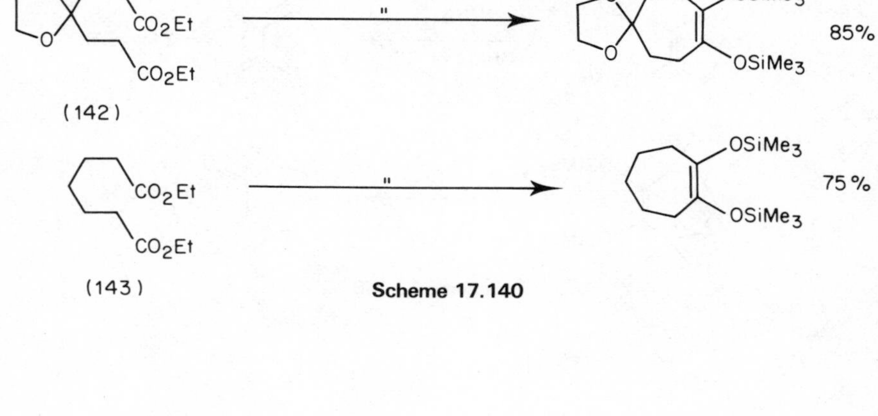

Na, $PhCH_3$, Me_3SiCl, reflux — (141) 60 – 95%

(142) — 85%

(143) — 75%

Scheme 17.140

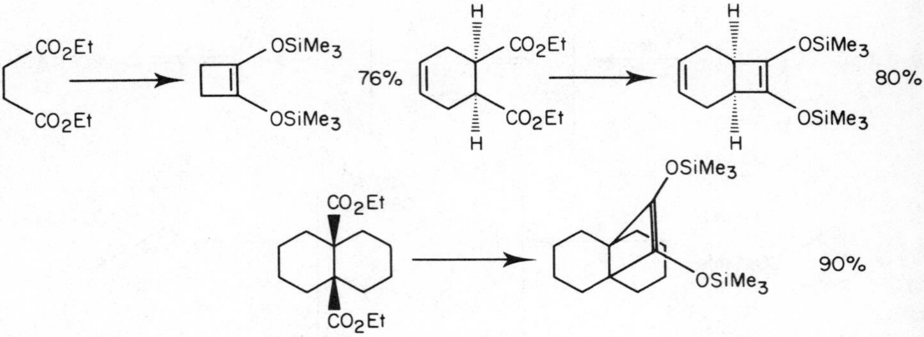

76% 80% 90%

Scheme 17.141

Modified conditions of this type have proved to be of striking value[216] in the synthesis of functionalized cyclobutanes, as illustrated in *Scheme 17.141.*

1,2-Bis(trimethylsilyloxy)alkenes are stable under non-hydrolytic conditions, but are readily cleaved[165] to the corresponding free acyloins by acid-catalysed hydrolysis or alcoholysis; mild basic methanolysis[217] permits isolation of the protected acyloins (144) (*Scheme 17.142*).

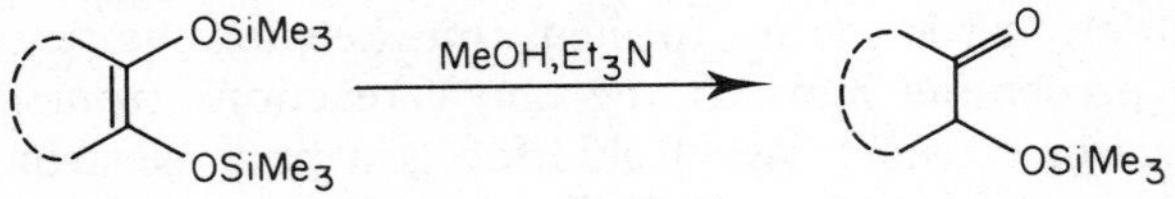

MeOH, Et_3N

(144)

Scheme 17.142

Treatment with bromine causes oxidative cleavage[218], and produces 1,2-diketones (*Scheme 17.143*). This cleavage is most successful in those cases[219] where the product diketone is non-enolizable or a cyclobutane-1,2-dione derivative; enolizable ketones tend to undergo further halogenation[220], because the trimethylsilyl bromide generated reforms highly reactive silyl enol ethers.

By careful control of the reaction conditions, further halogenation can be encouraged or suppressed in those cases involving cyclobutane-1,2-diones. For example, cyclobutane-1,2-dione (145) itself is readily prepared[221] as shown in *Scheme 17.144*, as is the adamantane derivative[222] (146). On the

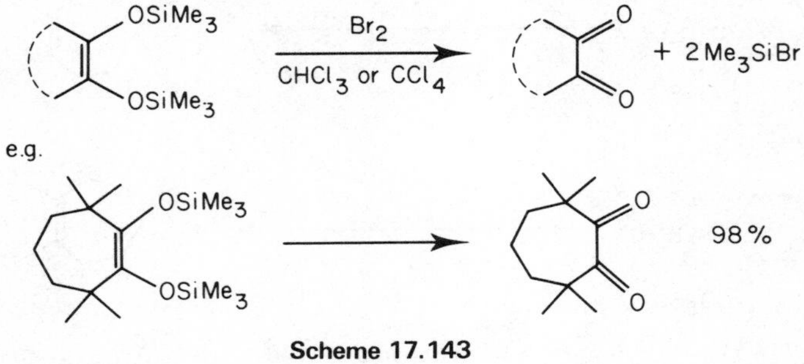

Br_2, $CHCl_3$ or CCl_4 + $2Me_3SiBr$

e.g.

98%

Scheme 17.143

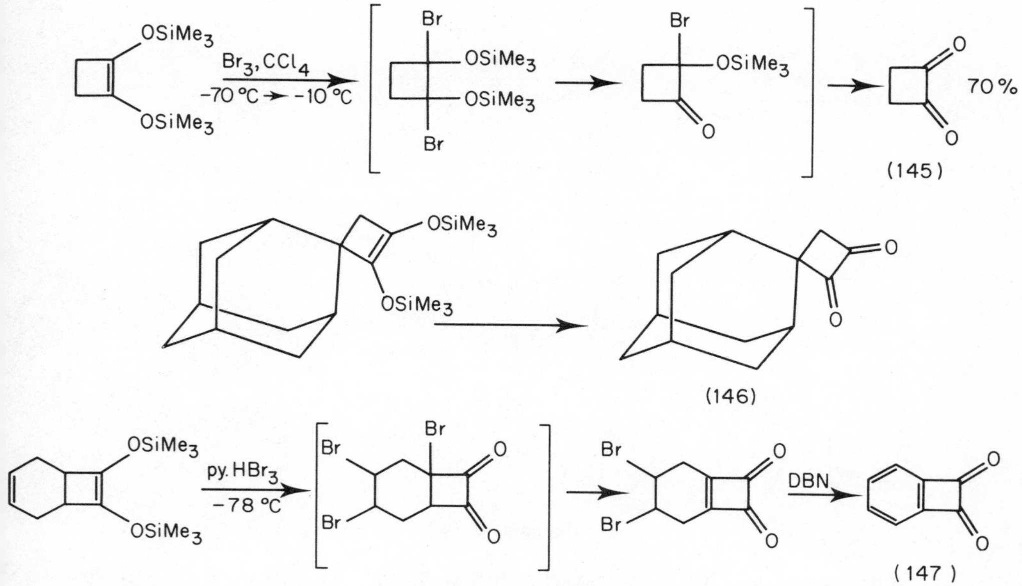

Scheme 17.144

other hand, further halogenation is encouraged by use of pyridine hydrobromide perbromide in a route[223] to benzocyclobutenedione (147).

1,2-Bis(trimethylsilyloxy)alkenes undergo cleavage when treated with two equivalents of methyl-lithium to produce 1,2-enediolates (*Scheme 17.145*), which by alkylation and further modification can lead to a variety of useful species.

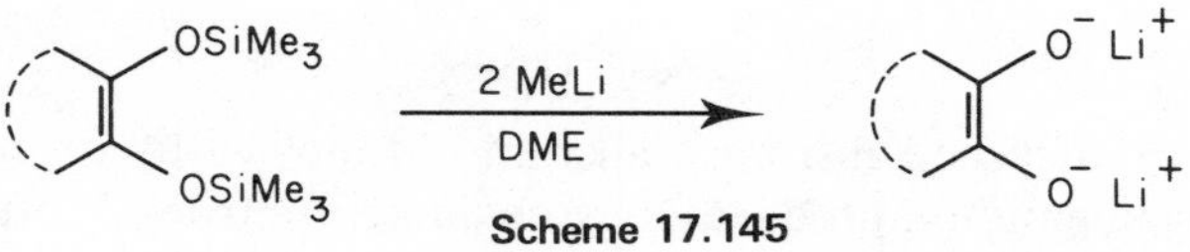

Scheme 17.145

Reaction of cyclopentenoid-derived substrates with ω-iodoesters produces functionalized cyclopentenones[224] of the type (148) (*Scheme 17.146*); such species are obviously of potential utility in prostanoid syntheses.

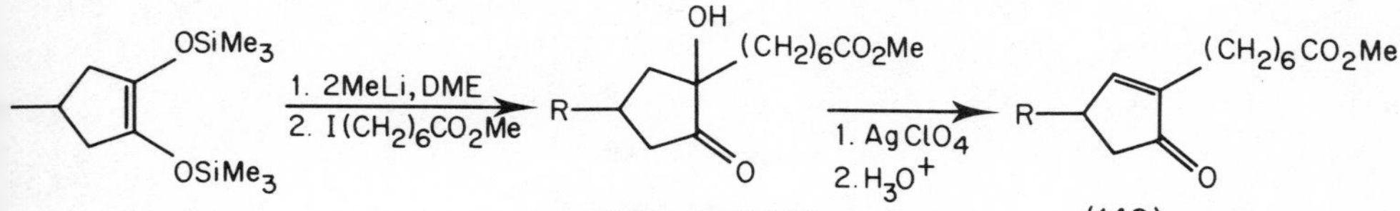

Scheme 17.146

The alkylation of alicyclic substrates with ω-bromoalkan-1-ols, followed by oxidative cleavage of the resulting diol hemiacetals (149), yields macrocyclic keto-lactones[225] (*Scheme 17.147*). This sequence of events has been applied to total syntheses[226] of ($\pm$)-diplodialide A and C (150), and of ($\pm$)-decan-9-olide (151).

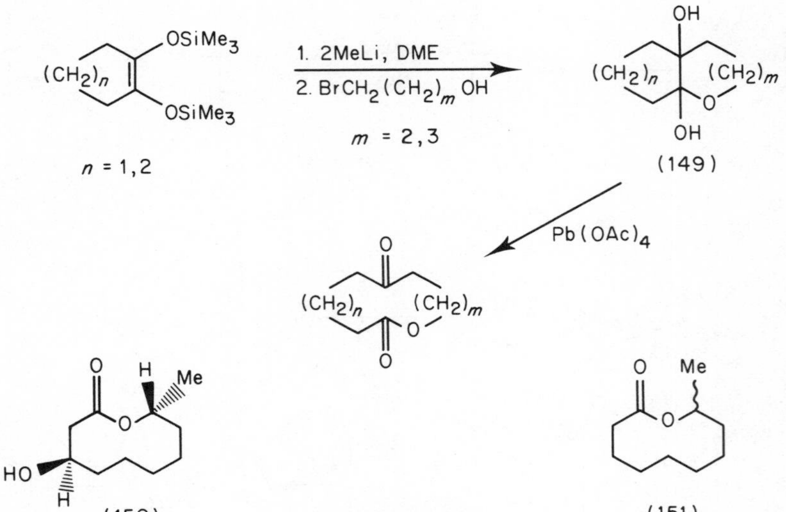

Scheme 17.147

With acyclic substrates, on the other hand, alkylation followed by reduction and oxidative cleavage provides a moderately efficient, if somewhat tortuous, route[227] to unsymmetrical ketones (*Scheme 17.148*).

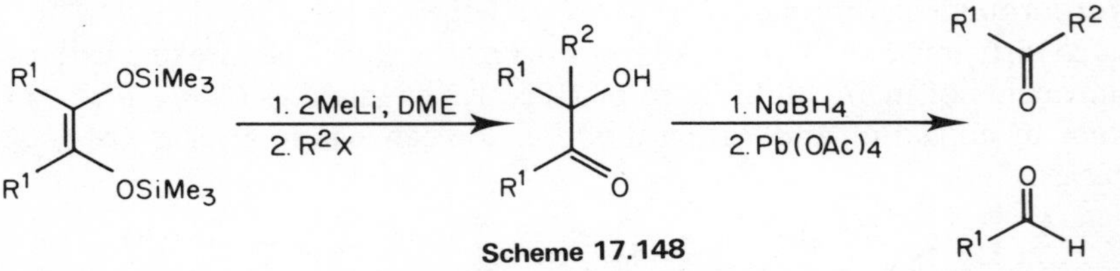

Scheme 17.148

Returning to alicyclic substrates, alkylation followed by oximation and Beckmann fragmentation produces long-chain keto-nitriles[228], often in very good yield (*Scheme 17.149*).

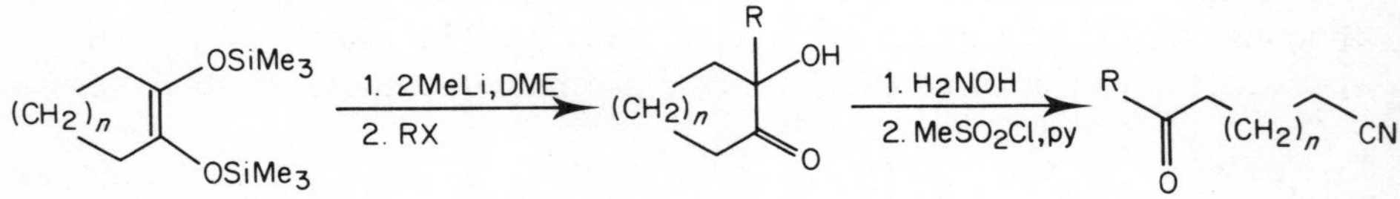

Scheme 17.149

Finally, both acyclic and large-ring alicyclic enediolates can be converted[229] into alkynes (*Scheme 17.150*), albeit in rather modest yields.

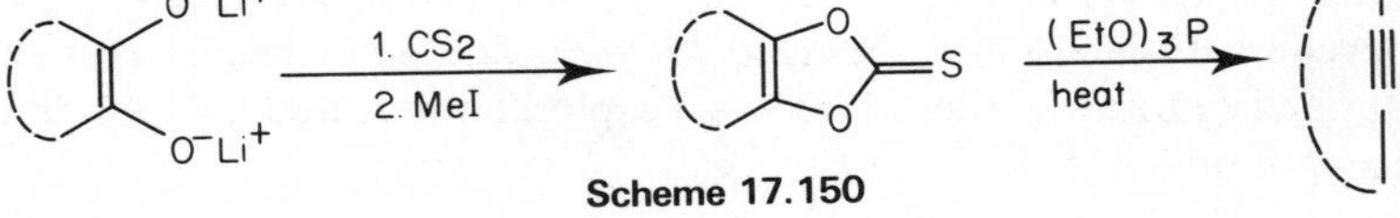

Scheme 17.150

Under certain conditions, 1,2-diesters and related species undergo reductive cleavage of the connecting σ-bond. An extension[230] to provide a method for the introduction of acetic acid fragments has also provided evidence for the mechanism of the reaction, which appears to proceed as shown in *Scheme 17.151*.

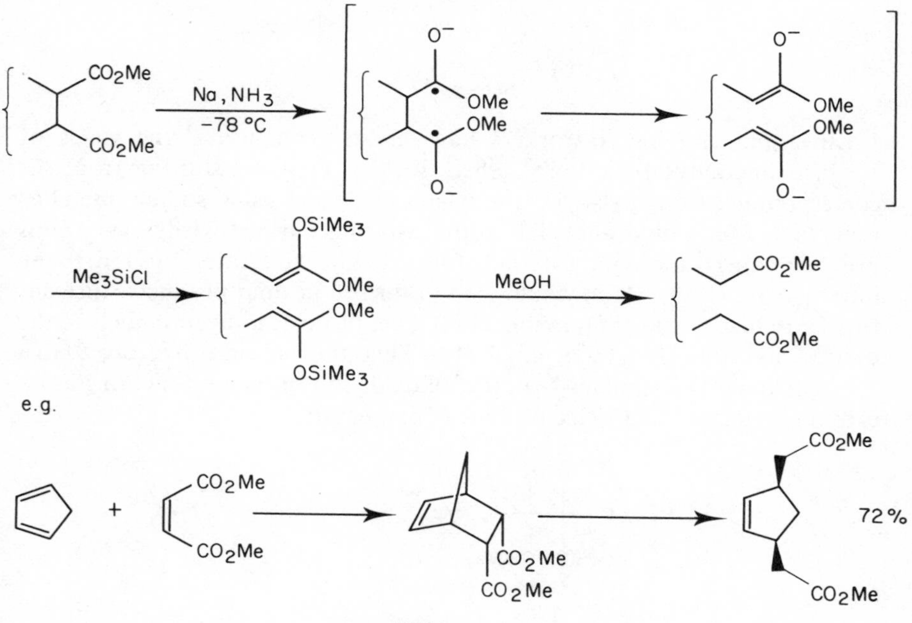
CO2Me
Na, NH3
−78 °C
CO2Me
O−
OMe
OMe
O−
O−
OMe
OMe
O−
OSiMe3
Me3SiCl
OMe
OMe
OSiMe3
MeOH
CO2Me
CO2Me
e.g.
CO2Me
+
CO2Me
CO2Me
CO2Me
CO2Me
72%
CO2Me

(152)

Scheme 17.151

The competitiveness and solvent dependence of this reductive process and the acyloin condensation can be exemplified by the contrasting acyloin reaction[231] of the diester (153) (*Scheme 17.152*) and the σ-cleavage reaction[230] of the diester (152) (*Scheme 17.151*). A further demonstration of the

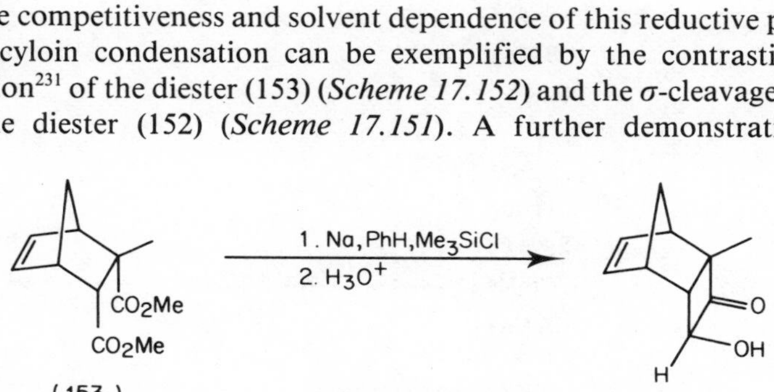
1. Na, PhH, Me3SiCl
2. H3O+
CO2Me
CO2Me
O
OH
H

(153)

Scheme 17.152

influence of solvent and/or metal on the course of reaction has been provided by Calas and his group. Carboxylic acid methyl esters react[232] with lithium metal and trimethylsilyl chloride in THF to give mixtures of *O*-silylated dimeric acyloin products and *C*-silylated monomers, as shown in *Scheme 17.153*.

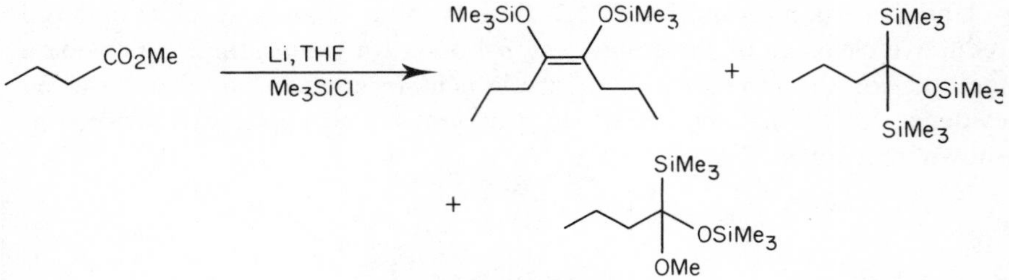

CO_2Me —(Li, THF; Me_3SiCl)→ Me_3SiO, $OSiMe_3$ + $SiMe_3$, $OSiMe_3$, $SiMe_3$ + $SiMe_3$, $OSiMe_3$, OMe

Scheme 17.153

Kuwajima and his co-workers have greatly enhanced[233] the yields of 1,1-bis(trimethylsilyl)alkyl silyl ethers such as (154) by starting from the corresponding trimethylsilyl carboxylates (155) and using sodium metal as reductant. Steric hindrance due to the two bulky trimethylsilyloxy groups prevents dimerization of the initially formed radical intermediate (156) to the intermediate (157), which would lead in turn to acyloin products. Instead, further reduction of the monomer (156) takes place, leading ultimately to the desired silyl ether (154) (*Scheme 17.154*). This ether is readily hydrolysed to a 1,1-bis(trimethylsilyl)alkan-1-ol, the alkoxide ion of which functions as an extremely strong chemoselective base (Chapter 20).

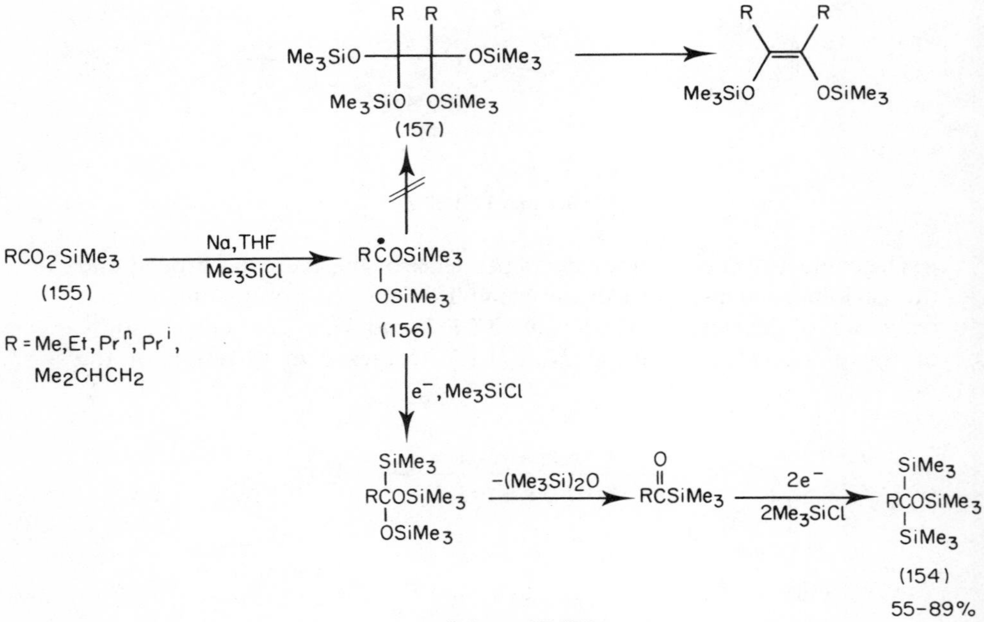

RCO_2SiMe_3 (155) —(Na, THF; Me_3SiCl)→ $R\dot{C}(OSiMe_3)OSiMe_3$ (156)

R = Me, Et, Pr^n, Pr^i, Me_2CHCH_2

(156) ⇸ (157) Me_3SiO–C(R)($OSiMe_3$)–C(R)($OSiMe_3$)–$OSiMe_3$ → R, R, Me_3SiO, $OSiMe_3$ (alkene)

(156) —(e^-, Me_3SiCl)→ $RC(SiMe_3)(OSiMe_3)OSiMe_3$ —($-(Me_3Si)_2O$)→ $RC(O)SiMe_3$ —($2e^-$, $2Me_3SiCl$)→ $RC(SiMe_3)_2OSiMe_3$ (154) 55–89%

Scheme 17.154

A further simplification[233] is possible when HMPA is present as co-solvent; under such conditions the free carboxylic acids themselves can be used, the required trimethylsilyl ester presumably being formed *in situ*, as shown in *Scheme 17.155*.

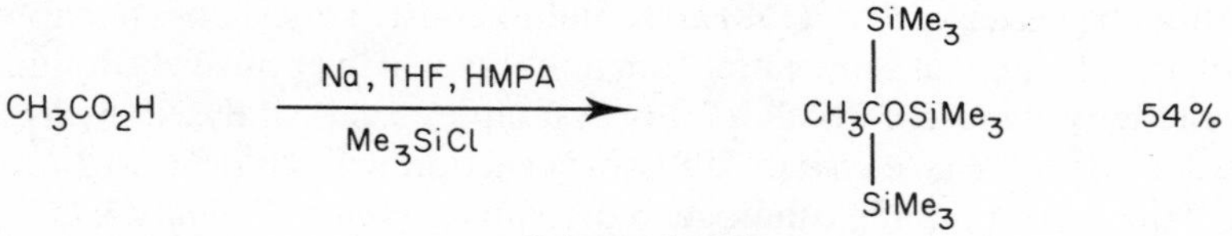

$$CH_3CO_2H \xrightarrow[Me_3SiCl]{Na,\ THF,\ HMPA} CH_3C(SiMe_3)_2OSiMe_3 \quad 54\%$$

Scheme 17.155

17.3.12 1-Trimethylsilyl trimethylsilyl enol ethers

The silyl enol ethers (158) formally derived from acylsilanes have considerable synthetic utility. Until recently, they were rather difficult[234] to prepare, but are now quite readily available by two main routes. 1,1-Bis(trimethylsilyl)alkan-1-ols, the preparation of which has just been discussed, undergo controlled oxidation[235] as their corresponding alkoxides to the requisite enol ethers (*Scheme 17.156*). An alternative, more flexible, method[235] starts from phenyl thiocarboxylic acid esters, and allows the preparation of, *inter alia*, the cyclohexane derivative (159).

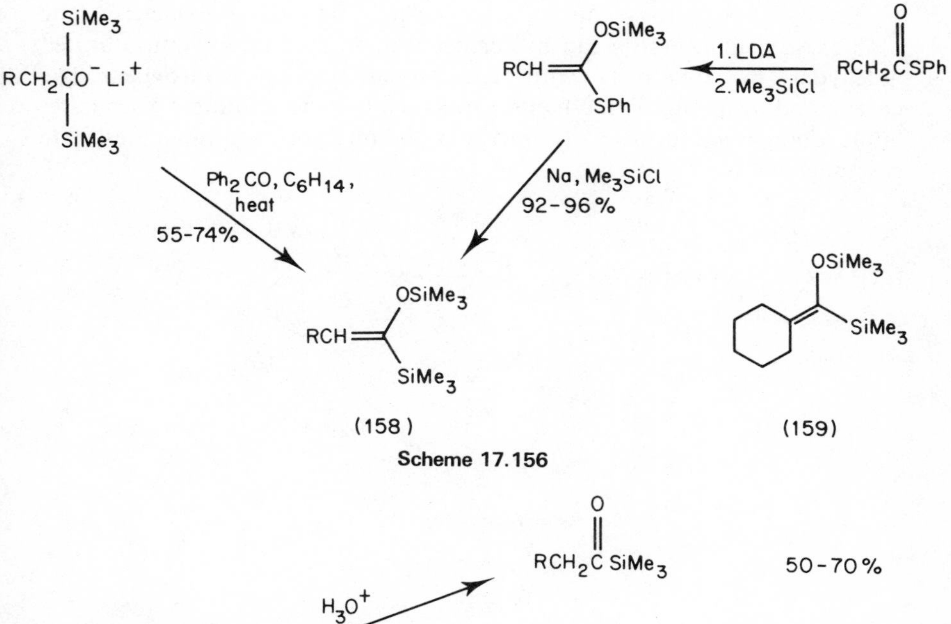

$$RCH_2C(SiMe_3)_2O^- Li^+ \xrightarrow[heat]{Ph_2CO,\ C_6H_{14},} RCH{=}C(OSiMe_3)SiMe_3 \ (158) \quad 55\text{–}74\%$$

$$RCH_2C(O)SPh \xrightarrow[2.\ Me_3SiCl]{1.\ LDA} RCH{=}C(OSiMe_3)SPh \xrightarrow{Na,\ Me_3SiCl} RCH{=}C(OSiMe_3)SiMe_3 \ (158) \quad 92\text{–}96\%$$

$$C_6H_{10}{=}C(OSiMe_3)SiMe_3 \ (159)$$

Scheme 17.156

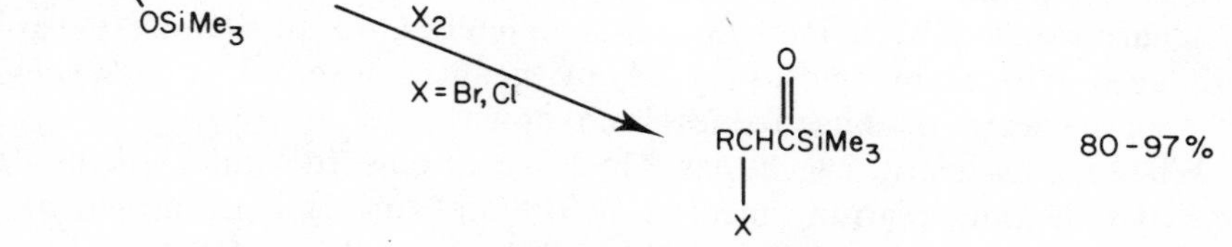

$$RCH{=}C(SiMe_3)OSiMe_3 \xrightarrow{H_3O^+} RCH_2C(O)SiMe_3 \quad 50\text{–}70\%$$

$$RCH{=}C(SiMe_3)OSiMe_3 \xrightarrow[X = Br,\ Cl]{X_2} RCHXC(O)SiMe_3 \ (160) \quad 80\text{–}97\%$$

Scheme 17.157

Although species such as (158) are simultaneously vinylsilanes (Chapter 7) and silyl enol ethers, it is the latter functionality which exclusively dominates the reactivity reported to date. For example, acid-catalysed hydrolysis yields[235] acylsilanes (*see also* p. 208), and reaction with chlorine or bromine gives[236] the corresponding α-halogeno derivatives (160) (*Scheme 17.157*).

Addition of phenyl sulphenyl chloride, followed by oxidative desulphurization, produces[237] $\alpha\beta$-unsaturated acylsilanes (*Scheme 17.158*).

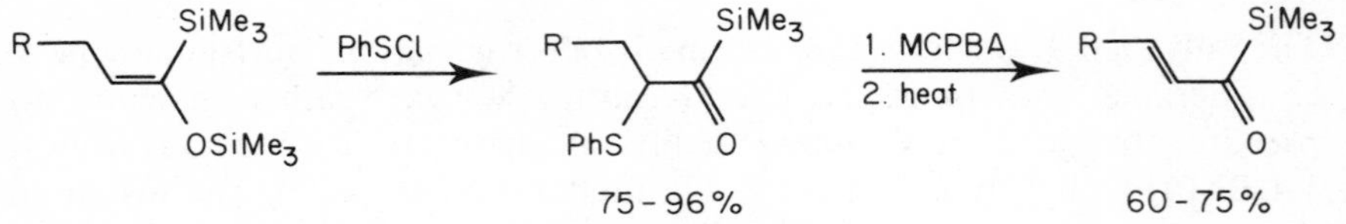

R, $SiMe_3$, $OSiMe_3$ —PhSCl→ R, $SiMe_3$, PhS, O (75-96%) —1. MCPBA 2. heat→ R, $SiMe_3$, O (60-75%)

Scheme 17.158

The parent compounds (158) react[238] with acetals in the presence of boron trifluoride etherate to give β-alkoxyacylsilanes (161) (*Scheme 17.159*); the use of stronger Lewis acids causes an additional process of decarbonylation, producing β-hydroxysilanes and the derived alkenes. The β-alkoxyacylsilanes (161) can be further transformed into $\alpha\beta$-unsaturated aldehydes by treatment with tetrabutylammonium hydroxide. The corresponding ammonium *o*-methoxyphenoxide is more suitable when base-labile products are formed; tetrabutylammonium fluoride causes undesirable rearrangements.

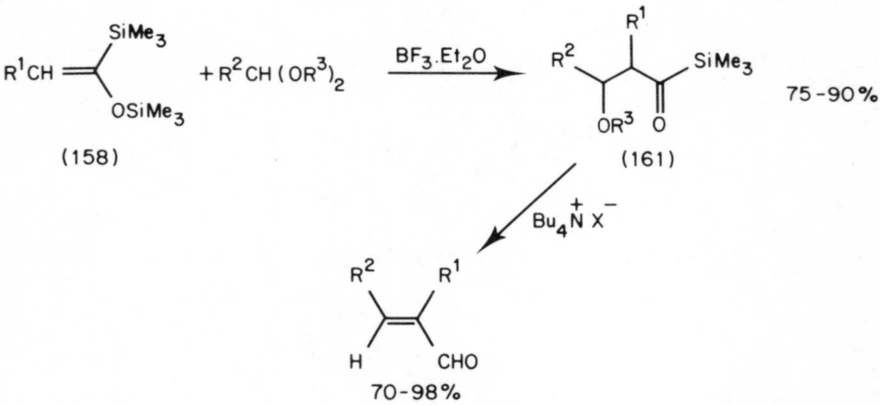

$R^1CH{=}C(SiMe_3)(OSiMe_3)$ (158) + $R^2CH(OR^3)_2$ —$BF_3.Et_2O$→ $R^2CH(OR^3)CH(R^1)COSiMe_3$ (161) 75-90%

(161) —$Bu_4\overset{+}{N}\overset{-}{X}$→ $R^2(H)C{=}C(R^1)CHO$ 70-98%

Scheme 17.159

α-Halogenoacyl silanes react in a most unexpected manner[236] with Grignard reagents. With methylmagnesium halides, β-keto-silanes (162) are produced in excellent yields, probably by the mechanism shown in *Scheme 17.160*; the degree of concertedness is unknown.

When the Grignard reagent has β-hydrogen atoms, the initial reaction is one of reduction; rearrangement as before and further reaction with the Grignard reagent ultimately produces β-hydroxysilanes (163) (*Scheme*

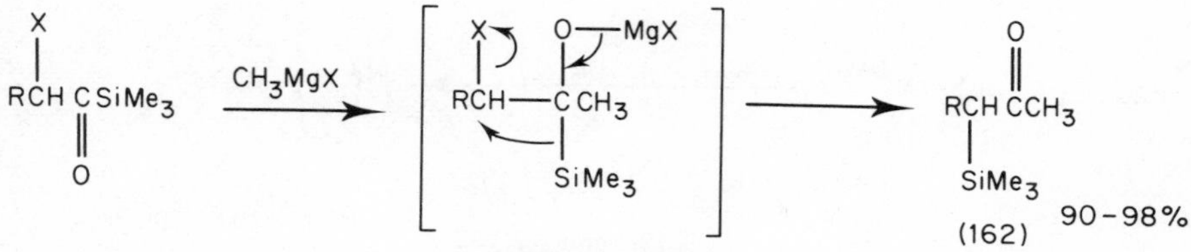

Scheme 17.160

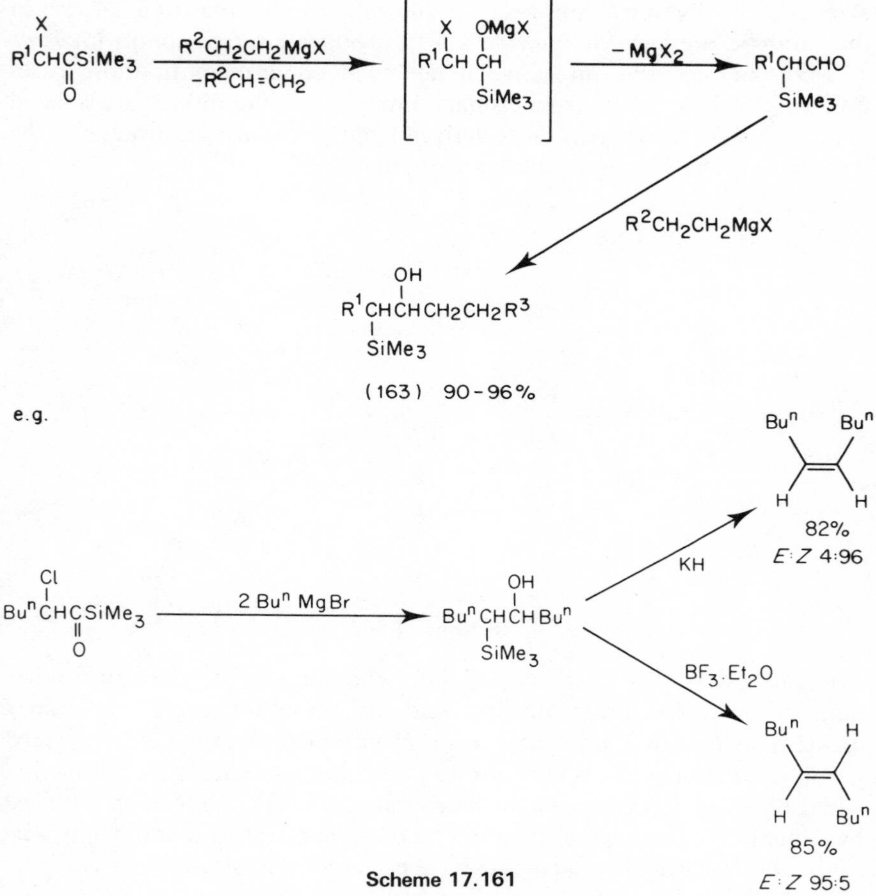

Scheme 17.161

17.161). As predicted by Cram's rule, *erythro* diastereoselectivity is observed in the second Grignard reaction, as demonstrated by the geometric isomer ratios of the derived alkenes (*see* Chapter 12).

17.3.13 Silyl nitronates (silyl esters of *aci*-nitroalkanes)

Although strictly not silyl enol ethers, silyl nitronates (164) will be discussed here for convenience. A variety of methods[239-242] has been delineated for their preparation, the most efficient and flexible[242] being shown in *Scheme 17.162.*

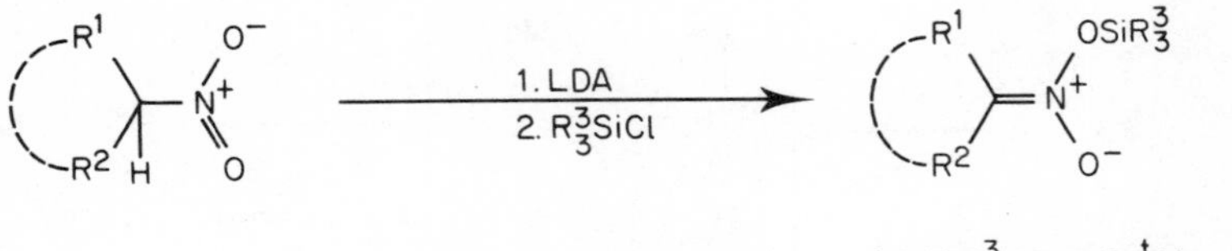

1. LDA
2. R^3_3SiCl

(164) R^3_3=Me$_3$, Bu^{t}Me$_2$

Scheme 17.162

In addition to acting as powerful 1,3-dipolar species in their reactions with alkenes[239,240], silyl nitronates react readily with a wide range of aldehydes in the catalytic presence of fluoride ion to produce 2-nitroalcohols (*Scheme 17.163*); ketones are unreactive under such conditions. In those cases involving silyl nitronates from primary nitroalkanes, the adducts are isolated in protected form, whereas those derived from secondary substrates give the free alcohols, which can be subsequently silylated.

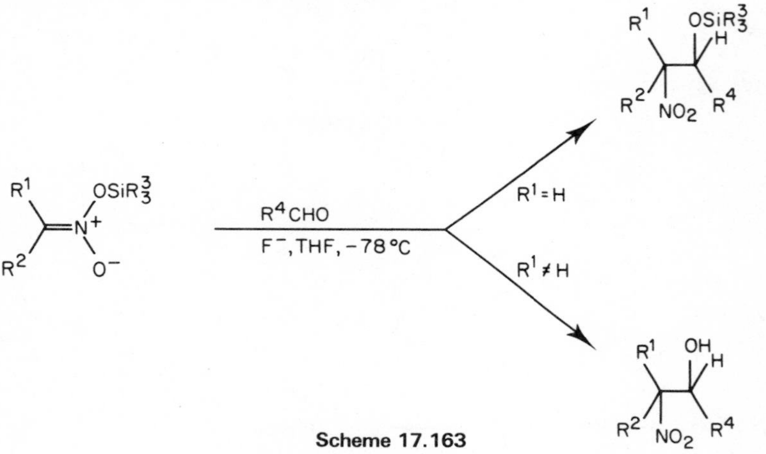

R^4CHO
F^-, THF, −78 °C
R^1 = H
R^1 ≠ H

Scheme 17.163

Regardless of the particular silyl functionality or of the substitution pattern, protected 2-nitroalcohols undergo smooth hydride[242] or Raney nickel reduction to 2-aminoalcohols (*Scheme 17.164*). Apparently the silyl group is not lost prior to nitro-group reduction, since attempted reduction with $LiAlH_4$ of unprotected 2-nitroalcohols results in bond scission followed by reduction of the original components of the substrate. Some other aspects of the chemistry of silyl nitronates have been reported briefly[243].

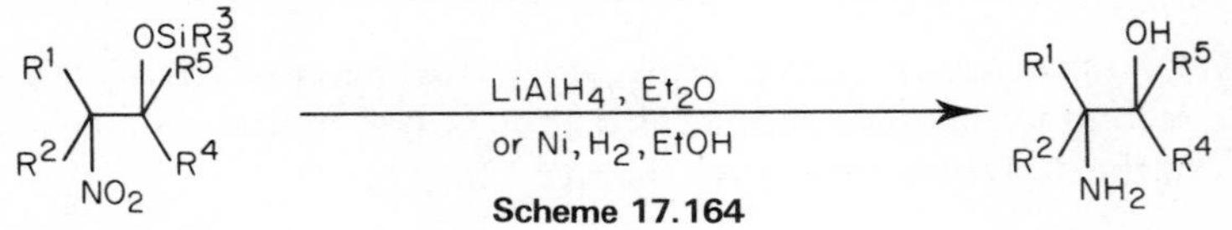

$LiAlH_4$, Et_2O
or Ni, H_2, EtOH

Scheme 17.164

17.4 Addendum

17.4.1 Preparation

Silylation of the adducts from diethylketene and organometallic reagents permits regioselective production[244] of certain silyl enol ethers

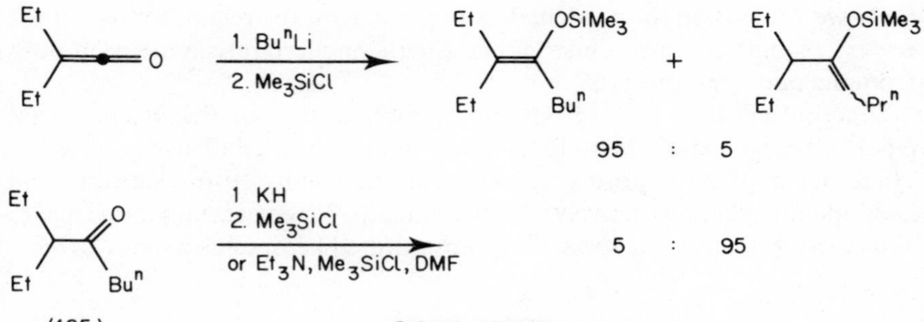

1. Bu^nLi
2. Me_3SiCl

95 : 5

1 KH
2. Me_3SiCl
or Et_3N, Me_3SiCl, DMF

5 : 95

(165)

Scheme 17.165

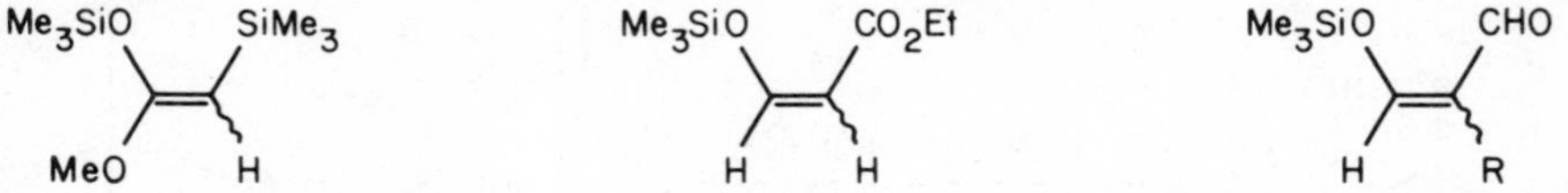

(166) (167) (168)

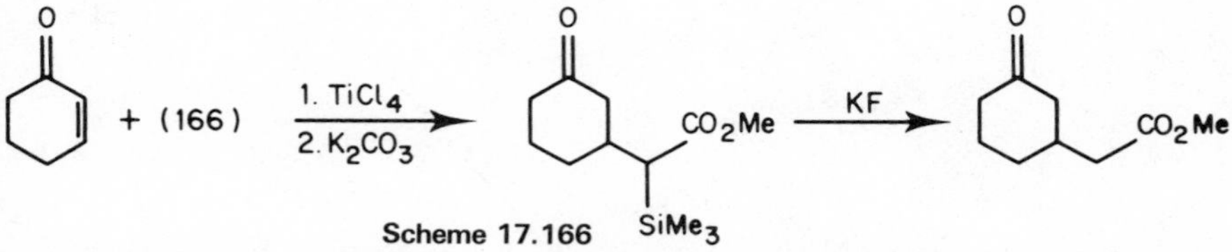

1. $TiCl_4$
2. K_2CO_3

KF

Scheme 17.166

Fe (III)

Ce (IV)

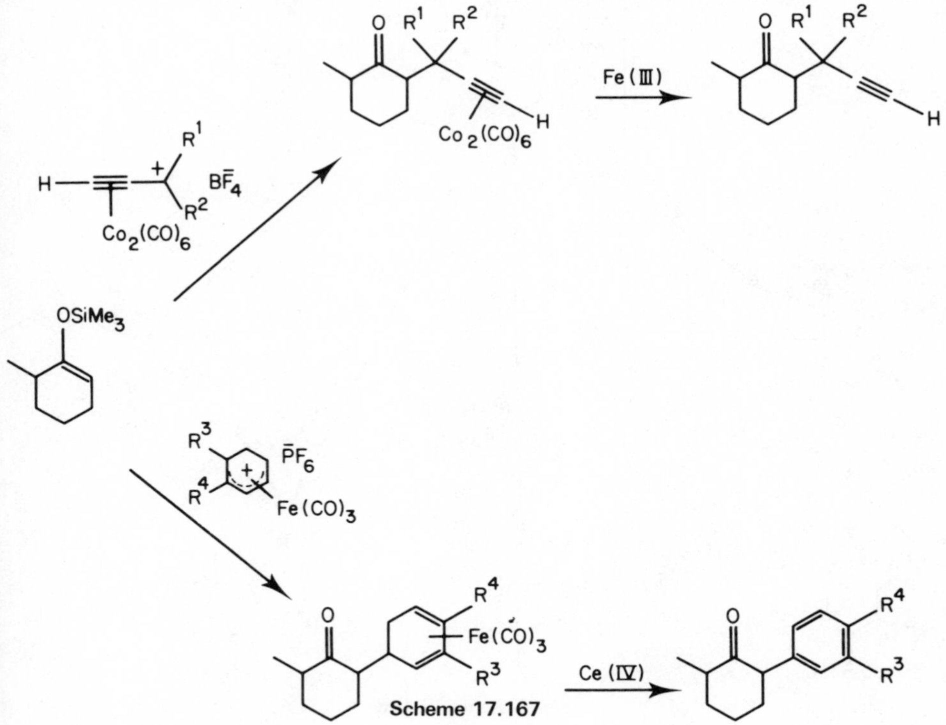

Scheme 17.167

(*Scheme 17.165*); in the particular example shown, the regioselectivity is the reverse of that obtained under either kinetic or thermodynamic conditions from the parent ketone (165).

Descriptions have been given for the preparation of the ketene acetal (166)[245], the silyl enol ether (167)[246], and the silyloxyacraldehydes (168)[247]. These act as masked forms of methyl acetate, ethyl formylacetate, and malondialdehyde, respectively; for example, the ketene acetal (166) behaves as a reactive nucleophile towards $\alpha\beta$-unsaturated ketones (*Scheme 17.166*).

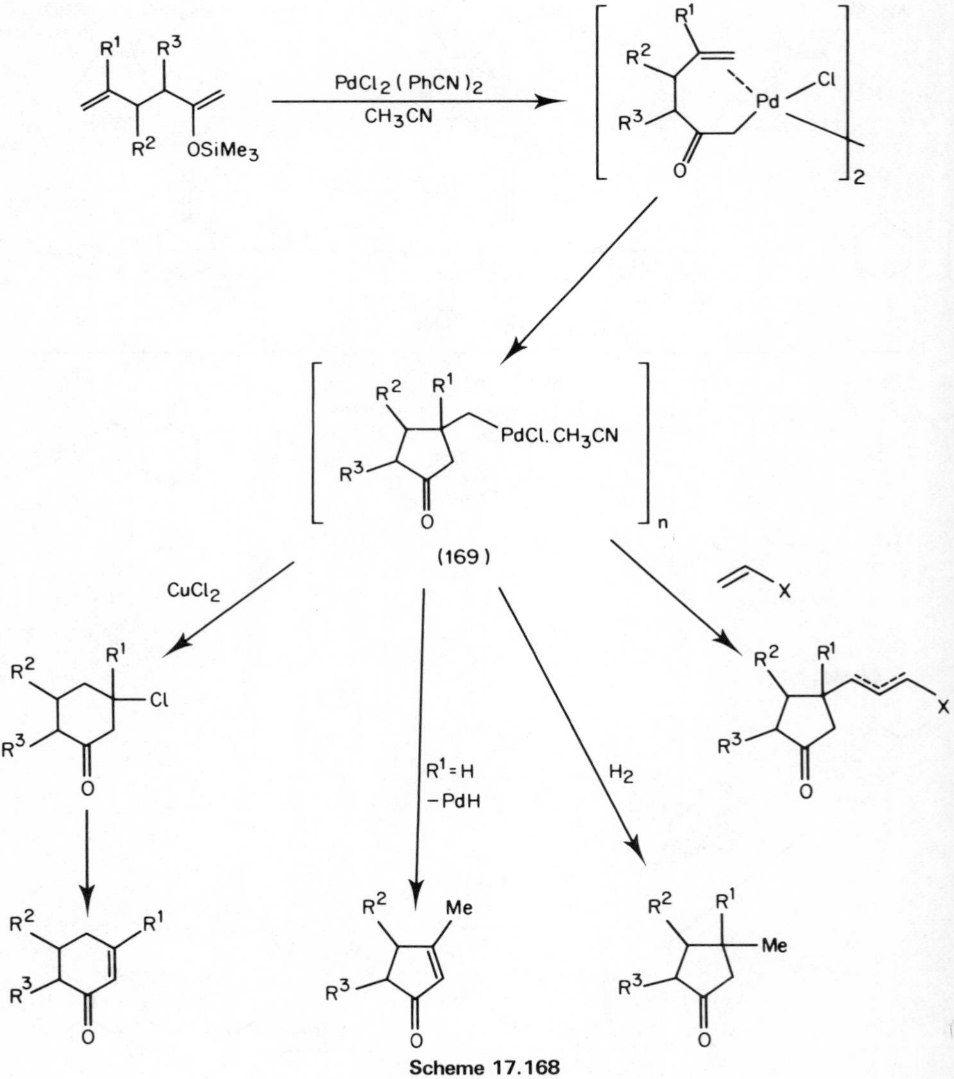

Scheme 17.168

17.4.2 Alkylation

Silyl enol ethers react with complexed carbocations[248, 249] as shown in *Scheme 17.167*; these reactions exemplify further the generalization that silyl enol ethers are comparable in reactivity with allylsilanes (p. 122).

γ-Unsaturated silyl enol ethers undergo a remarkable palladium(II)-induced cyclization[250]; appropriate modification of the intermediate complex (169) leads to functionalized cyclohexanone or cyclopentanone ring systems (*Scheme 17.168*).

17.4.3 Cycloaddition

Photo-induced [2 + 2]-cycloaddition has been employed in a facile route (*Scheme 17.169*) to angularly-substituted *cis*-hydrindanones and *cis*-decalones[251].

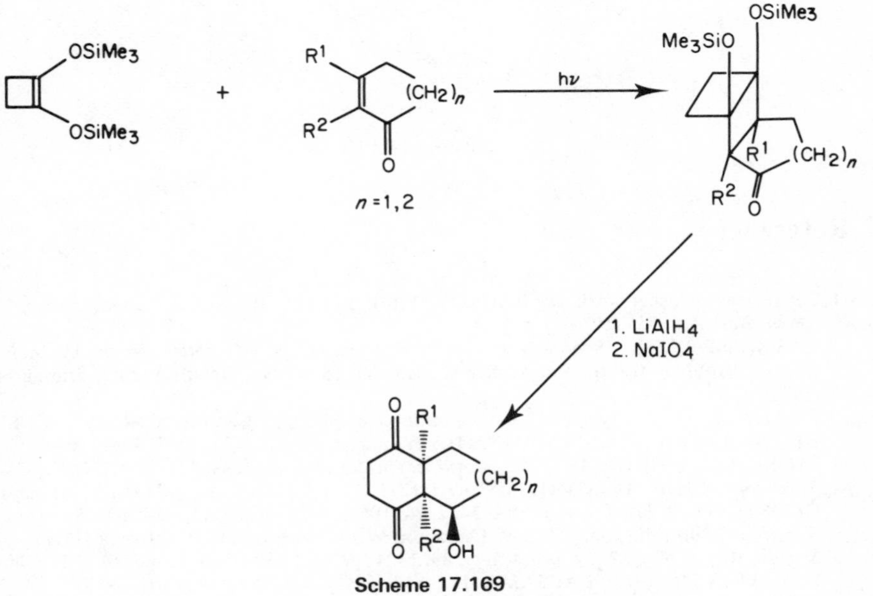

OSiMe3 OSiMe3 + R1 R2 (CH2)n O n = 1,2 hν Me3SiO OSiMe3 R1 (CH2)n R2 O

1. LiAlH4
2. NaIO4

O R1 (CH2)n O R2 OH

Scheme 17.169

The silyloxybutadienes (170)[252] and (171)[253] have found further use in natural product synthesis, acting as Diels–Alder dienes. An ingenious sequence of 'timed' Diels–Alder cycloadditions[254] which leads to the regiospecific formation of a tricyclic framework has been described. The silyloxydiene in the bis-diene (172) is the more reactive diene, as is the enone unit in the unsaturated ketone (173) the more reactive dienophile (*Scheme 17.170*).

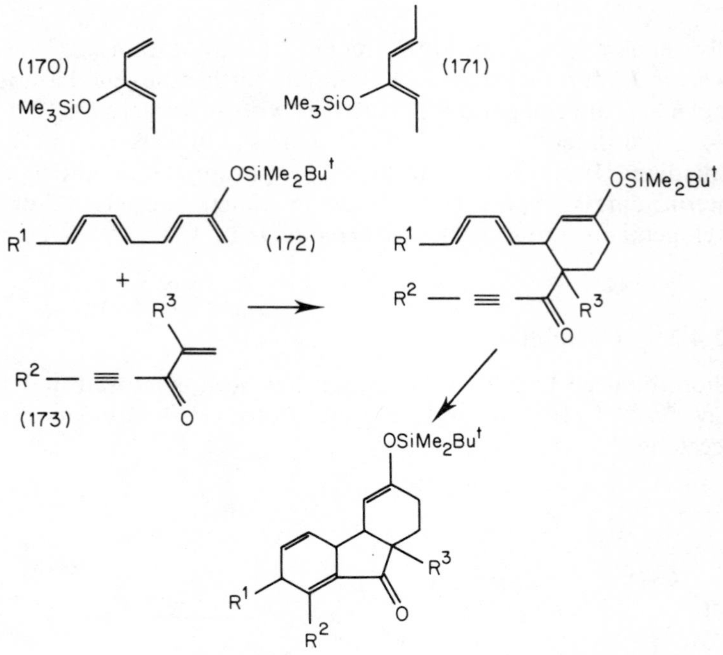

Scheme 17.170

References

1 For review of earlier work, *see* BAUKOV, Yu. I. and LUTSENKO, I. F., *Organometal. Chem. Rev. A* **6,** 355 (1970)

2 RASMUSSEN, J. K., *Synthesis* 91 (1977); HESSE, G., in *'Methoden der organischen Chemie (Houben–Weyl)'*, Ed. Müller E., vol. VI/1d, Georg Thieme Verlag, Stuttgart (1978)

3 CHAN, T. H., *'Synthesis in Organic Chemistry',* Cambridge (July 1979); BROWNBRIDGE, P. and CHAN, T. H., *Tetrahedron Lett.* 3423, 3427, 3431 (1980)

4 STORK, G. and HUDRLIK, P. F., *J. Am. chem. Soc.* **90,** 4462, 4464 (1968); STORK, G., *Pure appl. Chem.* **43,** 553 (1975); HOUSE, H. O., CZUBA, L. J., GALL, M. and OLMSTEAD, H. D., *J. org. Chem.* **34,** 2324 (1969); HOUSE, H. O., *'Modern Synthetic Reactions',* 2nd Edn., pp. 568–569, Benjamin, W. A., Menlo Park, California (1972)

5 *See also* BROWN, C. A., *J. org. Chem.* **39,** 1324 (1974); FLEMING, I. and PATERSON, I., *Synthesis* 736 (1979); LADJAMA, D. and RIEHL, J. J., *Synthesis* 504 (1979)

6 STORK, G. and SINGH, J., *J. Am. chem. Soc.* **96,** 6181 (1974); STORK, G. and d'ANGELO, J., *J. Am. chem. Soc.* **96,** 7114 (1974)

7 BOECKMAN, R. K., *J. Am. chem. Soc.* **96,** 6179 (1974); for other examples *see* PATTERSON, J. W. and FRIED, J. H., *J. org. Chem.* **39,** 2506 (1974); STOTTER, P. L. and HILL, K. A., *J. org. Chem.* **38,** 2576 (1973)

8 SAMSON, M. and VANDEWALLE, M., *Synth. Communs* **8,** 231 (1978); UTIMOTO, K., OBAYASHI, M., SHISHIYAMA, Y., INOUE, M. and NOZAKI, H., *Tetrahedron Lett.* 3389 (1980)

9 OJIMA, I., KOGURE, R. and NAGAI, Y., *Tetrahedron Lett.* 5035 (1972)

10 KAGAN, H. B., *Pure appl. Chem.* **43,** 401 (1975); HAYASHI, T., YAMAMOTO, K. and KUMADA, M., *Tetrahedron Lett.* 3 (1975)

11 OJIMA, I., NIHONYANAGI, M. and NAGAI, Y., *J. chem. Soc. chem. Communs* 938 (1972)

12 SAKURAI, H., MIYOSHI, K. and NAKADAIRA, Y., *Tetrahedron Lett.* 2671 (1977)

13 OJIMA, I. and NAGAI, Y., *J. Organometal. Chem.* **57,** C42 (1973)
14 SUZUKI, H., KOYAMA, Y., MORO-OKA, Y. and IKAWA, T., *Tetrahedron Lett.* 1415 (1979); for an extension to propargylic alcohol silyl ethers *see* HIRAI, K., SUZUKI, H., MORO-OKA, Y. and IKAWA, T., *Tetrahedron Lett.* 3413 (1980)
15 SEKI, Y., HIDAKA, A., MURAI, S. and SONODA, N., *Angew. Chem. int. Edn* **16,** 174 (1977); MURAI, S. and SONODA, N., *Angew. Chem. int. Edn* **18,** 837 (1979)
16 NAKAMURA, E., MUROFUSHI, T., SHIMUZU, M. and KUWAJIMA, I., *J. Am. chem. Soc.* **98,** 2346 (1976); NAKAMURA, E., HASHIMOTO, K. and KUWAJIMA, I., *Tetrahedron Lett.* 2079 (1978);
17 FESSENDEN, R. J. and FESSENDEN, J. S., *J. org. Chem.* **32,** 3535 (1967)
18 NAKAMURA, E., SHIMUZA, M. and KUWAJIMA, I., *Tetrahedron Lett.* 1699 (1976)
19 HEATHCOCK, C. H., PIRRUNG, M. C., BUSE, C. T., HAGEN, J. P., YOUNG, S. D. and SOHN, J. E., *J. Am. chem. Soc.* **101,** 7007 (1979); HEATHCOCK, C. H., BUSE, C. T., KLESCHICK, W. A., PIRRUNG, M. C., SOHN, J. E. and LAMPE, J., *J. org. Chem.* **45,** 1066 (1980) and references therein; BARTLETT, P. A., *Tetrahedron* **36,** 2 (1980)
20 (*a*) IRELAND, R. E., MUELLER, R. H. and WILLARD, A. K., *J. Am. chem. Soc.* **98,** 2868 (1976); (*b*) FATAFTAH, Z. A., KOPKA, I. E. and RATHKE, M. W., *J. Am. chem. Soc.* **102,** 3959 (1980)
21 RUBOTTOM, G. M., MOTT, R. C. and KRUEGER, D. S., *Synth. Communs* **7,** 327 (1977)
22 JOSHI, G. C. and PANDE, L. M., *Synthesis* 450 (1975)
23 STORK, G. and ISOBE, M., *J. Am. chem. Soc.* **97,** 4745 (1975)
24 HASHIMOTO, S., ITOH, A., KITAGAWA, Y., YAMAMOTO, H. and NOZAKI, H., *J. Am. chem. Soc.* **99,** 4192 (1977)
25 (*a*) POSNER, G. H., STERLING, J. J., WHITTEN, C. E., LENTZ, C. M. and BRUNELLE, D. J., *J. Am. chem. Soc.* **97,** 107 (1975); (*b*) SAKURAI, H., SHIRAHATA, A., ARAKI, Y. and HOSOMI, A., *Tetrahedron Lett.* 2325 (1980)
26 COATES, R. M., SANDEFUR, L. O. and SMILLIE, R. D., *J. Am. chem. Soc.* **97,** 1619 (1975); BLOCH, R., BOIVIN, F. and BORTOLUSSI, M., *J. chem. Soc. chem. Communs* 371 (1976)
27 SLUTSKY, J. and KWART, H., *J. Am. chem. Soc.* **95,** 8678 (1973)
28 KUO, Y.-N., YAHNER, J. A. and AINSWORTH, C., *J. Am. chem. Soc.* **93,** 6321 (1971)
29 DONALDSON, R. E. and FUCHS, P. L., *J. org. Chem.* **42,** 2032 (1977)
30 SIMCHEN, G. and KOBER, W., *Synthesis* 259 (1976); EMDE, H. and SIMCHEN, G., *Synthesis* 867 (1977)
31 TORKELSON, S. and AINSWORTH, C., *Synthesis* 431 (1977); 722 (1976)
32 *See* however, HENGGE, E. and PLETKA, H.-D., *Mh. Chem.* **104,** 1071 (1973)
33 BROOK, A. G., *J. Am. chem. Soc.* **80,** 1886 (1953); *Accts chem. Res.* **7,** 77 (1974); WRIGHT, A. and WEST, R., *J. Am. chem. Soc.* **96,** 3214 (1974); WEST, R., *Adv. organometal. Chem.* **16,** 1 (1977)
34 REICH, H. J., RUSEK, J. J. and OLSON, R. E., *J. Am. chem. Soc.* **101,** 2225 (1979)
35 (*a*) KUWAJIMA, I. and KATO, M., *J. chem. Soc. chem. Communs* 708 (1979); KUWAJIMA, I., KATO, M. and MORI, A., *Tetrahedron Lett.* 2745, 4291 (1980); (*b*) KUWAJIMA, I. and KATO, M., *Tetrahedron Lett.* 623 (1980); (*c*) REICH, H. J., OLSON, R. E. and CLARK, M. C., *J. Am. chem. Soc.* **102,** 1423 (1980)
36 BROOKS, A. G., DUFF, J. M., JONES, P. F. and DAVIS, N. R., *J. Am. chem. Soc.* **89,** 431 (1967); COREY, E. J., SEEBACH, D. and FREEDMAN, R., *J. Am. chem. Soc.* **89,** 434 (1967)
37 KUWAJIMA, I., ABE, R. and MINAMI, N., *Chemy Lett.* 993 (1976)
38 (*a*) HASSNER, A. and SODERQUIST, J. A., *J. organometal. Chem.* **131,** C1 (1977); (*b*) SODERQUIST, J. A. and HASSNER, A., *J. Am. chem. Soc.* **102,** 1577 (1980); CLINET, J.-C. and LINSTRUMELLE, G., *Tetrahedron Lett.* 3987 (1980); (*c*) YAMAMOTO, K., SUZUKI, S. and TSUJI, J., *Tetrahedron Lett.* 1653 (1980); (*d*) *see also* DEXHEIMER, E. M. and SPIALTER, L., *J. organometal. Chem.* **107,** 229 (1976); PEDDLE, G. J. D., *J. organometal. Chem.* **14,** 139 (1968); BOURGEOIS, P., *J. organometal. Chem.* **76,** C1 (1976); DUNOGUÈS, J., BOLOURTCHIAN, M., CALAS, R., DUFFAUT, N. and PICARD, J.-P., *J. organometal. Chem.* **43,** 157 (1972); BOURGEOIS, P., DUNOGUÈS, J. and DUFFAUT, N., *J. organometal. Chem.* **80,** C25 (1974)
39 STILL, W. C., *J. org. Chem.* **41,** 3063 (1976)
40 HOSOMI, A., HASHIMOTO, H. and SAKURAI, H., *J. org. Chem.* **43,** 2551 (1978)
41 STILL, W. C. and MACDONALD, T. L., *J. Am. chem. Soc.* **96,** 5561 (1974); *see also* WRIGHT, A., LING, D., BOUDJOUK, P. and WEST, R., *J. Am. chem. Soc.* **94,** 4784 (1972)

42 For the corresponding alkyl ethers, *see* EVANS, D. A., ANDREWS, G. C. and BUCKWALTER, B., *J. Am. chem. Soc.* **96,** 5560 (1974)
43 STILL, W. C. and MACDONALD, T. L., *J. org. Chem.* **41,** 3620 (1976)
44 OPPOLZER, W. and SNOWDEN, R. L., *Tetrahedron Lett.* 4187 (1976)
45 OPPOLZER, W. and SNOWDEN, R. L., *Tetrahedron Lett.* 3505 (1978)
46 BASSINDALE, A. R., BROOK, A. G. and HARRIS, J., *J. organometal. Chem.* **90,** C6 (1975)
47 BROOK, A. G., MacRAE, D. M. and BASSINDALE, A. R., *J. organometal. Chem.* **86,** 185 (1975); *see also* LITVINOVA, O. V., BAUKOV, Yu. I. and LUTSENKO, I. F., *Dokl. Akad. Nauk SSSR* **173,** 578 (1967); *Chem. Abstr.* **67,** 32720 (1967), for a catalysed bimolecular version
48 BASSINDALE, A. R., BROOK, A. G., CHEN, P. and LENNON, J., *J. organometal. Chem.* **94,** C21 (1975); HUDRLIK, P. F., WAN, C.-N. and WITHERS, G. P., *Tetrahedron Lett.* 1449 (1976); HUDRLIK, P. F., SCHWARTZ, R. H. and KULKARNI, A. K., *Tetrahedron Lett.* 2233 (1979)
49 GARST, M. E., BONFIGLIO, J. N., GRUDOSKI, D. A. and MARKS, J., *Tetrahedron Lett.* 2671 (1978); *J. org. Chem.* **45,** 2307 (1980)
50 BATES, R. B., KROPOSKI, L. M. and POTTER, D. E., *J. org. Chem.* **37,** 560 (1972)
51 JUNG, M. E. and BLUM, R. B., *Tetrahedron Lett.* 3791 (1977)
52 BAUKOV, Yu. I., BURLACHENKO, G. S. and LUTSENKO, I. F., *J. organometal. Chem.* **3,** 478 (1965)
53 KUO, Y.-N., CHEN, F., AINSWORTH, C. and BLOOMFIELD, J. J., *J. chem. Soc. chem. Communs* 136 (1971); KUO, Y.-N., CHEN, F. and AINSWORTH, C., *J. chem. Soc. chem. Communs* 137 (1971); AINSWORTH, C., CHEN, F. and KUO, Y.-N., *J. Organometal. Chem.* **46,** 59 (1972); AINSWORTH, C. and KUO, Y.-N., *J. organometal. Chem.* **46,** 73 (1972)
54 RATHKE, M. W. and SULLIVAN, D. F., *Synth. Communs* **3,** 67 (1973)
55 RASMUSSEN, J. K. and HASSNER, A., *J. org. Chem.* **39,** 2558 (1974)
56 HOLY, N. L. and WANG, Y. F., *J. Am. chem. Soc.* **99,** 944 (1977)
57 PATERSON, I. and FLEMING, I., *Tetrahedron Lett.* 993 (1979)
58 CHEN, F. and AINSWORTH, C., *J. Am. chem. Soc.* **94,** 4037 (1972)
59 YOSHII, E., KOBAYASHI, Y., KOIZUMI, T. and ORIBE, T., *Chem. pharm. Bull. Japan* **22,** 2767 (1974)
60 DANISHEFSKY, S., KITAHAWA, T., McKEE, R. and SCHUDA, P. F., *J. Am. chem. Soc.* **98,** 6715 (1976); DANISHEFSKY, S., PRISBYLLA, M. and LIPISKO, B., *Tetrahedron Lett.* 805 (1980); *see also* DANISHEFSKY, S., GUINGANT, A. and PRISBYLLA, M., *Tetrahedron Lett.* 2033 (1980)
61 For an intramolecular example, *see* OPPOLZER, W., HAUTH, H., PFÄFFLI, P. and WENGER, R., *Helv. chim. Acta* **60,** 1801 (1977)
62 OJIMA, I., INABA, S. and NAGAI, Y., *Tetrahedron Lett.* 4271 (1973); *Chemy Lett.* 1069 (1974)
63 MURAI, S., KUROKI, Y., HASEGAWA, K. and TSUTSUMI, S., *J. chem. Soc. chem. Communs* 946 (1972); MURAI, S., KUROKI, Y., AYA, T., SONODA, N. and TSUTSUMI, S., *Angew. Chem. int. Edn* **14,** 741 (1975)
64 (*a*) BURLACHENKO, G. S., MAL'TSEV, V. V., BAUKOV, Yu. I. and LUTSENKO, I. F., *J. gen. Chem. U.S.S.R.* **43,** 1708 (1973); STACHEL, H. D. and DANDL, K., *Tetrahedron Lett.* 2891 (1980); (*b*) RATHKE, M. W. and SULLIVAN, D. F., *Synth. Communs* **3,** 67 (1973); *Tetrahedron Lett.* 1297 (1973)
65 SCHMIDT, U. and SCHWOCHAU, M., *Tetrahedron Lett.* 4491 (1967); *Mh. Chem.* **98,** 1492 (1967)
66 FRIEDRICH, E. and LUTZ, W., *Angew. Chem. int. Edn* **16,** 413 (1977)
67 EVANS, D. A., HART, D. J. and KOELSCH, P. M., *J. Am. chem. Soc.* **100,** 4593 (1978)
68 JUNG, M. E., PAN, Y.-G., RATHKE, M. W., SULLIVAN, D. F. and WOODBURY, R. P., *J. org. Chem.* **42,** 3961 (1977)
69 RYU, I., MURAI, S., HATAYAMA, Y. and SONODA, N., *Tetrahedron Lett.* 3455 (1978); LOTT, R. S., BREITHOLLE, E. G. and STAMMER, C. H., *J. org. Chem.* **45,** 1151 (1980)
70 ITO, Y., HIRAO, T. and SAEGUSA, T., *J. org. Chem.* **43,** 1011 (1978); *see also* ITO, Y., NAKATSUKA, M., KISE, N. and SAEGUSA, T., *Tetrahedron Lett.* 2873 (1980)
71 KUROKI, Y., MURAI, S., SONODA, N. and TSUTSUMI, S., *Organometal. chem. Synth.* **1,** 465 (1972)

72 REUSS, R. H. and HASSNER, A., *J. org. Chem.* **39,** 1785 (1974); BLANCO, L., AMICE, P. and CONIA, J. M., *Synthesis* 194 (1976); LAZUKINA, L. A., KUKHAR, V. P. and PESOTSKAYA, G. V., *J. gen. Chem. U.S.S.R.* **45,** 2065 (1975); ITOH, Y., NAKATSUKA, M. and SAEGUSA, T., *J. org. Chem.* **45,** 2022 (1980)

73 RUBOTTOM, G. M. and MOTT, R. C., *J. org. Chem.* **44,** 1731 (1979); TORII, S., INOKUCHI, T., MISIMA, S. and KOBAYASHI, T., *J. org. Chem.* **45,** 2731 (1980)

74 ZEMBAYASHI, M., TAMAO, K. and KUMADA, M., *Synthesis* 422 (1977)

75 BROOK, A. G. and MACRAE, D. A., *J. organometal. Chem.* **77,** C19 (1974); RUBOTTOM, G. M., VASQUEZ, M. A. and PELLEGRINA, D. R., *Tetrahedron Lett.* 4319 (1974); *see,* for example, PENNANEN, S. I., *Tetrahedron Lett.* 657 (1980)

76 RUBOTTOM, G. M. and MARRERO, R., *J. org. Chem.* **40,** 3783 (1975)

77 MAUME, G. M. and HORNING, E. C., *Tetrahedron Lett.* 343 (1969)

78 RUBOTTOM, G. M., GRUBER, J. M. and MONG, G. M., *J. org. Chem.* **41,** 1673 (1976); *see also* RUBOTTOM, G. M., GRUBER, J. M. and KINCAID, K., *Synth. Communs* **6,** 59 (1976)

79 HASSNER, A., REUSS, R. H. and PINNICK, H. W., *J. org. Chem.* **40,** 3427 (1975)

80 RUBOTTOM, G. M. and NIEVES, M. I. L., *Tetrahedron Lett.* 2423 (1972)

81 JEFFORD, C. W. and RIMBAULT, C. G., *J. Am. chem. Soc.* **100,** 6515 (1978)

82 ADAM, W., ALZÉRRECA, A., LIU, J.-C. and YANY, F., *J. Am. chem. Soc.* **99,** 5768 (1977)

83 HOUSE, H. O., *Rec. chem. Prog.* **28,** 99 (1967)

84 STORK, G. and HUDRLIK, P. F., *J. Am. chem. Soc.* **90,** 4464 (1968)

85 HOUSE, H. O., GALL, M. and OLMSTEAD, H. D., *J. org. Chem.* **36,** 2361 (1971)

86 ULISS, D. B., HANDRICK, G. R., DALZELL, H. C. and RAZDAN, R. K., *J. Am. chem. Soc.* **100,** 2929 (1978) but *see* HAYASHI, T., KATSURO, Y. and KUMADA, M., *Tetrahedron Lett.* 3915 (1980)

87 KUWAJIMA, I. and NAKAMURA, E., *J. Am. chem. Soc.* **97,** 3257 (1975)

88 BOROWITZ, I. J., CASPER, E. W. R., CROUCH, R. K. and YEE, K. C., *J. org. Chem.* **37,** 3873 (1972)

89 BINKLEY, E. S. and HEATHCOCK, C. H., *J. org. Chem.* **40,** 2156 (1975)

90 NOYORI, R., YOKOYAMA, K., SAKATA, J., KUWAJIMA, I., NAKAMURA, E. and SHIMUZU, M., *J. Am. chem. Soc.* **99,** 1265 (1977)

91 NOYORI, R., NISHIDA, I., SAKATA, J. and NISHIZAWA, M., *J. Am. chem. Soc.* **102,** 1223 (1980); NOYORI, R., NISHIDA, I. and SAKATA, J., *Tetrahedron Lett.* 2085 (1980); *see also* OLOFSON, R. A. and CUOMO, J., *Tetrahedron Lett.* 819 (1980)

92 (*a*) CHAN, T. H., PATTERSON, I. and PINSONNAULT, J., *Tetrahedron Lett.* 4183 (1977); REETZ, M. T. and MAIER, W. F., *Angew. Chem. int. Edn* **17,** 48 (1978); REETZ, M. T. and SCHWELLNUS, K., *Tetrahedron Lett.* 1455 (1978); REETZ, M. T., CHATZIIOSIFIDIS, I., LÖWE, U. and MAIER, W. F., *Tetrahedron Lett.* 1427 (1979); (*b*) REETZ, M. T., MAIER, W. F., SCHWELLNUS, K. and CHATZIIOSIFIDIS, I., *Angew. Chem. int. Edn* **18,** 72 (1979)

93 PATERSON, I., *Tetrahedron Lett.* 1519 (1979)

94 YAMAMOTO, H. and NOZAKI, H., *Angew. Chem. int. Edn* **17,** 169 (1978)

95 MURAI, S., KUROKI, Y., AYA, T., SONODA, N. and TSUTSUMI, S., *J. chem. Soc. chem. Communs* 741 (1972)

96 MIYASHITA, M., TANAMI, T. and YOSHIKOSHI, A., *J. Am. chem. Soc.* **98,** 4679 (1976); *Org. Synth.* **59,** procedure 2123 (1980)

97 For a good review, *see* GRIECO, P. A., *Synthesis* 67 (1975)

98 PATERSON, I. and FLEMING, I., *Tetrahedron Lett.* 993, 995 (1979)

99 *See,* for example, REICH, H. J. and RENGA, J. M., *J. chem. Soc. chem. Communs* 135 (1974); TROST, B. M. and SALZMANN, T. N., *J. Am. chem. Soc.* **95,** 5321 (1973)

100 STORK, G., ROSEN, P., COOMBS, R. V. and TSUJI, J., *J. Am. chem. Soc.* **87,** 275 (1965)

101 PATERSON, I. and FLEMING, I., *Tetrahedron Lett.* 2179 (1979)

102 NIELSEN, A. T. and HOULIHAN, W. J., *Org. React.* **16,** 1 (1968)

103 *See,* for example, COREY, E. J., ENDERS, D. and BOCK, M. G., *Tetrahedron Lett.* 7 (1976); COREY, E. J. and ENDERS, D., *Tetrahedron Lett.* 3, 11 (1976); WITTIG, G. and REIFF, H., *Angew. Chem. int. Edn* **7,** 7 (1968); STORK, G. and ISOBE, M., *J. Am. chem. Soc.* **97,** 4745 (1975); HOUSE, H. O., CRUMRINE, D. S., TERANISHI, A. Y. and OLMSTEAD, H. D., *J. Am. chem. Soc.* **95,** 3310 (1973); MUKAIYAMA, T., INOMATA, K. and MURAKI, M., *J. Am. chem. Soc.* **95,** 967 (1973)

104 MUKAIYAMA, T., *Angew. Chem. int. Edn* **16,** 817 (1977)
105 MUKAIYAMA, T., NARASAKA, K. and BANNO, K., *Chemy Lett.* 1011 (1973); *J. Am. chem. Soc.* **96,** 7503 (1974); BANNO, K. and MUKAIYAMA, T., *Chemy Lett.* 279 (1976). For the same reaction using *catalytic* amounts of trimethylsilyl trifluoromethanesulphonate, *see* MURATA, S., SUZUKI, M. and NOYORI, R., *J. Am. chem. Soc.* **102,** 3248 (1980); *Tetrahedron Lett.* 2527 (1980)
106 OKIMA, I., YOSHIDA, K. and INABA, S., *Chemy Lett.* 429 (1977)
107 RATHKE, M. W., *Org. React.* **22,** 423 (1975)
108 CREGER, P. L., *Tetrahedron Lett.* 79 (1972)
109 SAIGO, K., OSAKI, M. and MUKAIYAMA, T., *Chemy Lett.* 989 (1975)
110 MUKAIYAMA, T. and HAYASHI, M., *Chemy Lett.* 15 (1974)
111 SAIGO, K., OSAKI, M. and MUKAIYAMA, T., *Chemy Lett.* 769 (1976)
112 NAKAMURA, E. and KUWAJIMA, I., *J. Am. chem. Soc.* **99,** 961 (1977); for applications, *see* ANDERSON, W. K. and LEE, G. E., *J. org. Chem.* **45,** 501 (1980); OPPOLZER, W. and WYLIE, R. D., *Helv. chim. Acta* **63,** 1198 (1980)
113 NAKAMURA, E., HASHIMOTO, K. and KUWAJIMA, I., *J. org. Chem.* **42,** 4166 (1977)
114 KUWAJIMA, I. and AZEGAMI, I., *Tetrahedron Lett.* 2369 (1979)
115 INOUE, T. and KUWAJIMA, I., *J. chem. Soc. chem. Communs* 251 (1980)
116 BERGMANN, E. D., GINSBURG, D. and PAPPO, R., *Org. React.* **10,** 179 (1959)
117 NARASAKA, K., SOAI, K. and MUKAIYAMA, T., *Chemy Lett.* 1223 (1974); SAIGO, K., OSAKI, M. and MUKAIYAMA, T., *Chemy Lett.* 163 (1976); NARASAKA, K., SOAI, K., AIKAWA, Y. and MUKAIYAMA, T., *Bull. chem. Soc. Japan* **49,** 779 (1976); *see also* DANISHEFSKY, S., VAUGHAN, K., GADWOOD, R. C. and TSUZUKI, K., *J. Am. chem. Soc.* **101,** 4262 (1979); SCHULTZ, A. G. and GODFREY, J. D., *J. Am. chem. Soc.* **102,** 2414 (1980); but *see* KITA, Y., SEGAWA, J., HARUTA, J., FUJII, T. and TAMURA, Y., *Tetrahedron Lett.* 3779 (1980)
118 CASINOS, I. and MESTRES, R., *J. chem. Soc. Perkin I* 1651 (1978) and references therein
119 ISHIDA, A. and MUKAIYAMA, T., *Bull. chem. Soc. Japan* **51,** 2077 (1978); **50,** 1161 (1977); MUKAIYAMA, T. and ISHIDA, A., *Chemy Lett.* 1201 (1975); HAYASHI, Y., NISHIZAWA, M. and SAKAN, T., *Chemy Lett.* 387 (1975); ISHIDA, A. and MUKAIYAMA, T., *Chemy Lett.* 1127 (1976)
120 FLEMING, I., GOLDHILL, J. and PATERSON, I., *Tetrahedron Lett.* 3209 (1979)
121 FLEMING, I., GOLDHILL, J. and PATERSON, I., *Tetrahedron Lett.* 3205 (1979)
122 YAMAMOTO, K., SUZUKI, S. and TSUJI, J., *Bull. chem. Soc. Japan* 649 (1978)
123 CHAN, T. H. and BROWNBRIDGE, P., *J. chem. Soc. chem. Communs* 578 (1979)
124 HUCKIN, S. N. and WEILER, L., *J. Am. chem. Soc.* **96,** 1082 (1974)
125 MURAI, S., HASEGAWA, K. and SONODA, N., *Angew. Chem. int. Edn* **14,** 636 (1975)
126 KRAMAROVA, E. P., BAUKOV, Yu. I. and LUTSENKO, I. F., *J. gen. Chem. U.S.S.R.* **43,** 1843 (1973)
127 KRAMAROVA, E. P., BAUKOV, Yu. I. and LUTSENKO, I. F., *J. gen. Chem. U.S.S.R.* **45,** 469 (1975)
128 WISSNER, A., *Tetrahedron Lett.* 2749 (1978)
129 LARSON, G. L., HERNÁNDEZ, D. and HERNÁNDEZ, A., *J. organometal. Chem.* **76,** 9 (1974)
130 LARSON, G. L. and HERNÁNDEZ, A., *J. organometal. Chem.* **102,** 123 (1975)
131 KLEIN, J., LEVENE, R. and DUNKELBLUM, E., *Tetrahedron Lett.* 2845 (1972)
132 LARSON, G. L., HERNÁNDEZ, E., ALONSO, C. and NIEVES, I., *Tetrahedron Lett.* 4005 (1975)
133 RUBOTTOM, G. M. and GRUBER, J. M., *J. org. Chem.* **43,** 1599 (1978)
134 STILL, W. C., *J. Am. chem. Soc.* **101,** 2493 (1979)
135 CLARK, R. D. and HEATHCOCK, C. H., *Tetrahedron Lett.* 2027 (1974); *see also* CREARY, X., *J. org. Chem.* **45,** 2419 (1980)
136 STORK, G. and NAIR, V., *J. Am. chem. Soc.* **101,** 1315 (1979)
137 ADAM, W. and LIU, J.-C., *J. Am. chem. Soc.* **94,** 2894 (1972); ADAM, W., del FIERRO, J., QUIROZ, F. and YANY, F., *J. Am. chem. Soc.* **102,** 2127 (1980)
138 WASSERMAN, H. H., LIPSHUTZ, B. H. and WU, J. S., *Heterocycles* **7,** 321 (1977)
139 ITO, Y., KONOIKE, T. and SAEGUSA, T., *J. Am. chem. Soc.* **97,** 649 (1975)
140 KOBAYASHI, Y., TAGUCHI, T. and TOKUNO, E., *Tetrahedron Lett.* 3741 (1977)
141 ITO, Y., KONOIKE, T., HARADA, T. and SAEGUSA, T., *J. Am. chem. Soc.* **99,** 1487 (1977)
142 INABA, S. and OJIMA, I., *Tetrahedron Lett.* 2009 (1977)

143 RUBOTTOM, G. M. and LOPEZ, M. I., *J. org. Chem.* **38,** 2097 (1973)
144 DENIS, J. M. and CONIA, J. M., *Tetrahedron Lett.* 4593 (1972)
145 MURAI, S., AYA, T. and SONODA, N., *J. org. Chem.* **38,** 4354 (1973)
146 DENIS, J. M., GIRARD, C. and CONIA, J. M., *Synthesis* 549 (1972)
147 RYU, I., MURAI, S., OTANI, S. and SONODA, N., *Chemy Lett.* 93 (1976)
148 For a review of some of these transformations, *see* CONIA, J. M., *Pure appl. Chem.* **43,** 317 (1975)
149 CONIA, J. M. and GIRARD, C., *Tetrahedron Lett.* 2767 (1973)
150 MURAI, S., SEKI, Y. and SONODA, N., *J. chem. Soc. chem. Communs* 1032 (1974)
151 MURAI, S., AYA, T., RENGE, T., RYU, I. and SONODA, N., *J. org. Chem.* **39,** 858 (1974); RYU, I., MURAI, S., OTANI, S. and SONODA, N., *Tetrahedron Lett.* 1995 (1977)
152 RYU, I., MURAI, S. and SONODA, N., *Tetrahedron Lett.* 4611 (1977)
153 GIRARD, C. and CONIA, J. M., *Tetrahedron Lett.* 3327 (1974)
154 SALAUN, J., GARNIER, B. and CONIA, J. M., *Tetrahedron* **30,** 1413 (1974)
155 TROST, B. M. and BOGDANOWICZ, M. J., *J. Am. chem. Soc.* **95,** 289, 2038, 5311 (1973); TROST, B. M. and KUROZUMI, S., *Tetrahedron Lett.* 1929 (1974); TROST, B. M. and SCUDDER, P. H., *J. Am. chem. Soc.* **99,** 7601 (1977); TROST, B. M., NISHIMURA, Y., YAMAMOTO, K. and McELVAIN, S. S., *J. Am. chem. Soc.* **101,** 1328 (1979); *see also* TROST, B. M. and MELVIN, L. S., *'Sulphur Ylides—Emerging Synthetic Intermediates',* Academic Press, New York and London (1975)
156 GIRARD, C., AMICE, P., BARNIER, J. P. and CONIA, J. M., *Tetrahedron Lett.* 3329 (1974)
157 TORII, S., OKAMOTO, T. and UENO, N., *J. chem. Soc. chem. Communs* 293 (1978)
158 WENKERT, E., GOODWIN, T. E. and RANU, B. C., *J. org. Chem.* **42,** 2137 (1977)
159 WENKERT, E., BUCKWALTER, B. L., CRAVEIRO, A. A., SANCHEZ, E. L. and SATHE, S. S., *J. Am. chem. Soc.* **100,** 1267 (1978); WENKERT, E., *Accts chem. Res.* **13,** 27 (1980)
160 AMICE, P., BLANCO, L. and CONIA, J. M., *Synthesis* 196 (1976)
161 ITO, Y., FUJII, S. and SAEGUSA, T., *J. org. Chem.* **41,** 2073 (1976); *Org. Synth.* **59,** 113 (1980)
162 ITO, Y., FUJII, S. and SAEGUSA, T., *J. org. Chem.* **42,** 2326 (1977)
163 ITO, Y., SUGAYA, T., NAKATSUKA, M. and SAEGUSA, T., *J. Am. chem. Soc.* **99,** 8366 (1977)
164 RUBOTTOM, G. M., MARRERO, R., KRUEGER, D. S. and SCHREINER, J. L., *Tetrahedron Lett.* 4013 (1977)
165 RÜHLMANN, K., *Synthesis* 236 (1971)
166 NAKAMURA, E. and KUWAJIMA, I., *J. Am. chem. Soc.* **99,** 7360 (1977)
167 MIZUNO, K., OKAMOTO, H., PAC, C., SAKURAI, H., MURAI, S. and SONODA, N., *Chemy Lett.* 237 (1975)
168 de MAYO, P. and TAKESHITA, H., *Can. J. Chem.* **41,** 440 (1963)
169 HALL, H. K. and YKMAN, P., *J. Am. chem. Soc.* **97,** 800 (1975)
170 CLARK, R. D. and UNTCH, K. G., *J. org. Chem.* **44,** 248, 253 (1979)
171 SNIDER, B. B., RODINI, D. J., CONN, R. S. E. and SEALFON, S., *J. Am. chem. Soc.* **101,** 5283 (1979); FIENMANN, H. and HOFFMANN, H. M. R., *J. org. Chem.* **44,** 2803 (1979)
172 OJIMA, I., INABA, S. and YOSHIDA, K., *Tetrahedron Lett.* 3643 (1977); OJIMA, I. and INABA, S., *Tetrahedron Lett.* 2077, 2081 (1980)
173 KREPSKI, L. R. and HASSNER, A., *J. org. Chem.* **43,** 3173 (1978); BRADY, W. T. and LLOYD, R. M., *J. org. Chem.* **44,** 2560 (1979); **45,** 2025 (1980); DEPRÉS, J.-P. and GREENE, A. E., *J. org. Chem.* **45,** 2036 (1980)
174 WOHL, R. A., *Helv. chim. Acta* **56,** 1826 (1973); *Tetrahedron Lett.* 3111 (1973); for a related ring-expansion of methylene cycloalkanes, *see* WOHL, R. A., *J. org. Chem.* **38,** 3862 (1973)
175 CITERIO, L. and STRADI, R., *Tetrahedron Lett.* 4227 (1977)
176 FLEMING, I., *'Frontier Orbitals and Organic Chemical Reactions',* Wiley (1976)
177 E.g. CASEAU, P. and FRAINNET, E., *Bull. Soc. chim. Fr.* 1658 (1972)
178 *Org. Synth.* **58,** 163 (1978)
179 JUNG, M. E. and McCOMBS, C. A., *Tetrahedron Lett.* 2935 (1976)
180 DANISHEFSKY, S., PRISBYLLA, M. P. and HINER, S., *J. Am. chem. Soc.* **100,** 2918 (1978)
181 JUNG, M. E. and McCOMBS, C. A., *J. Am. chem. Soc.* **100,** 5207 (1978)

182 RUBOTTOM, G. M. and KRUEGER, D. S., *Tetrahedron Lett.* 611 (1977); RUBOTTOM, G. M. and GRUBER, J. M., *J. org. Chem.* **42,** 1051 (1977); *see also* YAMAMOTO, H. and SHAM, H. L., *J. Am. chem. Soc.* **101,** 1609 (1979)

183 SYMMES, C. and QUIN, L. D., *J. org. Chem.* **41,** 238 (1976)

184 DANISHEFSKY, S., SINGH, R. K. and KITAHARA, T., *J. Am. chem. Soc.* **99,** 5180 (1977)

185 DANISHEFSKY, S. and KITAHARA, T., *J. Am. chem. Soc.* **96,** 7807 (1974); DANISHEFSKY, S., KITAHARA, T., YAN, C. F. and MORRIS, J., *J. Am. chem. Soc.* **101,** 6996 (1979); DANISHEFSKY, S., YAN, C. F., SINGH, R. K., GAMMILL, R. B., McCURRY, P. M., FRITSCH, N. and CLARDY, J., *J. Am. chem. Soc.* **101,** 7001 (1979)

186 DANISHEFSKY, S. and HIRAMA, M., *J. Am. chem. Soc.* **99,** 7740 (1977)

187 DANISHEFSKY, S., GOMBATZ, K., HARAYAMA, T., BERMAN, E. and SCHUDA, P., *J. Am. chem. Soc.* **100,** 6536 (1978); DANISHEFSKY, S., HIRAMA, M., GOMBATZ, K., HARAYAMA, T., BERMAN, E. and SCHUDA, P. F., *J. Am. chem. Soc.* **101,** 7020 (1979); *see also* DANISHEFSKY, S., MORRIS, J., MULLEN, G. and GAMMILL, R., *J. Am. chem. Soc.* **102,** 2438 (1980)

188 BOECKMAN, R. K., DOLAK, T. M. and CULOS, K. O., *J. Am. chem. Soc.* **100,** 7098 (1978)

189 KEANA, J. F. W. and ECKLER, P. E., *J. org. Chem.* **41,** 2850 (1976)

190 BANVILLE, J. and BRASSARD, P., *J. chem. Soc. Perkin I* 1852 (1976)

191 DANISHEFSKY, S., SINGH, R. K. and GAMMILL, R. B., *J. org. Chem.* **43,** 379 (1978)

192 DANISHEFSKY, S., YAN, C. F. and McCURRY, P. M., *J. org. Chem.* **42,** 1819 (1977)

193 YAMAMOTO, K., SUZUKI, S. and TSUJI, J., *Chemy Lett.* 649 (1978)

194 REETZ, M. T. and NEUMEIER, G., *Chem. Ber.* **112,** 2209 (1979)

195 IBUKA, T., MORI, Y. and INUBUSHI, Y., *Tetrahedron Lett.* 3169 (1976)

196 HOFFMANN, H. M. R., *Angew. Chem. int. Edn* **12,** 819 (1973)

197 NOYORI, R., *Accts chem. Res.* **12,** 61 (1979)

198 SAKURAI, H., SHIRAHATA, A. and HOSOMI, A., *Angew. Chem. int. Edn* **18,** 163 (1979)

199 ARNOLD, R. T. and HOFFMAN, C., *Synth. Communs* **2,** 27 (1972); JULIA, S., JULIA, M. and LINSTRUMELLE, G., *Bull. Soc. chim. Fr.* 2693 (1964)

200 IRELAND, R. E. and MUELLER, R. H., *J. Am. chem. Soc.* **94,** 5897 (1972); for an example involving competitive [1,3]-sigmatropic rearrangement *see* ARNOLD, R. T. and KULENOVIĆ, S. T., *J. org. Chem.* **45,** 891 (1980); for a lactone-based variant, *see* DANISHEVSKY, S., FUNK, R. L. and KERWIN, J. F., *J. Am. chem. Soc.* **102,** 6889 (1980)

201 IRELAND, R. E., MUELLER, R. H. and WILLARD, A. K., *J. org. Chem.* **41,** 986 (1976)

202 KATZENELLENBOGEN, J. A. and CHRISTY, K. J., *J. org. Chem.* **39,** 3315 (1974)

203 For other solutions to this general problem, *see* HIRAMA, M., GARVEY, D. S., LU, L. D.-L. and MASAMUNE, S., *Tetrahedron Lett.* 3937 (1979); EVANS, D. A., VOGEL, E. and NELSON, J. V., *J. Am. chem. Soc.* **101,** 6120 (1979); HEATHCOCK, C. H., BUSE, C. T., KLESCHICK, W. A., PIRRUNG, M. C., SOHN, J. E. and LAMPE, J., *J. org. Chem.* **45,** 1066 (1980); HOFFMANN, R. W. and ZEISS, H.-J., *Angew. Chem. int. Edn* **19,** 218 (1980)

204 IRELAND, R. E. and WILCOX, C. S., *Tetrahedron Lett.* 2839 (1977); IRELAND, R. E., THAISRIVONGS, S. and WILCOX, C. S., *J. Am. chem. Soc.* **102,** 1155 (1980)

205 BOYD, J., EPSTEIN, W. and FRÁTER, G., *J. chem. Soc. chem. Communs* 380 (1976)

206 HILL, R. K., SOMAN, R. and SAWADA, S., *J. org. Chem.* **37,** 3737 (1972); STORK, G. and RAUCHER, S., *J. Am. chem. Soc.* **98,** 1583 (1976)

207 STILL, W. C. and SCHNEIDER, M. J., *J. Am. chem. Soc.* **99,** 948 (1977)

208 KACHINSKI, J. L. C. and SALOMON, R. G., *Tetrahedron Lett.* 3235 (1977)

209 REUTER, J. M. and SALOMON, R. G., *Tetrahedron Lett.* 3199 (1978)

210 BLOCH, R. and DENIS, J. M., *J. organometal. Chem.* **90,** C9 (1975)

211 *See,* for example, SCHAEFER, J. P. and BLOOMFIELD, J. J., *Org. React.* **15,** 24 (1967)

212 SCHRÄPLER, U. and RÜHLMANN, K., *Chem. Ber.* **97,** 1383 (1964)

213 *See,* e.g., CORRIU, R., DABOSI, G. and MARTINEAU, M., *J. Chem. Soc. chem. Communs* 457 (1979)

214 BLOOMFIELD, J. J., OWSLEY, D. C., AINSWORTH, C. and ROBERTSON, R. E., *J. org. Chem.* **40,** 393 (1975)

215 BLOOMFIELD, J. J., *Tetrahedron Lett.* 591 (1968)

216 BLOOMFIELD, J. J., *Tetrahedron Lett.* 587 (1968); BLOOMFIELD, J. J. and NOLKE, J. M., *Org. React.* **23,** 259 (1976)

217 CREARY, X. and ROLLIN, A. J., *J. org. Chem.* **44,** 1017 (1979)
218 RÜHLMANN, K. and POREDDA, S., *J. prakt. Chem.* [4] **12,** 18 (1960)
219 STRATING, J., REIFFERS, S. and WYNBERG, H., *Synthesis* 209 (1971)
220 STRATING, J., REIFFERS, S. and WYNBERG, H., *Synthesis* 211 (1971)
221 HEINE, H.-G., *Chem. Ber.* **104,** 2869 (1971); HEINE, H.-G. and WENDISCH, D., *Justus Liebigs Annln Chem.* 463 (1976); CONIA, J. M. and DENIS, J. M., *Tetrahedron Lett.* 2845 (1971)
222 WYNBERG, H. W., STRATING, J. and REIFFERS, S., *Recl. Trav. chim. Pays-Bas Belg.* **89,** 982 (1970)
223 KOWAR, T. and LeGOFF, E., *Synthesis* 212 (1973); *see also* CARPINO, L. A. and TSAO, J.-H., *J. org. Chem.* **44,** 2564 (1979)
224 WAKAMATSU, T., HASHIMOTO, K., OGURA, M. and BAN, Y., *Synth. Communs* **8,** 319 (1978)
225 WAKAMATSU, T., AKASAKA, K. and BAN, Y., *Tetrahedron Lett.* 2751 (1977)
226 WAKAMATSU, T., AKASAKA, K. and BAN, Y., *Tetrahedron Lett.* 2755 (1977); *J. org. Chem.* **44,** 2008 (1979)
227 WAKAMATSU, T., AKASAKA, K. and BAN, Y., *Tetrahedron Lett.* 3879 (1974)
228 WAKAMATSU, T., FUKUI, M. and BAN, Y., *Synthesis* 341 (1976)
229 BAUER, D. P. and MACOMBER, R. S., *J. org. Chem.* **41,** 2640 (1976)
230 GASSMAN, P. G. and CREARY, X., *J. chem. Soc. chem. Communs* 1214 (1972)
231 JUNG, M. E., *J. chem. Soc. chem. Communs* 956 (1974)
232 PICARD, J. P., EKOUYA, A., DUNOGUÈS, J., DUFFAUT, N. and CALAS, R., *J. organometal. Chem.* **93,** 51 (1975)
233 KUWAJIMA, I., SATO, T., MINAMI, N. and ABE, T., *Tetrahedron Lett.* 1591 (1976)
234 *See,* e.g., BOURGEOIS, P., DUNOGUÈS, J. and DUFFAUT, N., *J. organometal. Chem.* **80,** C25 (1974)
235 KUWAJIMA, I., SATO, T., MINAMI, N. and ABE, T., *Tetrahedron Lett.* 1591 (1976); KUWAJIMA, I., ARAI, M. and SATO, T., *J. Am. chem. Soc.* **99,** 4181 (1977); KUWAJIMA, I., KATO, M. and SATO, T., *J. chem. Soc. chem. Communs* 478 (1978); *see also* COHEN, T. and MATZ, J. R., *J. Am. chem. Soc.* **102,** 6900 (1980)
236 SATO, T., ABE, T. and KUWAJIMA, I., *Tetrahedron Lett.* 259 (1978)
237 MINAMI, N., ABE, T. and KUWAJIMA, I., *J. organometal. Chem.* **145,** C1 (1978)
238 SATO, T., ARAI, M. and KUWAJIMA, I., *J. Am. chem. Soc.* **99,** 5827 (1977)
239 KASHUTINA, M. V., IOFFE, S. L. and TARTAKOVSKII, V. A., *Dokl. Akad. Nauk SSSR* **218,** 109 (1974); English version p. 607
240 TORSELL, K. and ZEUTHEN, O., *Acta chem. scand.* **B32,** 118 (1978); SHARMA, S. C. and TORSELL, K., *Acta chem. scand.* **B33,** 379 (1979)
241 OLAH, G. A., GUPTA, B. G. B., NARANG, S. C. and MALHOTRA, R., *J. org. Chem.* **44,** 4272 (1979); OLAH, G. A. and GUPTA, B. G. B., *Synthesis* 44 (1980)
242 COLVIN, E. W. and SEEBACH, D., *J. chem. Soc. chem. Communs* 689 (1978); COLVIN, E. W., BECK, A. K., BASTANI, B., SEEBACH, D., KAI, Y. and DUNITZ, J. D., *Helv. chim. Acta* **63,** 697 (1980)
243 SEEBACH, D., COLVIN, E. W., LEHR, F. and WELLER, T., *Chimia* **33,** 1 (1979)
244 TIDWELL, T. T., *Tetrahedron Lett.* 4615 (1979)
245 MATSUDA, I., MURATA, S. and IZUMI, Y., *J. org. Chem.* **45,** 287 (1980)
246 FOHLISCH, B. and GIERING, W., *Synthesis* 231 (1980)
247 REICHARDT, C. and RUST, C., *Synthesis* 232 (1980)
248 NICHOLAS, K. M., MULVANEY, M. and BAEYER, M., *J. Am. chem. Soc.* **102,** 2508 (1980)
249 KELLY, L. F., NARULA, A. S. and BIRCH, A. J., *Tetrahedron Lett.* 2455 (1980); *see also* BIRCH, A. J., NARULA, A. S., DAHLER, P., STEPHENSON, G. R. and KELLY, L. F., *Tetrahedron Lett.* 979 (1980); BIRCH, A. J., DAHLER, P., NARULA, A. S. and STEPHENSON, G. R., *Tetrahedron Lett.* 3817 (1980)
250 ITOH, Y., AOYAMA, H. and SAEGUSA, T., *J. Am. chem. Soc.* **102,** 4519 (1980); ITOH, Y., AOYAMA, H., MOCHIZUKI, A. and SAEGUSA, T., *J. Am. chem. Soc.* **101,** 494 (1979)
251 Van AUDENHOVE, M., De KEUKELEIRE, D. and VANDEWALLE, M., *Tetrahedron Lett.* 1979 (1980)
252 DANISHEFSKY, S., ZAMBONI, R., KAHN, M. and ETHEREDGE, S. J., *J. Am. chem. Soc.* **102,** 2097 (1980)
253 KRAUS, G. A. and TASCHNER, M. J., *J. org. Chem.* **45,** 1174 (1980)
254 KRAUS, G. A. and TASCHNER, M. J., *J. Am. chem. Soc.* **102,** 1974 (1980)

Chapter 18

Trimethylsilyl-based reagents

Formal replacement of the proton of a range of inorganic acids by the trimethylsilyl moiety gives a group of reagents which can behave as hard acids, particularly when silicon–oxygen bond formation takes place. When the anion of the parent acid is a soft base/good nucleophile, it can attack, for example, an ether molecule at the carbon atom carrying the now complexed oxygen, with consequent dealkylation or de-oxygenation, depending upon the fragment being followed; related processes can take place on carboxylic or phosphorus acid esters. With aldehydes and ketones as substrates, carbonyl addition products can result. These sequences are summarized in

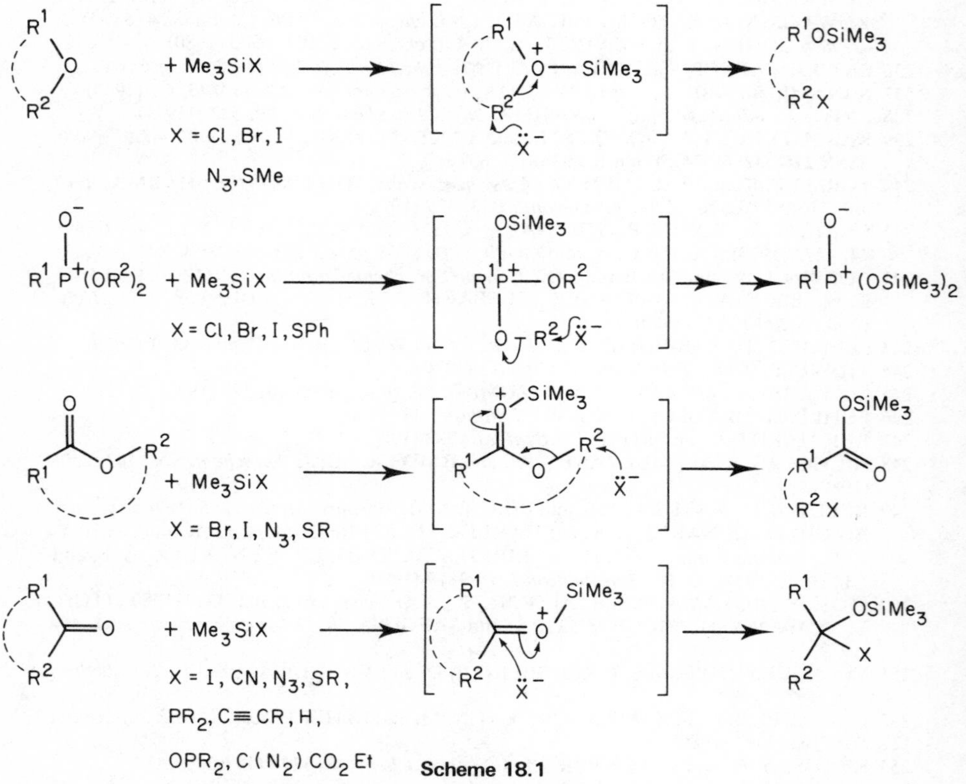

$+ Me_3SiX$

X = Cl, Br, I, N_3, SMe

$+ Me_3SiX$

X = Cl, Br, I, SPh

$+ Me_3SiX$

X = Br, I, N_3, SR

$+ Me_3SiX$

X = I, CN, N_3, SR, PR_2, C≡CR, H, OPR_2, $C(N_2)CO_2Et$

Scheme 18.1

Scheme 18.1; although drawn in a stepwise manner, they can, of course, be more or less concerted.

Two of the carbonyl addition processes outlined in *Scheme 18.1*, those of addition of an alkynyltrialkylsilane and of a trialkylsilane, are discussed in other Chapters. For the rest of the reactions shown, a brief description of preparative routes to some of the reagents (most of which are now commercially available, if somewhat expensive) will be followed by a selection of their applications. Virtually all of these reagents are moisture-sensitive, and appropriate handling precautions must be taken.

18.1 Reagent preparation

18.1.1 Bromotrimethylsilane, Me_3SiBr

This can be conveniently prepared[1] by stirring hexamethyldisiloxane with phosphorus(III) bromide in the presence of a catalytic amount of iron(III) chloride; the product is distilled directly from the reaction mixture (*Scheme 18.2*).

$$(Me_3Si)_2O + PBr_3 \xrightarrow{FeCl_3} 2Me_3SiBr$$

Scheme 18.2

18.1.2 Iodotrimethylsilane, Me_3SiI

This reagent can be obtained in a variety of ways, a simple one being reductive cleavage of hexamethyldisiloxane with iodine and aluminium powder[2,3]; once again, the product is isolated by direct distillation (*Scheme 18.3*). Alternatively, *in situ* generation of this highly reactive species often suffices. This can involve cleavage of 1,4-bis(trimethylsilyl)cyclohexa-2,5-diene[4], or of hexamethyldisilane[5] with iodine. Even more simply, an effective reagent can be obtained[6,7] by the *in situ* interaction between chlorotrimethylsilane and sodium iodide.

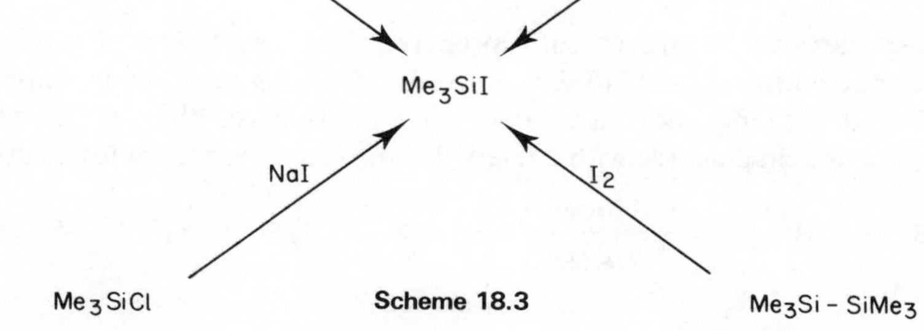

$$(Me_3Si)_2O \xrightarrow{I_2, Al} Me_3SiI$$

$$\text{1,4-bis(trimethylsilyl)cyclohexa-2,5-diene} \xrightarrow{I_2} Me_3SiI$$

$$Me_3SiCl \xrightarrow{NaI} Me_3SiI$$

$$Me_3Si-SiMe_3 \xrightarrow{I_2} Me_3SiI$$

Scheme 18.3

18.1.3 Cyanotrimethylsilane (trimethylsilyl cyanide), Me_3SiCN

Many preparations[8-11] of this reagent have been described, all involving reaction of chlorotrimethylsilane with some source of cyanide ion. The procedure outlined in *Scheme 18.4* has been described[11] in detail; the product is isolated by direct distillation.

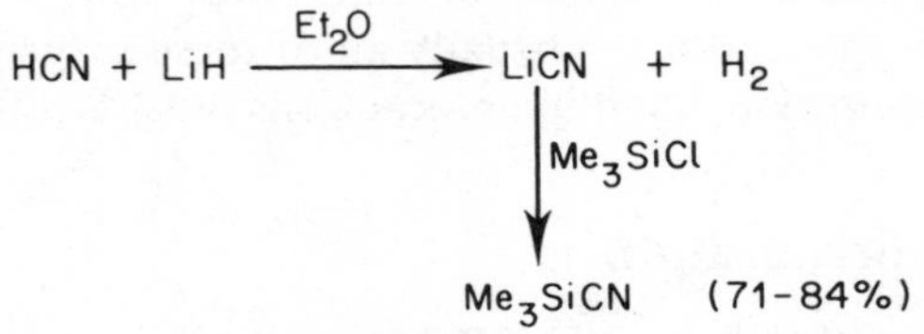

$$HCN + LiH \xrightarrow{Et_2O} LiCN + H_2$$

$$LiCN \xrightarrow{Me_3SiCl} Me_3SiCN \quad (71\text{-}84\%)$$

Scheme 18.4

18.1.4 Azidotrimethylsilane (trimethylsilyl azide), Me_3SiN_3

This species is prepared[12] by reaction of chlorotrimethylsilane with sodium azide in a high-boiling aprotic solvent (*Scheme 18.5*), the product being distilled directly; this procedure has been described[13] in full detail.

$$Me_3SiCl + NaN_3 \longrightarrow Me_3SiN_3 \quad 85\%$$

Scheme 18.5

18.1.5 Alkylthio- and arylthio-trimethylsilanes, $Me_3Si\text{–}SR$

These species[14] are conveniently obtained by silylation of the corresponding alkyl and aryl thiols[15-17], largely by methods similar to those employed for the preparation of alcohol- and phenol-derived silyl ethers (Chapter 15).

18.1.6 Bis(trimethylsilyl) sulphide (hexamethyldisilathiane), $(Me_3Si)_2S$

This useful silylating[18] reagent can also be used for the direct transfer of sulphur[14,19] to carbon. It is readily prepared by an imidazole-catalysed reaction[20,21] between hexamethyldisilazane and hydrogen sulphide (*Scheme 18.6*).

$$(Me_3Si)_2NH + H_2S \xrightarrow{1\%\ ImH} (Me_3Si)_2S \quad 96\%$$

Scheme 18.6

18.1.7 Tervalent phosphorus–oxygen–silicon reagents, $R^1_2POSiR^2_3$

Such species can be conveniently prepared[22-25] by silylation of a deprotonated phosphorus acid (*Scheme 18.7*), utilizing the wide range of chlorotrialkylsilanes now available. The alternative, that of treating a phosphorus acid chloride with a trialkylsilanoxide anion, is of lesser utility.

$$R^1_2P(O)H \xrightarrow[2.\ R^2_3SiCl]{1.\ Base} R^1_2P\text{-}OSiR^2_3 \quad 45\text{-}80\%$$

Scheme 18.7

18.2 Ester and ether cleavage

The halogenotrimethylsilanes will be grouped together here, since in broad terms their reactivity can be summarized by the first three processes shown in *Scheme 18.1.* The cleavage of cyclic and acyclic ethers employing halogenotrialkylsilanes has been known[26] for some time, although it is only recently that it has been studied systematically. For example, oxiranes are smoothly transformed into the corresponding chlorohydrins (*Scheme 18.8*) by treatment with chlorotrimethylsilane[27].

Bromo-, iodo-, and azido-trimethylsilane[28] react analogously, the last requiring catalysis by zinc chloride. If two equivalents of iodotrimethylsilane are employed, deoxygenation occurs, and alkenes[29] are produced. Methylthiotrimethylsilane[30] similarly opens oxiranes, oxetanes and tetrahydrofuran.

Two other routes to chlorohydrins are also shown in *Scheme 18.8.* The first of these involves the *in situ* generation[31] of trimethylsilyl hydroperoxide, Me_3SiOOH; the corresponding triphenylsilyl hydroperoxide, Ph_3SiOOH, is an isolable crystalline solid[32a] which converts alkenes into oxiranes[32b] in reasonable yields. t-Butyldimethylsilyl hydroperoxide has been used[33] for the conversion of a sensitive polyolefinic allylic alcohol, via its mesylate, into the corresponding hydroperoxide. The second route to chlorohydrins utilizes chlorotrimethylsilane to cleave[34] cyclic orthoacetates to chlorohydrin acetates; with suitable 1,2-diol precursors, a high degree of regioselectivity and of stereoselectivity[35] can be attained. In contrast, such orthoesters react with azidotrimethylsilane[36] to give mainly cyclic products, as shown in *Scheme 18.8.*

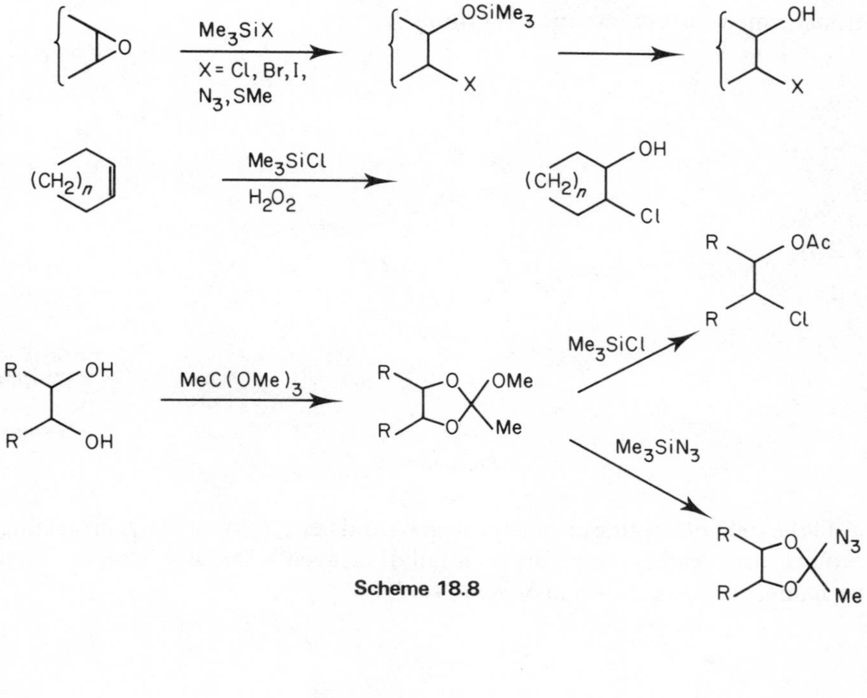

Scheme 18.8

In recent years, greater emphasis has been given to the use of the much more reactive bromo- and iodo-trimethylsilanes. For example, phosphorus(IV) acid dialkyl esters and related species are converted into the corresponding bis(trimethylsilyl) esters on treatment with halogenotrialkylsilanes, as shown in the third reaction in *Scheme 18.1*; mild hydrolysis[37] then provides the parent phosphorus(IV) acids. When chlorotrimethylsilane is used, the initial dealkylation process can require several days at reflux[38]. In contrast, bromotrimethylsilane can achieve the same transformation in one or two hours at room temperature[39], bromide ion being both a better leaving group and a better nucleophile than chloride ion. Iodotrimethylsilane[6,40] reacts even more rapidly, in some cases exothermically; the speed and selectivity of cleavage allows the presence of additional functional groups to be tolerated. Phenylthiotrimethylsilane[41] has been used for the stepwise selective dealkylation of phosphorus acid triesters, the alkyl portions of the esters reacting in the order of their S_N2 reactivity.

Iodotrimethylsilane appears to be the reagent of choice for the cleavage of simple esters, lactones, ethers and acetals, and the following few paragraphs will deal with this reagent almost exclusively. A variety of carboxylic acid esters are smoothly transformed[3,42] into the corresponding silyl ester (1) on treatment with iodotrimethylsilane (*Scheme 18.9*). Little selectivity is observed between methyl, ethyl, and isopropyl esters; t-butyl and benzyl esters, on the other hand, are cleaved much more rapidly, presumably by S_N1 cleavage of the probable intermediate (2). The initial reversible complexation can be catalysed[43] by iodine. The silyl ester (1) undergoes hydrolysis under very mild conditions (Chapter 16) to the parent carboxylic acid. If a considerable excess of iodotrimethylsilane is employed, the acid is converted in time into the corresponding acyl iodide; silyl esters of aromatic acids can be transformed further into methyl groups[43].

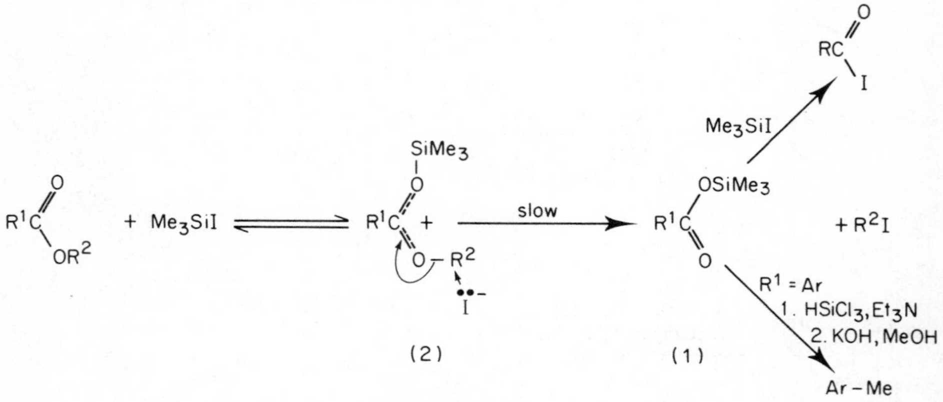

Scheme 18.9

In the field of peptide chemistry, benzyl and benzyloxycarbonyl protecting groups are readily (but differentially) cleaved[44], as are simple alkyl carbamates[45,46], as shown in *Scheme 18.10*.

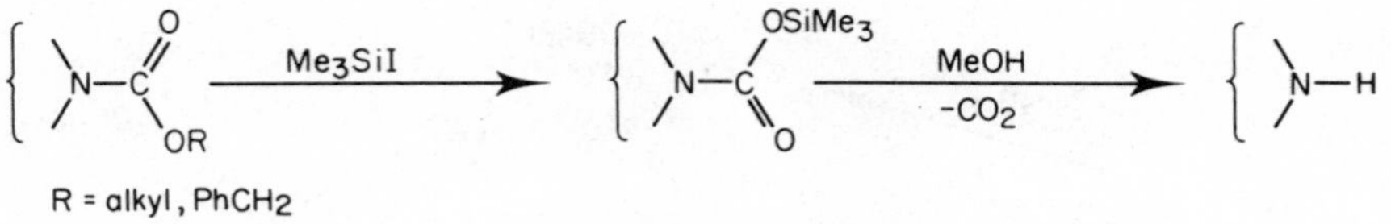

R = alkyl, $PhCH_2$

Scheme 18.10

β-Ketoesters and *gem*-diesters undergo dealkylation/decarboxylation[47], and lactones[48] are ring-opened to give *ω*-iodo silyl esters (*Scheme 18.11*); bromotrimethylsilane also cleaves lactones, whereupon *ω*-bromo silyl esters are produced. Similar reactions of azidotrimethylsilane[49] and of alkylthiotrialkylsilanes[14] have been observed to occur with *β*-lactones.

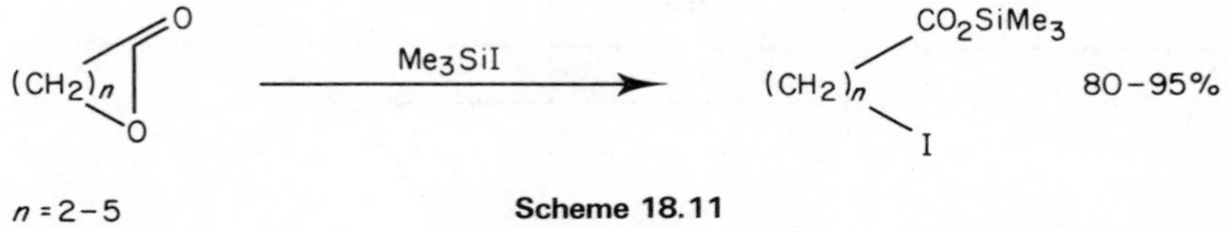

n = 2–5

Scheme 18.11

An interesting regioselective route[50] to *α*-methylene-*γ*-lactones is illustrated in *Scheme 18.12*. The key step of ester demethylation and elimination of amine probably proceeds as shown, the more substituted cyclopropane bond cleaving selectively.

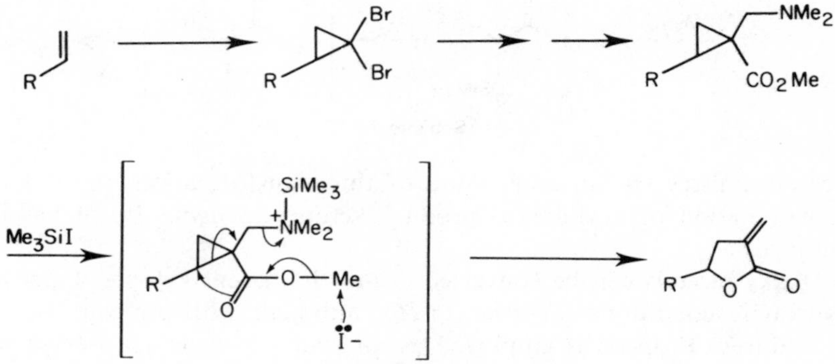

Scheme 18.12

Iodotrimethylsilane cleaves ethers to iodides and trimethylsilyl ethers[3,42]. With near-stoichiometric amounts of reagent, dialkyl ethers are cleaved to give a mixture of products (*Scheme 18.13*); when excess of reagent is employed, both alkyl fragments are obtained as alkyl iodides. Aralkyl ethers, on the other hand, undergo regiospecific cleavage; furthermore, aromatic methyl ethers can be cleaved selectively[51] in the presence of aromatic methylenedioxy groups.

Alcohols themselves[52] can be converted into the corresponding iodides, with the intermediacy of alkyl silyl ethers, which do not need to be isolated (*Scheme 18.14*); as expected, this reaction goes with inversion of

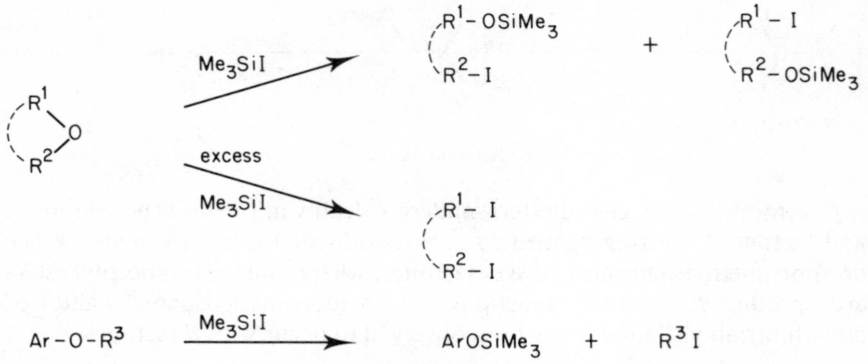

Scheme 18.13

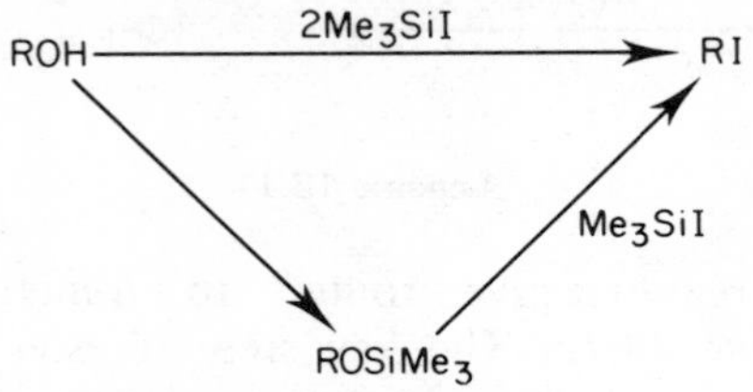

Scheme 18.14

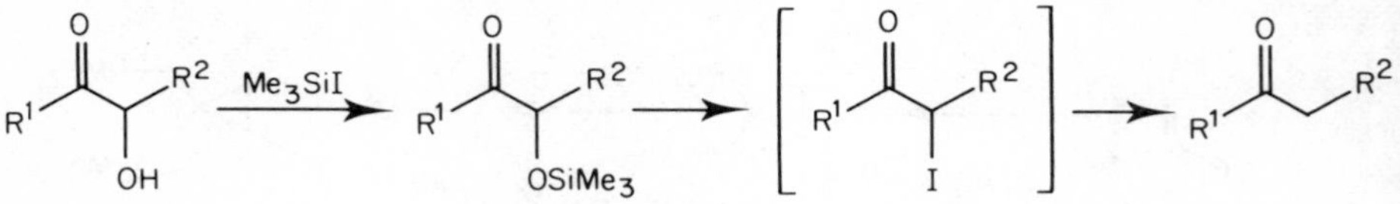

Scheme 18.15

stereochemistry. In an application of this transformation, the reductive deoxygenation of acyloins to produce ketones (*Scheme 18.15*) has been outlined[53].

Dialkyl acetals can be converted[54a] into the parent ketones under non-hydrolytic conditions (*Scheme 18.16*), although with aqueous isolation procedures. Propene is employed as an acid scavenger to prevent acid-catalysed aldol condensation. Application of this cleavage technique to ethylene acetals[54b] produces iodoalkyl (trimethylsilyloxy)ethyl ethers, which act as efficient alkylating agents.

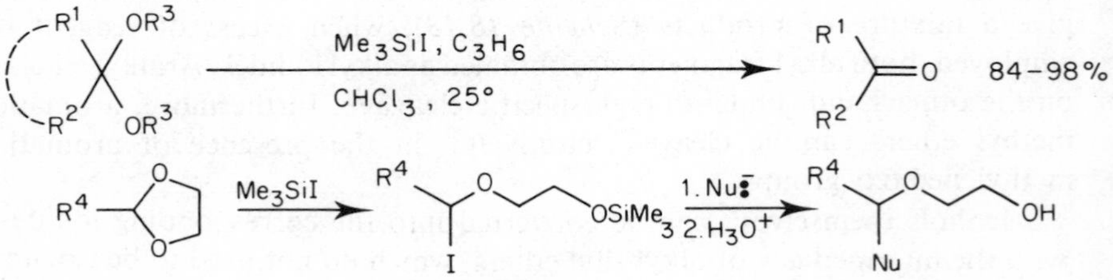

Scheme 18.16

Formaldehyde dimethylacetal similarly affords iodomethyl methyl ether[55] (*Scheme 18.17*) in good yield; some tetrahydrofuran derivatives (3) give[56] ω-iodo-alcohol trimethylsilyl ethers (4).

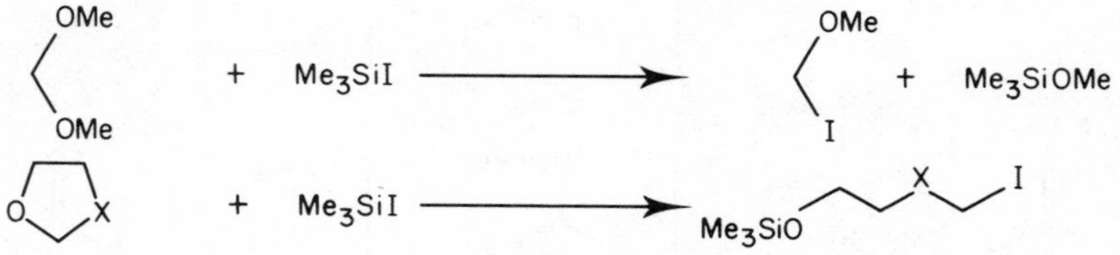

(3), X = O, S

(4)

Scheme 18.17

Returning to bromotrialkylsilanes, Masamune has described[57] their utility for the cleavage of methoxymethyl alcohol and acid-protecting groups. The selective cleavage of methoxymethyl esters in the presence of methoxymethyl ethers can be effected by bromotributylsilane and a trace amount of methanol, followed by mild hydrolysis (*Scheme 18.18*). Methoxymethyl ethers can themselves be cleaved by this reagent under somewhat different conditions, but often to give a mixture of products; in ideal circumstances, bromomethyl ethers are produced, ready for a variety of subsequent transformations.

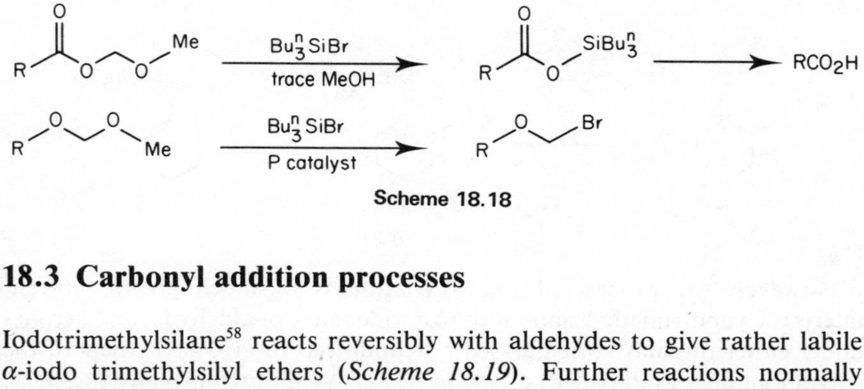

Scheme 18.18

18.3 Carbonyl addition processes

Iodotrimethylsilane[58] reacts reversibly with aldehydes to give rather labile α-iodo trimethylsilyl ethers (*Scheme 18.19*). Further reactions normally ensue: benzaldehyde, for instance, is converted into $\alpha\alpha$-diiodotoluene, whereas phenylacetaldehyde gives the dimeric structure (5). $\alpha\beta$-Unsaturated ketones undergo 1,4-addition[59] to produce β-iodoketones (*Scheme 18.20*); cyclopropyl ketones analogously give γ-iodoketones.

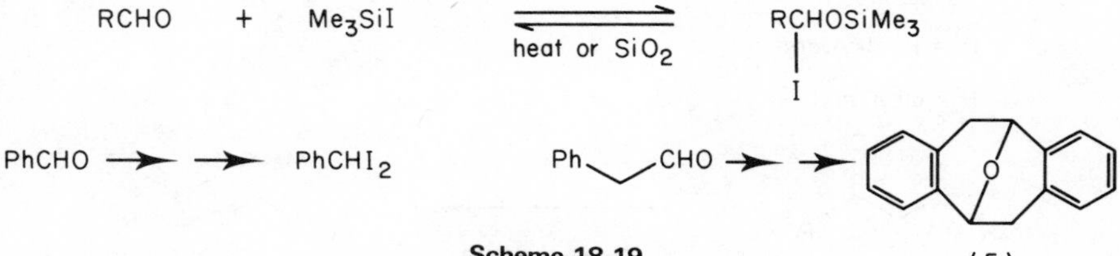

Scheme 18.19

(5)

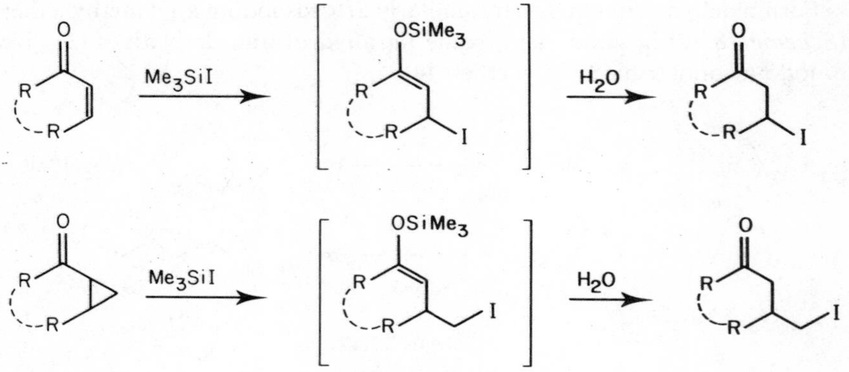

Scheme 18.20

In contrast, and in the presence of hexamethyldisilazane, both aldehydes and ketones are converted into their corresponding silyl enol ethers[60], thermodynamic product ratios being observed (*Scheme 18.21*). The reaction conditions are relatively mild; remarkably, enol ether formation takes precedence over ester cleavage in bifunctional substrates.

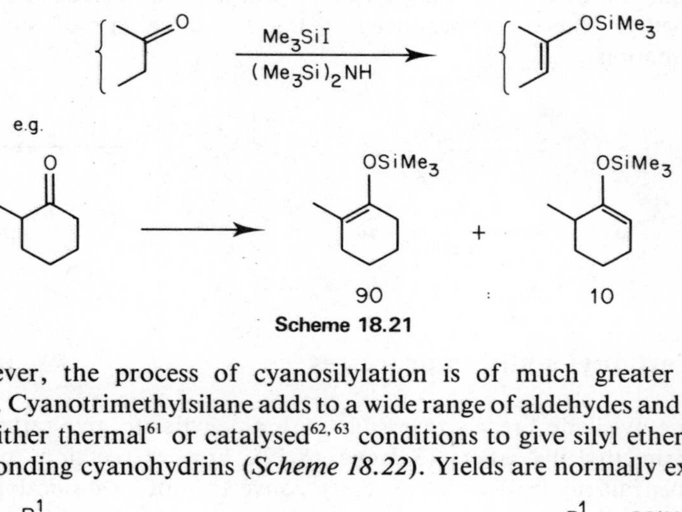

Scheme 18.21

However, the process of cyanosilylation is of much greater current interest. Cyanotrimethylsilane adds to a wide range of aldehydes and ketones under either thermal[61] or catalysed[62,63] conditions to give silyl ethers of the corresponding cyanohydrins (*Scheme 18.22*). Yields are normally excellent,

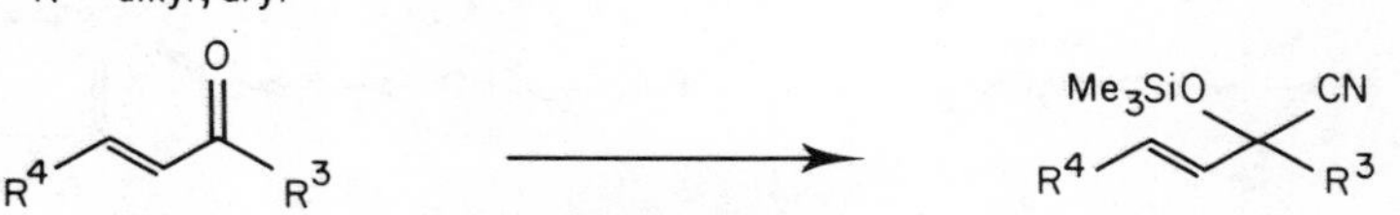

R^1 = H, alkyl, aryl

R^2 = alkyl, aryl

Scheme 18.22

even in cases of ketones which do not form cyanohydrins under the normal equilibrating conditions, and $\alpha\beta$-unsaturated carbonyl compounds undergo regioselective 1,2-addition. *In situ* generation[10] of the reagent from KCN and chlorotrimethylsilane is satisfactory only for reaction with aldehydes. Effective catalysts include ZnI_2, triphenylphosphine, and cyanide ion solubilized either as KCN-18-crown-6 or as tetrabutylammonium cyanide.

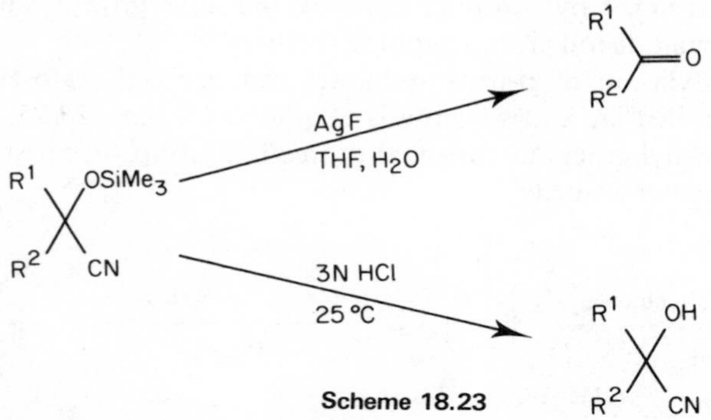

AgF, THF, H_2O; 3N HCl, 25 °C

Scheme 18.23

Cyanohydrin silyl ethers are distinguished from free cyanohydrins both in their ease of preparation and isolation and in their relative reluctance to undergo cleavage and reversion to the parent carbonyl compounds under basic or nucleophilic conditions. Liberation of the carbonyl group is readily achieved by treatment with AgF in aqueous THF. Alternatively, acid-catalysed hydrolysis[64] yields the parent cyanohydrins in almost quantitative yield (*Scheme 18.23*), even with highly hindered ketones such as benzophenone and t-butyl phenyl ketone.

Three major areas of application of cyanosilylation have been revealed so far. These are nucleophilic alkylation and acylation, carbonyl protection, and chain homologation.

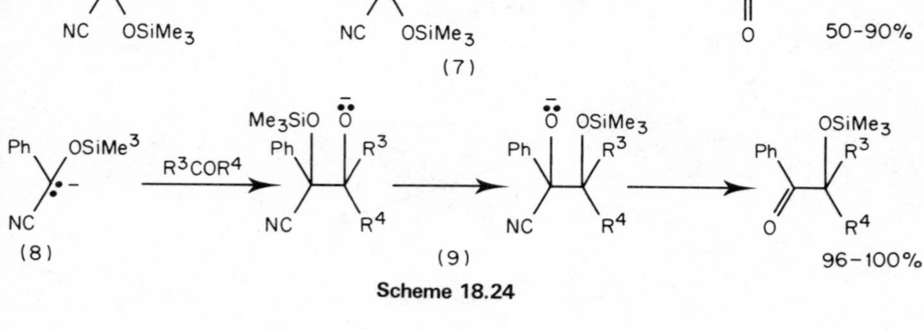

(6) 60–95%

(7) 50–90%

(8) (9) 96–100%

Scheme 18.24

The cyanohydrin silyl ethers derived from aryl, heteroaryl[65], and $\alpha\beta$-unsaturated aldehydes[66] can be deprotonated with LDA. The so-formed anions (6) and (7) react smoothly with a range of alkylating agents, the latter anion reacting regiospecifically; subsequent fluoride ion-induced hydrolysis produces ketones[67] in high yields (*Scheme 18.24*). The anion (8) derived from benzaldehyde adds quantitatively to aldehydes and ketones. The primary adduct (9) rearranges by 1,4-migration of the silyl group; subsequent elimination of cyanide ion gives acyloin silyl ethers[68].

Monocyanosilylation of *p*-benzoquinones can proceed regioselectively, and has been applied[69] to a new synthesis of quinols (*Scheme 18.25*). The use of cyanohydrin silyl ethers as carbonyl-protecting groups in prostaglandin synthesis has been described[70].

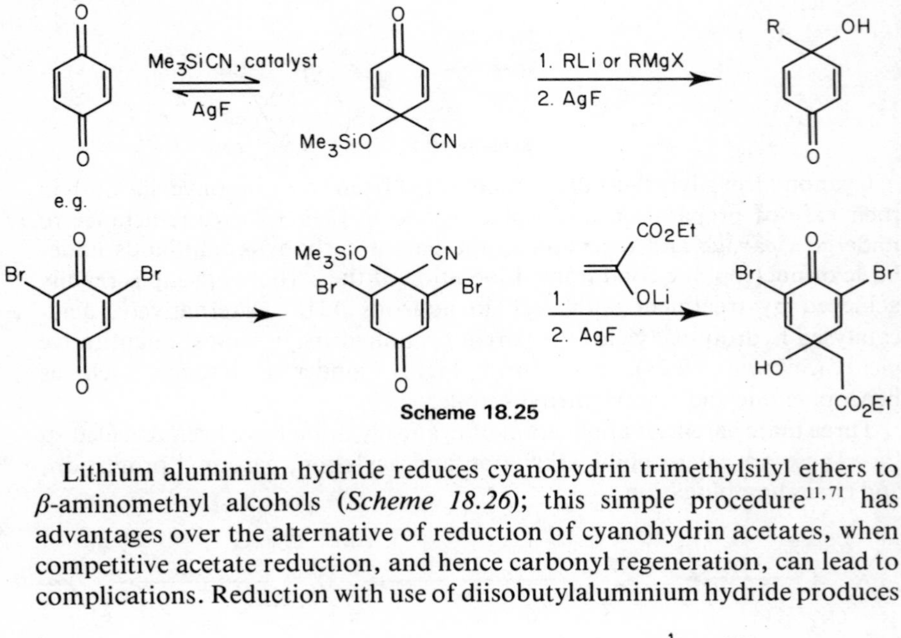

Me_3SiCN, catalyst; AgF; Me_3SiO; CN; 1. RLi or RMgX; 2. AgF; R; OH; e.g.; Br; CO_2Et; OLi; 1.; 2. AgF; HO; CO_2Et

Scheme 18.25

Lithium aluminium hydride reduces cyanohydrin trimethylsilyl ethers to β-aminomethyl alcohols (*Scheme 18.26*); this simple procedure[11,71] has advantages over the alternative of reduction of cyanohydrin acetates, when competitive acetate reduction, and hence carbonyl regeneration, can lead to complications. Reduction with use of diisobutylaluminium hydride produces

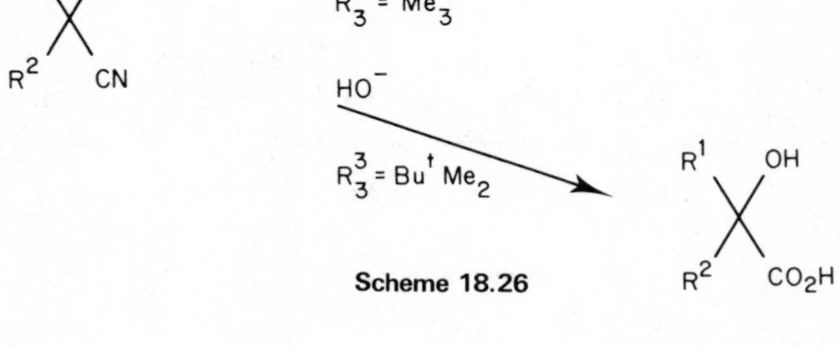

R^1; $OSiR^3_3$; R^2; CN; $LiAlH_4$; $R^3_3 = Me_3$; HO^-; $R^3_3 = Bu^tMe_2$; OH; NH_2; CO_2H

Scheme 18.26

α-trimethylsilyloxy aldehydes[72a]. Cyanohydrin t-butyldimethylsilyl ethers can be hydrolysed[72b] to α-hydroxy amides, and thence to α-hydroxy acids, without competitive desilylation in the initial hydrolysis process.

Azidotrimethylsilane[62, 73] adds readily to saturated aldehydes (*Scheme 18.27*), the reaction being catalysed either by $ZnCl_2$ or by azide ion solubilized with 18-crown-6; ketonic and αβ-unsaturated carbonyl compounds are unsatisfactory substrates. The product α-azido trimethylsilyl ethers (10) have been further transformed in two general ways. Thermolysis[73] leads directly to the corresponding *N*-trimethylsilyl amides, whereas reduction[74] with lithium aluminium hydride produces primary amines.

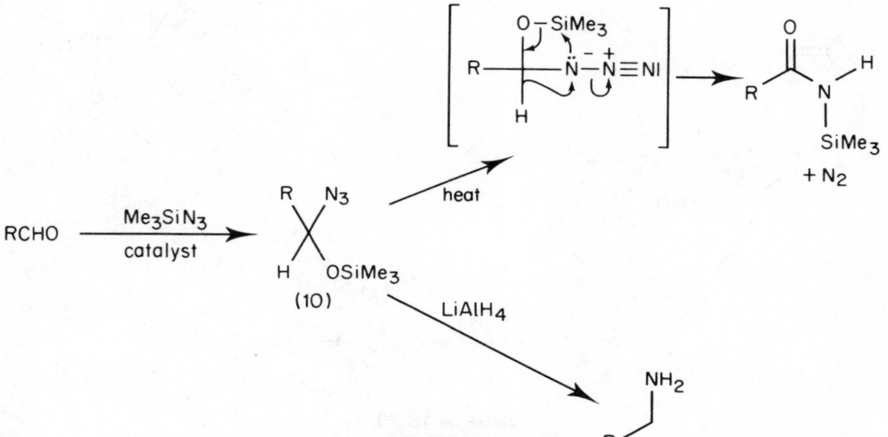

Scheme 18.27

An attractive route[49, 75] to isocyanates of diverse types is provided by the reaction of azidotrimethylsilane with acid chlorides, esters and anhydrides; the so-produced acyl azides (11) then undergo Curtius rearrangement as shown in *Scheme 18.28.* Halogenoethanoates, on the other hand, are converted[76] into azidoethanoates, which can act as precursors to alkoxycarbonyl nitrenes.

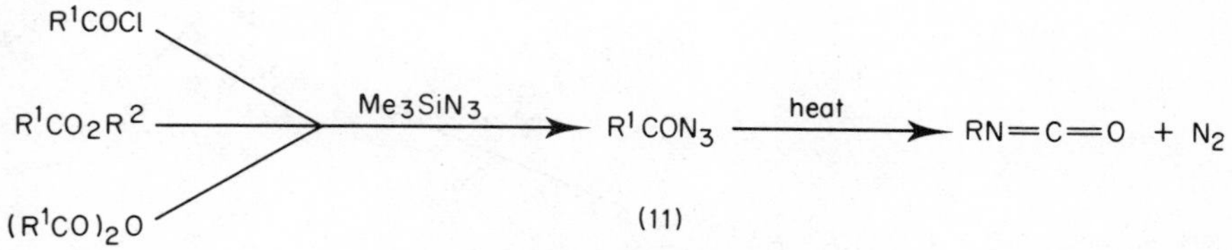

Scheme 18.28

A variety of tervalent phosphorus *O*-trialkylsilyl esters (12) add[77–80] spontaneously to carbonyl compounds, by a mechanism which has been shown to be intramolecular (*Scheme 18.29*). An alternative approach[81] which involves a two-reagent system is also shown; the latter method does not appear to proceed via generation of an intermediate silyl phosphite, but rather as depicted in *Scheme 18.29.*

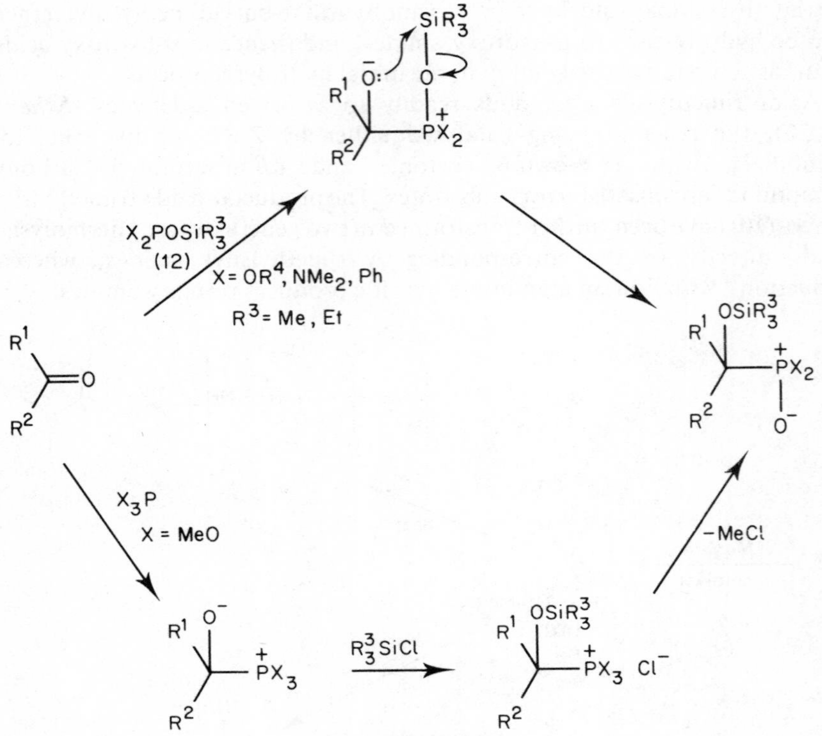

Scheme 18.29

With $\alpha\beta$-unsaturated substrates, either 1,2- or 1,4-addition, or a mixture of both, takes place, the particular outcome depending on the structure of the substrate and, to a lesser extent, that of the phosphorus reagent. In general terms, $\alpha\beta$-unsaturated aldehydes undergo preferential 1,2-addition, whereas $\alpha\beta$-unsaturated ketones normally give 1,4-adducts as shown in *Scheme 18.30.*

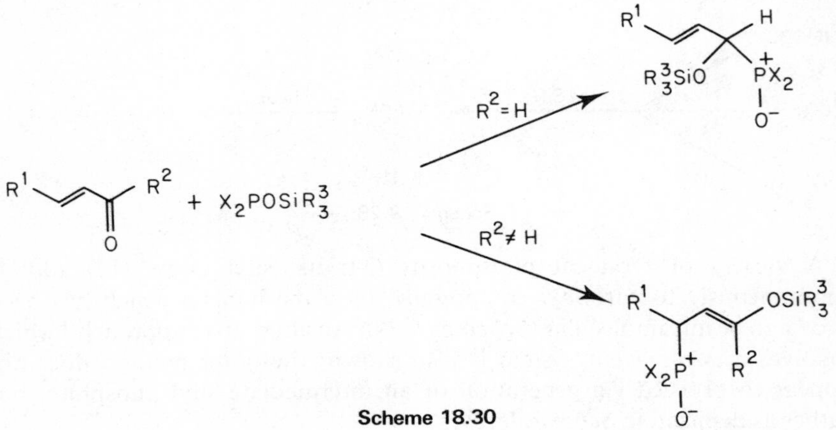

Scheme 18.30

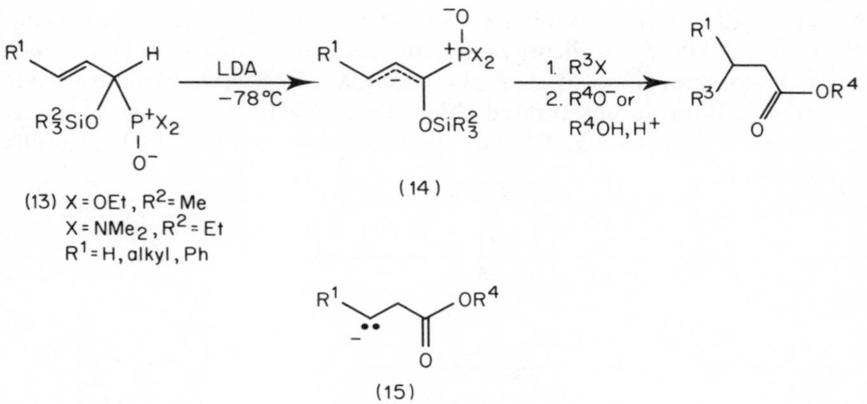

(13) X = OEt, R^2 = Me
X = NMe_2, R^2 = Et
R^1 = H, alkyl, Ph

(14)

(15)

Scheme 18.31

It is the products of addition to $\alpha\beta$-unsaturated aldehydes which have shown greatest potential so far. Such species (13) undergo ready deprotonation[82, 83] to give the stabilized anion (14). This anion reacts regioselectively with carbon electrophiles at what was originally the β-position; solvolysis then produces β-substituted esters or related species. In other words, the anion (14) is acting as the synthetic equivalent of the homoenolate anion (15) (*Scheme 18.31*). The adduct (16) derived from benzaldehyde reacts in a similar manner to the related adduct produced by cyanosilylation. Deprotonation gives the masked acyl anion (17), which leads on alkylation[82, 84] to alkyl phenyl ketones, and on reaction with carbonyl compounds[85] to acyloin silyl ethers (*Scheme 18.32*).

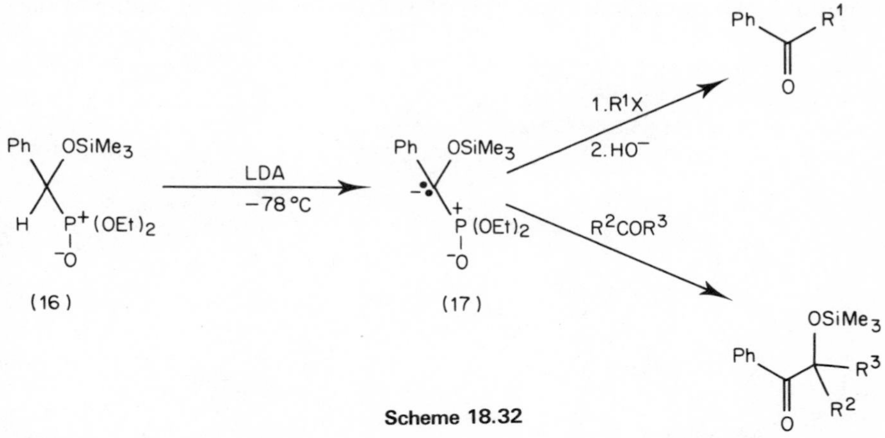

(16)

(17)

Scheme 18.32

Earlier reports[86] on the carbonyl-insertion reactions of alkylthiotrialkylsilanes indicated rather sluggish reactivity with all but highly electropositive carbonyl compounds. Additionally, enolizable ketones[87] gave rise to silyl enol ethers. In contrast, alkylthio- and arylthio-trimethylsilanes[88] react smoothly with aldehydes under the catalytic influence of anions such as cyanide or fluoride ions, or of imidazole[89], to produce *O*-silyl hemithioacetals (*Scheme 18.33*) in excellent yields. Ketones are unreactive under such catalytic conditions, so offering the potential of selective carbonyl group protection. Both $\alpha\beta$-unsaturated aldehydes and ketones undergo exclusive 1,4-addition to generate silyl enol ethers, a structural type of high utility (Chapter 17).

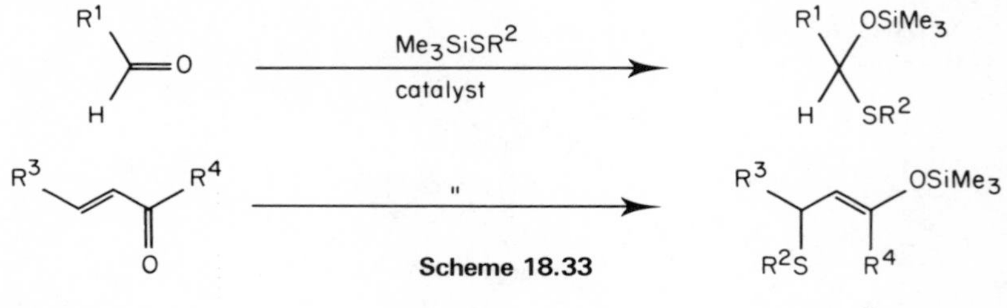

Scheme 18.33

When mild Lewis acids such as ZnI_2 are employed as catalyst, and when at least two equivalents of thiosilane are present, both aldehydes *and* ketones[88] are transformed into dithioacetals or dithioketals, respectively (*Scheme 18.34*); such a mild reagent system can be highly discriminating in selective carbonyl protection. The reaction can be stopped, if desired, at the intermediate hemithioacetal or hemithioketal stage. These species can also be produced[90] by use of the combination of chlorotrimethylsilane and a thiol in the presence of pyridine; it has been suggested[88] that the reactive thiosilane is formed *in situ*, and that the co-produced pyridine hydrochloride is then acting as an acid catalyst. Indeed, it has proven possible[88] to control acid-catalysed thiosilane carbonyl insertion processes to give either *O*-silyl hemithioketals or dithioketals simply by varying the degree of acidity of the

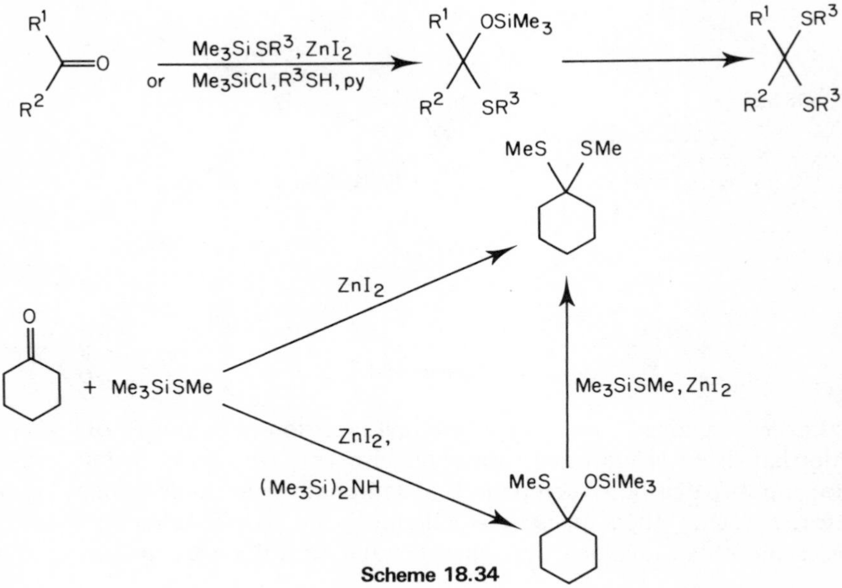

Scheme 18.34

medium. This can be achieved by having a suitable amine either present or absent, and an example is shown in *Scheme 18.34*.

It is thus possible to convert a wide variety of aldehydes and ketones into their corresponding protected forms under conditions which permit a high degree of selectivity. Additionally, *O*-silyl hemithioacetals undergo clean reduction[89] to sulphides as shown in *Scheme 18.35*.

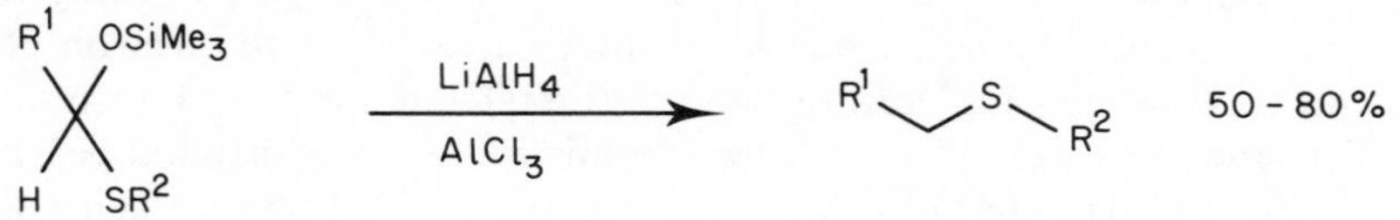

$$R^1CH(OSiMe_3)SR^2 \xrightarrow[AlCl_3]{LiAlH_4} R^1CH_2SR^2 \quad 50-80\%$$

R^1, R^2 = alkyl, aryl

Scheme 18.35

Parenthetically, it should be noted that unsymmetrical sulphides are readily obtained[90] by reaction of alkylthiotrimethylsilanes with alkyl halides (*Scheme 18.36*), the reaction being driven to completion by slow distillation of the halogenotrimethylsilane. Similarly, reaction with hexa-

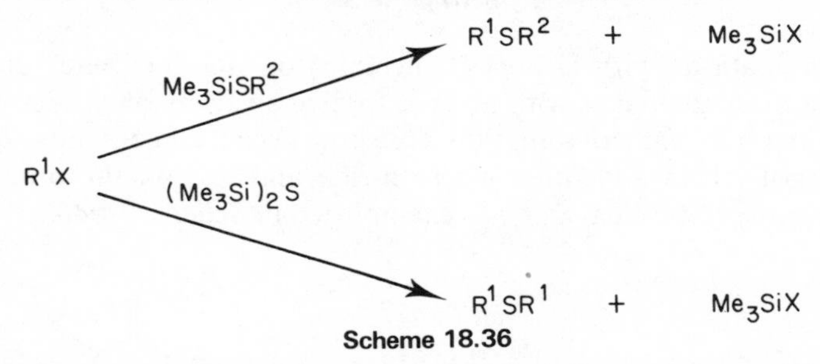

Scheme 18.36

methyldisilathiane leads to symmetrical sulphides. Phenylthiotrimethylsilane catalyses[91] the conjugate addition of benzenethiol to a variety of $\alpha\beta$-unsaturated systems. Unsymmetrical disulphides can be obtained from thiosilanes by reaction[92] with sulphenyl chlorides (*Scheme 18.37*).

$$R^1SCl + Me_3SiSR^2 \longrightarrow R^1SSR^2 + Me_3SiCl$$

Scheme 18.37 70-85%

The final anion-initiated carbonyl insertion process to be discussed here is the addition[93] of ethyl trimethylsilyldiazoacetate (18) to aldehydes, to give the aldol products[94] (19) (*Scheme 18.38*), even with $\alpha\beta$-unsaturated aldehydes. Treatment of such species as the free alcohols with Lewis acids produces acetylenic esters[95].

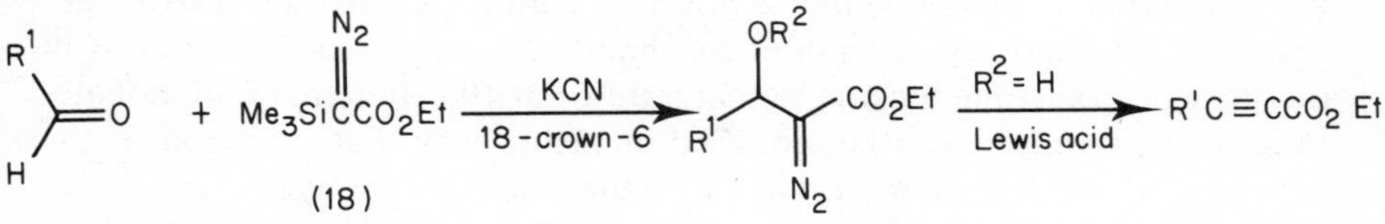

$$R^1CHO + Me_3SiC(=N_2)CO_2Et \ (18) \xrightarrow[18\text{-crown-6}]{KCN} R^1CH(OR^2)C(=N_2)CO_2Et \xrightarrow[\text{Lewis acid}]{R^2 = H} R^1C\equiv CCO_2Et$$

(19), $R^2 = SiMe_3$

Scheme 18.38

18.4 Some other addition processes

Azidotrimethylsilane is a reactive 1,3-dipolar species, capable of adding to both alkynes and alkenes. Its use is preferable to that of the highly explosive hydrazoic acid for the synthesis of 1,2,3-triazoles (*Scheme 18.39*) and related heterocyclic systems, the silyl group in the product being hydrolytically labile. It complements sodium azide in such cycloadditions, working best with electron-rich alkynes; azide ion is more effective with electron-poor substrates. This general area (amongst others) has been well reviewed[96] by Birkofer, a specialist in the chemistry of azidotrimethylsilane. Azidotrimethylsilane has also been used for the conversion of alkenes into aziridines[97], it being much safer to handle than the alternative arylsulphonyl azides.

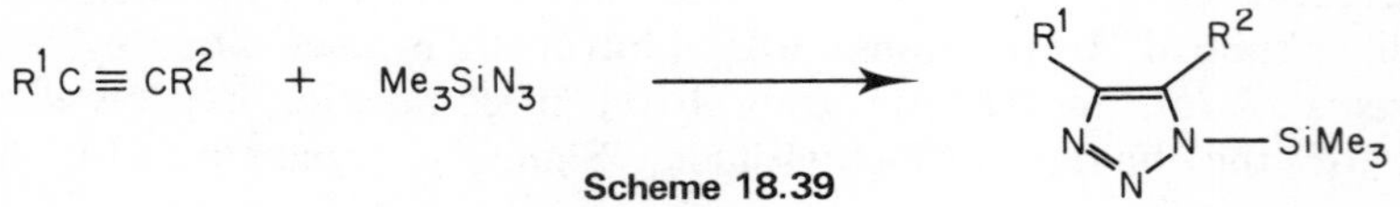

$$R^1C\equiv CR^2 \quad + \quad Me_3SiN_3 \longrightarrow$$

Scheme 18.39

In combination with lead(IV) acetate or iodobenzene diacetate, azidotrimethylsilane reacts with alkenes to give a variety of products[98]. The latter, milder reagent combination converts cyclic alkenes into α-azido-ketones; enol ethers and other electron-rich alkenes, on the other hand, undergo regiospecific cleavage[99], as exemplified in *Scheme 18.40*.

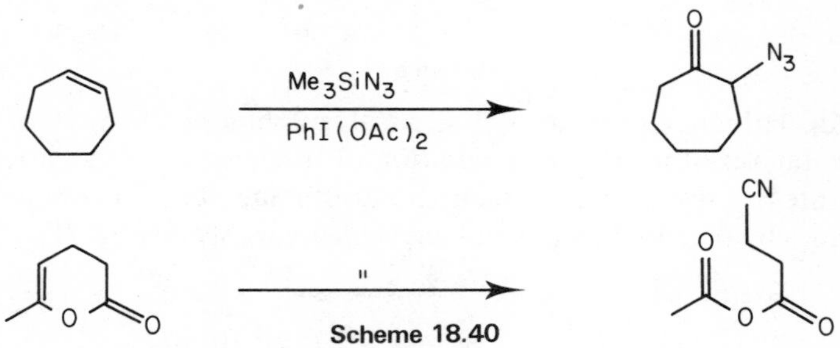

Scheme 18.40

18.5 Some other functionalized silane reagents

A wide range of nucleosides and nucleotides is synthesized[100] efficiently by interaction between *O*-silylated heterocycles and protected sugar derivatives. For spontaneous reaction, halogenoses or similar reactive derivatives are required. Alternatively, the more easily handled acetates react[101] smoothly in the presence of trimethylsilyl perchlorate or trifluoromethanesulphonate. Indeed, the sequence can be modified in such a way that silylation of the heterocyclic base, preparation of the active catalyst (20), and carbon–nitrogen bond formation with a per-acylated sugar can all be performed in one synthetic operation[102]. Sufficient amounts of chlorotrimethylsilane and hexamethyldisilazane are employed such as to ensure full silylation of all reactive groupings and also silylation of the requisite anions for production of the catalyst (20) (*Scheme 18.41*).

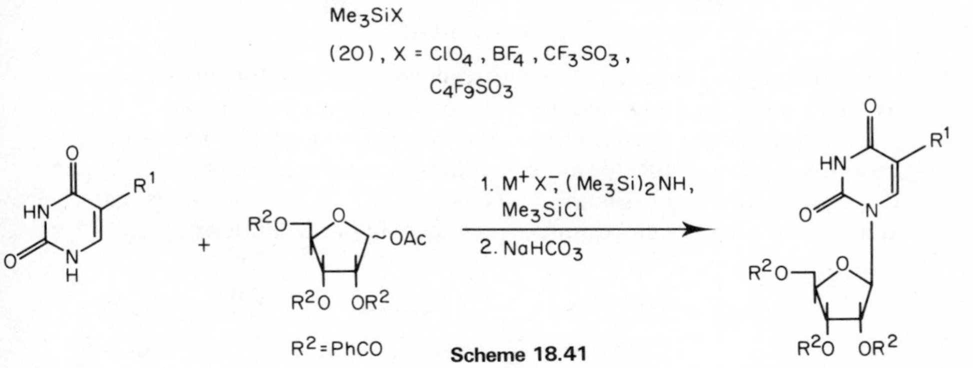

Me_3SiX

(20), X = ClO_4, BF_4, CF_3SO_3, $C_4F_9SO_3$

1. M^+X^-, $(Me_3Si)_2NH$, Me_3SiCl
2. $NaHCO_3$

R^2 = PhCO

Scheme 18.41

It is not normally possible[103] to convert an inactive, ester-protected acid directly and non-hydrolytically into an activated acid derivative under mild conditions. By considering the equilibrium shown in *Scheme 18.42*, Masamune[104] reasoned that if R^2O^- could be removed by reaction with MY, M^+ being a relatively hard acid with a strong affinity for oxygen and Y^- being a relatively soft base, then such a desirable sequence might become feasible. *N*-Trimethylsilylimidazole[105] fills such a role admirably, converting both phenyl and trifluoroethyl esters rapidly into acid imidazolides[106].

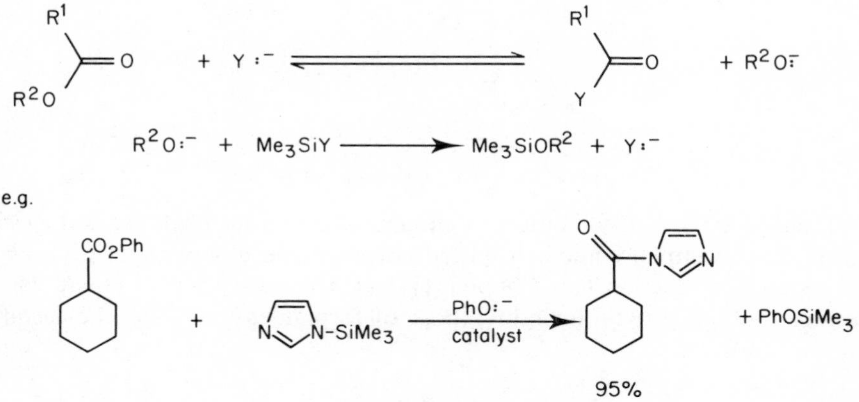

$R^2O:^- + Me_3SiY \longrightarrow Me_3SiOR^2 + Y:^-$

e.g.

PhO:⁻ catalyst; 95%; + $PhOSiMe_3$

Scheme 18.42

N-Trimethylsilylimidazole reacts with sulphur dichlorides to give efficient sulphur transfer[107, 108] reagents (21) (*Scheme 18.43*).

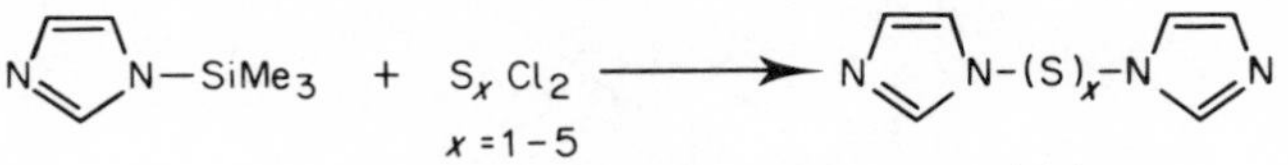

S_xCl_2, x = 1–5; (21)

Scheme 18.43

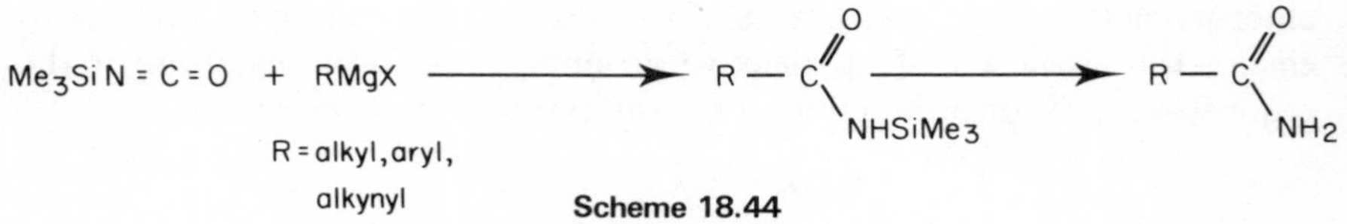

$Me_3SiN{=}C{=}O + RMgX \longrightarrow R{-}C(=O)NHSiMe_3 \longrightarrow R{-}C(=O)NH_2$

R = alkyl, aryl, alkynyl

Scheme 18.44

A wide range of Grignard reagents react with isocyanatotrimethylsilane to give homologous primary amides[109] in good yields (*Scheme 18.44*).

In a heavily patented process, tertiary ethynyl carbinols rearrange[110] smoothly to $\alpha\beta$-unsaturated aldehydes in the catalytic presence of silyl esters of vanadic acid and silanol (*Scheme 18.45*). Allylic alcohols can be equilibrated[111] with their regioisomers under similar conditions.

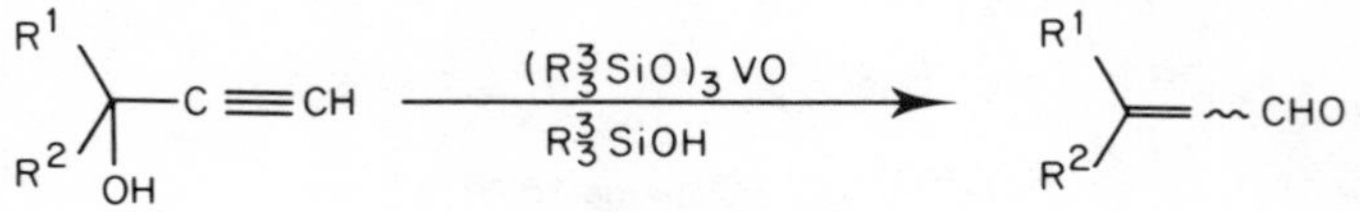

$R^1R^2C(OH){-}C{\equiv}CH \xrightarrow[R^3_3SiOH]{(R^3_3SiO)_3VO} R^1R^2C{=}CH{\sim}CHO$

Scheme 18.45

$(MeO)_2C< \xrightarrow[\text{Catalytic}]{Me_3SiOSO_2CF_3} [MeO^+(SiMe_3){-}C(OMe)< \longleftrightarrow MeO(SiMe_3)\;C{=}O^+Me] \xrightarrow{>C{=}C(OSiMe_3){-}} O{=}C{-}C{-}C(OMe)<$

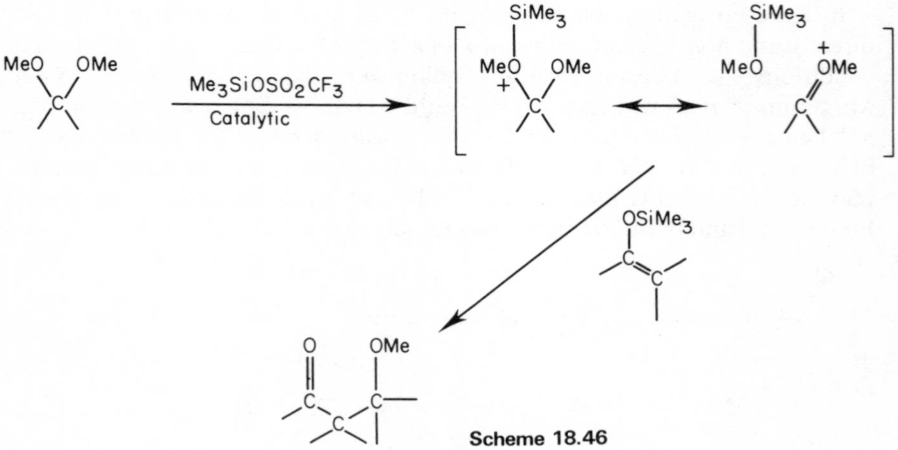

Scheme 18.46

Trialkylsilyl trifluoromethanesulphonates act as hard but weak Lewis acids. For example, trimethylsilyl trifluoromethanesulphonate catalyses[112] an aldol-type reaction (*see* Chapter 17.3.4) between silyl enol ethers and acetals (*Scheme 18.46*), including those of formaldehyde. Here, the acetal

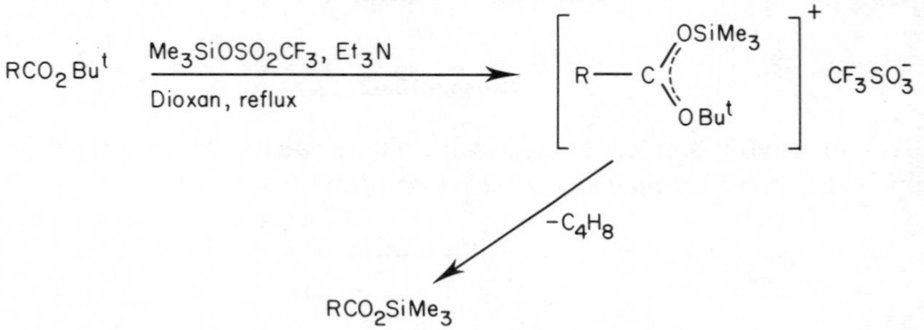

$RCO_2Bu^t \xrightarrow[\text{Dioxan, reflux}]{Me_3SiOSO_2CF_3,\ Et_3N} \left[R{-}C(OSiMe_3)(OBu^t)\right]^+ CF_3SO_3^- \xrightarrow{-C_4H_8} RCO_2SiMe_3$

Scheme 18.47

moiety is acting as an activating group rather than as a protecting group, since the reaction fails with aldehydes or ketones themselves. In appropriate cases, a high degree of *erythro* diastereoselectivity is observed. The same reagent transforms[113] carboxylic acid t-butyl esters into the corresponding trimethylsilyl esters (*Scheme 18.47*); benzyl esters survive these conditions. Trialkylsilyl trifluoromethanesulphonates have also been used[114] to generate the highly reactive heterodiene (22) from the epoxynitrone (23) (*Scheme 18.48*).

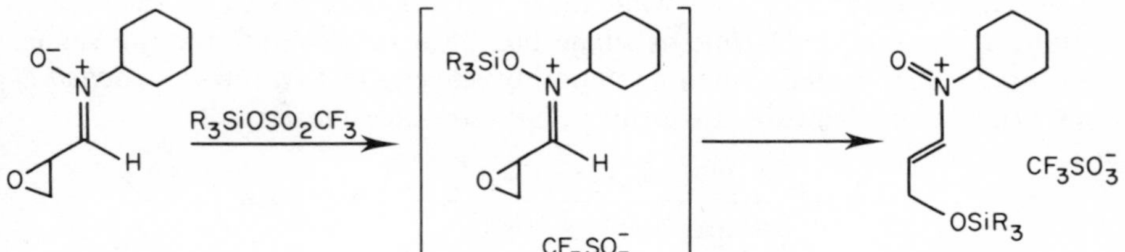

$R_3SiOSO_2CF_3$

(23) (22)

Scheme 18.48

Other applications of these sulphonate esters, including their use for the conversion of oxiranes into allylic alcohols and their general utility as silylating agents, are discussed in Chapters 15 and 17.

18.6 Addendum

18.6.1 Reagent preparation

Bis(trimethylsilyl) sulphate is readily available[115], and it can act[116] as a simple source of certain silyl reagents (*Scheme 18.49*). Phenylthio- and phenylseleno-trimethylsilane are conveniently prepared[117] by silylation of the thallium(I) salts of thiophenol and selenophenol, respectively.

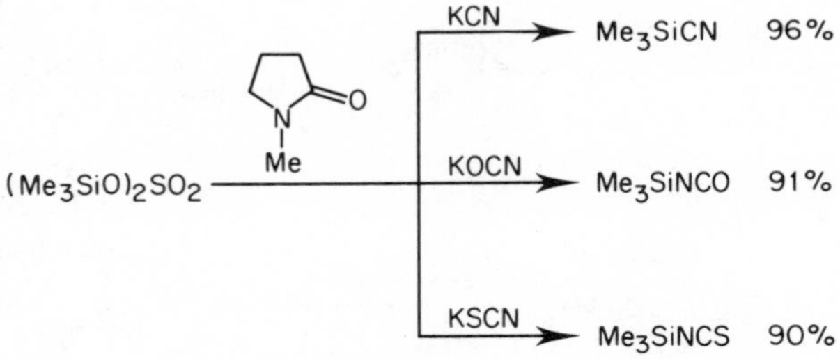

$(Me_3SiO)_2SO_2$ → KCN → Me_3SiCN 96%; KOCN → Me_3SiNCO 91%; KSCN → Me_3SiNCS 90%

Scheme 18.49

18.6.2 Cleavage reactions

Iodo-t-butyldimethylsilane (24), generated *in situ* as shown in *Scheme 18.50*, converts oxiranes into allylic alcohols[118]. The regioselectivity displayed is broadly similar to that observed with trimethylsilyl trifluoromethanesulphonate (p. 183), although here endocyclic alkene formation is preferred.

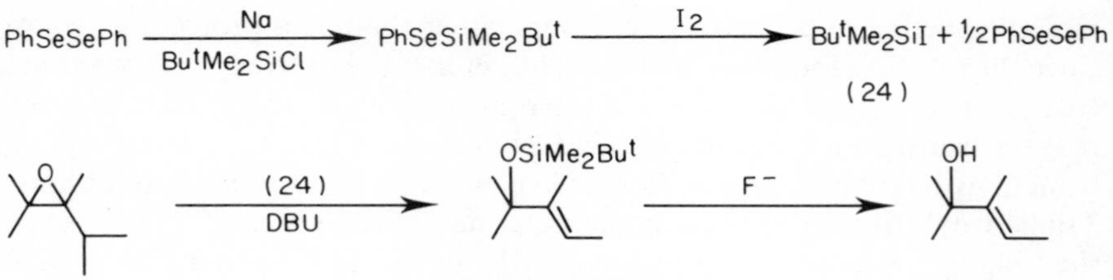
PhSeSePh
Na
ButMe2 SiCl
PhSeSiMe2 But
I2
ButMe2SiI + ½PhSeSePh
(24)
(24)
DBU
OSiMe2But
F⁻
OH
H

Scheme 18.50

The scope seems to be somewhat more general in terms of oxirane substitution pattern. Iodotrimethylsilane has been employed[119] for the same purpose. Thiotrimethylsilanes smoothly cleave methyl ethers (*Scheme 18.51*); benzyl ethers are also cleaved, but esters are unaffected[120].

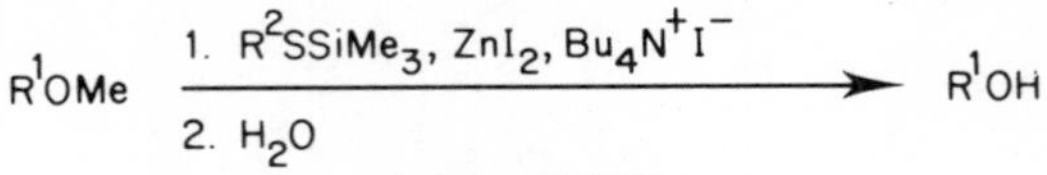
R1OMe
1. R2SSiMe3, ZnI2, Bu4N+I−
2. H2O
R1OH

Scheme 18.51

The reagent systems of chlorotrimethylsilane/lithium bromide and hexamethyldisilane/pyridinium hydrobromide perbromide convert alkyl, cycloalkyl and aralkyl alcohols into the corresponding bromides (*Scheme 18.52*). Although both combinations presumably generate bromotrimethylsilane, the latter system shows a remarkably high degree of selectivity for reaction with tertiary alcohols[121].

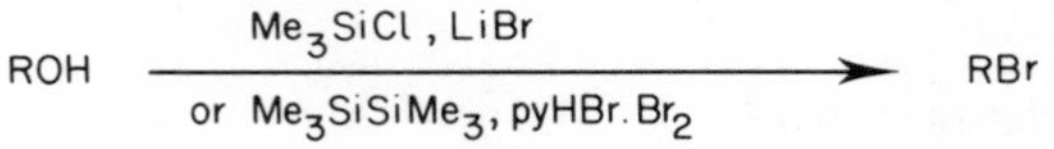
ROH
Me3SiCl, LiBr
or Me3SiSiMe3, pyHBr.Br2
RBr

Scheme 18.52

An impressive sequence[122] of silyl-based reactions leading from pyruvic acid to the trisodium salt of phosphoenol pyruvate is outlined in *Scheme 18.53*.

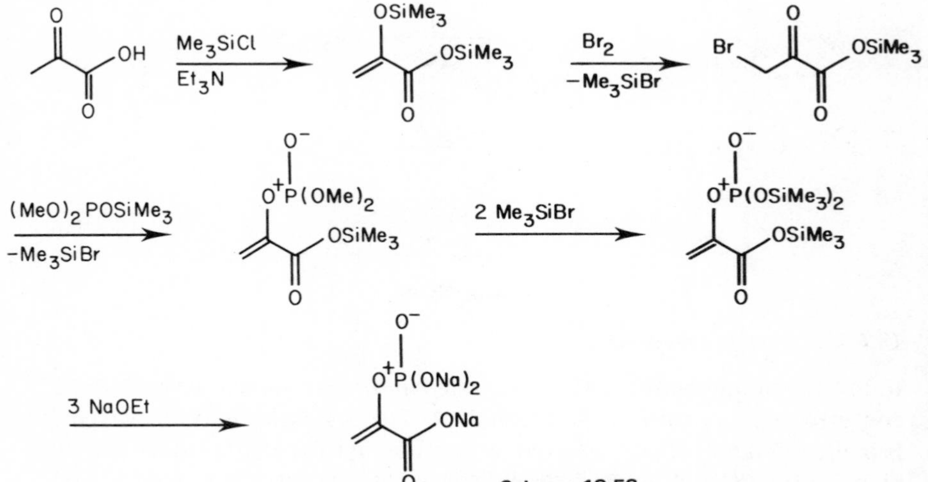
Me3SiCl
Et3N
OSiMe3
OSiMe3
Br2
−Me3SiBr
Br
OSiMe3
(MeO)2 POSiMe3
−Me3SiBr
O+P(OMe)2
2 Me3SiBr
O+P(OSiMe3)2
3 NaOEt
O+P(ONa)2
ONa

Scheme 18.53

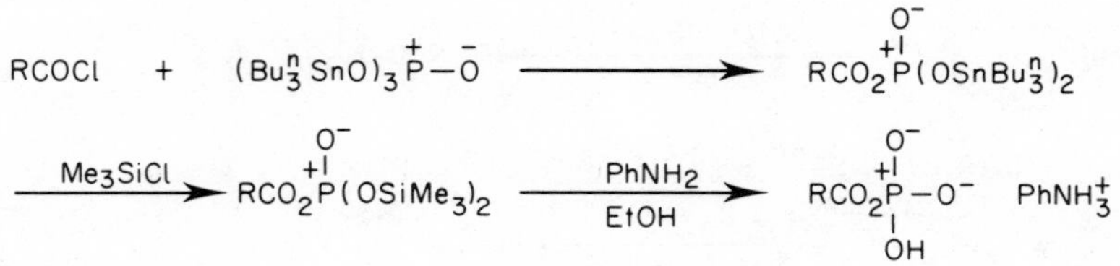

$$\mathrm{RCOCl} + (\mathrm{Bu^n_3SnO})_3\overset{+}{\mathrm{P}}-\overset{-}{\mathrm{O}} \longrightarrow \mathrm{RCO_2}\overset{+}{\mathrm{P}}(\mathrm{O^-})(\mathrm{OSnBu^n_3})_2$$

$$\xrightarrow{\mathrm{Me_3SiCl}} \mathrm{RCO_2}\overset{+}{\mathrm{P}}(\mathrm{O^-})(\mathrm{OSiMe_3})_2 \xrightarrow[\mathrm{EtOH}]{\mathrm{PhNH_2}} \mathrm{RCO_2}\overset{+}{\mathrm{P}}(\mathrm{O^-})(\mathrm{OH})-\mathrm{O^-}\ \ \mathrm{PhNH_3^+}$$

Scheme 18.54

Finally, two cleavage processes have been described in which carbon–oxygen fission is not involved. Unesterified acyl phosphates have been prepared[123] from stannyl phosphate intermediates by tin–oxygen bond fission (*Scheme 18.54*).

Iodotrimethylsilane cleaves aminals to produce Mannich salts[124], as exemplified in *Scheme 18.55* for the preparation of dimethyl(methylene)-ammonium iodide.

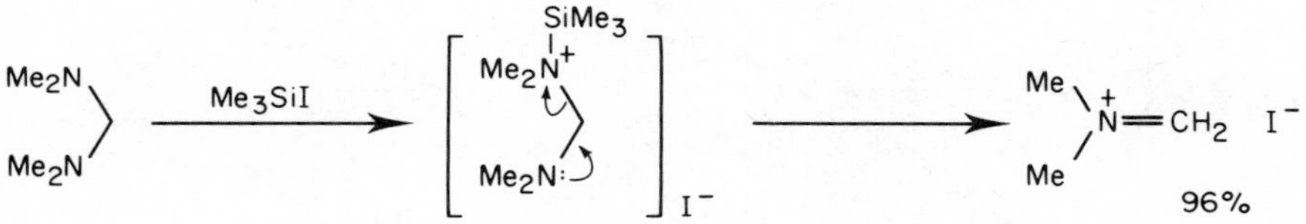

$$(\mathrm{Me_2N})_2\mathrm{CH_2} \xrightarrow{\mathrm{Me_3SiI}} \left[\mathrm{Me_2}\overset{+}{\mathrm{N}}(\mathrm{SiMe_3})-\mathrm{CH_2}-\mathrm{NMe_2}\right]\mathrm{I^-} \longrightarrow \mathrm{Me_2}\overset{+}{\mathrm{N}}=\mathrm{CH_2}\ \ \mathrm{I^-}\quad 96\%$$

Scheme 18.55

18.6.3 Carbonyl addition processes

Phenylselenotrimethylsilane adds to $\alpha\beta$-unsaturated carbonyl compounds to give products of 1,4-addition (*Scheme 18.56*), paralleling the reactivity of thiotrialkylsilanes; products of 1,2-addition are formed from saturated aldehydes only, ketones reacting too sluggishly. Suitable catalysts for the additions are triphenylphosphine[125] or iodotrimethylsilane[126].

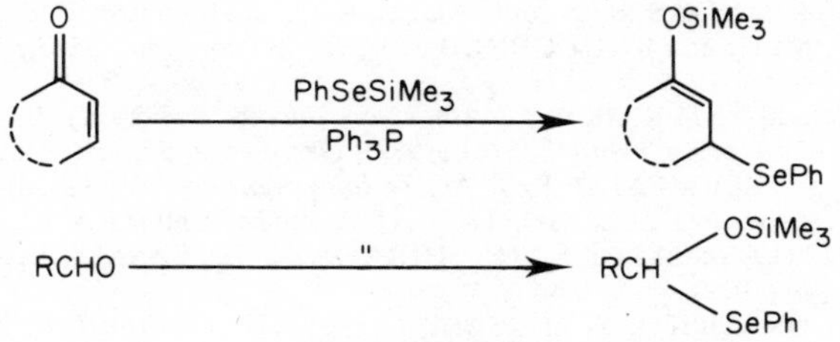

$$\text{cyclic enone} \xrightarrow[\mathrm{Ph_3P}]{\mathrm{PhSeSiMe_3}} \text{cyclic } \mathrm{OSiMe_3} \text{ enol ether bearing SePh}$$

$$\mathrm{RCHO} \xrightarrow{''} \mathrm{RCH(OSiMe_3)(SePh)}$$

Scheme 18.56

Cyanohydrins of saturated aldehydes are oxidized to acyl cyanides and thence to carboxylic acids when treated with pyridinium dichromate. If, however, the parent aldehyde is $\alpha\beta$-unsaturated, the β-carbon is disubstituted, and the γ-carbon carries at least one hydrogen, then the derived cyanohydrin trimethylsilyl ethers are oxidized to Δ^2-butenolides[127]. Preferential γ-methyl rather than γ-methylene or γ-methine oxidation is observed, but the original stereochemistry of the double bond is

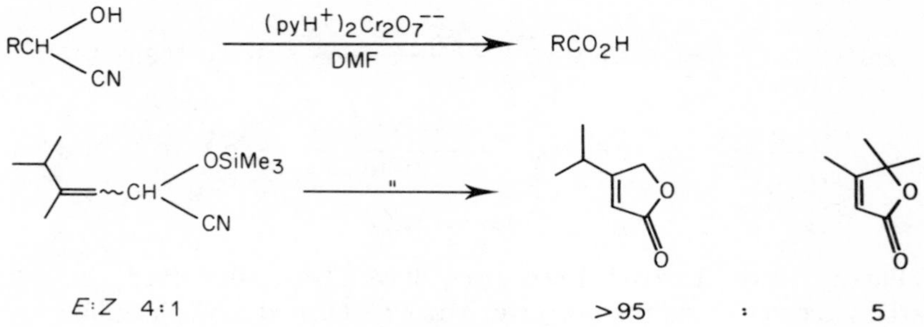

Scheme 18.57

unimportant, as exemplified in *Scheme 18.57*. Suitably substituted cyanohydrin trimethylsilyl ethers undergo Cope rearrangement[128] (*Scheme 18.58*) to give products of a formal homologous Claisen ester rearrangement.

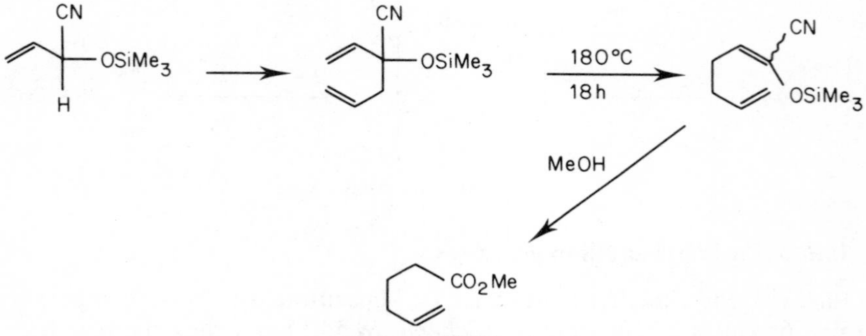

Scheme 18.58

References

1 GILLIAM, W. F., MEALS, R. N. and SAUER, R. O., *J. Am. chem. Soc.* **68,** 1161 (1946)
2 VORONKOV, M. G. and KHUDOBIN, Yu. I., *Izv. Akad. Nauk SSSR, Ser. khim.* 713 (1956)
3 JUNG, M. E. and LYSTER, M. A., *J. Am. chem. Soc.* **99,** 968 (1977); JUNG, M. E. and LYSTER, M. A., *J. org. Chem.* **42,** 3761 (1977); *Org. Synth.* **59,** 35 (1980)
4 JUNG, M. E. and BLUMENKOPF, T. A., *Tetrahedron Lett.* 3657 (1978)
5 OLAH, G. A., NARANG, S. C., GUPTA, B. G. B. and MALHOTRA, R., *Angew. Chem. int. Edn* **18,** 612 (1979); SAKURAI, H., SHIRAHATA, A., SASAKI, K. and HOSOMI, A., *Synthesis* 740 (1979)
6 MORITA, T., OKAMOTO, Y. and SAKURAI, H., *Tetrahedron Lett.* 2523 (1978); *J. chem. Soc. chem. Communs* 874 (1978)
7 OLAH, G. A., NARANG, S., GUPTA, B. G. B. and MALHOTRA, R., *J. org. Chem.* **44,** 1247 (1979); *Synthesis* 61 (1979)
8 BITHER, T. A., KNOTH, W. H., LINDSEY, R. V. and SHARKEY, W. H., *J. Am. chem. Soc.* **80,** 4151 (1958); EVERS, E. C., FREITAG, W. O., KEITH, J. N., KRINER, W. A., MacDIARMID, A. G. and SUJISHI, S., *J. Am. chem. Soc.* **81,** 4493 (1959); ZUBRICK, J. W., DUNBAR, B. I. and DURST, H. D., *Tetrahedron Lett.* 71 (1975); UZNANSKI, B. and STEC, W. J., *Synthesis* 154 (1978); RYU, I., MURAI, S., HORIIKE, T., SHINONAGI, A. and SONODA, N., *Synthesis* 154 (1978); HÜNIG, S. and WEHNER, G., *Synthesis* 522 (1979)
9 TAYLOR, E. C., ANDRADE, J. G., JOHN, K. C. and McKILLOP, A., *J. org. Chem.* **43,** 2280 (1978)
10 RASMUSSEN, J. K. and HEILMANN, S. M., *Synthesis* 523 (1979)

11 EVANS, D. A., CARROLL, G. L. and TRUESDALE, L. K., *J. org. Chem.* **39,** 914 (1974)
12 BIRKOFER, L., RITTER, A. and RICHTER, P., *Chem. Ber.* **96,** 2750 (1963)
13 *Org. Synth.* **50,** 107 (1970); *see also* WASHBURNE, S. S. and PETERSON, W. R., *J. Organometal. Chem.* **33,** 153 (1971)
14 ABEL, E. W. and ARMITAGE, D. A., *Adv. organometal. Chem.* **5,** 2 (1967)
15 GLASS, R. S., *J. Organometal. Chem.* **61,** 83 (1973)
16 OJIMA, I., NIHONYANAGI, M. and NAGAI, T., *J. organometal. Chem.* **50,** C26 (1973); OJIMA, I. and NAGAI, Y., *J. organometal. Chem.* **57,** C42 (1973); KUWAJIMA, I. and ABE, T., *Bull. chem. Soc. Japan* **51,** 2183 (1978)
17 EVANS, D. A., TRUESDALE, L. K., GRIMM, K. G. and NESBITT, S. L., *J. Am. chem. Soc.* **99,** 5009 (1977)
18 ABEL, E. W., *J. chem. Soc.* 4406 (1960); 4933 (1961)
19 ABEL, E. W., ARMITAGE, D. A. and BUSH, R. P., *J. chem. Soc.* 2455 (1964)
20 HARPP, D. N. and STELIOU, K., *Synthesis* 721 (1976); LOUIS, E. and URRY, G., *J. Inorg. Chem.* **7,** 1253 (1968)
21 *See also* OLAH, G. A., GUPTA, B. G. B., NARANG, S. C. and MALHOTRA, R., *J. org. Chem.* **44,** 4272 (1979)
22 BUGERENKO, E. F., CHERNYSHEV, E. A. and POPOV, E. M., *Izv. Akad. Nauk SSSR, Ser. khim.* 1391 (1966); *Chem. Abstr.* **66,** 76078 (1967)
23 ORLOV, N. F. and SUDAKOVA, E. V., *J. gen. Chem. U.S.S.R.* **39,** 211 (1969)
24 NESTEROV, L. V., KREPYSHEVA, N. E., SABIROVA, R. A. and ROMANOVA, G. N., *J. gen. Chem. U.S.S.R.* **41,** 2449 (1971)
25 EVANS, D. A., HURST, K. M. and TAKACS, J. M., *J. Am. chem. Soc.* **100,** 3467 (1978)
26 *E.g.* VORONKOV, M. G. and KHUDOBIN, Yu. I., *Zh. obshch. Khim.* **26,** 584 (1956); *Chem. Abstr.* **50,** 13729 (1956); ZAKHARKIN, L. I., STANKO, V. I. and BRATTSEV, V. A., *Izv. Akad. Nauk SSSR, Ser. khim.* 2069 (1961); *Chem. Abstr.* **56,** 10177 (1962); ADRIANOV, K. A., KURAKOV, G. A. and KHANANASHVILI, L. M., *J. gen. Chem. U.S.S.R.* **35,** 395 (1965)
27 LAVIGNE, A. A., TANCREDE, J., PIKE, R. M. and TABIT, C. T., *J. organometal. Chem.* **15,** 57 (1968); CEDER, O. and HANSSON, B., *Acta chem. scand.* **B30,** 574 (1976)
28 BIRKOFER, L. and KAISER, W., *Justus Liebigs Annln Chem.* 266 (1975)
29 DENIS, J. N., MAGNANE, R., Van EENOO, M. and KRIEF, A., *Nouveau J. Chim.* **3,** 705 (1979)
30 ABEL, E. W. and WALKER, D. J., *J. chem. Soc. (A)* 2338 (1968)
31 HO, T.-L., *Synth. Commun.* **9,** 37 (1979)
32 (*a*) DANNLEY, R. and JALICS, G., *J. org. Chem.* **30,** 2417 (1965); (*b*) REBEK, J. and McCREADY, R., *Tetrahedron Lett.* 4337 (1979)
33 COREY, E. J., MARFAT, A., FALCK, J. R. and ALBRIGHT, J. O., *J. Am. chem. Soc.* **102,** 1433 (1980)
34 NEWMAN, M. S. and OLSON, D. R., *J. org. Chem.* **38,** 4203 (1973)
35 HARTMANN, W., HEINE, H.-G. and WENDISCH, D., *Tetrahedron Lett.* 2263 (1977)
36 HARTMANN, W. and HEINE, H.-G., *Tetrahedron Lett.* 513 (1979)
37 *See,* for example, SEKINE, M. and HATA, T., *J. chem. Soc. chem. Communs* 285 (1978)
38 RABINOWITZ, R., *J. org. Chem.* **28,** 2975 (1963); VORONKOV, M. S. and ZAGONNIK, W. N., *Zh. obshch. Khim.* **27,** 2975 (1957)
39 McKENNA, C. E., HIGA, M. T., CHEUNG, N. H. and McKENNA, M.-C., *Tetrahedron Lett.* 155 (1977); RUDINSKAS, A. J. and HULLAR, T. L., *J. mednl. pharm. Chem.* **19,** 1367 (1976); GROSS, H., BOCK, C., COSTISELLA, B. and GLOEDE, J., *J. prakt. Chem.* **320,** 344 (1978)
40 ZYGMUNT, J., KAFARSKI, P. and MASTALERZ, P., *Synthesis* 609 (1978); CHOJNOWSKI, J., CYPRYK, M. and MICHALSKI, J., *Synthesis* 777 (1978); BLACKBURN, G. M. and INGLESON, D., *J. chem. Soc. chem. Communs* 870 (1978)
41 TAGEUCHI, Y., DEMACHI, Y. and YOSHII, E., *Tetrahedron Lett.* 1231 (1979)
42 HO, T.-L. and OLAH, G. A., *Angew. Chem. int. Edn* **15,** 774 (1976)
43 BENKESER, R. A., MOZDZEN, E. C. and MUTH, C. L., *J. org. Chem.* **44,** 2185 (1979)
44 LOTT, R. S., CHAUHAN, V. S. and STAMMER, C. H., *J. chem. Soc. chem. Communs* 495 (1979)
45 JUNG, M. E. and LYSTER, M. A., *J. chem. Soc. chem. Communs* 315 (1978)
46 *See also* PIRKLE, W. H. and HAUSKE, J. R., *J. org. Chem.* **42,** 2781 (1977)
47 HO, T.-L., *Synth. Communs* **9,** 233 (1979)
48 KRICHELDORF, H. R., *Angew. Chem. int. Edn* **18,** 689 (1979)
49 KRICHELDORF, H. R., *Chem. Ber.* **106,** 3765 (1973)

50 HIYAMA, T., SAIMOTO, H., NISHIO, K., SHINODA, M., YAMAMOTO, H. and NOZAKI, H., *Tetrahedron Lett.* 2043 (1979)
51 MINAMIKAWA, J. and BROSSI, A., *Tetrahedron Lett.* 3085 (1978)
52 JUNG, M. E. and ORNSTEIN, P. L., *Tetrahedron Lett.* 2659 (1977)
53 HO, T.-L., *Synth. Communs* **9,** 665 (1979)
54 (*a*) JUNG, M. E., ANDRUS, W. A. and ORNSTEIN, P. L., *Tetrahedron Lett.* 4175 (1977); (*b*) BRYANT, J. D., KEYSER, G. E. and BARRIO, J. R., *J. org. Chem.* **44,** 3733 (1979)
55 JUNG, M. E., MAZUREK, M. A. and LIM, R. M., *Synthesis* 588 (1978)
56 KEYSER, G. E., BRYANT, J. D. and BARRIO, J. R., *Tetrahedron Lett.* 3263 (1979)
57 MASAMUNE, S., *Aldrichim. Acta* **11,** 23 (1978)
58 JUNG, M. E., MOSSMAN, A. B. and LYSTER, M. A., *J. org. Chem.* **43,** 3698 (1978)
59 MILLER, R. D. and McKEAN, D. R., *Tetrahedron Lett.* 2305 (1979); 2639 (1980)
60 MILLER, R. D. and McKEAN, D. R., *Synthesis* 730 (1979)
61 NEEF, H. and MÜLLER, R., *J. prakt. Chem.* **315,** 367 (1973)
62 EVANS, D. A., TRUESDALE, L. K. and CARROLL, G. L., *J. chem. Soc. chem. Communs* 55 (1973); EVANS, D. A. and TRUESDALE, L. K., *Tetrahedron Lett.* 4929 (1973)
63 LIDY, W. and SUNDERMEYER, W., *Chem. Ber.* **106,** 587 (1973)
64 GASSMAN, P. A. and TALLEY, J. J., *Tetrahedron Lett.* 3773 (1978); *Org. Synth.* **59,** procedure 2115 (1980)
65 DEUCHERT, K., HERTENSTEIN, U., HÜNIG, S. and WEHNER, G., *Chem. Ber.* **112,** 2045 (1979)
66 HERTENSTEIN, U., HÜNIG, S. and ÖLLER, M., *Synthesis* 416 (1976); for an extension, *see* JACOBSON, R. M., LAHM, G. P. and CLADER, J. W., *J. org. Chem.* **45,** 395 (1980)
67 For other acyl anion equivalents, *see* LEVER, O. W., *Tetrahedron* **32,** 1943 (1976); SEEBACH, D., *Angew. Chem. int. Edn* **18,** 239 (1979)
68 HÜNIG, S. and WEHNER, G., *Chem. Ber.* **112,** 2062 (1979)
69 EVANS, D. A., HOFFMAN, J. M. and TRUESDALE, L. K., *J. Am. chem. Soc.* **95,** 5822 (1973); EVANS, D. A. and WONG, R. Y., *J. org. Chem.* **42,** 350 (1977)
70 STORK G. and KRAUS, G., *J. Am. chem. Soc.* **98,** 6747 (1976)
71 EVANS, D. A., CARROLL, G. L. and TRUESDALE, L. K., *J. org. Chem.* **39,** 914 (1974); TAKAISHI, N., FUGIKURA, Y., INAMOTO, Y. and AIGAMI, K., *J. org. Chem.* **42,** 1737 (1977)
72 (*a*) COREY, E. J., TIUS, M. A. and DAS, J., *J. Am. chem. Soc.* **102,** 1742 (1980); (*b*) COREY, E. J., CROUSE, D. N. and ANDERSON, J. E., *J. org. Chem.* **40,** 2140 (1975); *see also* GRUNEWALD, G. L., BROUILLETTE, W. J. and FINNEY, J. A., *Tetrahedron Lett.* 1219 (1980)
73 BIRKOFER, L., MÜLLER, F. and KAISER, W., *Tetrahedron Lett.* 2781 (1967); BIRKOFER, L. and KAISER, W., *Justus Liebigs Annln Chem.* 266 (1975)
74 KYBA, E. P. and JOHN, A. M., *Tetrahedron Lett.* 2737 (1977)
75 MacMILLAN, J. H. and WASHBURNE, S. S., *J. org. Chem.* **38,** 2982 (1973) and references therein
76 KRICHELDORF, H. R., *Synthesis* 695 (1972)
77 NOVIKOVA, Z. S., MASHOSHINA, S. N., SAPOZHNIKOVA, T. A. and LUTSENKO, I. F., *J. gen. Chem. U.S.S.R.* **41,** 2655 (1971) and references therein
78 KONOVALOVA, I. V., BURNAEVA, L. A., SAIFULLINA, N. S. and PUDOVIK, A. N., *J. gen. Chem. U.S.S.R.* **46,** 17 (1976) and references therein
79 SEKINE, M., OKIMOTO, K. and HATA, T., *J. Am. chem. Soc.* **100,** 1001 (1978) and references therein
80 EVANS, D. A., HURST, K. M. and TAKACS, J. M., *J. Am. chem. Soc.* **100,** 3647 (1978)
81 BIRNUM, G. H. and RICHARDSON, G. A., *Chem. Abstr.* **60,** 5551d (1964)
82 EVANS, D. A., TAKACS, J. M. and HURST, K. M., *J. Am. chem. Soc.* **101,** 371 (1979)
83 HATA, T., NAKAJIMA, M. and SEKINE, M., *Tetrahedron Lett.* 2047 (1979)
84 HATA, T., HASHIZUME, A., NAKAJIMA, M. and SEKINE, M., *Tetrahedron Lett.* 363 (1978)
85 KOENIGKRAMER, R. E. and ZIMMER, H., *Tetrahedron Lett.* 1017 (1980)
86 ABEL, E. W., WALKER, D. J. and WINGFIELD, J. N., *J. chem. Soc. (A)* 1814 (1968)
87 OJIMA, I. and NAGAI, Y., *J. organometal. Chem.* **57,** C42 (1973)
88 EVANS, D. A., TRUESDALE, L. K., GRIMM, K. G. and NESBITT, S. L., *J. Am. chem. Soc.* **99,** 5009 (1977)

89 GLASS, R. S., *Synth. Communs* **6,** 47 (1976)
90 ONG, B. S. and CHAN, T. H., *Synth. Communs* **7,** 283 (1977)
91 RICCI, A., DANIELI, R. and PIRAZZINI, G., *J. chem. Soc. Perkin I* 1069 (1977)
92 HARPP, D. N., FRIEDLANDER, B. T., LARSEN, C., STELIOU, K. and STOCKTON, A., *J. org. Chem.* **43,** 3481 (1978)
93 EVANS, D. A., TRUESDALE, L. K. and GRIMM, K. G., *J. org. Chem.* **41,** 3335 (1976)
94 For other routes to such products from both aldehydes and ketones, *see* WENKERT, E. and McPHERSON, C. A., *J. Am. chem. Soc.* **94,** 8084 (1972); SCHÖLLKOPF, U., BANHIDAI, B., FRASNELLI, H., MEYER, R. and BECKHAUS, H., *Justus Liebigs Annln Chem.* 1767 (1974)
95 WENKERT, E. and McPHERSON, C. A., *Synth. Communs* **2,** 331 (1972)
96 BIRKOFER, L. and STUHL, O., *Topics curr. Chem.* **88,** 33 (1980)
97 *E.g.,* STOUT, D. M., TAKAYA, T. and MEYERS, A. I., *J. org. Chem.* **40,** 563 (1975)
98 ZBIRAL, E., *Synthesis* 285 (1972)
99 EHRENFREUND, J. and ZBIRAL, E., *Justus Liebigs Annln Chem.* 290 (1973)
100 LUKEVICS, E., ZABLOTSKAYA, A. E. and SOLOMENNIKOVA, I. I., *Russ. chem. Revs* **43,** 140 (1974)
101 VORBRÜGGEN, H. and KROLIKIEWICZ, K., *Angew. Chem. int. Edn* **14,** 421 (1975)
102 VORBRÜGGEN, H. and BENNUA, B., *Tetrahedron Lett.* 1339 (1978)
103 *See,* however, ANDERSON, A. G. and KONO, D. H., *Tetrahedron Lett.* 5121 (1973); BURTON, D. J. and KOPPES, W., *J. chem. Soc. chem. Communs* 425 (1973)
104 BATES, G. S., DIAKUR, J. and MASAMUNE, S., *Tetrahedron Lett.* 4423 (1976)
105 BIRKOFER, L., RICHTER, P. and RITTER, A., *Chem. Ber.* **93,** 8204 (1960)
106 STAAB, H. and ROHR, W., in *'Newer Methods of Preparative Organic Chemistry',* Ed. Foerst, W., p. 61, Academic Press, New York (1968)
107 ZEHER, F. and DEGEN, B., *Angew. Chem. int. Edn* **6,** 703 (1967)
108 HARPP, D. N., STELIOU, K. and CHAN, T. H., *J. Am. chem. Soc.* **100,** 1222 (1978)
109 PARKER, K. A. and GIBBONS, E. G., *Tetrahedron Lett.* 981 (1975); *see also* BOURGEOIS, P., MERAULT, G. and CALAS, R., *J. organometal. Chem.* **59,** C4 (1973)
110 PAULING, H., ANDREWS, D. A. and HINDLEY, N. C., *Helv. chim. Acta* **59,** 1233 (1976); OLSON, G. L., MORGAN, K. D. and SAUCY, G., *Synthesis* 25 (1976); ERMAN, M. B., AUL'CHENKO, I. S., KHEIFITS, L. A., DULOVA, V. G., NOVIKOV, Yu. N. and VOL'PIN, M. E., *Tetrahedron Lett.* 2981 (1976)
111 CHABARDES, P., KUNTZ, E. and VARAGNAT, J., *Tetrahedron* **33,** 1775 (1977)
112 MURATA, S., SUZUKI, M. and NOYORI, R., *J. Am. chem. Soc.* **102,** 3248 (1980); *Tetrahedron Lett.* 2527 (1980)
113 BORGULYA, J. and BERNAUER, K., *Synthesis* 545 (1980)
114 RIEDIKER, M. and GRAF, W., *Helv. chim. Acta* **62,** 205 (1979)
115 SCHMIDT, M. and SCHMIDBAUER, H., *Angew. Chem.* **70,** 470 (1958); *Chem. Ber.* **99,** 2446 (1961); DUFFAUT, N., CALAS, R. and DUNOGUÈS, J., *Bull. Soc. chim. Fr.* 512 (1963)
116 KANTLEHNER, W., HAUG, E. and MERGEN, W. W., *Synthesis* 460 (1980)
117 DETTY, M. P. and WOOD, G. P., *J. org. Chem.* **45,** 80 (1980)
118 DETTY, M. R., *J. org. Chem.* **45,** 924 (1980)
119 SAKURAI, H., SASAKI, K. and HOSOMI, A., *Tetrahedron Lett.* 2329 (1980); KRAUS, G. A. and FRAZIER, K., *J. org. Chem.* **45,** 2579 (1980)
120 HANESSIAN, S. and GUINDON, Y., *Tetrahedron Lett.* 2305 (1980)
121 OLAH, G. A., GUPTA, B. G. B., MALHOTRA, R. and NARANG, S. C., *J. org. Chem.* **45,** 1638 (1980)
122 SEKINE, M., FUTATSUGI, T., YAMADA, K. and HATA, T., *Tetrahedron Lett.* 371 (1980); *see also* DAVIDSON, R. M. and KENYON, G. L., *J. org. Chem.* **45,** 2698 (1980)
123 YAMAGUCHI, K., KAMIMARA, T. and HATA, T., *J. Am. chem. Soc.* **102,** 4534 (1980)
124 BRYSON, T. A., BONITZ, G. H., REICHEL, C. J. and DARDIS, R. E., *J. org. Chem.* **45,** 524 (1980)
125 LIOTTA, D., PATY, P. B., JOHNSTON, J. and ZIMA, G., *Tetrahedron Lett.* 5091 (1978)
126 DETTY, M., *Tetrahedron Lett.* 5087 (1978); 4189 (1979)
127 COREY, E. J. and SCHMIDT, G., *Tetrahedron Lett.* 731 (1980)
128 ZIEGLER, F. E., NELSON, R. V. and WANG, T., *Tetrahedron Lett.* 2125 (1980)

Chapter 19

Nitrogen-substituted silanes

After the halogenosilanes, the amino- and amido-silanes[1-3] are the most reactive classes of organosilane in which silicon is bonded to a more electronegative element. The silicon–nitrogen bond is readily formed by methods similar to those detailed in Chapter 15; an auxiliary tertiary amine is normally added to preclude any loss of starting material by salt formation. The bond is readily cleaved (*Scheme 19.1*) under mild electrophilic conditions; indeed, a range of amino- and amido-silanes is described in Chapter 15 as potent silyl-transfer reagents. This chapter will explore the fate of the nitrogen fragment.

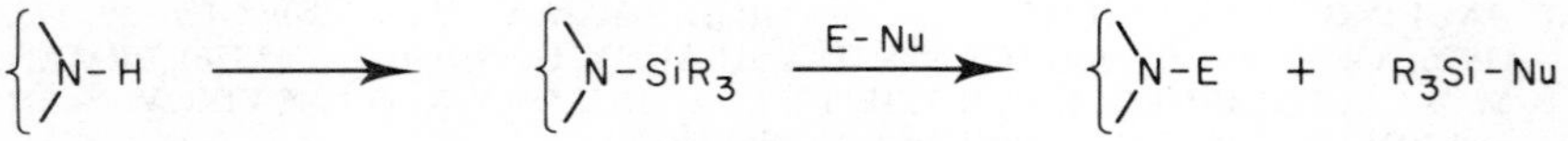

$$\rangle N-H \longrightarrow \rangle N-SiR_3 \xrightarrow{E-Nu} \rangle N-E + R_3Si-Nu$$

Scheme 19.1

19.1 Amino-silanes

The protection afforded to a primary or secondary amine rests on the temporary replacement of an N–H bond with an N–Si bond. Such protection can withstand a variety of (hindered) nucleophilic conditions. For example, *NN*-bis(trimethylsilyl) glycinate esters (1) can be deprotonated[4] and then alkylated on carbon to give homologous α-amino-acid derivatives (*Scheme 19.2*). Alkylation of the anions with chlorotrimethylsilane gives products of *O*-substitution, and reaction with aldehydes leads to serine derivatives.

NO-Bis(trimethylsilyl) hydroxylamine (2) rearranges on deprotonation to the oxygen anion (3). This species reacts with arylsulphonyl (and acyl) chlorides to give, after hydrolysis, the corresponding *O*-substituted hydroxylamine (*Scheme 19.3*), for example *O*-mesitylenesulphonyl hydroxylamine[5].

Deprotonated imines undergo silylation on nitrogen[6a] or carbon[6b] (*Scheme 19.4*) depending largely on the steric bulk of the nitrogen substituent. *N*-Silylated enamines react with methyl-lithium in a manner reminiscent of silyl enol ethers (Chapter 17), regenerating the deprotonated imine[7]. Enamines themselves are easily prepared[8] by use of *N*-trimethylsilyl secondary amines.

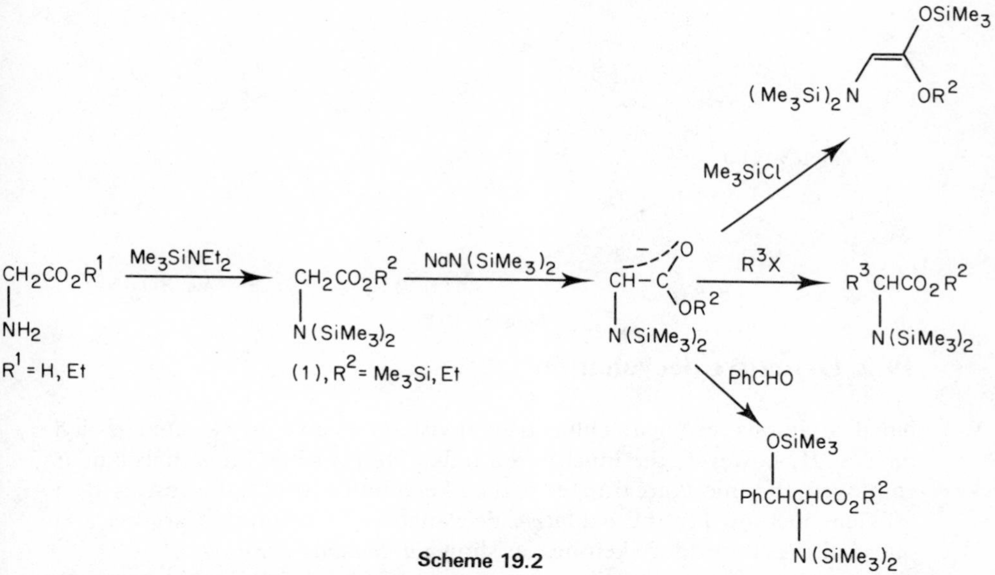

Scheme 19.2

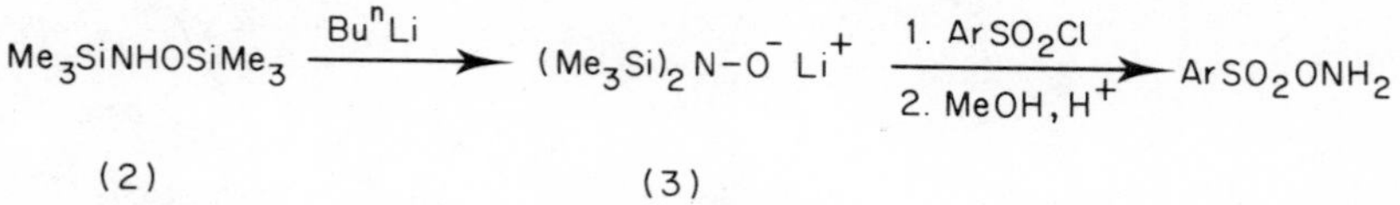

Scheme 19.3

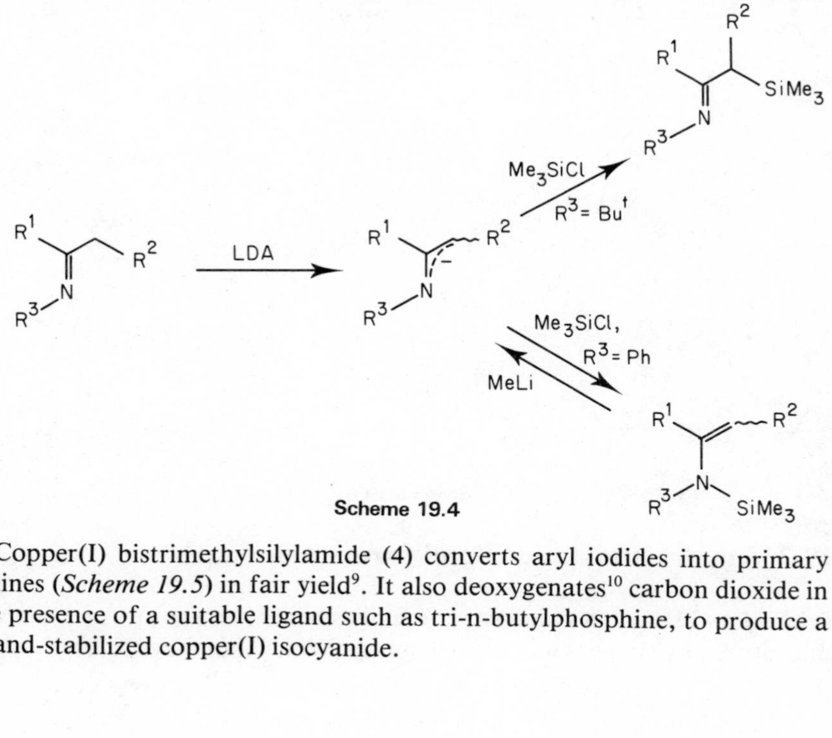

Scheme 19.4

Copper(I) bistrimethylsilylamide (4) converts aryl iodides into primary amines (*Scheme 19.5*) in fair yield[9]. It also deoxygenates[10] carbon dioxide in the presence of a suitable ligand such as tri-n-butylphosphine, to produce a ligand-stabilized copper(I) isocyanide.

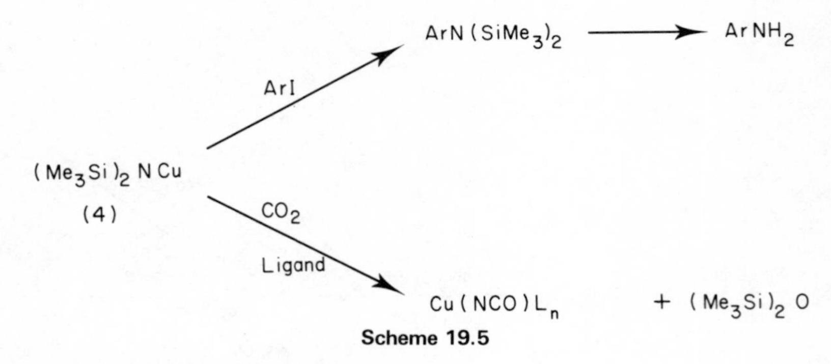

Scheme 19.5

19.2 Oxidative decyanation

Nitrile α-anions react with chlorotrimethylsilane to give, as expected, α-silyl nitriles. If, however, the much more bulky chloro-t-butyldimethylsilane is employed, the anions are trapped in their ketenimine form[11]; this results in an efficient method for the oxidative decyanation of secondary aralkyl and diaryl nitriles to produce ketones, as shown in *Scheme 19.6.*

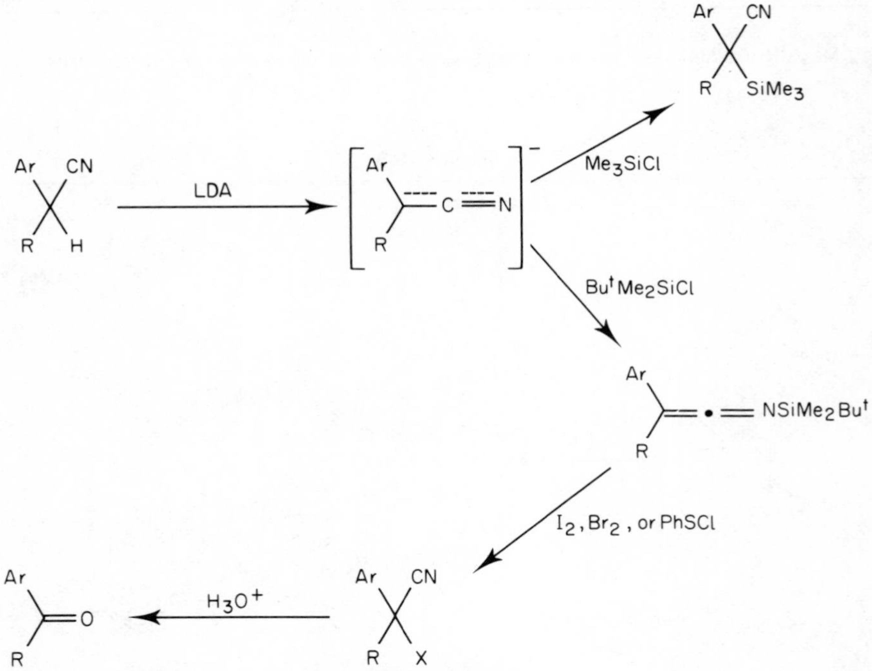

Scheme 19.6

19.3 Amides and amide bond formation

Primary and secondary amides are readily silylated by use of, for example, chlorotrimethylsilane and triethylamine. With both acyclic and cyclic amides, monosilylation takes place at nitrogen; the few exceptions to this generality are those cyclic cases where *O*-substitution will result in heteroaromatization[1]. Primary amides can, of course, undergo bis-silylation, to give either amide (5) or imidate (6) structures. A careful ^{15}N NMR investigation[12] of such bis(trimethylsilyl) amides revealed that whereas bis(trimethylsilyl)formamide[13] has the amide structure (5), all other bis(trimethylsilyl) amides studied are in the imidate form (6).

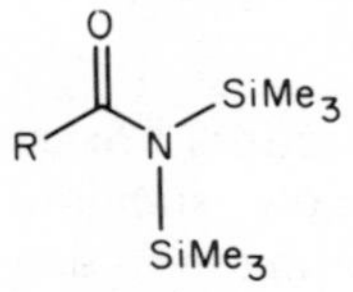

(5) 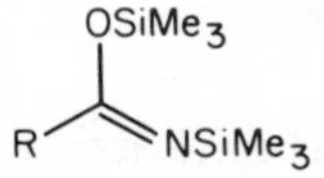 (6)

The amide functionality of substituted azetidin-2-ones can be readily masked by silylation, permitting anion formation and subsequent condensation at the 3-position. The most important applications of this technique use aldehydes as electrophiles, giving synthetic entry to the thienamycin and olivanic series of potent β-lactamase-resistant antibiotics[14]. The particular synthesis of thienamycin (7) shown in *Scheme 19.7* is of considerable interest both in its intensive use of organosilicon chemistry and in its commercial practicability[15].

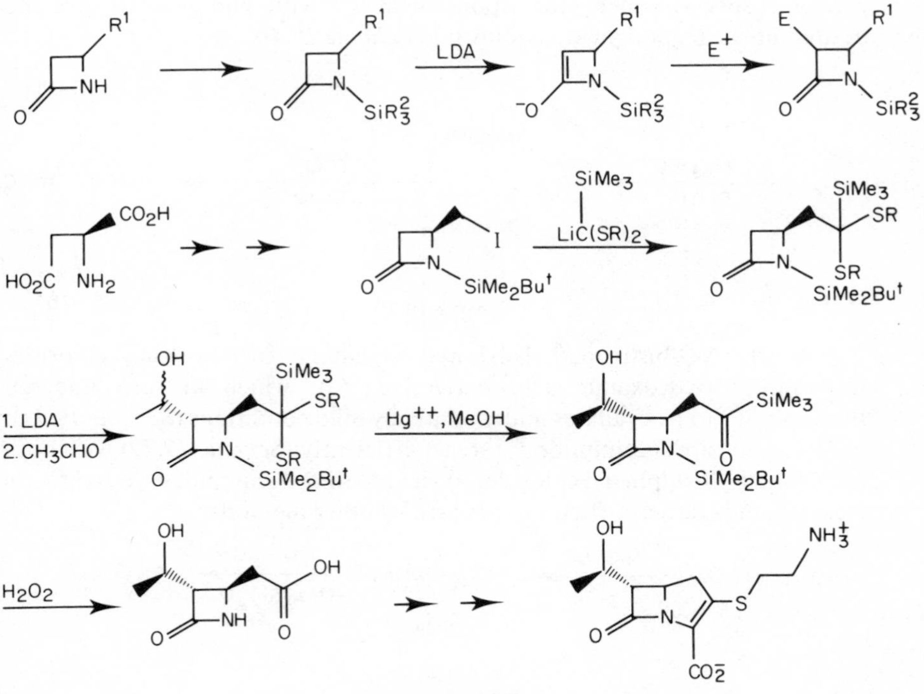

Scheme 19.7 (7)

Other uses of trialkylsilyl amides include the preparation of hydroxamic acids[16] by oxidation of *N*-trimethylsilyl secondary amides by a molybdenum oxide–HMPA complex (*Scheme 19.8*); yields are modest.

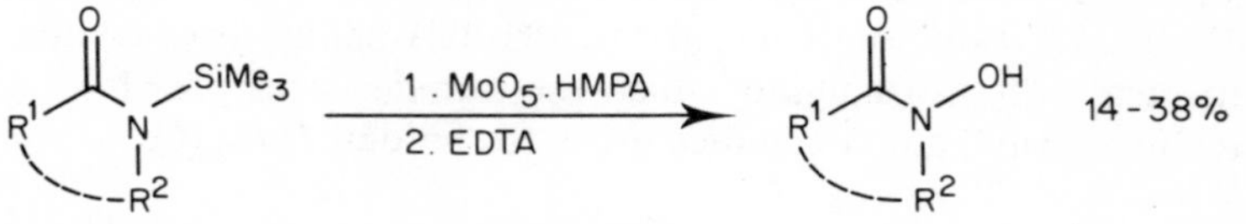

$$R^1C(=O)N(SiMe_3)-R^2 \xrightarrow[2.\ EDTA]{1.\ MoO_5.HMPA} R^1C(=O)N(OH)-R^2 \quad 14-38\%$$

Scheme 19.8

Carboxyl-activated amino-acids and peptides react readily with *N*-trimethylsilylamino-acids, providing a now standard method[1,2,3,17] (*Scheme 19.9*) for amide/peptide bond formation; depending on the activation provided, the silicon–nitrogen bond may or may not be preserved in the immediate product.

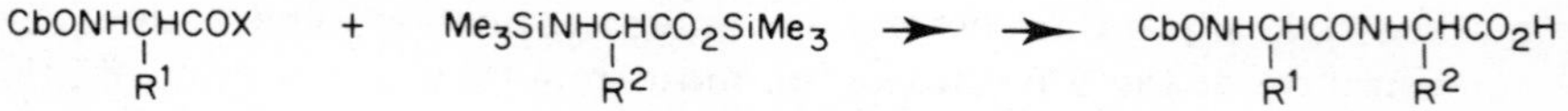

$$CbONHCH(R^1)COX + Me_3SiNHCH(R^2)CO_2SiMe_3 \rightarrow \rightarrow CbONHCH(R^1)CONHCH(R^2)CO_2H$$

Scheme 19.9

Monosilyl primary amines undergo acylation with acyl chlorides with selective loss of a proton, the silicon–nitrogen bond being preserved. For example, $\alpha\omega$-diamines do not react cleanly with phosgene to produce cyclic ureas. Conversion into the *NN'*-bis(trimethylsilyl) derivatives, however, produces species which react spontaneously[18] with phosgene to give the desired ureas in good yield, as shown in *Scheme 19.10*.

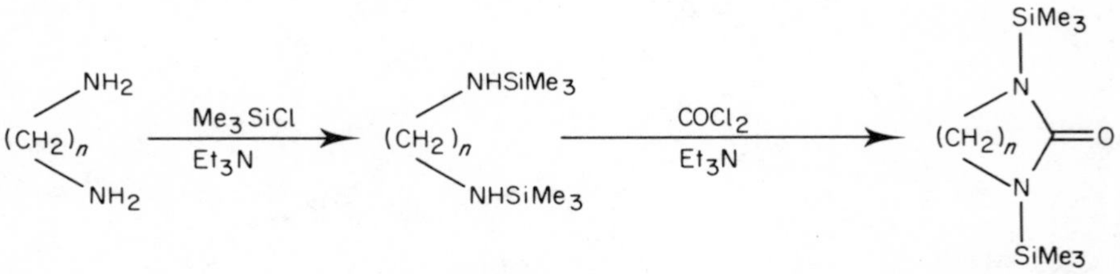

$$(CH_2)_n(NH_2)_2 \xrightarrow[Et_3N]{Me_3SiCl} (CH_2)_n(NHSiMe_3)_2 \xrightarrow[Et_3N]{COCl_2} (CH_2)_n[N(SiMe_3)]_2C{=}O \quad 60-75\%$$

$n = 3, 4$

Scheme 19.10

Similarly, *NO*-bis(trimethylsilyl) hydroxylamine reacts with acyl chlorides to produce hydroxamic acid derivatives (8), which in turn undergo thermolysis[19] to isocyanates under relatively mild conditions (*Scheme 19.11*).

N-Trialkylsilylsuccinimides[20,21] react efficiently (*Scheme 19.12*) with both alkyl- and aryl-sulphenyl chlorides to yield *N*-thiosuccinimides[22], a structural class which is rather difficult to prepare by other methods.

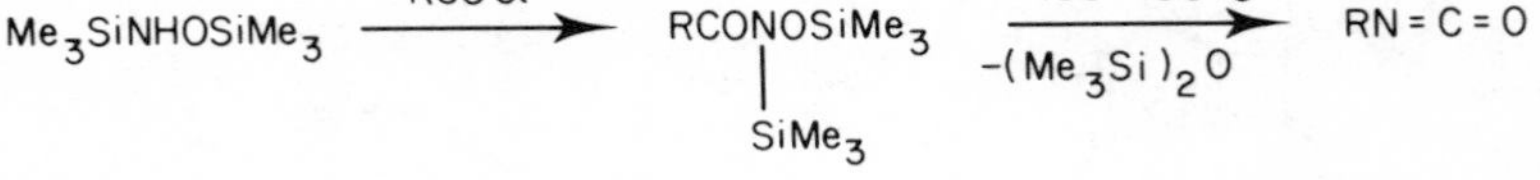

$$Me_3SiNHOSiMe_3 \xrightarrow{RCOCl} RCON(SiMe_3)OSiMe_3 \xrightarrow[-(Me_3Si)_2O]{130-150\,^{\circ}C} RN{=}C{=}O$$

(8)

Scheme 19.11

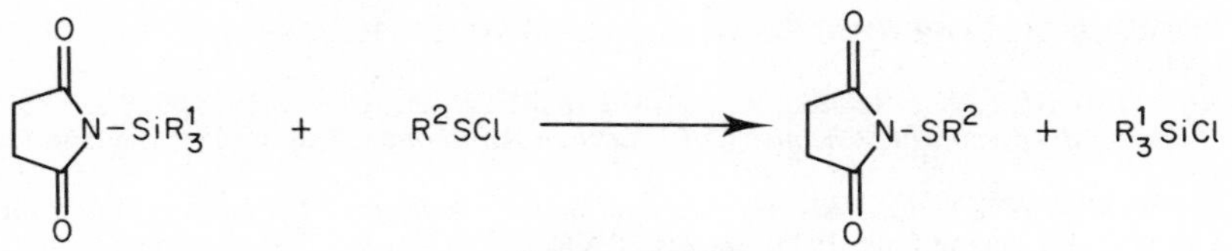

$$\text{(succinimido)N–SiR}^1_3 + \text{R}^2\text{SCl} \longrightarrow \text{(succinimido)N–SR}^2 + \text{R}^1_3\text{SiCl}$$

Scheme 19.12

N-Silylated amines can be induced to undergo carbon–nitrogen fission, and hence deamination. For example, treatment with dinitrogen tetroxide produces nitrates[23], whereas ruthenium-catalysed reaction with phenylselenide anion yields phenyl selenoalkanes[24]. Both of these processes are exemplified in *Scheme 19.13*.

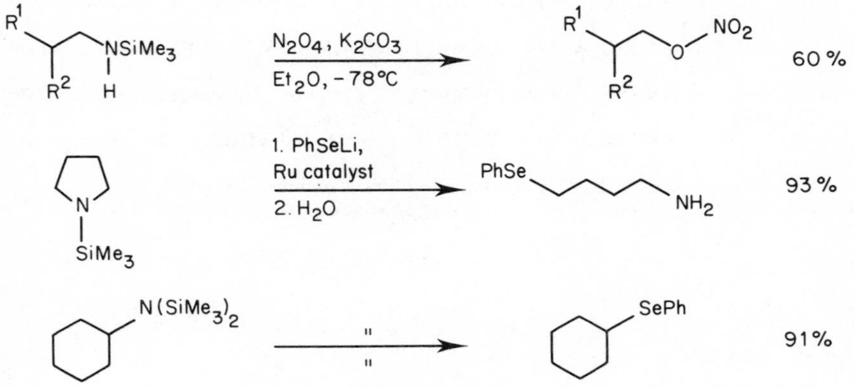

$$\text{R}^1\text{CH(R}^2\text{)CH}_2\text{NHSiMe}_3 \xrightarrow[\text{Et}_2\text{O}, -78\,°\text{C}]{\text{N}_2\text{O}_4,\ \text{K}_2\text{CO}_3} \text{R}^1\text{CH(R}^2\text{)CH}_2\text{ONO}_2 \quad 60\,\%$$

$$\text{N-(SiMe}_3\text{)pyrrolidine} \xrightarrow[\text{2. H}_2\text{O}]{\text{1. PhSeLi, Ru catalyst}} \text{PhSe(CH}_2)_4\text{NH}_2 \quad 93\,\%$$

$$\text{C}_6\text{H}_{11}\text{N(SiMe}_3)_2 \xrightarrow[\text{"}]{\text{"}} \text{C}_6\text{H}_{11}\text{SePh} \quad 91\,\%$$

Scheme 19.13

References

1 BIRKOFER, L. and RITTER, A., in *'Newer Methods of Preparative Organic Chemistry'*, Ed. Foerst, W., vol. 5, p. 211, Academic Press, New York (1968)

2 KLEBE, J., *Adv. org. Chem.* **8,** 97 (1972); *Accts chem. Res.* **3,** 299 (1970)

3 PIERCE, A. E., *'Silylation of Organic Compounds',* Pierce Chemical Co., Rockford, Illinois (1968)

4 RÜHLMANN, K. and KUHRT, G., *Angew Chem.* **80,** 797 (1968)

5 KING, F. D. and WALTON, D. R. M., *Synthesis* 788 (1975)

6 (*a*) AHLBRECHT, H. and LIESCHING, D., *Synthesis* 746 (1976); *see also* WALTER, W. and LUKE, H. W., *Angew Chem. int. Edn* **16,** 535 (1977); (*b*) COREY, E. J., ENDERS, D. and BOCK, M. G., *Tetrahedron Lett.* 7 (1976)

7 AHLBRECHT, H., DÜBER, E. O., ENDERS, D., EICHENAUER, H. and WEUSTER, P., *Tetrahedron Lett.* 3691 (1978)

8 COMI, R., FRANCK, R. W., REITANO, M. and WEINREB, S. M., *Tetrahedron Lett.* 3107 (1973); *see,* however, HELLBERG, L. H. and JUAREZ, A., *Tetrahedron Lett.* 3553 (1974)

9 KING, F. D. and WALTON, D. R. M., *J. chem. Soc. chem. Communs* 256 (1974); *Synthesis* 40 (1976); for a different method, *see* BARTON, D. H. R., BREWSTER, A. G., LEY, S. V. and ROSENFELD, M. N., *J. chem. Soc. chem. Communs* 147 (1977)

10 TSUDA, T., WASHITA, H. and SAEGUSA, T., *J. chem. Soc. chem. Communs* 468 (1977)

11 WATT, D. S., *J. org. Chem.* **39,** 2799 (1974); SELIKSON, S. J. and WATT, D. S., *Tetrahedron Lett.* 3029 (1974)

12 YODER, C. H., COPENHAFER, W. C. and DuBESHTER, B., *J. Am. chem. Soc.* **96,** 4283 (1974)
13 KANTLEHNER, W., KUGEL, W. and BREDERECK, H., *Chem. Ber.* **105,** 2264 (1972)
14 *'Recent Advances in the Chemistry of β-Lactam Antibiotics',* Cambridge, England (July 1980)
15 CHRISTENSEN, B. G., *'Recent Advances in the Chemistry of β-Lactam Antibiotics',* Cambridge, England (July 1980); *see also* RATCLIFFE, R. W., SALZMANN, T. N. and CHRISTENSEN, B. G., *Tetrahedron Lett.* 31 (1980); SALZMANN, T. N., RATCLIFFE, R. W., CHRISTENSEN, B. G. and BOUFFARD, F. A., *J. Am. chem. Soc.* **102,** 6161 (1980)
16 MATLIN, S. A. and SAMMES, P. G., *J. chem. Soc. chem. Communs* 1222 (1972)
17 *See,* for example, KOPPEL, G. A., McSHANE, L., JOSE, F. and COOPER, R. D. G., *J. Am. chem. Soc.* **100,** 3933 (1978)
18 BIRKOFER, L., KÜHLTHAU, H. P. and RITTER, A., *Chem. Ber.* **93,** 2810 (1960); MIRONOV, V. F., SHELUDYAKOV, V. D. and KOZYUKOV, V. P., *Russ. chem. Revs* **48,** 473 (1979)
19 KING, F. D., PIKE, S. and WALTON, D. R. M., *J. chem. Soc. chem. Communs* 351 (1978); *see also* RIGAUDY, J., LYTWYN, E., WALLACH, P. and CUONG, N. K., *Tetrahedron Lett.* 3367 (1980)
20 SAKURAI, H., HOSOMI, A., NAKAJIMA, J. and KUMADA, M., *Bull. chem. Soc. Japan* **39,** 2263 (1966)
21 CALAS, R., FRAINNET, E. and DENTONE, Y., *C.r. hebd. Séanc Acad. Sci., Paris* **259,** 3377 (1964)
22 HARPP, D. N., FRIEDLANDER, B., MULLINS, D. and VINES, S. M., *Tetrahedron Lett.* 963 (1977)
23 WUDL, F. and LEE, T. B. K., *J. chem. Soc. chem. Communs* 490 (1970)
24 MURAHASHI, S.-I. and YANO, T., *J. Am. chem. Soc.* **102,** 2456 (1980)

Chapter 20

Silicon-substituted bases and ligands

Bis(trimethylsilyl)amine, or hexamethyldisilazane (1), is a commercially available derivative of ammonia with manifold applications, a major one being as a silyl-transfer reagent (Chapter 15). It also provides an extremely bulky ligand[1-3] which can stabilize metals in low coordinative environments. An example of such a stabilizing effect can be seen in the isolation[4] of the first two-coordinate phosphorus- and arsenic-centred radicals (2); bis-(trimethylsilyl)methyl ligands provide similar stabilization to the radicals (3).

$(Me_3Si)_2NH$ (1) $[(Me_3Si)_2N]_2M^{\bullet}$ (2), M = P, As $[(Me_3Si)_2CH]_2M^{\bullet}$ (3), M = P, As

Turning to alkali metal derivatives, lithium[5], sodium[6], and potassium[7] bis(trimethylsilyl)amides (*Scheme 20.1*) have all found extensive use as exceptionally strong, non-nucleophilic bases. Many such applications have occurred in earlier Chapters, and only some further selected examples will be discussed here.

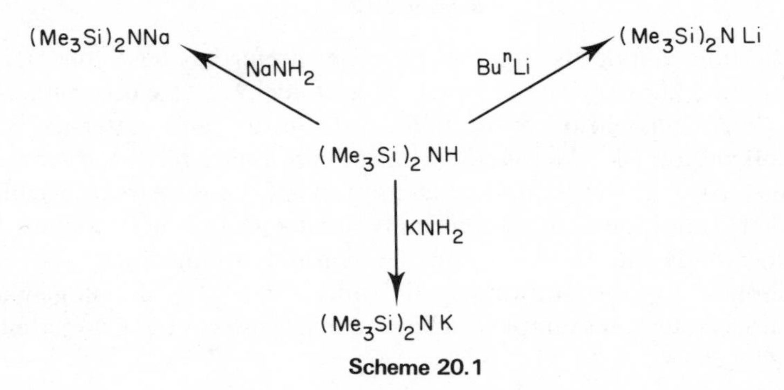

Scheme 20.1

The lithium amide has been recommended[8] for the generation of kinetic enolates; the sodium amide[9] can also be employed for this purpose, but the resulting enolates are, as expected, less regiostable. With conjugated dienones, a γ-rather than an ε-proton is removed[10], to give cross-conjugated enolate ions (*Scheme 20.2*).

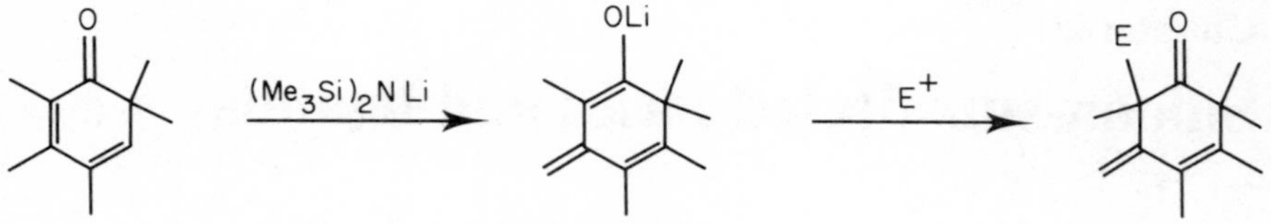

$(Me_3Si)_2NLi$ E^+

Scheme 20.2

Enolate ions prepared with such non-nucleophilic bases have been employed[11] as protecting groups for ketones during hydride reduction; protection is also afforded against Grignard and Wittig reaction conditions. An example is shown in *Scheme 20.3*.

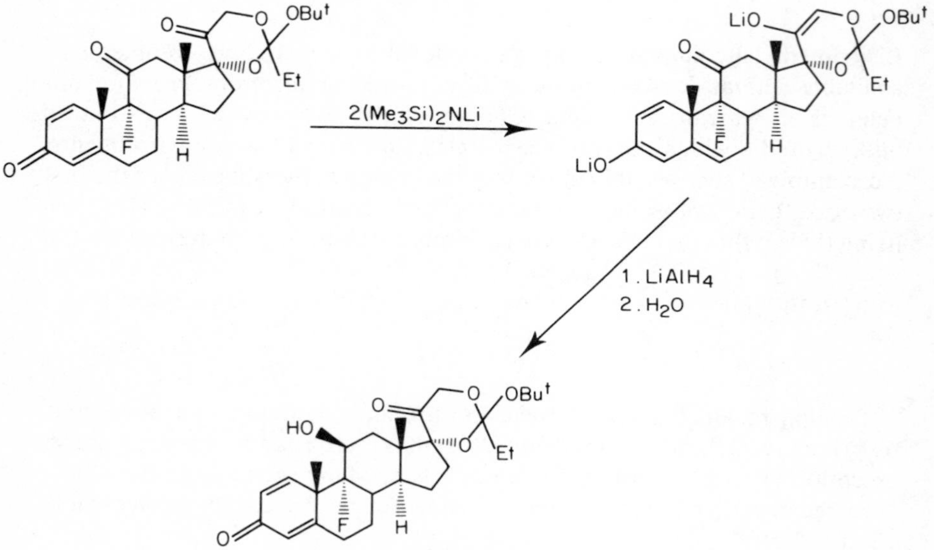

$2(Me_3Si)_2NLi$ 1. $LiAlH_4$ 2. H_2O

Scheme 20.3

The lithium amide is reportedly[12] the preferred base in Darzens condensations, allowing the use even of acetaldehyde as electrophile. The lithium base has also been used for intra- and inter-molecular hydroxyalkylation of γ-lactones[13]. The sodium amide has been employed advantageously[14] in Dieckmann condensations of $\alpha\omega$-diesters, especially in cases where additional nucleophilically labile groups are present. The potassium amide has found use in the synthesis of medium- and large-ring lactones by the intramolecular alkylation[15] of ω-halogenoalkyl phenylthioacetates; an example can be seen in a synthesis of (±)-recifeiolide (4) (*Scheme 20.4*).

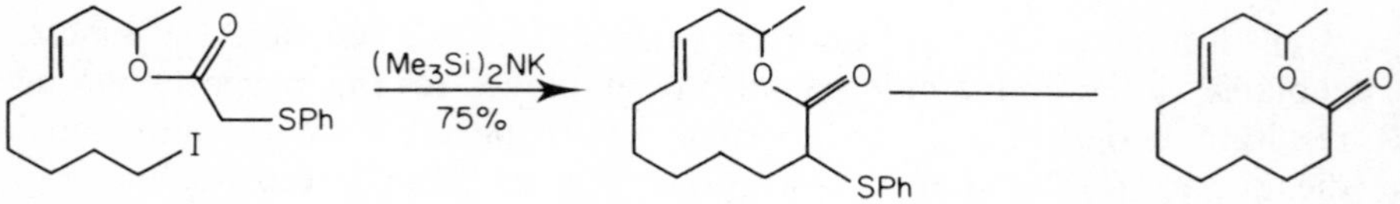

$(Me_3Si)_2NK$ 75%

Scheme 20.4 (4)

Intramolecular displacement of halide ions from halogeno-acetals provides a synthesis[16] of functionalized bicyclic diketones; when the lithium amide is used as base, the product is 95 per cent *trans*-nitrile (5) whereas, remarkably, if the metal ion is potassium, the stereochemistry is completely reversed, giving 95 per cent *cis*-nitrile (5), as shown in *Scheme 20.5*.

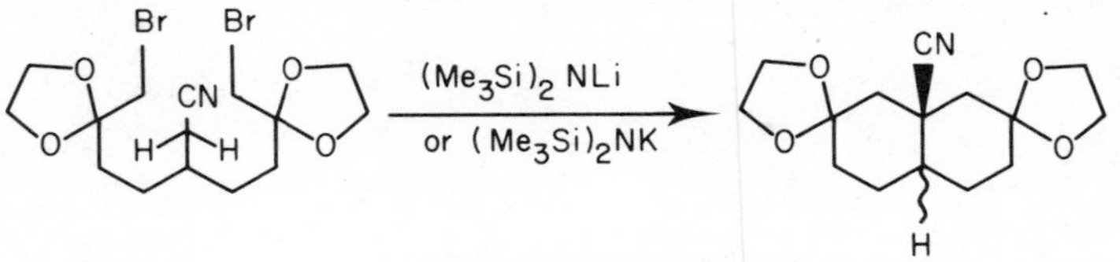

Br Br CN H H

$(Me_3Si)_2NLi$ or $(Me_3Si)_2NK$

CN H (5)

Scheme 20.5

Cyclizations involving attack of a carbanion on an electrophilic carbon atom usually result in the formation of a five- rather than of a six-membered ring, and rarely of a four-membered ring. Stork[17] has described a process of 'epoxynitrile cyclization', in which these tendencies are reversed. The reversal is ascribed to the geometric constraints imposed by the oxiran ring in each case, making it difficult for the nitrile anion and the oxiran C–O bond to come into line for formation of a five-membered ring (*Scheme 20.6*). The second reaction shown is highly stereoselective, and has been employed[18] in a synthesis of (±)-grandisol (6).

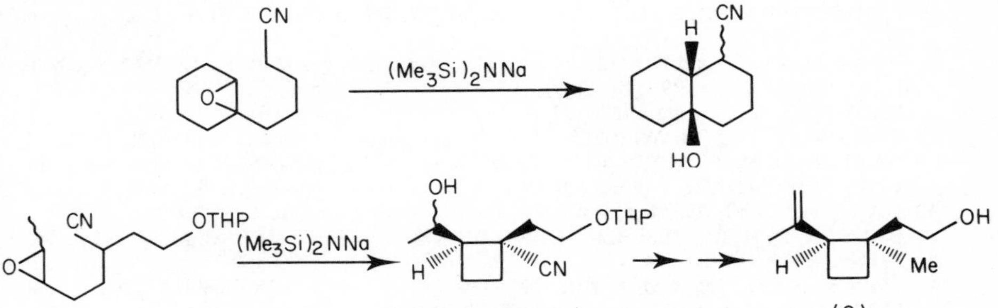

CN O

$(Me_3Si)_2NNa$

H CN HO

CN OTHP O

$(Me_3Si)_2NNa$

OH OTHP H CN

H Me OH (6)

Scheme 20.6

Finally, the sodium amide has been recommended for the generation[19] of halogenocarbenes from dihalogenomethanes, and for the preparation[20] of alkylidenephosphoranes free from lithium salts.

Lithium 1,1-bistrimethylsilyl-3-methylbutoxide (7) is an exceptionally selective hindered strong base[21]. It regiospecifically removes methyl protons from acetate esters[22] and methyl ketones[23], even in the simultaneous presence of enolizable aldehydes; the latter then trap the so-formed enolate anions and thus provide a new range of regiospecific aldol condensations (*Scheme 20.7*).

Trimethylsilylmethylpotassium (8) is a powerful reagent[24] for the deprotonation/metallation of allylic and benzylic hydrocarbons; however, the combined reagent n-butyl lithium/potassium t-butoxide usually works equally well.

Me_3SiCH_2K

(8)

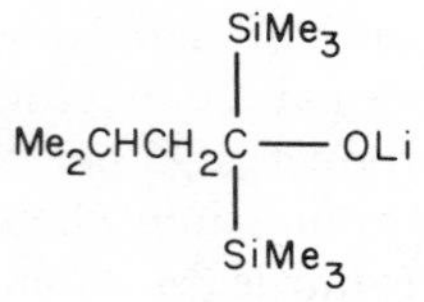

$$Me_2CHCH_2C(SiMe_3)_2OLi$$

(7)

$$R^1CHO + MeCOR^2 \xrightarrow{(7)} R^1CH(OH)CH_2COR^2$$

$R^2 = CH_2R^3$ or OR^4

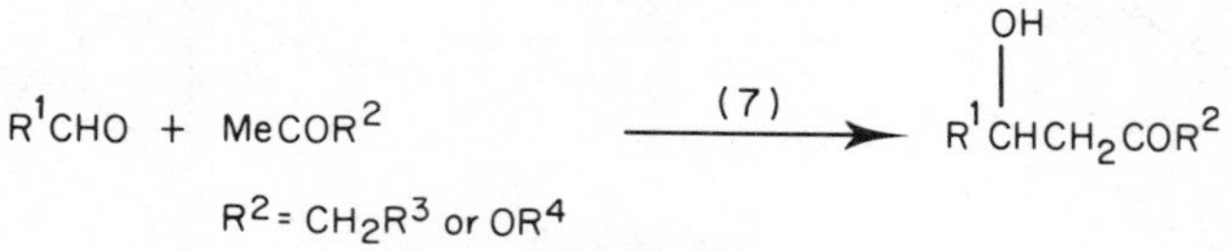

Scheme 20.7

References

1 HARRIS, D. H. and LAPPERT, M. F., *Organometal. Chem. Rev.* **2,** 13 (1976)
2 LAPPERT, M. F., POWER, P. P., SANGER, A. R. and SRIVASTAVA, R. C., *'Metal and Metalloid Amides',* Chap. 5, Ellis Horwood, Chichester (1980)
3 BRADLEY, D. C. and CHISHOLM, M. H., *Accts chem. Res.* **9,** 273 (1976)
4 GYNANE, M. J. S., HUDSON, A., LAPPERT, M. F., POWER, P. P. and GOLDWHITE, H., *J. chem. Soc. chem. Communs* 623 (1976); *see also* GRILLER, D. and INGOLD, K. U., *Accts chem. Res.* **9,** 13 (1976)
5 AMONOO-NEIZER, E. H., SHAW, R. A., SKOVLIN, D. O. and SMITH, B. C., *J. chem. Soc.* 2997 (1965)
6 WANNAGAT, U. and NEIDERPRÜM, H., *Chem. Ber.* **94,** 1540 (1961); WANNAGAT, U., *Pure appl. Chem.* **19,** 329 (1969)
7 BROWN, C. A., *Synthesis* 427 (1974)
8 TANABE, M. and CROWE, D. F., *J. chem. Soc. chem. Communs* 564 (1973)
9 BARTON, D. H. R., HESSE, R. H., TARZIA, G. and PECHET, M. M., *Chem. Communs* 1497 (1969); TANABE, M. and CROWE, D. F., *Chem. Communs* 1498 (1969)
10 HART, H., LOVE, G. M. and WANG, I. C., *Tetrahedron Lett.* 1377 (1973)
11 BARTON, D. H. R., HESSE, R. H., WILSHIRE, C. and PECHET, M. M., *J. chem. Soc. Perkin I* 1075 (1977)
12 BORCH, R. F., *Tetrahedron Lett.* 3761 (1972); *see*, however, KYRIAKAKOU, G. and SEYDEN-PENNE, J., *Tetrahedron Lett.* 1737 (1974)
13 BROWN, E., DHAL, R. and ROBIN, J.-P., *Tetrahedron Lett.* 733 (1979)
14 HURD, R. N. and SHAH, D. H., *J. org. Chem.* **38,** 390 (1973)
15 TAKAHASHI, T., HASHIGUCHI, S., KASUGA, K. and TSUJI, J., *J. Am. chem. Soc.* **100,** 7424 (1978)
16 STORK, G., GARDNER, J. O., BOECKMAN, R. K. and PARKER, K. A., *J. Am. chem. Soc.* **95,** 2014 (1973); STORK, G. and BOECKMAN, R. K., *J. Am. chem. Soc.* **95,** 2016 (1973)
17 STORK, G., CAMA, L. D. and COULSON, D. R., *J. Am. chem. Soc.* **96,** 5268 (1974)
18 STORK, G. and COHEN, J. F., *J. Am. chem. Soc.* **96,** 5270 (1974)
19 MARTEL, B. and HIRIART, J. M., *Synthesis* 201 (1972)
20 BESTMANN, H. J., STRANSKY, W. and VOSTROWSKY, O., *Chem. Ber.* **109,** 1694 (1976)
21 KUWAJIMA, I., SATO, T., MINAMI, N. and ABE, T., *Tetrahedron Lett.* 1591 (1976); KUWAJIMA, I., ARAI, M. and SATO, T., *J. Am. chem. Soc.* **99,** 4181 (1977)
22 KUWAJIMA, I., MINAMI, N. and SATO, T., *Tetrahedron Lett.* 2253 (1976); MINAMI, N. and KUWAJIMA, I., *Tetrahedron Lett.* 1423 (1977)
23 KUWAJIMA, I., SATO, T., ARAKI, M. and MINAMI, N., *Tetrahedron Lett.* 1817 (1976)
24 STÄHLE, M., HARTMANN, J. and SCHLOSSER, M., *Helv. chim. Acta* **60,** 1730 (1977); SCHLOSSER, M. and HARTMANN, J., *J. Am. chem. Soc.* **98,** 4674 (1976); SCHLOSSER, M. and RAUCHSCHWALBE, G., *J. Am. chem. Soc.* **100,** 3258, 6544 (1978)

Chapter 21

Silanes as reducing agents

The addition of the Si–H linkage to multiply-bonded substrates is important not only as a method of reduction, but also as a major route to complex organosilanes such as vinylsilanes. Such addition can be brought about under catalytic or 'ionic' conditions, the latter also being used for hydride transfer to carbonium ions. A distinctly different method of multiple-bond reduction employs chlorotrimethylsilane–metal systems. Finally, certain silanes can smoothly deoxygenate phosphine and amine oxides and sulphoxides, and can reduce carbonyl and carbinol derivatives to hydrocarbons.

21.1 Hydrosilylation

The term 'hydrosilylation' is used to describe the transition metal-catalysed addition of a hydrosilane to a multiply-bonded system. The area has been extensively reviewed[1-3] recently, particular emphasis having been justifiably given to contributions made by Russian workers. In view of such recent compilations, only a limited discussion of the more salient features is appropriate here.

21.1.1 Alkynes and alkenes

The catalyst most frequently employed in hydrosilylation of alkynes and alkenes is hexachloroplatinic acid[4]. Under such conditions, alkynes undergo

Scheme 21.1

cis-addition[5,6] (*Scheme 21.1*), and so provide one of the major routes to vinylsilanes (Chapter 7). Peroxide-initiated addition[5], on the other hand, gives mainly the products of *trans*-addition, whereas nickel(II) catalyses[7] a stereoselective double *cis*-addition (*Scheme 21.1*).

Terminal alkynes react with a high degree of regioselectivity[8], with lower reaction temperatures[9] favouring a greater proportion of the terminal isomer (*Scheme 21.2*); not unexpectedly, this regioselectivity is reversed[10] when alkynylsilanes are employed as substrates.

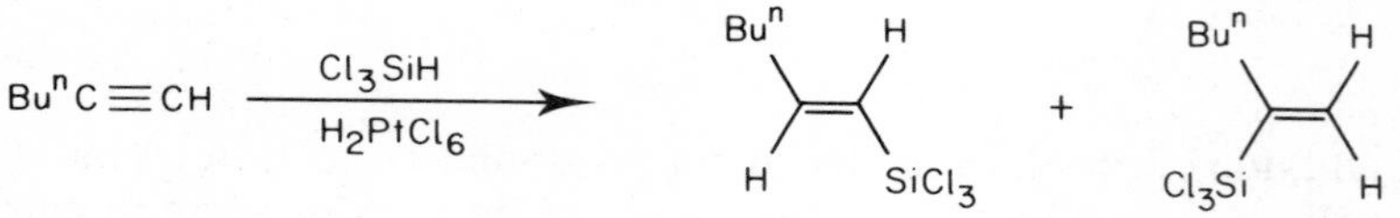

$$Bu^nC{\equiv}CH \xrightarrow[H_2PtCl_6]{Cl_3SiH} (E)\text{-}Bu^nCH{=}CHSiCl_3 + Bu^n(Cl_3Si)C{=}CH_2$$

reaction temp.			
(i) reflux	78	:	22
(ii) 5 °C	95	:	5

Scheme 21.2

In comparison with alkynes, simple alkenes undergo hydrosilylation much more sluggishly[1-3]. The relative reactivity[11] of a particular alkene reflects both its electron density and its degree of substitution. Indeed, the comparative order[12] of reactivity of some alkynes and alkenes has been established as:

$$HC{\equiv}C{-} \;>\; H_2C{=}CH{-} \;>\; H_2C{=}C(CH_3){-}$$

Alkenes react with trialkylsilanes and carbon monoxide[13] in the presence of transition metal complexes to produce homologous silyl enol ethers (Chapter 17) (*Scheme 21.3*). The reactions of various other organic substrates with this reagent system have also been described[13].

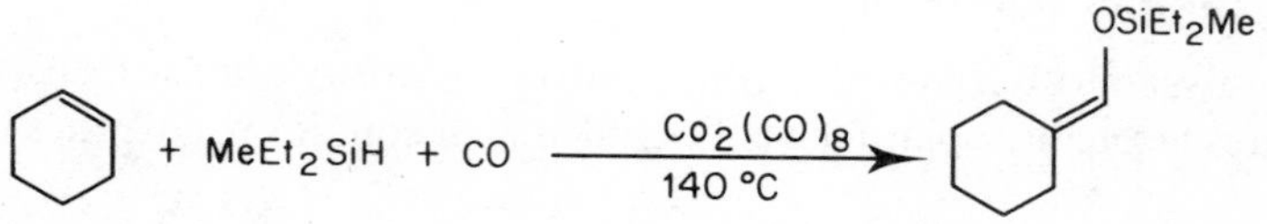

$$\text{cyclohexene} + MeEt_2SiH + CO \xrightarrow[140\,°C]{Co_2(CO)_8} \text{cyclohexylidene}{=}CH{-}OSiEt_2Me$$

Scheme 21.3

21.1.2 Carbonyl compounds

Hydrosilanes add to the carbonyl group of saturated aldehydes and ketones to produce alkoxysilanes[1-3]. A wide variety of catalysts, including tris(triphenylphosphine)rhodium chloride[14], is effective in promoting this reaction. The addition of tertiary amines or of certain thiols alters its course to become one of dehydrogenative silylation, producing silyl enol ethers (Chapter 17) from readily enolizable ketones.

$\alpha\beta$-Unsaturated carbonyl compounds undergo a rhodium-catalysed

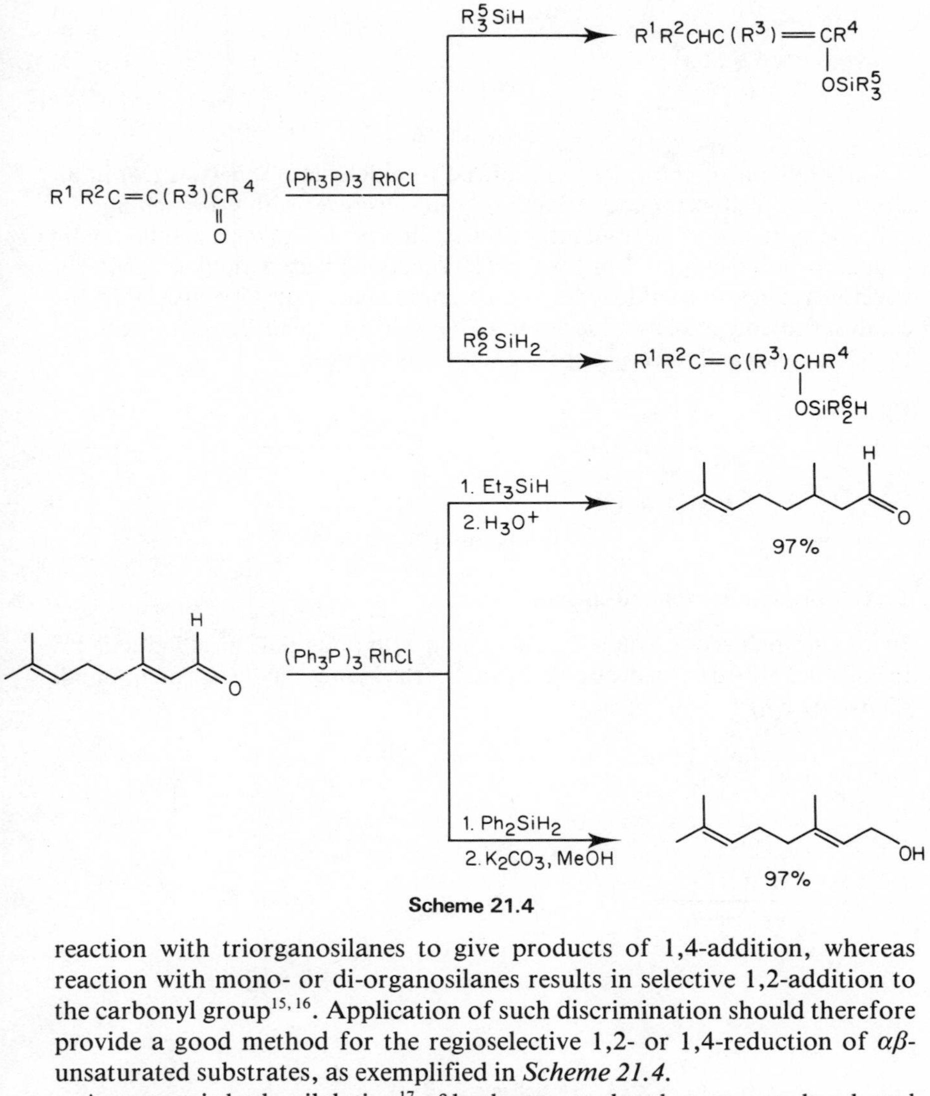

Scheme 21.4

reaction with triorganosilanes to give products of 1,4-addition, whereas reaction with mono- or di-organosilanes results in selective 1,2-addition to the carbonyl group[15,16]. Application of such discrimination should therefore provide a good method for the regioselective 1,2- or 1,4-reduction of $\alpha\beta$-unsaturated substrates, as exemplified in *Scheme 21.4.*

Asymmetric hydrosilylation[17] of both saturated and unsaturated carbonyl substrates[18,19] (*Scheme 21.5*) including oxo-esters[19,20], can be achieved by use of chiral rhodium catalysts, a moderate to high degree of chirality normally being induced.

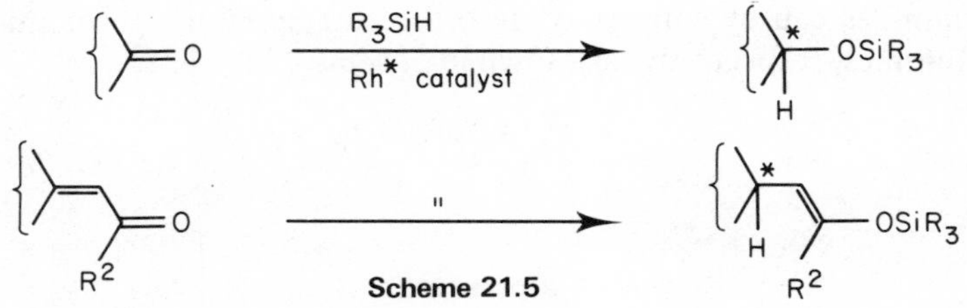

Scheme 21.5

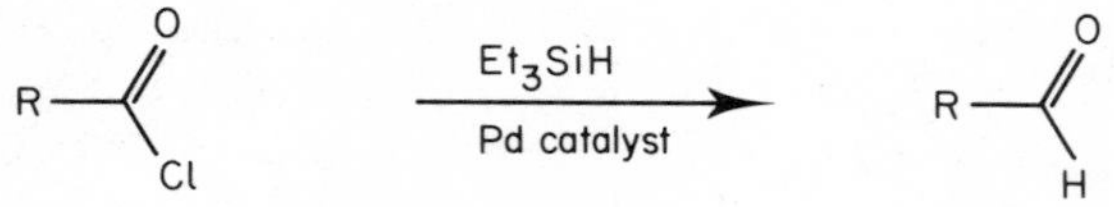

$$\mathrm{RCOCl} \xrightarrow[\text{Pd catalyst}]{\mathrm{Et_3SiH}} \mathrm{RCHO}$$

Scheme 21.6

Carboxylic acid chlorides are reduced to aldehydes (*Scheme 21.6*) in an alternative[21] to Rosenmund reduction; yields decrease with α-branching.

In the presence of an organotin catalyst in a protic solvent, commercially available polymethylhydrosiloxane (1) functions[22] as a mild reagent for selective reduction of aldehydes and ketones to carbinols (*Scheme 21.7*); the catalyst provides a tin hydride as the active reducing agent. In the presence of Pd/C, alkenes and nitro-groups are smoothly reduced.

$$R^1R^2C{=}O + \frac{1}{n}[\mathrm{MeHSiO}]_n \ (1) \xrightarrow[\mathrm{EtOH}]{(R^3_3\mathrm{Sn})_2\mathrm{O}} R^1R^2CH\text{–}OH$$

R^2 = H, alkyl

Scheme 21.7

21.1.3 Imines and related species

Imines are reduced to amines[23], once again with potential chiral induction[24], in what is claimed to be the best method for the reduction of such compounds (*Scheme 21.8*).

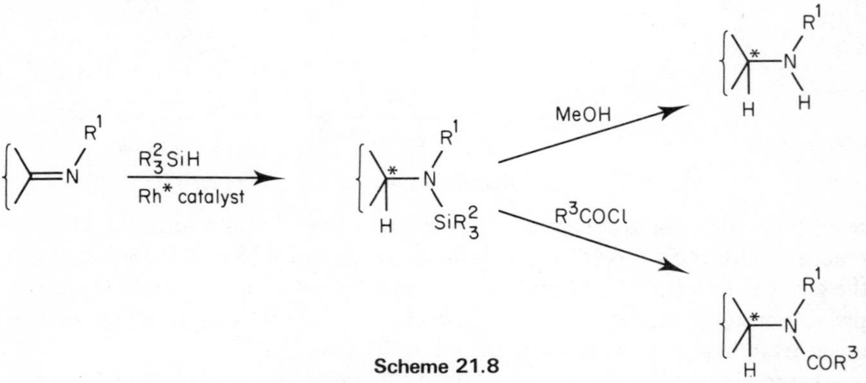

R^2_3SiH, Rh* catalyst; MeOH; R^3COCl

Scheme 21.8

In the presence of rhodium catalysts, pyridines undergo 1,4-addition with trimethylsilane to give *N*-silyl species, which can then be converted into the parent 1,4-dihydropyridines by controlled hydrolysis[25]. Isocyanates and carbodiimides can be converted[26] into the corresponding formamides and formamidines, respectively, as shown in *Scheme 21.9*.

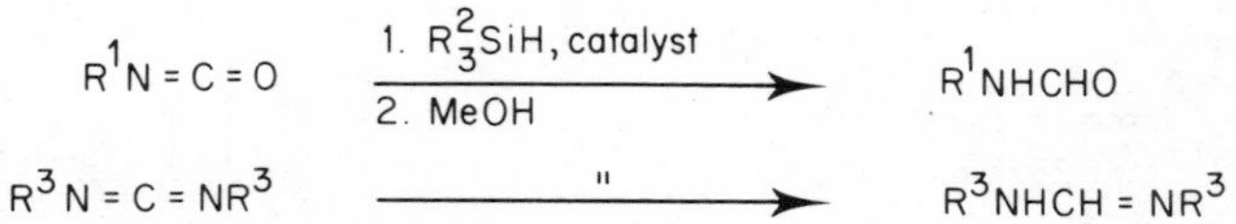

$$R^1N{=}C{=}O \xrightarrow[\text{2. MeOH}]{\text{1. } R^2_3SiH\text{, catalyst}} R^1NHCHO$$

$$R^3N{=}C{=}NR^3 \xrightarrow{\quad '' \quad} R^3NHCH{=}NR^3$$

Scheme 21.9

21.2 Ionic hydrogenation

In common with the similarly polarized boron and aluminium hydrides, silicon hydrides (hydrosilanes) can transfer hydride ions to electropositive carbon centres. However, unlike the first two reducing agents, hydrosilanes require additional activation of the carbon centre by Lewis or protic acids before such hydride transfer can take place. This most useful process[27], summarized in *Scheme 21.10* for protic acids, is known accordingly as 'ionic hydrogenation'.

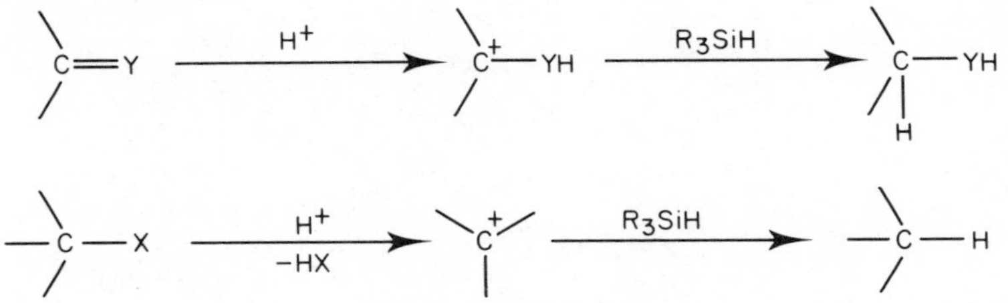

$$>C{=}Y \xrightarrow{H^+} >\overset{+}{C}{-}YH \xrightarrow{R_3SiH} >C(H){-}YH$$

$$-\!\!\!>C{-}X \xrightarrow[-HX]{H^+} >\overset{+}{C}< \xrightarrow{R_3SiH} -\!\!\!>C{-}H$$

Scheme 21.10

Of all the hydrosilanes, triethylsilane seems to be the most useful in terms of both rate of reaction and of handling. Choice of the other member of the reagent pair, the protic or Lewis acid, is governed by the substrate and by the particular product desired.

21.2.1 Alkynes and alkenes

Triethylsilane–trifluoroacetic acid will reduce alkenes to alkanes only if a relatively stable, e.g., tertiary, carbonium ion can be formed on protonation of the substrate[28]. The effect of variation of the bulk of the hydrosilane on the stereoselectivity of this process has been investigated[29]. Terminal arylalkynes are reduced only slowly to arylalkanes by this reagent system[30]. In short, it is not a particularly attractive method for the production of alkanes from such unsaturated substrates. With non-conjugated polyolefins, however, advantage can sometimes be taken of the fact that ionic hydrogenation will preferentially reduce the more substituted double bonds; this is the reverse preference to that observed under normal catalytic hydrogenation conditions.

21.2.2 Carbonyl compounds, carbinols and related species

In dramatic contrast, ionic hydrogenation of aldehydes and ketones can be a most useful process, permitting selective production of alcohols, symmetrical ethers, carboxylate esters, or acetamides (*Scheme 21.11*); the competitive reactions leading to these species can be manipulated by subtle variation of conditions[31]. A stereochemical study with cyclic ketone substrates has shown[32] that use of large Lewis acids and/or of large organosilanes favours equatorial attack by hydride.

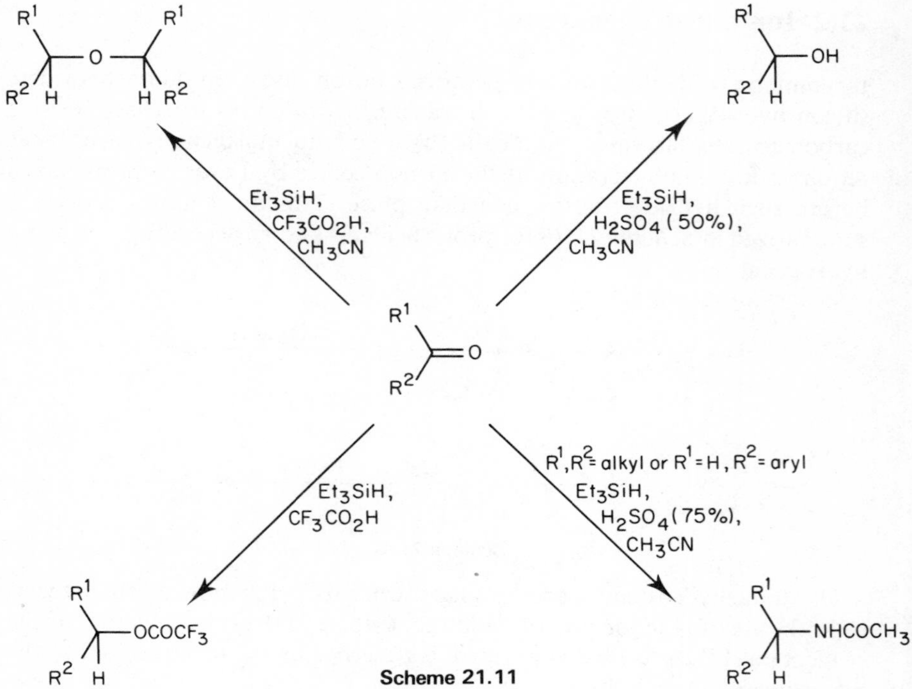

R^1 R^1 O H R^2 H R^2 — Et_3SiH, CF_3CO_2H, CH_3CN
R^1 OH R^2 H — Et_3SiH, H_2SO_4 (50%), CH_3CN
R^1 R^2 C=O
R^1 OCOCF$_3$ R^2 H — Et_3SiH, CF_3CO_2H
R^1 NHCOCH$_3$ R^2 H — R^1,R^2 = alkyl or R^1 = H, R^2 = aryl; Et_3SiH, H_2SO_4 (75%), CH_3CN

Scheme 21.11

Under certain conditions, carbinols can be deoxygenated to hydrocarbons. Tertiary alcohols are reduced by triethylsilane–trifluoroacetic acid[28]. The scope of this transformation can be increased by use of boron trifluoride gas[33] as Lewis acid, when even simple secondary alcohols are cleanly reduced; under these conditions other reducible groups, such as aromatic nitro functions, are unaffected. The latter reagent system can also reduce[34] aldehydes and ketones directly (*Scheme 21.12*); oxiranes[35] are also reduced to hydrocarbons, but at a much slower rate.

The reagent combination of triethylsilane–trifluoroacetic acid can reduce aralkyl ketones to aralkanes[36], and has been employed in the reductive

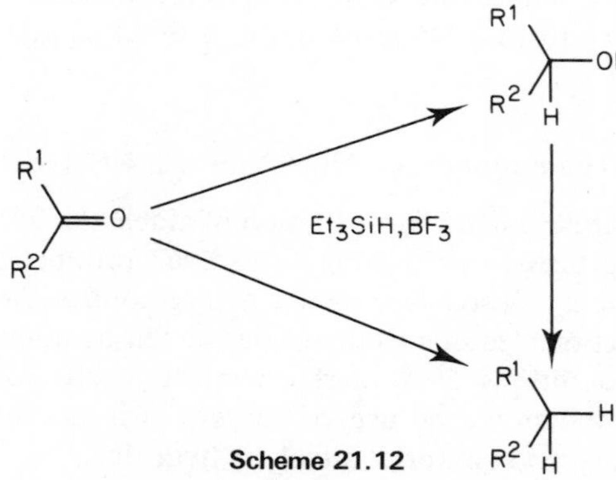

R^1 R^2 C=O — Et_3SiH, BF_3 → R^1 R^2 C(H)OH → R^1 R^2 C(H)H

Scheme 21.12

$$RCONH_2 + HCHO \rightleftharpoons RCONHCH_2OH \xrightarrow[CF_3CO_2H]{Et_3SiH} RCONHCH_3$$

Scheme 21.13

methylation[37] of amides and related compounds shown in *Scheme 21.13*. Alkyl chlorides and bromides can be reduced to hydrocarbons, catalysis by aluminium(III) chloride being required (*Scheme 21.14*). This efficient process[38] seems to be quite general for primary, secondary, and tertiary alkyl halides; rearrangement of the initially formed carbonium ion to a more stable ion can occur prior to hydride transfer.

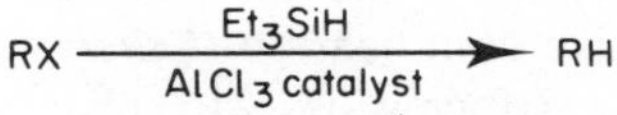

$$RX \xrightarrow[AlCl_3 \text{ catalyst}]{Et_3SiH} RH$$

Scheme 21.14

21.2.3 Miscellaneous

A variety of other unsaturated systems[27], such as Schiff bases and certain heterocyclic ring systems[39], undergo ionic hydrogenation. Nitriles can be reduced as the derived nitrilium ions, leading to aldimines[40] and thence to aldehydes (*Scheme 21.15*); this complements the known reduction of such ions to amines by use of sodium borohydride.

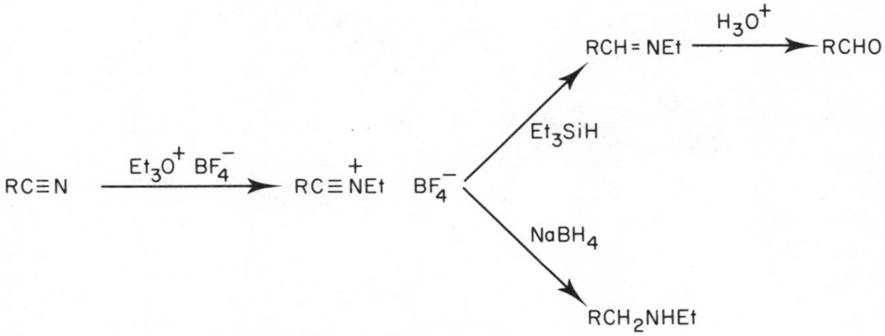

$$RC{\equiv}N \xrightarrow{Et_3O^+ \ BF_4^-} RC{\equiv}\overset{+}{N}Et \ BF_4^-$$

$$RC{\equiv}\overset{+}{N}Et \ BF_4^- \xrightarrow{Et_3SiH} RCH{=}NEt \xrightarrow{H_3O^+} RCHO$$

$$RC{\equiv}\overset{+}{N}Et \ BF_4^- \xrightarrow{NaBH_4} RCH_2NHEt$$

Scheme 21.15

21.3 Reductive silylation

This process involves treatment of an organic substrate with a chlorotrimethylsilane–metal reducing system. The French team of Calas, Dunoguès and their co-workers have carried out exhaustive studies on the reactions of such systems with a wide variety of organic structural types. Concentrating particularly on the two combinations of chlorotrimethylsilane–Mg–HMPA and chlorotrimethylsilane–Li–THF, they have shown that these reagents can provide access to many new *C*-silylated compounds. Additionally, the products are frequently those arising from one-electron transfer processes, thus introducing a new method of reductive dimerization. Since the area has been thoroughly reviewed[41] by the French investigators themselves, only a brief outline will be given here.

21.3.1 Unsaturated hydrocarbons

A major area of application is the reductive silylation of arenes and conjugated dienes, discussed in Chapter 9 as preparative routes to allylsilanes.

21.3.2 *αβ*-Unsaturated carbonyl compounds

Another fruitful area of application of these reagent systems can be seen in their reaction with *αβ*-unsaturated ketones and esters. Extensive investigation[42] has shown that the major processes involved can be either 1,4-disilylation or reductive dimerization (*Scheme 21.16*), depending upon the substrate and upon the reducing system. The latter pathway can produce 1,6-dicarbonyl compounds in synthetically useful yields, especially when catalytic amounts of titanium(IV) or iron(III) salts are present.

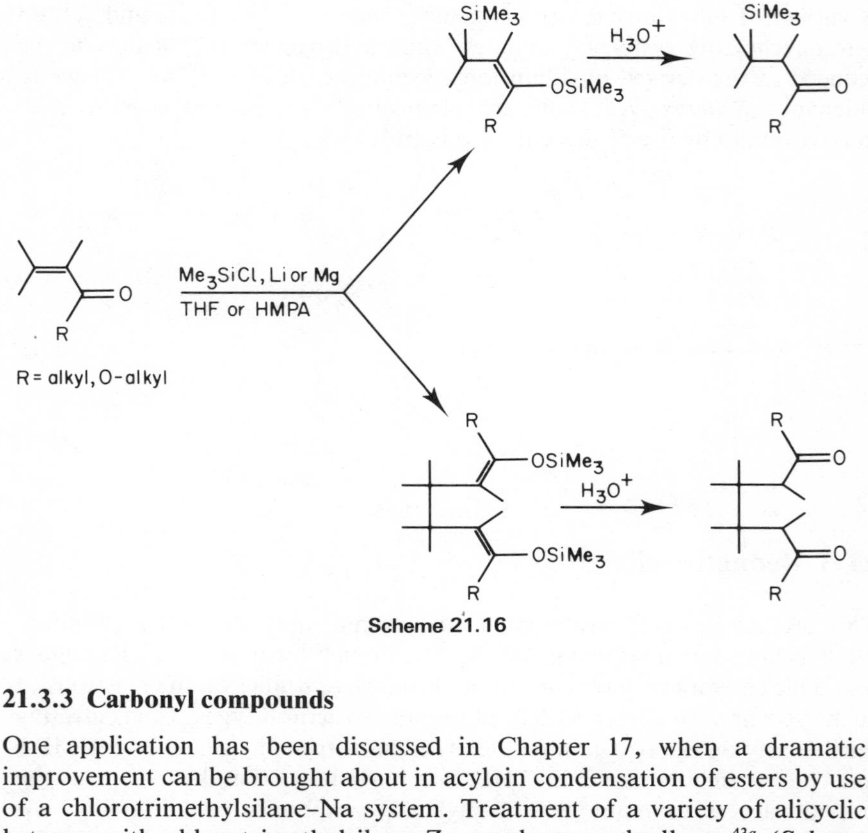

SiMe$_3$ OSiMe$_3$ R H_3O^+ SiMe$_3$ O   R

O   R   Me$_3$SiCl, Li or Mg   THF or HMPA

R = alkyl, O-alkyl

R   OSiMe$_3$ H_3O^+ OSiMe$_3$ R   R   O   O   R

Scheme 21.16

21.3.3 Carbonyl compounds

One application has been discussed in Chapter 17, when a dramatic improvement can be brought about in acyloin condensation of esters by use of a chlorotrimethylsilane–Na system. Treatment of a variety of alicyclic ketones with chlorotrimethylsilane–Zn produces cycloalkenes[43a] (*Scheme 21.17*). The proposed involvement of carbenes or metal carbenoids in this reaction has been substantiated by showing[43b] that phenylcarbene can be generated from benzaldehyde under similar conditions.

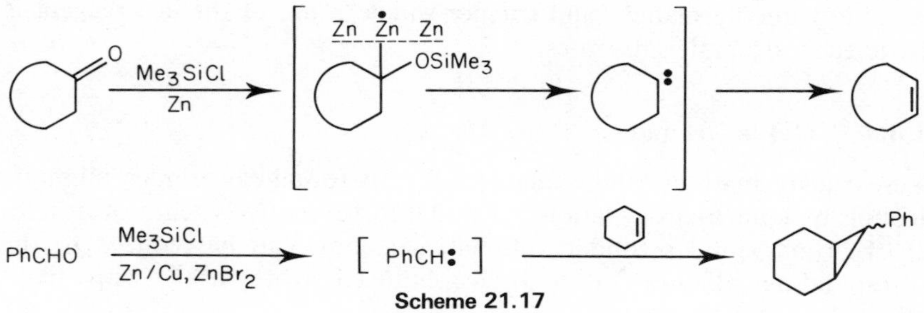

Scheme 21.17

21.4 Deoxygenation processes

21.4.1 $\overset{+}{P}-\overset{-}{O}$ to P

Phosphine oxides are smoothly deoxygenated by a variety of functionalized silanes. With phosphorus–chiral substrates, the stereochemical outcome can easily be controlled by suitable choice of reagent[44–49] (*Table 21.1*). Of these reagents, it should be noted that trichlorosilane–triethylamine will carry out this process selectively[50] in the presence of an aromatic carbonyl group.

Table 21.1 Deoxygenation of acyclic and cyclic phosphine oxides

Reagent	*Phosphine oxide*	*Stereochemical outcome*	*References*
Cl_3SiH	Acyclic	Retention	44
Cl_3SiH–Et_3N	Acyclic	Inversion	44, 45
	Cyclic	Retention	46, 47
Si_2Cl_6	Acyclic	Inversion	45
	Cyclic	Retention	48
$PhSiH_3$	Acyclic	Retention	49
	Cyclic	Retention	49

21.4.2 $\overset{+}{N}-\overset{-}{O}$ to N

Amine oxides[45] and related species are deoxygenated by hexachlorodisilane. The conditions employed are mild, allowing preparation of, for example, the isoindene[51] (2) (*Scheme 21.18*).

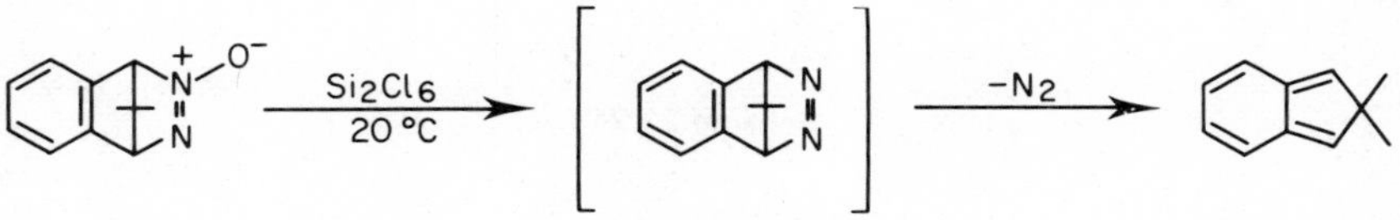

(2)

Scheme 21.18

21.4.3 $\overset{+}{S}-\overset{-}{O}$ to S

Sulphoxides can be reduced to sulphides by a variety of functionalized silanes, including hexachlorodisilane[45], hexamethyldisilathiane[52], bromo-

and iodo-trimethylsilane[53], and trichlorosilane[54]; use of the last reagent is restricted to aromatic substrates.

21.4.4 C–OH to CH and C=O to CH_2

These transformations can be achieved directly for all bar primary aliphatic alcohols by ionic hydrogenation (*see* p. 329). Alternatively, chloroformates (3) of primary and secondary aliphatic alcohols can be reduced to the corresponding alkanes by a radical-induced reaction[55,56] with tri-*n*-propylsilane (*Scheme 21.19*).

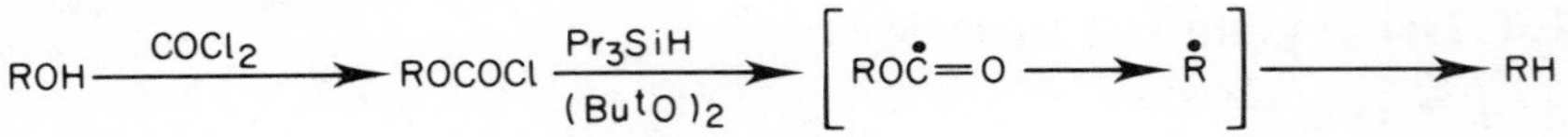

$$\text{ROH} \xrightarrow{\text{COCl}_2} \underset{(3)}{\text{ROCOCl}} \xrightarrow[(\text{Bu}^t\text{O})_2]{\text{Pr}_3\text{SiH}} \left[\text{RO}\dot{\text{C}}{=}\text{O} \longrightarrow \dot{\text{R}}\right] \longrightarrow \text{RH}$$

Scheme 21.19

In the presence of triethylamine, trichlorosilane reduces a wide range of aromatic carbonyl compounds[57], particularly carboxylic acids (*Scheme 21.20*) to hydrocarbons. The same reagent system reduces tertiary amides to amines.

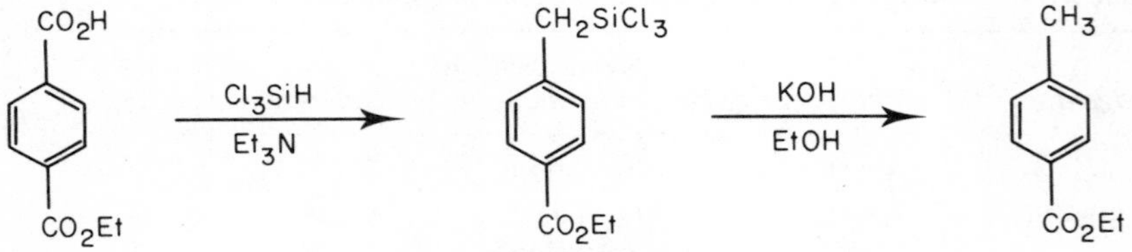

$$p\text{-}(\text{CO}_2\text{H})\text{C}_6\text{H}_4(\text{CO}_2\text{Et}) \xrightarrow[\text{Et}_3\text{N}]{\text{Cl}_3\text{SiH}} p\text{-}(\text{CH}_2\text{SiCl}_3)\text{C}_6\text{H}_4(\text{CO}_2\text{Et}) \xrightarrow[\text{EtOH}]{\text{KOH}} p\text{-}(\text{CH}_3)\text{C}_6\text{H}_4(\text{CO}_2\text{Et})$$

Scheme 21.20

On the other hand, and by use of the chlorotrimethylsilane–Mg–HMPA system[58], a variety of substituted methyl benzoates are reductively silylated to aroylsilanes (*Scheme 21.21*).

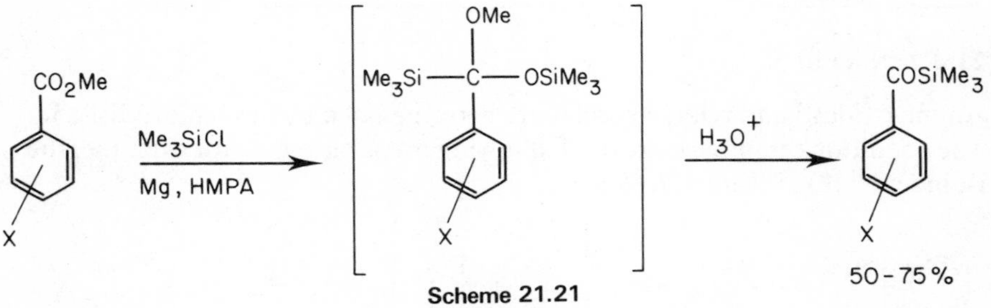

$$\text{X-C}_6\text{H}_4\text{CO}_2\text{Me} \xrightarrow[\text{Mg, HMPA}]{\text{Me}_3\text{SiCl}} \left[\text{Me}_3\text{Si-C(OMe)(OSiMe}_3\text{)-C}_6\text{H}_4\text{X}\right] \xrightarrow{\text{H}_3\text{O}^+} \underset{50\text{-}75\%}{\text{X-C}_6\text{H}_4\text{COSiMe}_3}$$

Scheme 21.21

References

1 LUKEVICS, E., *Russ. chem. Revs* **46,** 264 (1977)
2 LUKEVICS, E., BELYAKOVA, Z. V., POMERANTSEVA, M. G. and VORONKOV, M. G., *Organometal. chem. Rev.* **5,** 1 (1977)
3 *See also* EABORN, C. and BOTT, R. W., *'Organometallic Compounds of the Group IV Elements',* Ed. MacDiarmid, A. G., vol. 1, part 1, p. 105, Marcel Dekker, New York (1968)

4 SPEIER, J. L., *Adv. organometal. Chem.* **17,** 407 (1979)
5 BENKESER, R. A., BURROUS, M. L., NELSON, L. E. and SWISHER, J. V., *J. Am. chem. Soc.* **83,** 4385 (1961)
6 For a mechanistic discussion, *see* CHALK, A. J. and HARROD, J. F., *J. Am. chem. Soc.* **87,** 16 (1965)
7 TAMAO, K., MIYAKE, N., KISO, Y. and KUMADA, M., *J. Am. chem. Soc.* **97,** 5603 (1975)
8 BENKESER, R. A., CUNICO, R. F., JONES, P. R. and NERLEKAR, P. G., *J. org. Chem.* **32,** 2634 (1967)
9 YOSHIDA, J., TAMAO, K., TAKAHASHI, M. and KUMADA, M., *Tetrahedron Lett.* 2161 (1978)
10 DUNOGUÈS, J., BOURGEOIS, P., PILLOT, J.-P., MERAULT, G. and CALAS, R., *J. organometal. Chem.* **87,** 169 (1975)
11 BENKESER, R. A. and MUENCH, W. C., *J. Am. chem. Soc.* **95,** 285 (1973)
12 GAR, T. K., BUYAKOV, A. A., KISIN, A. V. and MIRONOV, V. F., *J. gen. Chem. U.S.S.R.* **41,** 1596 (1971)
13 MURAI, S. and SONODA, N., *Angew. Chem. int. Edn* **18,** 837 (1979) and references therein
14 OJIMA, I., NIHONYANAGI, M. and NAGAI, Y., *J. chem. Soc. chem. Communs* 938 (1972); OJIMA, I., NIHONYANAGI, M., KOGURE, T., KUMAGAI, M., HORIUCHI, S., NAKATSUGAWA, K. and NAGAI, Y., *J. organometal. Chem.* **94,** 449 (1975)
15 OJIMA, I., KOGURE, T. and NAGAI, Y., *Tetrahedron Lett.* 5035 (1972)
16 OJIMA, I., *Strem Chemiker* **8,** 1 (1980) and references therein; OJIMA, I., YAMAMOTO, K. and KUMADA, M., in *'Aspects of Homogeneous Catalysis',* Ed. Ugo, R., vol. 3, chap. 3, Reidel, Dordrecht, Nederland (1977)
17 VALENTINE, D. and SCOTT, J. W., *Synthesis* 329 (1978)
18 KAGAN, H. B., *Pure appl. Chem.* **43,** 401 (1975) and references therein
19 HAYASHI, T., YAMOMOTO, K. and KUMADA, M., *Tetrahedron Lett.* 3 (1975); OJIMA, I., KOGURE, T. and NAGAI, Y., *Chemy Lett.* 985 (1975); OJIMA, I., KOGURE, T. and KUMAGAI, M., *J. org. Chem.* **42,** 1671 (1977) and references therein
20 CORRIU, R. J. P. and MOREAU, J. J. E., *J. organometal. Chem.* **91,** C7 (1975); **85,** 19 (1975); **64,** C51 (1974)
21 CITRON, J. D., *J. org. Chem.* **34,** 1977 (1969); *see also* DENT, S. P., EABORN, C. and PIDCOCK, A., *Chem. Communs* 1703 (1970)
22 LIPOWITZ, J. and BOWMAN, S. A., *J. org. Chem.* **38,** 162 (1973)
23 OJIMA, I., KOGURE, T. and NAGAI, Y., *Tetrahedron Lett.* 2475 (1973)
24 LANGLOIS, N., DANG, T.-P. and KAGAN, H. B., *Tetrahedron Lett.* 4865 (1973); KAGAN, H. B., LANGLOIS, N. and DANG, T.-P., *J. organometal. Chem.* **90,** 353 (1975)
25 COOK, N. C. and LYONS, J. E., *J. Am. chem. Soc.* **87,** 3283 (1965); **88,** 3396 (1966)
26 OJIMA, I. and INABA, S., *J. organometal. Chem.* **140,** 97 (1977)
27 KURSANOV, D. N., PARNES, Z. N. and LOIM, N. M., *Synthesis* 633 (1974)
28 CAREY, F. A. and TREMPER, H. S., *J. org. Chem.* **36,** 758 (1971) and references therein; *see,* for example, SEMMELHACK, M. F. and YAMASHITA, A., *J. Am. chem. Soc.* **102,** 5924 (1980)
29 DOYLE, M. P. and McOSKER, C. C., *J. org. Chem.* **43,** 693 (1978)
30 ZDANOVICH, V. I., KUDRYAVTSEV, R. V. and KURSANOV, D. N., *Izv. Akad. Nauk SSSR., Ser. khim.* 472 (1970)
31 DOYLE, M. P., DeBRUYN, D. J., DONNELLY, S. J., KOOISTRA, D. A., ODUBELA, A. A., WEST, C. T. and SONNEBELT, S. M., *J. org. Chem.* **39,** 2740 (1974)
32 DOYLE, M. P., McOSKER, C. C., BALL, N. and WEST, C. T., *J. org. Chem.* **42,** 1922 (1977); DOYLE, M. P., WEST, C. T., DONNELLY, S. J. and McOSKER, C. C., *J. organometal. Chem.* **117,** 129 (1976)
33 ADLINGTON, M. G., ORFANOPOULOS, M. and FRY, J. L., *Tetrahedron Lett.* 2955 (1976)
34 FRY, J. L., ORFANOPOULOS, M., ADLINGTON, M. G., DITTMAN, W. and SILVERMAN, S. B., *J. org. Chem.* **43,** 374 (1978); *Org. Synth.* **59,** procedure 2134 (1980)
35 FRY, J. L. and MRAZ, T. J., *Tetrahedron Lett.* 849 (1979)
36 RAYNOLDS, P. W., MANNING, M. J. and SWENTON, J. S., *Tetrahedron Lett.* 2383 (1977)
37 AUERBACH, J., ZAMORE, M. and WEINREB, S., *J. org. Chem.* **41,** 725 (1976)
38 DOYLE, M. P., McOSKER, C. C. and WEST, C. T., *J. org. Chem.* **41,** 1393 (1976)
39 *See,* for example, GUILLERM, G., FRAPPIER, F., TABET, J.-C. and MARQUET, A., *J. org. Chem.* **43,** 3776 (1977)

40 FRY, J. L., *J. chem. Soc. chem. Communs* 45 (1974)
41 CALAS, R. and DUNOGUÈS, J., *Organometal. Chem. Rev.* **2,** 277 (1976); CALAS, R., *J. organometal. Chem.* **200,** 11 (1980)
42 DUNOGUÈS, J., CALAS, R., BOLOURTCHIAN, M., BIRAN, C. and DUFFAUT, N., *J. organometal. Chem.* **57,** 55 (1973); PICARD, J.-P., DUNOGUÈS, J. and CALAS, R., *J. organometal. Chem.* **77,** 167 (1974)
43 (*a*) MOTHERWELL, W. B., *J. chem. Soc. chem. Communs* 935 (1973); HODGE, P. and KHAN, M. N., *J. chem. Soc. Perkin I* 809 (1975); (*b*) SMITH, C. L., ARNETT, J. and EZIKE, J., *J. chem. Soc. chem. Communs* 653 (1980)
44 HORNER, L. and BALZER, W. D., *Tetrahedron Lett.* 1157 (1965); FRITZSCHE, H., HASSERODT, U. and KORTE, F., *Chem. Ber.* **97,** 1988 (1964); **98,** 171, 1681 (1965)
45 NAUMANN, K., ZON, G. and MISLOW, K., *J. Am. chem. Soc.* **91,** 7012 (1969)
46 HAWES, W. and TRIPPET, S., *J. chem. Soc. (C)* 1465 (1969)
47 HALL, C. R. and SMITH, D. J. H., *Tetrahedron Lett.* 1693 (1974)
48 DeBRUIN, K. E., ZON, G., NAUMANN, K. and MISLOW, K., *J. Am. chem. Soc.* **91,** 7027 (1969)
49 MARSI, K. L., *J. org. Chem.* **39,** 265 (1974)
50 SEGALL, Y., GRANOTH, I. and KALIR, A., *J. chem. Soc. chem. Communs* 501 (1974)
51 DOLBIER, W. R., MATSUI, K., DEWEY, H. J., HORÁK, D. V. and MICHL, J., *J. Am. chem. Soc.* **101,** 2136 (1979)
52 SOYSA, H. S. D. and WEBER, W. P., *Tetrahedron Lett.* 235 (1978)
53 OLAH, G. A., GUPTA, B. G. B. and NARANG, S. C., *Synthesis* 583 (1977)
54 CHAN, T. H., MELNYK, A. and HARPP, D. N., *Tetrahedron Lett.* 201 (1969)
55 BILLINGHAM, N. C., JACKSON, R. A. and MALEK, F., *J. chem. Soc. chem. Communs* 344 (1977)
56 *See also* BARRETT, A. G. M., PROKOPIOU, P. A. and BARTON, D. H. R., *J. chem. Soc. chem. Communs* 1175 (1979) and references therein
57 BENKESER, R. A., *Accts chem. Res.* **4,** 94 (1971); BENKESER, R. A. and EHLER, D. F., *J. org. Chem.* **38,** 3660 (1973); *Org. Synth.* **53,** 159 (1973)
58 PICARD, J.-P., CALAS, R., DUNOGUÈS, J., DUFFAUT, N., GERVAL, J. and LAPOUYADE, P., *J. org. Chem.* **44,** 420 (1979)

Index

Abbreviations, vii
γ-Acetoxyalkynylsilanes, precursors of allenylsilanes, 170–171
Acetals,
cleavage processes with Me_3SiX reagents, 294–295
reaction with allylsilanes, 110, 114
reaction with silyl enol ethers and ketene acetals, 227–229, 230, 306
Acyl azides, 299
Acyl cyanides, 309–310
Acyl phosphates, 309
Acylation,
of alkynylsilanes, 167–168
of allylsilanes, 97, 105, 109, 110, 112, 113, 121
of arylsilanes, 128
of silyl enol ethers, 232–233
of silyl ethers, 185
of silyl ketene acetals, 233–235
of vinylsilanes, 70–72, 74
Acyloin condensation, Rühlmann modification, 267–272
Acyloin silyl ethers, 297–298, 301
Acyloxysilanes (silyl carboxylates),
hydrolysis of, 193–194, 292
in ester cleavage by Me_3SiX reagents, 292, 293, 306–307
preparation and reactions, 193–196
single electron reduction of, 272–273
Acylsilanes (α-ketosilanes),
in routes to β-hydroxysilanes, 274–275
in routes to silyl enol ethers, 9, 207–209
preparation, 137–138, 208–209, 273–275, 317, 334
reaction with Wittig ylides, 31–32
thermal rearrangement, 212
Addition,
of halogens to vinylsilanes, stereochemistry of, 67, 69–70
of hydrosilanes to multiple bonds, 46–47, 325–331
of Me_3SiX reagents to carbonyl compounds, 295–303, 309
Alcohols,
deoxygenation of, 330, 334
silylation of, 179–180, 184–189
Alcohols (*cont.*)
source of alkyl iodides, 293–294
Aldehydes *see also* Carbonyl compounds *and* Silyl enol ethers,
αβ-epoxysilanes as precursors of, 84
hydrosilylation of, 326–328
ionic hydrogenation of, 329–331
reactions with ethyl trimethylsilylacetate, 202
reaction with organosilyl anions, 137–138
in routes to vinylsilanes, 54–59
αβ-unsaturated, from tertiary ethynyl carbinols, 306
αβ-unsaturated, reaction with Me_3SiX reagents, 296–297, 300–302
Aldol reactions *see also* Hydroxyalkylation,
of silyl enol ethers, 219–220, 226–227, 231
of silyl nitronates, 276
Alkanethiols,
silylation, 202, 290
hemithioacetal formation, 302
Alkene geometry,
inversion, 67, 69, 73, 147–148
retention, 64–66, 69, 73
Alkenes,
cyclic, reaction with Et_2MeSiH/CO, 200–201
heteroatom-substituted, from αβ-epoxysilanes, 88–91
hydrosilylation, 325–326
ionic hydrogenation, 329
photochemical addition to vinylsilanes, 76–77
preparation from allylsilanes, 104–106, 109, 115, 117–118, 121
preparation from vinylsilanes, 63–66, 72–73, 75
silyl-Wittig (Peterson olefination) routes to, 141–161
strained bridgehead, 156
β-Alkoxyacylsilanes, 274
Alkoxysilanes *see also* Alkyl silyl ethers,
by hydrosilylation of carbonyl compounds, 326
Alkyl bromides,
from alcohols, 308
reduction to hydrocarbons, 331

Alkyl carbamates,
cleavage, 160, 292–293
precursors of isocyanates, 160
Alkyl chlorides, reduction to hydrocarbons, 331
Alkyl iodides, from alcohols, 293–294
Alkyl silyl ethers,
methanolysis, 178–179
preparation, cleavage and applications, 181–189
Alkyl trimethylsilyl ketene acetals *see* Silyl ketene acetals
Alkyl trimethylsilyl malonates, 195
Alkylation, of organosilyl anions, 136
Alkylation, Lewis acid catalysed,
of allylsilanes, 109, 122
of silyl enol ethers and ketene acetals, 221–226, 279
Alkylative reduction, route to vinylsilanes, 50
Alkylthiotrimethylsilanes *see* Thiosilanes
Alkynes,
hydrosilylation, 325–326
in vinylsilane preparation, 45–54, 325–326
ionic hydrogenation, 329
synthesis, 148, 270, 303
terminal, protection of, 165–167
Alkynyl ketones, from alkynylsilanes, 167–168
Alkynylsilanes,
acylation, 167–168
applications, 165–170
catalytic hydrogenation, route to vinylsilanes, 48
cleavage, 165, 170
cycloaddition reactions, 130–131
hydrometallation, in routes to vinylsilanes, 48–54
intramolecular carbonium ion attack, 169–170
preparation, 165
source of alkynyl anions for addition to carbonyl compounds, 170
Alkynyl trialkylborates, 53
Allene oxides, generation of, 154
Allenes, by silyl-Wittig elimination, 152–154
Allenyl ketones, synthesis, 265
Allenylsilanes,
preparation, 170–171
reactions, 170–171
Allyl alcohol esters *see also* Silyl ketene acetals,
Claisen rearrangement and stereochemical control, 261–265
Allyl ketones, synthesis, 265
Allyl-metal species, in routes to allylsilanes, 98–100
Allyl silyl ethers, deprotonation and alkylation, 112, 209–211
Allyl silylmethyl ethers, 1,2-rearrangements of, 31
Allyl sulphides,
from allylsilanes, 106–107
other routes, 117
Allyl phosphine oxides, preparation, 117
Allylsilanes,
as allyl anion equivalents, 114–115
ene reactions of, 119–120
lithio anions of, 118–119
preparation, 98–104, 120
reactions with electrophiles, 104–116, 121–122
regiostability and rearrangement, 35–36, 98
relative reactivity, 97
stereochemistry of reaction with electrophiles, 107
trimethylenemethane from, 115
Amides,
bond formation, 196, 316–318
N-methylation of, 331
primary, synthesis of, 306
silylation of, 316–317
tertiary, reduction of, 334
Amine oxides, deoxygenation of, 333
Amines,
N-silylated, 314–316, 318–319
synthesis of chiral, 328
α-Amino acids, silyl derivatives, 194, 314–315
2-Aminoalcohols, synthesis of, 276
α-Aminoketones, enol silylation of, 212
Aminosilanes, 314–316, 319–319
Annelation reactions,
using silylated vinyl ketones, 76–77
via $\alpha\beta$-epoxysilanes, 85
Anthraquinones, synthesis of, 256–257
Anti-addition, of halogens to vinylsilanes, 67
Anti-elimination,
in electrophile-induced desilylation of vinylsilanes, 62, 67–74
in silyl-Wittig reaction, 88, 90, 145–147
Anti-solvolysis, of $\alpha\beta$-epoxysilanes, 89–90
Aralkyl ketones, ionic hydrogenation, 330
Arenes, reductive silylation of, 103, 332
Arylalkynes, silylation of, 165, 202
Arylsilanes,
electronic effects of silyl substitution, 15, 16
electrophile-induced desilylation, 16, 126–132
preparation, 125–126
Aryl silyl ethers, 179, 202, 205
1-Aryl-1-silyloxycyclopropanes, oxidative cleavage of, 249
Arylsulphonylhydrazones, in route to vinylsilanes, 59
Arylthiotrimethylsilanes *see* Thiosilanes
Aziridines, synthesis of, 304
Azidotrimethylsilane (Trimethylsilyl azide),
alkene cleavage with, 304
carbonyl addition reactions of, 299
cycloaddition reactions of, 304
dealkylation with, 293
oxirane cleavage with, 291
preparation of, 290
α-Azidotrimethylsilyl ethers, preparation and reactions, 299

Baeyer–Villiger oxidation,
of dichlorocyclobutanones, 107, 108
of γ-ketosilanes, 158–159
Bases, silicon-substituted, 321–324
Benzoquinones, monocyanosilylation of, 298
Benzyl ethers, cleavage of, 308
Benzyloxysilanes, anionic 1,2-rearrangement of, 31
Benzylthiosilanes, anionic 1,2-rearrangement of, 31
Bis-silylated arenes, 126, 128–131
Bis(silylenol) ethers, preparation and reactions, 229–232
N,O-Bis(trimethylsilyl)acetamide, silylating agent, 180
Bis(trimethylsilyl)acetic acid esters, 56
1,1-Bis(trimethylsilyl)alkan-1-ols, 272–273, 323–324
1,1-Bis(trimethylsilyl)alkyl silyl ethers, 272–273
Bis(trimethylsilyl)amine *see* Hexamethyldisilazane
1,2-Bis(trimethylsilyl)benzene, rearrangement of, 32
1,4-Bis(trimethylsilyl)benzene, synthesis of, 126
3,4-Bis(trimethylsilyl)benzocyclobutenes, synthesis of, 130
N,O-Bis(trimethylsilyl)carbamate, silylating agent, 180
Bis(trimethylsilyl)ethyne, 129–130, 167–168
N,N-Bis(trimethylsilyl)glycinate esters, 314–315
N,O-Bis(trimethylsilyl)hydroxylamine, 314–315, 318
Bis(trimethylsilyl)ketene, 176
Bis(trimethylsilyl) malonate, 195–196, 266–267
N,O-Bis(trimethylsilyl)sulphamic acid, silylating agent, 180
Bis(trimethylsilyl)sulphate, as source of Me_3SiX reagents, 307
Bis(trimethylsilyl)sulphide *see* Hexamethyldisilathiane
Bis(trimethylsilyl)thioketene, preparation and reactions, 176–177
N,O-Bis(trimethylsilyl)trifluoroacetamide, silylating agent, 180
N,N′-Bis(trimethylsilyl)urea, silylating agent, 180
1,2-Bis(trimethylsilyloxy)cycloalkenes,
preparation by modified acyloin condensation, 267–268
source of 1,2-enediolates, 269–270
1,2-Bis(trimethylsilyloxy)cyclobutenes,
Lewis acid catalysed reaction with acetals, 228
preparation of, 268
source of cyclobutane-1,2-diones, 268–269
1,2-Bis(trimethylsilyloxy)cyclopentenes, 269
1,2-Bis(trimethylsilyloxy)cyclopropanes, oxidative cleavage, 249
σ-Bonds, reductive cleavage of, 271
Bond strengths, selected dissociation energies, 4–5
ω-Bromo silyl esters, from lactones, 293
1-Bromoalkenylsilanes, preparation, 58
Bromodesilylation of vinyl silanes, 63–73
β-Bromoethylsilanes, partial solvolysis of, 19
Bromotributylsilane, in selective ester cleavage, 295
α-Bromotrialkylsilylethenes, 57, 58, 160
Bromotrimethylsilane (Trimethylsilyl bromide),
dealkylation with, 292–293
deoxygenation with, 333–334
oxirane cleavage with, 291
preparation of, 289
Brook rearrangement, 30, 31, 137, 162, 207, 209
t-Butyldimethylsilyl chloride *see* Chloro-t-butyldimethylsilane
t-Butyldimethylsilyl enol ethers *see also* Silyl enol ethers
ester-derived, 214, 234, 261–265
ketone-derived, 199, 251
nitroalkane-derived, 276
t-Butyldimethylsilyl ethers,
applications, 185–189, 205
cleavage of, 185
preparation of, 184–185
t-Butyldimethylsilyl hydroperoxide, in hydroperoxide formation, 291
t-Butyldimethylsilyl ketene acetals *see also* Silyl ketene acetals
preparation, 214, 261–265
reactions, 234, 261–265
t-Butyldimethylsilyl perchlorate, silylating agent, 184
t-Butyldiphenylsilyl chloride, silylating agent, 189
t-Butyldiphenylsilyl ethers, preparation and cleavage, 189
t-Butyl group, alkylation of silyl enol ethers, 221
t-Butyl trimethylsilyl carbonate, nucleophiles with, 193–194

Carbinols *see* Alcohols
Carbodemetallation, in routes to vinylsilanes, 48–54
Carbon–silicon bonds,
stability, 5–7
summary of some cleavage processes, 6
Carbonyl compounds *see also* Aldehydes, Ketones *and* Silyl enol ethers
addition processes, with Me_3SiX reagents, 295–303, 326–327
aldol condensation, with silyl enol ethers, 219–221, 226
aldol condensation, with silyl nitronates, 276

Carbonyl compounds (*cont.*)
allylsilanes from, 101-102, 109
alkenes from, by silyl-Wittig (Peterson olefination), 54-56, 143-152
alkynes from, 148
cycloalkenes from, 332-333
αβ-epoxysilanes as precursors of, 84-86
α-halogeno-αβ-unsaturated, 248
hydrosilylation of, 326-328
ionic hydrogenation of, 329-331
reactions with allylsilanes, 110-111, 114-115
reactions with organosilyl anions, 137-138
unsaturated *see* αβ-Unsaturated carbonyl compounds
vinylsilanes from, 54-59, 80
Carbonyl groups, 1,2-transposition of, 92-93
Carboxylic acid chlorides, reduction to aldehydes, 328
Carboxylic acid esters,
cleavage reactions, with Me_3SiX reagents, 292-293, 295, 307
unsaturated *see* αβ-Unsaturated carboxylic acid esters
Carboxylic acids,
by oxidation of aldehyde cyanohydrins, 309-310
protection of, 159, 194-195
reduction of aromatic to hydrocarbons, 334
Chirality, in reactions at silicon, 11, 31-37, 42
Chloro-t-butyldimethylsilane (t-Butyldimethylsilyl chloride) in routes to:
alkyl silyl ethers, 184
silyl enol ethers, 199, 251
silyl ketene acetals, 214, 260-265
silyl keteneimines, 316
silyl nitronates, 276
Chlorohydrins, from oxiranes, 291
Chloromethyltrimethylsilane,
Grignard reagent from, 26, 144, 149
preparation of, 41
Wittig ylide from 80, 141-142
Chlorosilanes, 41
Chlorotriethylsilane, 186-187, 265
Chlorotrimethylsilane (Trimethylsilyl chloride) in routes to:
acyloxysilanes, 194-196
alkyl silyl ethers, 178-179
alkynylsilanes, 165
allylsilanes, 98-100, 103-104, 120
amidosilanes, 316-317
aminosilanes, 318
arylsilanes, 125-126
chlorohydrins from oxiranes, 291
hexamethyldisilane, 134
isocyanates from amides, 160
Me_3SiX reagents, 289-290, 305, 308
silyl enol ethers, 199, 203, 204, 206, 210, 213, 267-268, 273, 277
silyl hemithioacetals, 302
silyl ketene acetals, 214-216, 277-278
silyl nitronates, 276
vinylsilanes, 48, 57, 59, 60, 79
Chlorotrimethylsilane-Li-THF, reductive silylation with, 103-104, 331-333
Chlorotrimethylsilane-Mg-HMPA, reductive silylation with, 331-333
Chlorotrimethylsilane-Zn, ketone deoxygenation with, 332-333
Claisen rearrangements, 260-265, 310
Cope rearrangements, 181-182, 310
Copper(I) bistrimethylsilylamide, 315-316
α-Cuparenone, synthesis of, 260
Curtius rearrangement of acyl azides, 299
Cyanohydrin t-butyldimethylsilyl ethers, 299
Cyanohydrin trimethylsilyl ethers, 296-299, 310
Cyanotrimethylsilane (Trimethylsilyl cyanide),
carbonyl addition reactions, of, 296-298
preparation of, 290, 307
Cycloaddition reactions,
[2 + 1], with silyl enol ethers, 243-247
[2 + 1], with vinylsilanes, 117, 155
[2 + 2], with allylsilanes, 101, 107, 108
[2 + 2], with silyl enol ethers, 250-253, 279
[2 + 2], with vinylsilanes, 76-77
[3 + 2], with silyl enol ethers, 253-254
[3 + 2], with trimethylenemethane, 115
[4 + 2], with alkynylsilanes, 130-131
[4 + 2], with allylsilanes, 100-101
[4 + 2], with silyl-1,3-butadienes, 102-103, 106
[4 + 2], with silyloxy-1,3-butadienes, 254-259, 279-280
[4 + 3] and [2 + 3], with silyl enol ethers, 259-260
Cyclobutane-1,2-diones, synthesis of, 268-269
Cycloheptatrienylidene, generation of, 132
Cyclopentenones, synthesis of, 247, 269
Cyclopropanes, by cyclopropanation of vinylsilanes, 117, 155
Cyclopropane silyl ethers,
by cyclopropanation of silyl enol ethers, 243-247
homoenolate anions from, 250
ring opening reactions of, 244-250
Cyclopropanols, vinyl-substituted, 245-246
Cyclopropenes, generation from β-halogenosilanes, 155
Cyclopropyl ketones,
preparation from cyclopropyltrimethylsilane, 117
source of γ-iodoketones, 295
Cyclopropyltrimethylsilane,
acylation of, 117
preparation of, 117

Darzens condensation,
preferred base in, 322
route to αβ-epoxysilanes, 83, 86
(±)-Decan-9-olide, synthesis of, 269-270
Decyanation, oxidative, 316
Deoxygenation processes, 332-334
Deuterio-desilylation of vinylsilanes, 64-66

Dialkyl acetals,
preparation of, 190
cleavage by Me_3SiX reagents, 294–295
Dialkyl ethers, cleavage by Me_3SiX reagents, 293–294
Dichlorocyclobutanones,
Baeyer–Villiger oxidation of, 107, 108
from allylsilanes, 101, 107
Dichlorodimethylsilane *see* Dimethyldichlorosilane
Dieckmann condensation, 267, 322
Diels–Alder cycloaddition *see* Cycloaddition reactions, [4+2]
Dienes,
cycloaddition *see* Cycloaddition reactions
reductive silylation of, route to allylsilanes, 103–104, 332
silyloxy-substituted, synthesis of, 254–259
silyl-substituted, synthesis of, 47, 100, 106
1,2-Diesters,
from silyl ketene acetals, 243
reductive σ-bond cleavage of, 271
Dihalogenocarbenes,
addition to silyl enol ethers, 247
addition to vinylsilanes, 155
1,2-Dihydroxysilanes,
base-induced reactions of, 90–91
from αβ-epoxysilanes, 90
β-Diketone silyl enol ethers, 1,5-rearrangement of, 37–38
Diketones,
by halogenoacetal cyclization, 323
from silyl enol ethers, 228, 231, 233, 242, 247, 268–269
Dimethyl acetylenedicarboxylate, 252
Dimethyldichlorosilane (Dichlorodimethylsilane),
as trap in pinacol cyclization, 184
silylating agent for diols, 183
Dimethyl fumarate, 251
Dimethyl(methylene)ammonium iodide,
Mannich reaction with silyl enol ethers, 216, 223–224
preparation of, 309
1,1-Dimethylsilaethene, 7,8
(±)-Diplodiolides A and C, syntheses of, 269–270
1,2-Disilacyclobutanes, 9
1,2-Disilylethenes, synthesis of, 46–47, 60
1,3-Disilylpropene, stepwise sulphonation of, 97–98
1,2-Disilylstyrene, protiodesilylation of, 66
(±)-Disodium prephenate, synthesis of, 256
1,3-Dithianes,
in routes to acylsilanes, 208-209
in silyl-Wittig, 151
Dithioketals, preparation of, 302–303
Djerassi–Prelog lactone, syntheses of, 186, 238

β-Effect, 12, 15–20, 62, 87, 89, 98
Electronegativity, selected values, 5–7
Elimination reactions,
geometry of, in silyl-Wittig, 89, 145–148
of β-functionalized silanes, 55–58, 141–148
Enamines, *N*-silylated, 314–315
Ene reaction,
of allylsilanes, 119–120
of silyl enol ethers, 239–240
1,2-Enediolates, from 1,2-bis(trimethylsilyloxy)alkenes, 269
Enol carboxylates, 233
Enolate geometry, control of, 202, 262–265
Enolate ions, generation of, 198–215, 296, 321–322
Enones *see* αβ-Unsaturated carbonyl compounds
Epoxynitrile cyclization, 323
αβ-Epoxysilanes,
allene oxides from, 154
desilylation of, 155
general ring opening reactions, 86–89
isomerization to carbonyl compounds, 84–94
preparation of, 83–84, 86
pyrolytic rearrangement of, 93–94, 212
ring opening reactions with hydride reagents, 86, 92–93
ring opening reactions with organocuprates, 142–143, 146
βγ-Epoxysilanes, 95
(±)-Eriolanin, synthesis of, 186
Erythronolides A and B, synthesis of, 186
Esters *see* Carboxylic acid esters
Ethers,
cleavage of by Me_3SiX reagents, 292–295, 308
reactions with organosilyl anions, 138, 147–148
1-Ethoxy-1-trimethylsilyloxycyclopropane,
preparation of, 250
synthetic equivalent of β-anion of ethyl propionate, 249–250
Ethyl propiolate, 251–252
Ethylene acetals, cleavage of by Me_3SiX reagents, 294
Ethyl trimethylsilylacetate,
preparation of, 201
silylating agent, 201–202, 219
Ethyl trimethylsilyldiazoacetate, in carbonyl insertion, 303
Ethynyl carbinols, αβ-unsaturated aldehydes from, 306

Fluoride ion induced desilylation,
of alkynylsilanes, 165–166, 170
of allylsilanes, 114–115
of arylsilanes, 128
of αβ-epoxysilanes, 155
of β-functionalized silanes, 77, 152–156, 159–160, 277
of silyl enol ethers, 205, 218–220, 236, 238

Fluoride ion induced desilylation (*cont.*)
of silyl ethers, 185–186, 189
of silyl nitronates, 276
via pentafluorosilicates, 138–139
Fluorosilanes, 40–41
(±)-Frullanolide, synthesis of, 232
2,3-Furanediones, from silyl enol ethers, 232

Geometric differentiation, of vinylsilanes, 60–61
Geometry, control of *see also* Stereochemistry,
in electrophile-induced desilylation of vinylsilanes, 63–76
in enolate ion formation, 201, 202, 262–265
in syntheses of vinylsilanes, 45–60
(±)-Grandisol, synthesis of, 323
Grignard reagents *see* Silyl Grignard reagents

Halodemetallation, in routes in vinylsilanes, 50–51, 53
α-Halogeno-αβ-unsaturated carbonyl compounds, 248
Halogenoacetals, cyclization of, 323
α-Halogenoacylsilanes,
preparation of, 208–209, 273
reaction with Grignard reagents, 274–275
α-Halogenoalkyl ketones,
from silyl enol ethers, 216
in routes to silyl enol ethers, 203–204
ω-Halogenoalkyl phenylthioacetates, lactones from, 322
β-Halogenoalkylsilanes,
and the β-effect, 18–19
1,2-elimination reactions of, 152–159
Halogenoalkyltrialkylsilanes, 15
Halogenocarbenes, generation of, 323
Halogenocyclopropenes, 155
Halogenotrimethylsilanes *see* appropriate halogen
Hexachlorodisilane, in deoxygenation processes, 333
Hexamethyldisilane,
in route to aroylsilanes, 210
preparation of, 134
source of bromotrimethylsilane, 308
source of iodotrimethylsilane, 289
source of trimethylsilyl anion, 135
Hexamethyldisilathiane,
in deoxygenation processes, 333
preparation of, 290
reactions as sulphur transfer reagent, 303
Hexamethyldisilazane [Bis(trimethylsilyl)amine],
deprotonation of, silicon-substituted bases from, 321
Hexamethyldisilazane [Bis(trimethylsilyl)amine] (*cont.*)
silylating agent for alcohols, 179–180, 305
silylating agent for carbonyl compounds, 206, 296
Hexaphenyldisiloxane, 12
Homoenolate anion equivalents, 211–212, 250
Hydroalumination, in routes to vinylsilanes, 49–51
Hydroboration, in routes to vinylsilanes, 51–52
Hydroboration-oxidation,
of silyl enol ethers, 235
of vinylsilanes, 74, 76
Hydrogenation,
catalytic, of alkynylsilanes, 48
ionic, 329–331
Hydrometallation, in routes to vinylsilanes, 48–52
Hydrosilanes,
hydrosilylation and, 325–328
ionic hydrogenation and, 329–331
organosilyl anions by deprotonation of, 135
Hydrosilylation,
general, 325–328
in dehydrogenative silylation, 200
of acyl halides, 328
of alkenes, 325–326
of alkynes, 46–48, 325–326
of carbonyl compounds, 326–327
of conjugated dienes, 102
of αβ-unsaturated carbonyl compounds, 200, 326–327
of αβ-unsaturated carboxylic acid esters, 214–215
Hydrostannylation, in routes to vinylsilanes, 54
Hydroxamic acids, preparation from amides, 318
Hydroxyalkylation, alkylsilane anions and, 118–119, 152
Hydroxyalkylation, fluoride ion induced,
allylsilanes and, 114–115
silyl enol ethers and, 219–221
Hydroxyalkylation, Lewis acid induced,
allylsilanes and, 110–111, 113–114, 121
silyl enol ethers and, 226–227, 231
β-Hydroxyalkylsilanes,
diastereoselective preparation of, 145–147, 275
1,2-elimination reactions of, 142–152, 160–162, 274–275
preparation of, 142–152, 160–162, 174–175
1-Hydroxyalkyl vinylsilanes
preparation of, 58, 153
in routes to allene oxides, 154
in routes to allenes, 153–154
α-Hydroxyamides, from cyanohydrins, 299
γ-Hydroxyesters, preparation of, 249–250
(2-Hydroxyethyl)allenylsilanes, oxidation of, 171–172

Hydroxyl groups *see* Alcohols *and* Alkyl silyl ethers
Hydroxylamine, *N,O*-bis(trimethylsilyl), 314–315, 318
α-Hydroxylation,
  of silyl enol ethers, 217, 236–237
  of silyl ketene acetals, 237
Hyperconjugation and the β-effect, 16–17

Imidazole,
  silylation catalyst, 184
  N-trimethylsilyl, 180, 305
Imidazoles, from silyl enol ethers, 254
Imidazolides, preparation of, 305
Imines,
  chiral amines from, 328
  silylation of anions from, 314–315
Inversion,
  of alkene geometry *see* Alkene geometry
  of chirality at silicon, 36, 42
ω-Iodoalcohol trimethylsilyl ethers, 295
Iodoalkyl (trimethylsilyloxy)ethyl ethers, 294
Iodo-t-butyldimethylsilane (t-Butyldimethylsilyl iodide),
  in conversion of oxiranes into allylic alcohols, 307–308
  preparation of, 307–308
β-Iodoketones, from αβ-unsaturated ketones, 295–296
γ-Iodoketones, from cyclopropyl ketones, 295–296
Iodomethyl methyl ether, preparation of, 295
Iodomethyltrimethylsilane (Trimethylsilylmethyl iodide),
  in alkylation of amines, etc., 160
  in alkylation of carbanions, 156–157
  Wittig ylide from, allylsilane preparation, 102
Iodosilanes, 41
ω-Iodo silyl esters, from lactones, 293
Iodotrimethylsilane (Trimethylsilyl iodide),
  acetal cleavage with, 294–295
  carbonyl addition reactions of, 295
  dealkylation with, 292–293, 309
  deoxygenation with, 291, 293–294, 334
  in preparation of silyl enol ethers, 296–297
  oxirane cleavage with, 291, 308
  preparation of, 116, 289
α-Iodo trimethylsilyl ethers, 295
Ionic hydrogenation,
  of alkenes and alkynes, 329
  of carbonyl compounds and carbinols, 329–330
  of nitriles, 331
  reductive methylation of amides, 330–331
Ipso-desilylation, of arylsilanes, 128–131
Isocyanates, routes to, 160, 299, 318
Isocyanatotrimethylsilane,
  preparation of, 307
Isocyanatotrimethylsilane (*cont.*)
  synthesis of primary amides, 306
Isopropyldimethylsilyl protecting groups, 186, 205

Karahanaenone, synthesis of, 260
Ketene silyl acetals *see* Silyl ketene acetals
β-Ketoacids,
  silatropic rearrangement and, 37, 203–204
  synthesis of, 266
β-Ketoamides, preparation of, 252
β-Ketoesters,
  bis(silylenol) ethers of, 231–232, 258–259
  dealkylation-decarboxylation of, 293
  synthesis of, 233–234, 247
Ketolactones, synthesis of macrocyclic, 269–270
Ketone carbonyl group, 1,2-transposition of, 92–93
Ketones *see also* Carbonyl compounds *and* Silyl enol ethers,
  allylsilanes from, 101–102, 109
  αβ-epoxysilanes as precursors of, 85–87, 90, 91–93
  ethylidene generation from, 160–161
  hydrosilylation of, 326–328
  ionic hydrogenation of, 329–331
  oxidative decyanation route to, 316
  reaction with alkyl silyl ethers, formation of ketals, 189
  reaction with Me_3SiX reagents, 295–303, 326–327
  reaction with organosilyl anions, 137–138
  αβ-unsaturated *see* αβ-Unsaturated carbonyl compounds
  unsymmetrical, synthesis of, 270
  vinylsilanes from, 54–55, 58–59, 80
Ketonitriles, synthesis of long-chain, 270
α-Ketosilanes *see* Acylsilanes
β-Ketosilanes,
  preparation of, 91–92, 145–147, 274–275
  route to β-hydroxyalkylsilanes, 144–145, 275
  thermal rearrangement of, 34, 212
γ-Ketosilanes,
  as masked αβ-unsaturated ketones, 157–158
  preparation of, 157
  silicon-assisted Baeyer–Villiger oxidation of, 158–159

Lactams, from aminoacids, 196
β-Lactams,
  preparation from allylsilanes, 107–108

β-Lactams (*cont.*)
preparation from silyl ketene acetals, 241, 252–253
N-protection of, 317
silyl-Wittig reaction and, 151
Lactones,
by Baeyer–Villiger oxidation, 107, 158–159
by cyclization, 322
by oxidation of cyanohydrins, 309–310
cleavage of, by Me_3SiX reagents, 292–293
Djerassi–Prelog, 186, 238
from allenylsilanes, 171–172
hydroxyalkylation of, 322
macrocyclic, 269–270
silylation of, 214
Lead(IV) carboxylates, in oxidation,
of arylsilanes, 128, 131
of silyl enol ethers, 238
Ligands, silicon-substituted, 321
α-Lithiosilyl halides, silaethenes from, 9
α-Lithio-α-trimethylsilyl species,
in routes to acylsilanes, 317, 209
in routes to vinylsilanes, 55–58
in silyl-Wittig, 149–151, 160–161
preparation of, 21–29
Lithium bis(trimethylsilyl)amide, 321–324
Lithium 1,1-bis(trimethylsilyl)-3-methylbutoxide, 323–324
Loganin, synthesis of, 107–108

Mannich,
reactions of silyl enol ethers, 216, 223–224
salts, 309
α-Metallated organosilanes,
factors in stabilization of, 12–13
preparation of, 21–29
Methoxymethyl esters, selective cleavage of, 295
Methoxymethyltrimethylsilane,
metallation of, 161–162
reductive nucleophilic acylation with, 161–162
Methyl acetoacetate, silylation of, 231–232
Methyl benzoates, aroylsilanes from, 334
Methyl ethers, cleavage by Me_3SiX reagents, 293, 295, 308
(±)-Methyl santolinate, synthesis of, 264–265
α-Methylation, of silyl enol ethers, 224–225
α-Methylenation, of silyl enol ethers, 223–226
α-Methylene-γ-lactones, synthesis of, 150, 224, 293
(±)-Methylmaysenine, synthesis of, 186
Methylsuccinic anhydride, silylation of, 230–231
Methylthiotrimethylsilane *see* Thiosilanes
Michael reaction,
with allylsilanes, 111
with silyl enol ethers and ketene acetals, 228–229, 250–251
Migrations of silyl groups, 30–39, 112, 186, 208–210, 298
(±)-Muscone, synthesis of, 168, 249

1-Naphthylsilanes, 1,2-rearrangements of, 32–33
2-Nitroalcohols, synthesis and reduction of, 276
Nucleosides,
hydroxyl protection of, 186
synthesis of, 304–305
Nucleophilic acylation/alkylation,
in preparation of acylsilanes, 208–209, 317
using aldehyde-derived Me_3SiX addition products, 297–298, 301
using 1-ethoxy-1-trimethylsilyloxycyclopropane, 249–250
using methoxymethyltrimethylsilane, 161–162

(±)-Oestrone, synthesis of, 131
Organodisilanes,
in route to organosilyl anions, 135
preparation of, 134
Organopentafluorosilicates,
preparation of, 136
reactions with electrophiles, 128, 138–139
Organosilicon compounds, major suppliers, 2
Organosilyl anions,
preparation of, 134–136
reactions of, 136–138, 147–148
Oxyallyl zwitterions, generation of, 154, 259–260
Oxidation *see also* Baeyer–Villiger oxidation,
of cyanohydrins, 309–310
of β-hydroxyalkylsilanes to β-ketosilanes, 145, 147
of silyl enol ethers and ketene acetals,
to α-hydroxycarbonyl compounds, 236–238
to αβ unsaturated carbonyl compounds, 239–240, 241–242
with singlet oxygen, 239–241
with ozone, 238–239
Oxidative coupling,
of silyl enol ethers, 242
of silyl ketene acetals, 243
Oxiranes, ring opening reactions of *see also* αβ-Epoxysilanes,
with Me_3SiX reagents,
isomerization to allylic alcohols, 183, 307–308
to produce derivatives of 1,2-diols, 291
with organosilyl anions, in alkene inversion, 147–148
Ozonolysis,
of silyl enol ethers, 238–239
of vinylsilanes, 78–79

Penicillin sulphoxides, thermal rearrangement of, 181–182
(±)-Pentalenolactone, synthesis of, 256
(±)-Periplanone, synthesis of, 186
Peterson olefination (silyl-Wittig reaction), 55–56, 143–152
Phenols,
from arylsilanes, 128, 131
silylation of, 178–179, 202, 205
Phenoxysilanes,
lithium/ammonia reduction of, 205
rates of methanolysis of, 178–179
Phenyldimethylsilyl lithium,
preparation of, 135
reaction with oxiranes, alkene inversion with, 147
Phenylselenotrimethylsilane,
preparation of, 307
carbonyl addition processes, 309
Phenylsulphenyl chloride, in sulphenylation,
of allylsilanes, 107
of silyl enol ethers, 230–231, 274
Phenylthiotrimethylsilane *see* Thiosilanes
Phenyltriethylsilane, protiodesilylation of, 127
Phenyltrimethylsilane,
electrophile-induced desilylation of, 126
nitration of, 16, 129
2-Phenylvinylsilane *see* β-Trimethylsilylstyrene
Phosphine oxides, deoxygenation of, 333
Phosphoenol pyruvate, synthesis of, 308–309
Phosphorus-oxygen-silicon reagents,
carbonyl addition processes, 299–301
preparation of, 290
Phosphorus(V) esters, dealkylation of by Me_3SiX reagents, 288, 292, 308, 309
Photo-oxygenation of silyl enol ethers, 240–241
Photochemical cycloaddition *see* Cycloaddition reactions
π-bonds,
α-anion stabilization and, 12–13
β-effect and, 12, 16–19
silabenzenes and, 9–10
silaethenes and, 7–9
siloxanes and, 11–12
Potassium bis(trimethylsilyl)amide, 321–324
Prostaglandins, synthesis of, 186–189, 298
Protiodemetallation, in routes to vinylsilanes, 48–54
Protiodesilylation,
of allylsilanes, 105–106, 109
of arylsilanes, 127–128
of vinylsilanes, 64–66, 72–73
Proton abstraction *see* α-Metallated organosilanes
(±)-Pumiliotoxin C, synthesis of, 259
Pummerer rearrangement *see also* Selenoxides and Sulphoxides, 33–36
(±)-Pyrenophorin, synthesis of, 186
Pyridines, hydrosilylation of, 328
Quinols, synthesis from quinones, 298
Quinones,
1,2-addition using allylsilanes, 111–112
1,2-addition using cyanotrimethylsilane, 298
cycloaddition with silyloxybutadienes, 256–258

Rearrangement reactions with migration of silicon, 30–39, 112, 186, 208–210, 298
(±)-Recifeiolide, synthesis of, 322
Reducing agents, silanes as, 325–336
Reductive cleavage of σ-bonds, 271
Reductive nucleophilic acylation, 161–162
Reformatsky reactions, 196, 202, 227
Retention,
of alkene geometry *see* Alkene geometry
of chirality at silicon, 31, 34, 36, 37, 42

Selenoethers, α-silylation of, 34
Selenoxides, Pummerer rearrangement of, 34–35
(±)-Serratenediol, synthesis of, 186
(±)-Seychellene, synthesis of, 254–255
Showdomycin, synthesis of, 228
σ-bonds,
and hypervalency, 11
reductive cleavage of, 271
Sigmatropic rearrangements,
general, 30–38
of allylsilanes, 98, 100
of silyl enol ethers, 260–267
Silabenzenes, 9–10
Silaethenes, 7–9
Silanes, as reducing agents, 325–334
Silicon–hydrogen bonds,
cleavage of, 135, 184
polarity reversal, 6
Siloxanes, 11–12
α-Silyl alkoxides, Brook rearrangement of, 30–32, 112, 207–211
Silyl anions *see* Organosilyl anions
Silyl carboxylates *see* Acyloxysilanes
Silylcarboxylic acids, synthesis of, 136
Silyl cuprates, reactions of, 79, 86, 137–138
Silyl dienes *see* Dienes
Silyl dienol ethers, preparation and reactions, 229–232
Silyl enol ethers,
acylation of, 232–235
aldol reactions of, fluoride ion induced, 219–221
aldol reactions of, Lewis acid induced, 226–228, 231
alkylation of, direct, 277, 279
alkylation of, fluoride ion induced, 218–219
alkylation of, Lewis acid induced, 221–226, 229–230

Silyl enol ethers (*cont.*)
  cleavage of, to enolate ions, 217–218, 220–221, 269–270
  cyclization of, γ-unsaturated, 278–279
  cycloaddition reactions of,
    [2 + 1], 243–250
    [2 + 2], 250–253, 279
    [3 + 2], 253–254
    [4 + 2], 254–259, 279–280
    [4 + 3] and [2 + 3], 259–260
  geometric control in preparation of, 201–202
  hydroboration-oxidation of, 235
  Michael additions and, 228–229, 232
  of acylsilanes, 273–275
  of β-ketoesters, 231, 258
  oxidation of, to α-hydroxycarbonyl compounds, 236–238
  oxidative coupling of, 242–243
  ozonolysis of, 238–239
  preparation of, 91–92, 93–94, 198–213, 259, 267–268, 276–278, 296
  ring contraction of cyclic, 253
  [3,3]-sigmatropic rearrangements of, 265
  uncatalysed reactions, summary of, 215–217
  αβ-unsaturated carbonyl compounds from, 240–242
Silyl esters,
  of carboxylic acids *see* Acyloxysilanes
  of *aci*-nitroalkanes, 275–276
  of phosphorus(V) acids, 288, 292
Silyl ethers *see* Alkyl silyl ethers
α-Silyl Grignard reagents,
  in routes to vinylsilanes, 60
  in silyl-Wittig reaction, 144, 149
  preparation of, 21, 25–27
O-Silyl hemithioacetals, preparation and reactions, 301–303
Silyl ketene acetals *see also* Silyl enol ethers,
  geometric control in preparation of, 262–264
  inter- and intramolecular rearrangements of, 266–267
  preparation of, 213–215, 260–265, 276–278
  [3,3]-sigmatropic rearrangements involving, 261–265
Silyl nitronates, 36, 275–276
Silyl perchlorates, silylating agents, 184
α-Silyl selenoxides,
  aldehydes from, 34–35
  Pummerer rearrangement of, 34
Silyl-Wittig reaction (Peterson olefination), 55–56, 142–151
β-Silylated ynamines, 169–170
Silylating agents, for alcohols, 179–181, 184–189
Silylation,
  of acids, 193–196
  of alcohols, 179–181, 184–189
  of allyl-metal species, 98–100
Silylation (*cont.*)
  of amines, 314–318
  of aryl-metal species, 125
  of enolate ions, 198–215
  of terminal alkynes, 165–167
  of vinyl-metal species, 59–60
  reductive, 103–104, 331–333
α-Silylcarbanions,
  in routes to β-hydroxysilanes, 142–143, 149–152
  preparation of, 21–27
Silylketenes, preparation and reactions, 174–177
Silylmethylallyl cation, trimethylenemethane from, 115
Silyloxy dienes *see* Dienes
Silyloxy Claisen rearrangement, 260–265
Silyloxy Cope rearrangement, 181–182, 310
Silyloxyallylsilanes,
  as homoenolate anion equivalents, 211–212
  preparation of, 211–212
1-Silyloxybicyclo[*n*.1.0]alkanes, preparation and reactions, 248–250
β-Silyloxyborane intermediates, 235
Silyloxycyclopropanes, preparation and reactions, 243–250
1-Silyloxy-2,2-dichlorobicyclo[*n*.1.0]alkanes, electrolysis of, 247
α-Silyloxyketones, by oxidation of silyl enol ethers, 236
Simmons–Smith cyclopropanation, 243
Sodium bis(trimethylsilyl)amide, 321–324
Solvolysis of silyl ethers, 178–179
β-Stannyl silyl enol ethers, in protection of enones, 237
Stereochemistry *see also* Geometry,
  of electrophilic attack on alkynylsilanes, 169
  of electrophilic attack on allylsilanes, 107
  of electrophilic attack on vinylsilanes, 63–75
  of silyl-Wittig (Peterson olefination), 145–147
  of substitution at silicon, 42–43
  of vinylsilanes, geometric differentiation, 60–61
Substitution types, in vinylsilane preparation, 62
Succinic anhydride, silylation of, 230–231
Sulphenylation,
  of allylsilanes, 106–107
  of silyl enol ethers, 216, 230–231, 274
Sulphides, synthesis of, 303
β-Sulphonylsilanes,
  fluoride ion induced 1,2-elimination of, 156
  preparation of, 156
Sulphoxides,
  deoxygenation of, 333–334
  sila-Pummerer rearrangement of, 33–34
  α-silylation of, 33–34
Sulphur transfer reagents, 305
Syn-elimination, in silyl-Wittig, 89, 145–148

(±)-Temsin, synthesis of, 186
Tetrachlorosilane, 1, 160
Tetraethylsilane, 1
Tetraketones, synthesis of macrocyclic, 249
Tetramethylsilane, 2
Thienamycin, synthesis of, 317
Thiosilanes,
as sulphur transfer reagents, 303
carbonyl addition reactions of, 301–302
dealkylation with, 292, 293, 308
oxirane ring opening with, 291
preparation of, 290, 307
Thiotrimethylsilane *see* Thiosilanes
Tri-n-propylsilane, in reduction of chloroformates, 334
N-Trialkylsilylamides, 180, 299, 317–318
α-Trialkylsilylvinyl ketones,
annelation with, 76–77
photocycloaddition with, 76–77
preparation of, 58
1,2,3-Triazoles, 304
Trichlorosilane,
in conversion of amides into isocyanates, 160
in deoxygenation processes, 333–334
in hydrosilylation, 326
Triethylsilane-trifluoroacetic acid, in ionic hydrogenation, 329–331
Trimethylenemethane, 115
N-Trimethylsilylacetamide, silylating agent, 180
N-Trimethylsilyl aminoacids, 318
Trimethylsilyl azide *see* Azidotrimethylsilane
Trimethylsilyl bromide *see* Bromotrimethylsilane
Trimethylsilyl bromoacetate, 196
Trimethylsilylbromoketene, preparation and reactions, 174–175
1-Trimethylsilylbuta-1,3-diene *see* Dienes
Trimethylsilyl chloride *see* Chlorotrimethylsilane
Trimethylsilyl cyanide *see* Cyanotrimethylsilane
5-Trimethylsilylcyclopentadiene,
preparation of, 100
reactions of, 100–101, 106–107, 118
Trimethylsilyldiazomethane,
alkyne preparation with, 148
1,1,2-Trimethylsilaethene from, 7–8
N-Trimethylsilyldiethylamine, silylating agent, 180
Trimethylsilyl enol ethers *see* Silyl enol ethers
Trimethylsilyl esters *see* Acyloxysilanes
Trimethylsilyl ethers *see* Alkyl silyl ethers
Trimethylsilylethyne, 53–54, 172
Trimethylsilyl hydroperoxide, oxirane ring opening with, 291
N-Trimethylsilylimidazole,
in imidazole preparation, 305
silylating agent, 180
Trimethylsilyl iodide *see* Iodotrimethylsilane
Trimethylsilylketene, preparation and reactions, 174, 176
Trimethylsilyl ketene acetals *see* Silyl ketene acetals
Trimethylsilyl β-ketoesters, silatropic rearrangement of, 37–38, 203–204
Trimethylsilyl lithium,
addition to enones, 137
preparation of, 135
Trimethylsilylmethyl chloride *see* Chloromethyltrimethylsilane
Trimethylsilylmethyl group, electron-releasing effects of, 16–18
Trimethylsilylmethyl iodide *see* Iodomethyltrimethylsilane
Trimethylsilylmethylketene, preparation and reactions, 175
Trimethylsilylmethyl lithium, 22, 26, 27
Trimethylsilylmethyl magnesium chloride, 26, 144
Trimethylsilylmethyl phenyl selenide,
in route to homologous aldehydes, 35
preparation of, 35
Trimethylsilylmethyl phenyl sulphoxide, 33
Trimethylsilylmethyl potassium, 323
Trimethylsilylmethyl trifluorosulphonate, 160
α-Trimethylsilyloxy aldehydes, 299
Trimethylsilyloxybuta-1,3-dienes *see* Dienes
Trimethylsilyl perchlorate, 304
Trimethylsilyl potassium, oxirane ring opening with, 147
Trimethylsilyl pyruvate, 194
β-Trimethylsilylstyrene,
electrophile-induced desilylation of, 63–64, 69–70
preparation of, 45
Trimethylsilyl trifluoromethanesulphonate,
catalyst in alkyl silyl ether reactions, 189–190
catalyst in allylsilane reactions, 114
in acetal activation, aldol with silyl enol ethers, 306
in t-butyl ester cleavage, 306
in epoxynitrone activation, 307
in oxirane isomerization to allylic alcohols, 183
in preparation of silyl enol ethers, 205–206, 304–305
1-Trimethylsilyl trimethylsilyl enol ethers,
electrophile-induced desilylation of, 62–63, 273–274
preparation of, 273
Triphenylsilanol, 12
Triphenylsilyl hydroperoxide, in oxirane formation, 291
Trityldimethylsilyl groups, in alcohol protection, 189

αβ-Unsaturated carbonyl compounds,
1,4-addition of organosilyl anions to, 137–138

αβ-Unsaturated carbonyl compounds (*cont.*)
addition reactions with Me_3SiX reagents, 295–302, 309–310
from silyl enol ethers, 240–242
hydrosilylation of, 200, 326–327
precursors of silyl enol ethers, 199–200, 245–246, 309, 326–327
protection of, 137, 157–158, 237
reaction with allylsilanes, Lewis acid catalysed, 111
reductive silylation of, 332
via acylation of alkynylsilanes, 168
αβ-Unsaturated carboxylic acid esters,
as precursors of silyl ketene acetals, 214–215
hydrosilylation of, 214–215
Ureas, synthesis of, 318

Vinyl bromides, 63–64, 66, 67, 69–70, 73, 79–80, 88
Vinyl carbanions, silylation of, 59–60
Vinyl halides,
from vinylsilanes, 64, 66–74
Vinyl halides (*cont.*)
in routes to vinylsilanes, 60
Vinyl iodides, 67–68, 74
Vinyl ketones *see* α-Trialkylsilylvinyl ketones
Vinylcyclopropanols, 245–246
Vinylidene carbenes, generation of, 78–79, 148–149
Vinylsilanes,
conversion into αβ-epoxysilanes, 83
disubstituted, 50–56
electrophile-induced desilylation of, 62–82
geometric differentiation of, 60–61
monosubstituted, 45–49
organometallic addition to, 21
ozonolysis of, 78, 79
preparation, 44–61, 79–80, 326
silyl-Wittig reaction and, 153
substitution types and availability, 44
trisubstituted, 58, 79

Wittig reaction, in route to allylsilanes, 101–102